Biological Effects and Physics of Solar and Galactic Cosmic Radiation

Part A

NATO ASI Series

Advanced Science Institutes Series

A series presenting the results of activities sponsored by the NATO Science Committee, which aims at the dissemination of advanced scientific and technological knowledge, with a view to strengthening links between scientific communities.

The series is published by an international board of publishers in conjunction with the NATO Scientific Affairs Division

A	**Life Sciences**	Plenum Publishing Corporation
B	**Physics**	New York and London
C	**Mathematical and Physical Sciences**	Kluwer Academic Publishers
D	**Behavioral and Social Sciences**	Dordrecht, Boston, and London
E	**Applied Sciences**	
F	**Computer and Systems Sciences**	Springer-Verlag
G	**Ecological Sciences**	Berlin, Heidelberg, New York, London,
H	**Cell Biology**	Paris, Tokyo, Hong Kong, and Barcelona
I	**Global Environmental Change**	

Recent Volumes in this Series

Volume 243A —Biological Effects and Physics of Solar and Galactic Cosmic Radiation, Part A
edited by Charles E. Swenberg, Gerda Horneck, and E. G. Stassinopoulos

Volume 243B —Biological Effects and Physics of Solar and Galactic Cosmic Radiation, Part B
edited by Charles E. Swenberg, Gerda Horneck, and E. G. Stassinopoulos

Volume 244 —Forest Development in Cold Climates
edited by John Alden, J. Louise Mastrantonio, and Søren Ødum

Volume 245 —Biology of *Salmonella*
edited by Felipe Cabello

Volume 246 —New Developments in Lipid–Protein Interactions and Receptor Function
edited by K. W. A. Wirtz, L. Packer, J. Å. Gustafsson, A. E. Evangelopoulos, and J. P. Changeux

Volume 247 —Bone Circulation and Vascularization in Normal and Pathological Conditions
edited by A. Schoutens, J. Arlet, J. W. M. Gardeniers, and S. P. F. Hughes

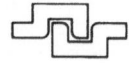

Series A: Life Sciences

Biological Effects and Physics of Solar and Galactic Cosmic Radiation

Part A

Edited by

Charles E. Swenberg

Armed Forces Radiobiology Research Institute
Bethesda, Maryland
and Complexity Incorporated
Potomac, Maryland

Gerda Horneck

DLR Institute of Aerospace Medicine
Cologne, Germany

and

E. G. Stassinopoulos

NASA-Goddard Space Flight Center
Greenbelt, Maryland

Springer Science+Business Media, LLC

Proceedings of a NATO Advanced Study Institute on
Biological Effects and Physics of Solar and Galactic Cosmic Radiation,
held October 13–23, 1991,
in Algarve, Portugal

NATO-PCO-DATA BASE

The electronic index to the NATO ASI Series provides full bibliographical references (with keywords and/or abstracts) to more than 30,000 contributions from international scientists published in all sections of the NATO ASI Series. Access to the NATO-PCO-DATA BASE is possible in two ways:

—via online FILE 128 (NATO-PCO-DATA BASE) hosted by ESRIN, Via Galileo Galilei, I-00044 Frascati, Italy

—via CD-ROM "NATO-PCO-DATA BASE" with user-friendly retrieval software in English, French, and German (©WTV GmbH and DATAWARE Technologies, Inc. 1989)

The CD-ROM can be ordered through any member of the Board of Publishers or through NATO-PCO, Overijse, Belgium.

Library of Congress Cataloging-in-Publication Data

NATO Advanced Study Institute on Biological Effects and Physics of Solar and Galactic Cosmic Radiation (1991 : Algarve, Portugal)
Biological effects and physics of solar and galactic cosmic radiation. Part A / edited by Charles E. Swenberg, Gerda Horneck, and E.G. Stassinopoulos.
p. cm. -- (NATO advanced science institutes series. Series A, Life sciences ; v. 243A)
"Proceedings of a NATO Advanced Study Institute on Biological Effects and Physics of Solar and Galactic Cosmic Radiation, held October 13-23, 1991 in Algarve, Portugal"--T.p. verso.
Includes bibliographical references and index.
ISBN 978-1-4613-6266-1 ISBN 978-1-4615-2918-7 (eBook)
DOI 10.1007/978-1-4615-2918-7
1. Radiation--Toxicology--Animal models--Congresses.
2. Radiation--Toxicology--Mathematical models--Congresses. 3. Outer space--Exploration--Health aspects--Congresses. I. Swenberg, Charles E. II. Horneck, G. (Gerda) III. Stassinopoulos, E. G. IV. Title. V. Series.
RC1151.R33N38 1991
616.9'897--dc20 93-8450
CIP

Additional material to this book can be downloaded from http://extra.springer.com.

Originally published by Plenum Press, New York in 1993
Softcover reprint of the hardcover 1st edition 1993

PREFACE

Space missions subject human beings or any other target of a spacecraft to a radiation environment of an intensity and composition not available on earth. Whereas for missions in low earth orbit (LEO), such as those using the Space Shuttle or Space Station scenario, radiation exposure guidelines have been developed and have been adopted by spacefaring agencies, for exploratory class missions that will take the space travellers outside the protective confines of the geomagnetic field sufficient guidelines for radiation protection are still outstanding. For a piloted Mars mission, the whole concept of radiation protection needs to be reconsidered. Since there is an increasing interest of many nations and space agencies in establishing a lunar base and /or exploring Mars by manned missions, it is both, timely and important to develop appropriate risk estimates and radiation protection guidelines which will have an influence on the design and structure of space vehicles and habitation areas of the extraterrestrial settlements.

This book is the result of a multidisciplinary effort to assess the state of art in our knowledge on the radiation situation during deep space missions and on the impact of this complex radiation environment on the space traveller. It comprises the lectures by the faculty members as well as short contributions by the students given at the NATO Advanced Study Institute "Biological Effects and Physics of Solar and Galactic Cosmic Radiation" held in Armacao de Pera, Portugal, 12-23 October, 1991. The following scientists served on the Organizing Committee:

C. E. Swenberg, Armed Forces Radiobiology Research Institute, Bethesda, Maryland, USA

G. Horneck, Deutsche Forschungsanstalt für Luft- und Raumfahrt, Köln, Germany

E.G. Stassinopoulos, NASA Goddard Space Flight Center, Greenbelt, Maryland, USA

P.D. McCormack, US Naval Medical Center, Washington D.C., USA

The participants, coming from various countries including Russia, Ukraine, Czechoslovakia, Bulgaria are listed at the end of this book. The event is in many respects a sequel to the NATO Advanced Study Institute "Terrestrial Space Radiation and its Biological Effects", Corfu, Greece, October 1987, which was mainly concerned with radiation problems for manned missions in Low Earth Orbit (LEO).

During this meeting, it was emphasised that in order to safeguard future human enterprises in space, especially those of very long duration or beyond our geomagnetic shield, an intense research program has to be initiated. The program objectives should include:

(1) with respect to the radiation environment in deep space missions: development of a better physical model of the galactic cosmic radiation modulation as a function of solar cycle observables; development of models of HZE particles propagation in the interplanetary medium; a better understanding on the periodicity (magnitude, duration) of a solar cycle, in order to make a prediction several years in advance; a standardized approach on an international scale for predicting/forecasting solar flares that give rise to proton events and an

efficient warning system; microdosimetric approaches in future development of space dosimetry; determination of the dose contributing products as a function of shielding; testing of new space technologies with respect to their vulnerability to space radiation;

(2) with respect to biological responses to the radiation in space: selection of appropriate biological test systems for radiobiological space experiments in order to quantify and qualify various long-term biological radiation effects; analysis of the biological responses to single particle traversals; development of biological dosimeters; a better understanding of the radiobiological chain of events as a function of radiation characteristics from the initial interactions altering the essential chemical processes, such as DNA strand breaks to the biological response, e.g. cellular lethality, mutagenesis, transformation; investigation of transeffects, i. e. radiation lesions in the DNA that may lead to changes at remote sites; development of radiobiological models appropriate for determining biological effects of HZE particles; international cooperation in the analysis of radiation effects in higher organisms and humans including data obtained in space;

(3) with respect to risk estimates for deep space missions: development of new and more appropriate concepts to quantitate the radiation risk in space missions; development of shielding concepts; development of countermeasures including radioprotectants, nutritional supplements; determination of the radiation tolerances of different individuals.

This list is by no means complete. It reflects the need for a long-term program where ground based studies will be augmented by flight experiments, especially in high inclination orbits or on precursor missions to Moon and Mars. It was considered to be extremely important to reach a standardisation on an international level with respect to data collection, protocol comparison and formulation of guidelines for future exploratory class missions.

The committee is most grateful to the North Atlantic Treaty Organization for the outstanding support provided for this meeting and for the production of this monograph. It also acknowledges substantial financial support provided by German Aerospace Research Establishment DLR, The US Armed Forces Radiobiology Research Institute, Bethesda MD, the US Department of Energy, and the Committee on Interagency Radiation Research and Policy Coordination in cooperation with Oak Ridge Associated Universities. We thank Lisa Steimel for her valuable and efficient assistance in assuring that the meeting was truly successful. The editors acknowledge the advice and guidance of Mr. Gregory Safford of Plenum Press and the assistance of Dr. Mei-Lie Swenberg in typing and reorganizing the manuscripts to the final version.

Charles E. Swenberg

Gerda Horneck

E. G. Stassinopoulos

CONTENTS

RADIATION EFFECTS ON BIOLOGICAL SYSTEMS

THEORETICAL MODELS OF BIOLOGICAL RADIATION ACTION

RADIATION-INDUCED DNA LESIONS IN EUKARYOTIC CELLS, THEIR REPAIR AND BIOLOGICAL RELEVANCE

M. Frankenberg-Schwager

GSF: Forschungszentrum für Umwelt und
Gesundheit, Institut für Biophysikalische
Strahlenforschung, Paul-Ehrlich-Str. 20
D-6000 Frankfurt 70, FRG

INTRODUCTION

A variety of lesions can be detected in the DNA of eukaryotic cells irradiated with ionizing radiation. These include DNA- protein crosslinks (DPCs), base alterations and base detachments, sugar alterations, bulky lesions, i.e. clusters of base damage, DNA single- and double-strand breaks. For most of these lesions, i. e. DNA-protein crosslinks, base damage, single- and double-strand breaks and bulky lesions, the kinetics of enzymatic repair has been studied. The influence of various factors, such as oxia/anoxia, linear energy transfer (LET) of the radiation used, incubation medium, cell-cycle stage, cell age, thiol content, hyperthermia,on the induction and repair of these lesions is described and their biological relevance is outlined. Radiation-sensitive cell lines are also included. This paper is an extended and updated version of two earlier publications (Frankenberg-Schwager, 1989 and 1990) with special emphasis on the effect of LET.

RADIATION-INDUCED DNA-PROTEIN CROSSLINKS

Induction of DNA- Protein Crosslinks

DNA-protein crosslinks are formed by covalent linkage between DNA and proteins of the nuclear matrix (e.g. Oleinick et al., 1986). Mainly those DNA regions which contain actively transcribed , and presumably also replicated, sequences are involved in the linkage to proteins, (e.g. Chiu et al., 1982, Oleinick et al., 1986).There is evidence that these DNA sequences are associated with the nuclear matrix (Ciejek et al., 1983, Robinson et al., 1983, Mc Cready et al., 1982). A relatively high amount of about 6000 DPCs is observed per normal, unirradiated cell of Chinese hamster V79 lung fibroblasts in exponential growth (Oleinick et al.,1986). DPCs are induced linearly with radiation dose in the range of 10 to 100 Gy as shown for exponentially growing leukemia mouse cells (Cress and Bowden ,1983)and Chinese hamsterV79 cells (Fig. 1) (Chiu et al. ,1984 , Oleinick et al., 1986).γ-rays increase the number of new linkages between DNA and protein at a frequency of 150 DPCs per Gy per V79 cell (Ramakrishnan et al. ,1987). Thus DPCs are induced at about a 4 time higher frequency than DNA double-strand breaks (Ramakrishnan et al., 1987). The same yield of radiation-induced DPCs is found in normal and radiation-sensitive cell lines (like Ataxia telangiectasia (AT) cells and the Chinese hamster CHO mutant xrs)(Oleinick et al., 1990). At higher doses the number of DPCs approaches a plateau value which corresponds to the

Biological Effects and Physics of Solar and Galactic Cosmic Radiation, Part A, Edited by C.E. Swenberg *et al.*, Plenum Press, New York, 1993

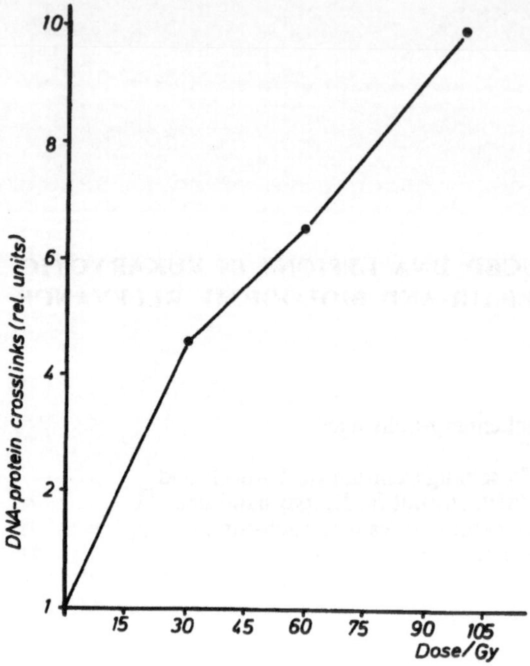

Figure 1. Induction of DNA-protein Crosslinks in Exponentially Growing Chinese Hamster V79 cells Exposed to ^{60}Co-γ rays (Redrawn from Chiu et al. 1984).

number of DNA attachment sites to the nuclear matrix (Ramakrishnan et al. ,1987). The yield of DPCs in irradiated V79 cells is reduced in the presence of cysteamine (Radford ,1986a). Based on the protective effect of the OH radical scavenger dimethylsulphicoxide (DMSO), which is more pronounced in active DNA than in bulk DNA(Chiu et al., 1986), it appears that OH radicals are responsible for the production of DPCs in V79 cells (Oleinick et al., 1986). The yield of DPCs is enhanced when cells are irradiated under hypoxic conditions as shown fora human fibroblast cell line (Fornace and Little, 1977), for CHO cells (Meyn and Jenkins, 1984) and for mouse L and Chinese hamster V79 cells (Radford, 1986a). Interestingly, DPCs are detected in cells from tissues irradiated in situ in acutely hypoxic (N_2 - asphyxiated) mice and even in air-breathing animals (Meyn et al., 1986), indicating the presence of radio-biological hypoxic cells in tissues of air-breathing animals. Preirradiation hyperthermia (43^{0}C for 15 min or longer) has no effect on the yield of radiation-induced DPCs in exponentially growing mouse leukemia cells (Cress and Bowden,1983)and in V79 cells (Radford, 1986a), suggesting that the increased amount of nuclear protein after hyperthermia (Roti-Roti and Winward, 1981) is not available for covalent bonding to the DNA damaged by ionizing radiation. In contrast, pre-irradiation hyperthermia increased the yield of DPCs in mouse L cells (Radford, 1986a).

Repair of DNA-Protein Crosslinks

Mammalian cells are capable of repairing radiation-induced DPCs. DNA is removed from protein with biphasic kinetics during postirradiation incubation of cells at 37^0 C under growth conditions. Human fibroblasts in stationary phase from a confluent monolayer irradiated with 50 Gy under oxic or anoxic conditions exhibit a rapid component of repair of DPCs with a half-time constant $t_{1/2}$ of about 2 h and a slow component with a $t_{1/2}$-value of 12-13 h (Fornace and Little ,1977). In actively transcribed DNA sequences, DPCs are not only preferentially induced but also the removal of DPCs in these regions is faster as compared to nontranscribed DNA (Oleinick et al.,1986). In agreement with these findings is the fast repair of DPCs in exponentially growing monolayer cultures of V79 cells irradiated with 60 Gy. The $t_{1/2}$- value reported for the rapid component is 1 h which is followed by a slower \component of repair (Chiu et al.,1984). Likewise, for growing mouse leukemia cells

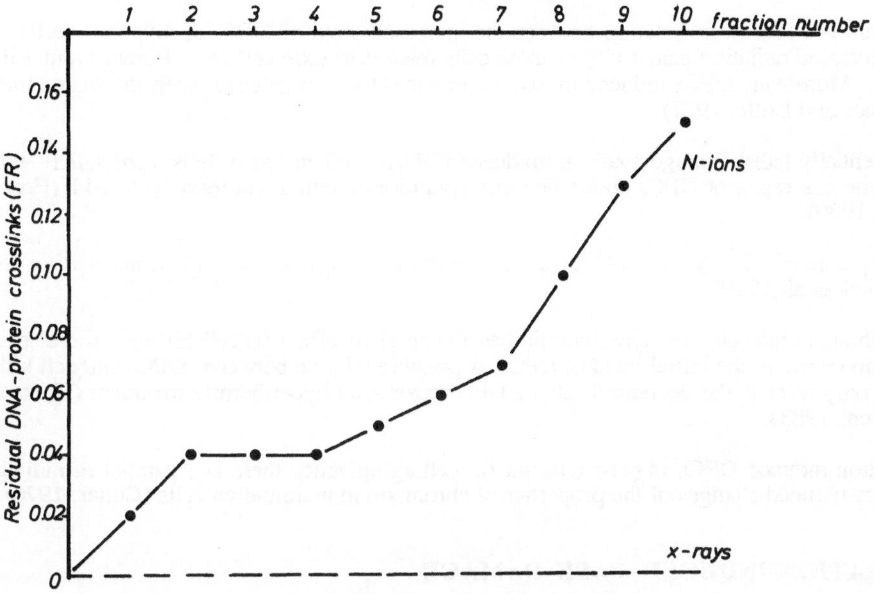

Figure 2. **DNA-protein Crosslinks Remaining after a Postirradiation Incubation for 6h at 37 °C of Human Melanoma Cells Exposed to 6 Gy of N-ions (LET = 530 keV/μm) or to 11 Gy of X-rays. Data Calculated from Eguchi et al.(1987). DNA-protein Crosslinks were Assayed by Alkaline Filter Elution with and without Proteinase K Treatment. FR: Fraction of DNA Bound to Protein.**

irradiated with 50 Gy, a rapid removal of DPCs with a $t_{1/2}$ -value of 20 min is followed by a slower component with a $t_{1/2}$ -value of 2 h (Cress and Bowden, 1983). After preirradiation hyperthermic treatment of these cells, only the slow component of the repair kinetics can be detected (Cress and Bowden , 1983). The slower removal of DPCs after preirradiation hyperthermia may be explained by the finding that the chemical nature of the bonding is changed with heat (Cress and Bowden, 1983). Removal of DPCs is also observed in mouse C3H/10T1/2 fibroblasts which are held after irradiation with 50 Gy under nongrowth conditions at 37°C; in this case DPC removal is relatively slow with a $t_{1/2}$ -value of 10-12 h (Fornace et al., 1980). DPCs are removed with the same efficiency in normal and radiation-sensitive mammalian cell lines (like AT and xrs_6 (Oleinick et al. ,1990). Interestingly, Chinese hamster V79 fibroblasts remove γ-ray induced DPCs but not UV-induced DPCs (Chiu et al..1984), which may point to a different structure of both types of DPCs. More residual DNA-protein crosslinks are observed in human melanoma cells incubated for 6 h after irradiation with high LET radiation (N ions, linear energy transfer (LET) = 530 keV/mm) compared to sparsely ionizing radiation (Fig.2) (Eguchi et al.. 1987). The mechanism of DPC removal appears to involve excision of the adduct, repair replication and a ligation step (Chiu et al.. 1984). Poly-(ADP-ribose) plays a role in the release of DPCs, since 3-aminobenzamide, an inhibitor of the Poly-(ADP- ribose) polymerase, slows down the second component of the DPC repair kinetics (Chiu et al..1984). Cycloheximide has no effect on the repair of DPCs, suggesting that no newly synthesized proteins are required for DPC release (Chiu et al.. 1984). Repair of DPCs seems to require an intact nuclear membrane, as indicated by its absence in metaphase cells (Oleinick et al..1986).

Biological Relevance of DNA-Protein Crosslinks

The biological relevance of DPCs may depend on their chemical structure. To the normally occurring 6000 DPCs per V79 cell in exponential growth phase, only 150 DPCs are added by a mean lethal radiation dose (Ramakrishnan et al., 1987). Radiation-induced DPCs must be structurally different from normally occurring DPCs if they are involved in cell killing. However, evidence suggests that radiation-induced DPCs do not play a role in radiation-induced cell killing:

a) There is a lack of correlation between the increased yield of DPCs in hypoxic cells and the decreased radiation sensitivity of these cells relative to oxic cells(e.g. Fornace and Little, 1977). Moreover, DPCs induced in oxic or anoxic cells are released with the same kinetics (Fornace and Little, 1977).

b) Potentially lethal damage repair in irradiated C3H/10T1/2 mouse cells is completed within 6 h, whereas repair of DPCs under the same conditions continues at least up to 24 h (Fornace et al., 1980).

c) The same efficiency for DPC release is observed in normal and radiationsensitive cells (Oleinick et al.,1990).

d) Although preirradiation hyperthermia has a synergistic effect on cell killing, there seems to be no effect on the initial yield of DPCs. A possible relation between DPCs and cell killing could only refer to the decreased rate of DPC repair after hyperthermic treatment (Cress and Bowden, 1983).

Radiation-induced DPCs may be relevant for cell aging, since there is a parallel in radiation- and age-induced changes of the properties of chromatin in mammalian cells (Cutler, 1976).

RADIATION-INDUCED BASE DAMAGE

Induction of Base Damage

Damage to the four DNA bases is a major type of lesion caused by ionizing radiation in mammalian cells (Cerutti ,1976, Hutchinson, 1985). It is estimated that after a given dose the total number of damaged bases in human lung fibroblasts may be twice the number of DNA single-strand (SSBs) breaks (Cerutti,1974). Various radiation-induced alterations to the thymine base have been characterized (Cerutti , 1974) which, in this paper, are all termed "thymine damage". A linear induction of thymine damage is observed for human lung fibroblasts irradiated with γ-rays up to 3000 Gy (Mattern and Cerutti, 1975), for EMT6 mouse sarcoma cells (up to 1000 Gy) (Dooley et al., 1984), for HeLa cells irradiated with carbon ions up to about 10 000 Gy and for CHO cells exposed to carbon ions up to 50 000 Gy (Fig. 3) (Mattern and Welch ,1979). Using a more sensitive method (i.e. a monoclonal antibody in combination with the ELISA-technique) to detect thymine damage (specifially

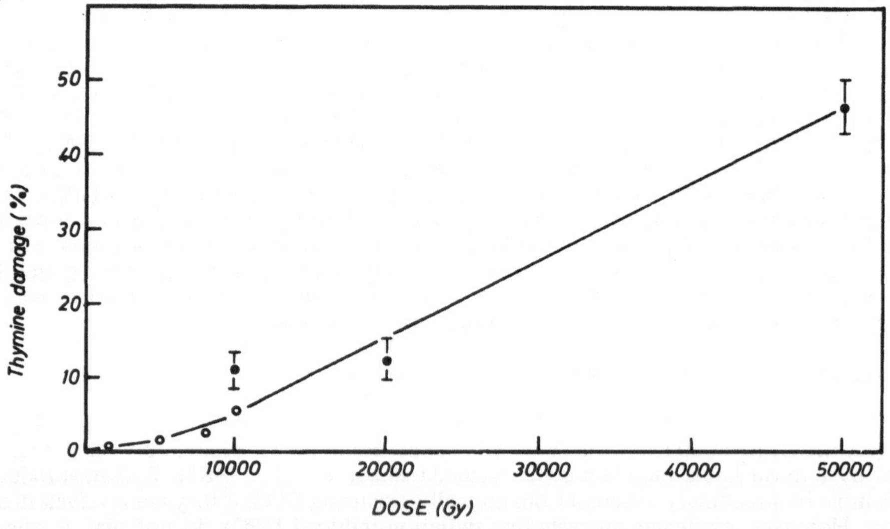

Figure 3. Induction of Thymine Damage in HeLa (o) and CHO (●) Cells Exposed to Carbon Ions (LET = 170 or 780 keV/μm) (Redrawn from Mattern and Welch, 1979).

thymine glycol) in the DNA of normal human fibroblasts, the linear induction of thymine damage by low doses (up to 5 Gy) of γ-rays was confirmed (Leadon, 1990). The efficiency of induction of thymine damage in HeLa and CHO cells exposed to carbon ions (LET 170 and 780 keV/mm) is decreased relative to X-rays (Mattern and Welch, 1979). The frequency of induction of thymine damage depends on the chromatin structure. Thymine damage is about three times more frequently induced in replicating chromatin (Warters and Childers , 1982), and a higher yield is also reported for actively transcribing chromatin of HeLa cells (Patil et al., 1985). Base damage induced by much lower doses can be monitored as endonuclease-sensitive sites (ESSs), i.e. base damaged sites incised by endonuclease yielding additional SSBs (Paterson ,1978). However, only about half of the amount of radiation-induced base damage can be detected as ESSs, deduced on the one hand from the twice as frequent induction of total base damage relative to SSBs (Cerutti, 1974) and on the other hand from the same frequency of ESSs and SSBs observed in human fibroblasts (Fornace et al., 1986). A linear relationship between induced ESSs and dose is found in mouse L 1210 cells (Fornace,1982), in CHO cells (Van der Schans et al., 1979), in mouse L and Chinese hamster V79 cells (Radford, 1986a). Thiol compounds have no effect on the yield of radiation-induced ESSs in CHO cells (Van der Schans et al., 1979) but cysteamine is reported to protect against ESSs in mouse L and V79 cells (Radford, 1986a). The yield of base damage measured as ESSs is about equal under oxic or hypoxic conditions in irradiated human fibroblasts (Paterson et al.,1976), in CHO cells (Van der Schans et al., 1982a) and in mouse L cells (Radford ,1986a). In contrast, the yield of ESSs in Chinese hamster cells is 2.5 times higher under oxic versus anoxic irradiation conditions (Skov et al., 1979). There is another paper reporting a high oxygen enhancement ratio (OER) of about 4 for the induction of ESSs in V79 cells, however, here also extremely high OER s for induced DNA single- (OER ~ 10) and double-strand breaks (OER ~ 6) have been measured (Kinsella et al., 1986). Fibroblasts from normal and Xeroderma pigmentosum individuals show the same efficiency of induction of ESSs (Fornace et al., 1986). Preirradiation hyperthermia has no effect on the induction of thymine damage in CHO cells (Warters and Roti Roti , 1978) and of ESSs in mouse L cells (Radford ,1986a). A powerful and sensitive method combining capillary gas chromatography with mass spectrometry with selected ion-monitoring was applied to detect base damage in cultured human lymphocytes irradiated under anoxic conditions (Dizdaroglu et al., 1987). At a dose range from 10 to 100 Gy the linear induction of a 8,5- cyclo-2-deoxyguanosine could be monitored. This type of damage arises by intramolecular cyclisation between guanine and the neighbouring sugar moiety and was not known before to occur in irradiated cells.

Repair of Base Damage

Thymine damage is removed from the DNA of cells during postirradiation incubation in growth medium at 37° C (Mattern et al. ,1973, 1975; Mattern and Welch, 1979, Paterson et al. 1976. Van der Schans et al., 1979). Repair of radiation-induced thymine damage is usually completed within 1 h and faster than repair of UV- induced thymine damage, which requires one to several hours for completion. For example, about 80 % of thymine damage induced by a dose of 2500 Gy is removed from the DNA of human lung fibroblasts, SV 40 transformed human fibroblasts and CHO cells within 15 min of postirradiation incubation (Mattern et al., 1975). Excision of thymine damage induced by ionizing radiation is equally efficient in human and rodent cells, in contrast to UV-induced thymine damage which seems to be reduced in the latter cells. This indicates that the endonucleolytic incision step differs for base damage induced by ionizing radiation and by UV-light. Endonucleolytic activity towards γ-irradiated DNA has been detected in human cells (Brent, 1976) and in calf thymus cells (Bacchetti and Benne , 1975). Radiation-induced base damage, monitored as ESS, is removed exponentially with time (first order reaction) with $t_{1/2}$-values of about 1 h in human fibroblasts (Paterson et al., 1976) and of about 1.5 h in Xeroderma pigmentosum fibroblasts (Fornace et al., 1986). The removal of ESS is impaired in a radiation-sensitive cell line of Ataxia telangiectasia patients (Paterson et al., 1976; Paterson and Smith, 1979) indicating a defective excision repair in these cells. Other AT cell lines, however, show normal excision repair (Hariharan et al., 1981; Van der Schans et al., 1981). The rate of ESS removal is not affected by the absence of oxygen and the presence of thiol compounds during irradiation of CHO cells (Van der Schans et al., 1979, 1982 a). Approximately 50% of the thymine glycols induced by 4 Gy of γ-rays were removed within 45 min from normal human fibroblasts and from human mammary epithelial cells, as detected by the monoclonal antibody

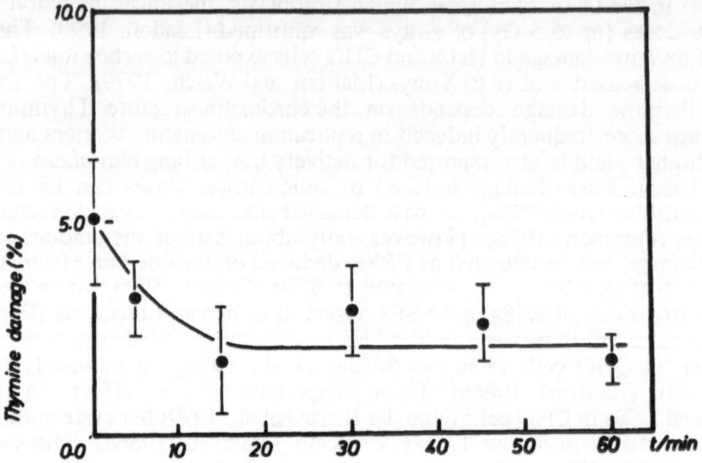

Figure 4. Removal of Thymine Damage During Incubation at 37 ° C from HeLa cells Exposed to Carbon Ions (780 keV/μm) at a Dose of 10 000 Gy (Redrawn from Mattern and Welch, 1979).

assay combined with ELISA (Leaper, 1990). The chromatin structure of cells is important not only for the frequency of induction of thymine damage but also for the rate of repair. In active chromatin of HeLa cells the rate of removal of thymine damage is higher than in inactive chromatin (Patil et al.,1985). Although high LET radiation reduces the yield of thymine damage, no effect on the rate of its removal is observed in HeLa cells exposed to carbon ions (Fig. 4) (Mattern and Welch, 1979). Base damage is removed from eukaryotic cells by two modes of excision repair. The "base excision repair" acts mainly on X-ray-induced base damage, whereas the "nucleotide excision repair" acts on damage induced by UV-light (e.g. Friedberg, 1985). The "base excision repair" starts with the removal of damaged bases from the DNA by glycosylases. Two DNA glycosylases (one specific for several 5-6 saturated thymine derivatives, the other specific for imidazole-ring opended purines) were found to be involved in the removal of radiation-induced base damage (Breimer and Lindahl, 1985). The resultant apurinic/apyrimidinic (AP) site intermediates are then substrates for AP-endonucleases which incise the DNA backbone (Lindahl, 1976). The formation and repair of such AP-sites was demonstrated in HeLa cells after irradiation with a low dose (3.75 Gy) of γ-rays (Moran and Ebisuzaki, 1987).

Biological Relevance of Base Damage

There is evidence against the involvement of base damage in radiation-induced cell killing:

a) Lack of correlation between cell killing and base damage induction independent of LET. For example, for carbon ions the efficiency for cell killing is maximum at a LET-value of about 200 keV /mm, yet the yield of thymine damage is reduced by a factor of 5 (Mattern and Welch, 1979). Furthermore, the rate and extent of repair is the same after high and low LET radiation (Mattern and Welch,1979).

b) Hypoxia protects against radiation-induced cell killing but, in most cases, not against base damage (Paterson et al., 1976; Radford, 1986a; Van der Schans et al., 1982a).

c) Some AT cell lines of similar radiosensitivity are capable of repairing ESSs (Hariharan et al. , 1981; Van der Schans et al., 1981), others are not (Paterson et al., 1976; Paterson and Smith, 1979).

d) Hyperthermia enhances radiation-induced cell killing but has apparently no effect on the yield of base damage, measured as ESSs (Radford, 1986a).

Radiation-induced base damage plays a major role in the induction of point mutations (reviewed by Breimer, 1988) in essential genes. For example, 84 % of the mutational events in the essential gene (APRT) of Chinese hamster cells are classified as "point mutations" (Grosovsky et al., 1988). In contrast, mainly large deletions rather than point mutations given rise to mutational events in a non-essential gene (HPRT) of V79 cells (Thacker, 1986, Nicklas et al.,1991). Thymine damage may play a role in the process of cellular aging. With increasing age of human WI-38 cells in culture, the efficiency of removal of radiation-induced thymine damage decreases and hence an accumulation of thymine damage in the DNA of old cells is observed (Mattern and Cerutti,1979).

RADIATION-INDUCED DNA SINGLE-STRAND BREAKS

Induction of DNA Single-Strand Breaks

Induction of SSBs has been studied for a large variety of mammalian cells and has been found to increase in proportion to the dose of radiation (e.g. Ahnström and Edvardsson , 1974, Coquerelle et al., 1973, Kohn et al., 1976, Roots et al., 1979). The same induction frequency for SSB is observed in radiation-resistant Chinese hamster and in radiation-sensitive mouse cells (Sakai and Okada, 1984, Radford, 1986a). Likewise, no difference in SSB induction is detected in normal CHO cells and their radiosensitive counterparts EM9 and NM2 (Giaccia et al.,1985, VanAnkeren et al.,1988), in radioresistant (L5178Y-R) and sensitive (L5178Y-S) murine leukemic lymphoblast cell lines (Wlodek and Hittelman,1987, Evans et al., 1987), and in normal human fibroblasts and radiosensitive cells from retinoblastoma patients (Woods et al. , 1982). The efficiency of SSB induction decreases with increasing LET of heavy ions as summarized in Fig. 5 (Roots et al., 1990). In agreement with this, the yield of SSB in human melanoma cells exposed to N-ions (530 keV/mm) was also smaller than after X-ray exposure (Eguchi et al., 1987). The efficiency of SSB induction by neutrons is decreased compared to sparsely ionizing radiation as shown for human skin fibroblasts (Van der Schans et al., 1983), for Chinese hamster V79 cells (Ahnström and

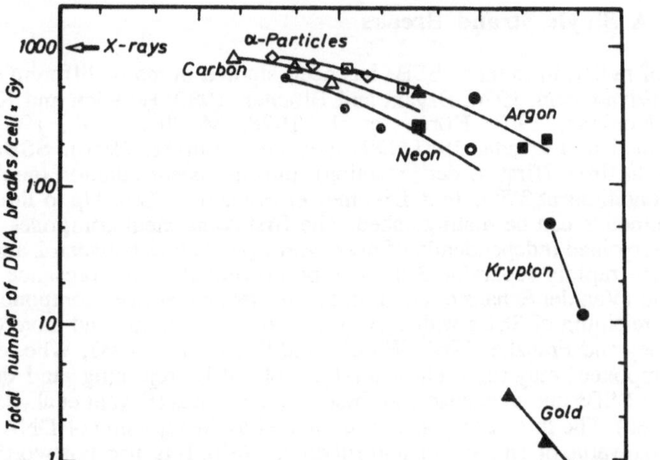

Figure 5. Induction Frequency of Total Number of DNA Strand Breaks per Mammalian Cell (i.e. Single-strand Breaks, Alkaline Labile Sites, Double-strand Breaks) as a Function of LET (Redrawn from Roots et al.,1990). (Δ), C, 18, 49, 88 keV/μm, human T1 Kidney Cells (Roots et al., 1979); (▲), C, 360 keV/μm, Chinese Hamster V79 Cells (Kampf and Eichhorn, 1983); (◘), C, 332 keV/μm, Chinese Hamster V79 Cells (Ritter et al., 1977); (●), Ne, 41, 109, 194 keV/μm, (Roots et al.,1979); (o), Ne, 900 keV/μm, (Kampf and Eichhorn, 1983); (O), Ne, 827 keV/μm, (Ritter et al. 1977); (◘), A, 112, 272 keV/μm, (Roots et al. 1979); (●), A, 1953, 2900 keV/μm, (Ritter et al. 1977); (◇),α-Particles 33, 66, 160 keV/μm, (Kampf and Eichhorn, 1983); (●), Kr, 3000, 5000 KeV/μm, (Rydberg, 1985); (▲), Au, 1500, 2700, 4400 keV/μm (Rydberg, 1985).

Edvardsson, 1974, Peak et al., 1989), for mouse cells (Sakai et al., 1987), and for human P3 epithelial teratocarcinoma cells (Peak et al., 1989). Hypoxia decreases the yield of radiation-induced SSBs relative to oxic conditions (Palcic and Skarsgard, 1972; Lennartz et al.,1975a).The decrease is by a factor of 3 for Chinese hamster cells V 79 fibroblasts (Koch and Painter, 1975) and for human fibroblasts (Revesz, 1985), but also a factor of about 4 has been reported for Chinese hamster cells (Hohman et al. 1976). Radiation-induced SSBs, assayed under alkaline conditions, comprise about 30%- 50% alkaline labile sites in cells irradiated under oxia (Lennartz et al., 1973, Matsudaira et al., 1977) and even more of these sites (about 50%) are produced under anoxic conditions (Lennartz et al. ,1973). Exogenously supplied thiol compounds reduce the yield of SSBs in human fibroblasts (Van der Schans et al., 1979) and in mouse and Chinese hamster cells (Radford , 1986a) irradiated under hypoxic conditions. In agreement with these findings is the increased yield of SSBs in irradiated anoxic human fibroblasts depleted in endogenous glutathione (Edgren et al., 1981). It was reported that sensitization to SSB induction occurs in HeLa and CHO cells depleted of intracellular thiol compounds at moderate oxygen concentrations (0.5-10mM), whereas sensitization under oxic or extreme hypoxic conditions (0.01mM oxygen) is marginal (Van der Schans et al., 1986). The radioprotective agent WR-1065 is reported to decrease the yield of radiation-induced SSBs (Grdina and Nagy, 1986). The efficiency of SSB induction does not vary during the cell cycle in CHO cells (Humphrey et al., 1968, Lett and Sun, 1970, Graubmann and Dikomey, 1983), in mouse mammary tumour cell lines (Sweigert et al., 1988) and in HeLaS3 cells (Watanabe and Horikawa , 1977). Preirradiation hyperthermia has no effect on the yield of SSB in CHO cells (Dewey et al., 1980, Iliakis et al., 1990), HeLa cells (Lunec et al., 1981) and mouse cells (Radford ,1986a). Contradictory to this, an increased frequency of radiation-induced SSBs following hyperthermic treatment of CHO cells is observed (Corry et al., 1977; Mills and Meyn , 1981). Depending on the assay applied, preirradiation hyperthermic treatment of CHO cells resulted in an enhanced yield of SSB (measured by alkaline filter elution) or had no effect on the induction of SSB (measured by alkaline sucrose gradient centrifugation) (Warters et al., 1987).SSBs are preferably induced by radiation in transcriptionally active DNA sequences as shown in mouse (Chiu et al., 1982) and Chinese hamster (Oleinick et al., 1985) cells. This is explained by the better accessibility of OH radicals to DNA during transcription.

Repair of DNA Single-Strand Breaks

Repair of radiation-induced SSBs has been studied in many different cell lines (e.g. Ahnström and Edvardsson, 1974, Bryant and Blöcher, 1980, Bowden and Kasunic, 1981, Jorritsma and Konings, 1983, Körner et al., 1978, McGhie et al., 1983, McWilliams et al., 1983, Sakai and Okada, 1981, Dikomey and Franzke, 1986). SSBs are rejoined exponentially with time (first order reaction) during postirradiation incubation of cells under growth conditions at 37^{0} C (e.g. Dikomey et Franzke ,1986). Up to three components of SSB repair kinetics can be distinguished. The first component comprises 70-90% of all SSBs which are rejoined independently of dose with $t_{1/2}$ -values between 2 and 10 min. The induction of these rapidly reparable SSBs can be prevented by the presence of cysteamine during irradiation (Van der Schans et al. ,1982a).The second, slower component represents a dose-dependent rejoining of SSBs with $t_{1/2}$ -values of about 10 min and more (e.g. Bryant et al.,1984, Dikomey and Franzke, 1986, Wheeler and Wierowski , 1983, Wheeler and Nelson, 1987). This component may represent a mixture of SSB rejoining and dose-dependent formation of new SSBs due to excision at base damage sites (Bryant et al., 1984; Dikomey and Franzke, 1986). The third component is attributed to the rejoining of DNA double-strand breaks with a $t_{1/2}$ -value of 1h (Bryant and Blöcher, 1980). It is also noteworthy that the rate of SSB rejoining - if measured under alkaline conditions - is slowed down due to the much slower repair of alkaline labile sites (Lennartz et al. ,1975b, Matsudaira et al., 1977). SSBs are rejoined with the same rate in radiation-resistant and -sensitive human cell lines (Paterson et al., 1976, Fornace and Little, 1980, Hariharan et al., 1981, Woods et al., 1982, Van der Schans et al., 1982a, 1983) and rodent cells (Sakai and Okada, 1984, Van Ankeren et al., 1988, Wlodek and Hittelman, 1987, Evans et al., 1987). However, a slower rate of SSB rejoining was reported for the radiosensitive CHO cell line EM9 (Thompson et al. ,1982, Schwartz et al.,1987, VanAnkeren et al., 1988). A higher fraction of unrejoined SSBs was found in some radiosensitive cells (Sakai and Okada, 1984, Van Ankeren et al., 1988). No difference in the rate of SSB rejoining is observed in undifferentiated and terminally differentiated rodent brain cells, yet the slower overall SSB rejoining rate in differentiated

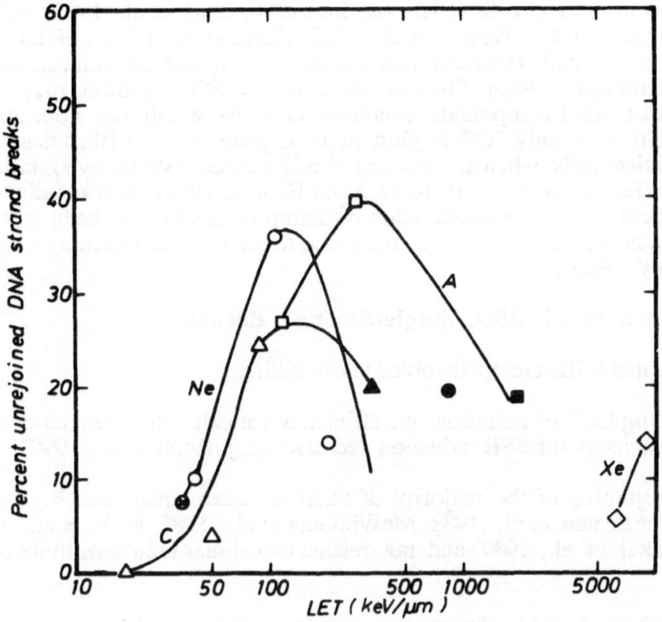

Figure 6. Unrejoined total DNA strand breaks as a function of LET (Redrawn from Roots et al. 1990; see ibid for further details) (▲), C, 17.8, 49.3, 88.4 keV/μm, human T1 kidney cells, 20 Gy (Roots et al., 1979); (▲), C, 332 keV/μm, Chinese hamster V79 cells, 300-450 Gy (Ritter et al.,1977); (●), Ne, 35 keV/μm, rabitt retina cells, 10 Gy (Lett et al., 1986); (○), Ne, 41, 109, 194 keV/μm (Roots et al.,1979); (●), Ne, 827 keV/μm (Ritter et al.,1977); (□), A, 112, 272.4 keV/μm (Roots et al., 1979); (■), A, 1953 keV/μm (Ritter et al., 1977); (◇), Xe, 6200, 8800 keV/μm, bovine lens epithelial cells, 72-200 Gy (Aufderheide et al., 1987).

neurons is mainly due to the increased fraction of slow reparable breaks in these cells (Wheeler and Nelson,1987). Contact inhibited mouse C3H/10T1/2 cells in stationary phase exhibit a slow rejoining of SSBs when cells are kept after irradiation under nongrowth conditions. 90% of the induced SSBs are rejoined during the first hour and most of the remaining SSBs disappear during the next 5h (Fornace et al., 1980). The rate of SSB rejoining after high LET irradiation (α-particles) in CHO cells is decreased compared to low LET irradiation (Cole et al., 1980). In contrast, a similar rate of SSB rejoining is observed after exposure to low LET radiation or to neutrons of human fibroblasts and of cells from AT patients (Van der Schans et al., 1983), of human lymphocytes (McWilliams et al., 1983), of HeLa cells (Maki et al., 1986), as well as of human melanoma cells exposed to N-ions (Eguchi et al., 1987). The fraction of nonrejoined SSBs increases with the LET of heavy ions (Fg. 6) (for review see Roots et al., 1990). An increased yield of unrejoined SSBs induced by neutrons is reported for V 79 cells (Ahnström and Edvardsson, 1974; Körner et al., 1978, Peak et al., 1989), for mouse cells (Furuno et al. ,1979; Sakai et al., 1987) and for human P3 epithelial teratocarcinoma cells (Peak et al., 1989). In CHO cells 20% of the SSBs induced by α-particles remained unrejoined after 6h compared to only 10% after low LET irradiation (Cole et al., 1980). There is no correlation of the rate of SSB rejoining with the cell cycle dependent variation in radiosensitivity (Sawada and Okada, 1970), but a similar cell cycle dependent variation of nonrejoined SSBs and cellular radiosensitivity is observed (Sakai and Okada, 1984). On the other hand, the rate of SSB rejoining was found to be dependent on the chromatin structure of irradiated cells. For example, a slower rate of rejoining is measured in Chinese hamster fibroblasts in metaphase compared to asynchronous cells (Oleinick et al., 1985). There is no effect of age on the capacity to rejoin SSBs, as observed for stimulated human peripheral blood lymphocytes (Turner et a., 1982). Radiation-induced SSBs are rejoined even under extreme hypoxic conditions in V79 cells, but the amount of SSBs left unrejoined after an incubation period of one hour is 20% in cells under extreme hypoxia and only 10% in cells kept under oxic conditions (Koch and Painter, 1975). Treatment of cells at temperatures of about 43°C prior to irradiation reduces the rate and the extent of SSB rejoining as shown for CHO cells (Clark and Lett,1976; Clark et al., 1981; Mills and Meyn, 1981; Mills and Meyn, 1983; Warters et al. ,1987; Iliakis et al., 1990), HeLa cells (Lunec et al. ,1981), and mouse cells (Bowden and Casunic, 1981) and Ehrlich ascites tumour cells (Jorritsma and Konings , 1983). The rate and extent of SSB rejoining may depend on the level of intracellular thiol compounds. Rejoining of SSBs within one hour after irradiation under oxic conditions is only 70% in glutathione-depleted human fibroblasts compared to glutathione-proficient cells, whereas rejoining of SSBs induced under hypoxia is about 100% in both cell lines (Edgren et al., 1981; Revesz and Edgren, 1984). In contradiction to this, no effect of thiol depletion on postirradiation rejoining of SSBs has been reported (Vos et al.,1986). The radioprotector WR-1065 was found to inhibit SSB rejoining in irradiated cells (Grdina and Nagy, 1986).

Biological Relevance of DNA Single-Strand Breaks

Evidence that SSBs are not involved in cell killing:

a) With increasing LET of radiation, the efficiency for cell killing increases (e.g. Fowler, 1981), yet the efficiency for SSB induction decreases (e.g. Roots et al., 1990).

b) The rate of rejoining of the majority of SSBs is about equal after high and low LET radiation (Van der Schans et al. ,1983, McWilliams et al., 1983, Maki et al., 1986, Eguchi et al., 1987, Sakai et al., 1987) and not related to cellular radiosensitivity (e.g. Szumiel , 1981).

c) A level of SSBs, induced by hydrogen peroxide via OH radicals, corresponding to 10 Gy does not result in killing of Chinese hamster cells and does not affect cell inactivation by γ-rays (Ward et al.,1985).

However, a correlation between unrejoined SSBs and cell killing is observed for radiation-resistant and -sensitive cell lines (Sakai and Okada, 1984), for the cell cycle dependency (Sakai and Okada , 1984), for the LET dependency (Cole et al., 1980; Ritter et al., 1977), and for preirradiation hyperthermia (Clark and Lett ,1976; Bowden and Casunic, 1981; Lunec et al., 1981; Mills and Meyn ,1981). It remains to be elucidated to which extent DNA double-strand breaks contribute to the amount of unrejoined SSBs (see Ritter et al., 1977; Dikomey and Franzke, 1986; Radford, 1986a; Sakai et al., 1987, Peak et al., 1991).

RADIATION-INDUCED DNA DOUBLE-STRAND BREAKS

Induction of DNA Double-Strand Breaks

After irradiation with (supra)lethal doses, a linear relationship between double-strand breaks (DSBs) and doses has been observed for various types of eukaryotic cells assayed by various methods (e.g. Corry and Cole,1968, Lehman and Ormerod, 1970, Blöcher, 1982; Van der Schans et al., 1982a, McMillan et al., 1990). Contradictory to this, in cases where neutral filter elution was used to measure DSBs, it was reported that a non-linear, shoulder-type induction curve for DSBs is found at low doses as applied in survival studies (Radford, 1985a, Prise et al., 1987, Kelland et al., 1988, Okayasu et al., 1988, Kosaka et al. ,1990, Peak et al. , 1991) and it was suggested that DSBs are induced in cells in proportion to the square of dose (Blazek et al., 1989). Controversial opinions exist about the possibility that this finding is a technical artifact depending on the cellular lysis conditions applied (Okayasu and Illiakis, 1989, Radford, 1990a, see also Hutchinson ,1989). For example, in three Chinese hamster ovary cell lines of varying radiosensitivities a non-linear DSB induction curve is observed by neutral filter elution at pH 9.6, whereas at pH 7.0 all three cell lines show a linear dose-response for the yield of DSBs (VanAnkeren et al., 1988). In yeast cells, DSBs are induced under various irradiation conditions linearly with dose in a dose range applied in survival assays (Fig. 7) (Frankenberg-Schwager et al., 1979; Frankenberg et al., 1986). Yeast mutants, exhibiting no detectable rejoining of DSBs, show an exponential survival curve where about one DSB per cell corresponds to one lethal event (Frankenberg et al., 1981). On the basis that one unrepaired DSB is lethal for a yeast cell, a curvi-linear induction of DSB with dose should result in a shoulder-type rather than an

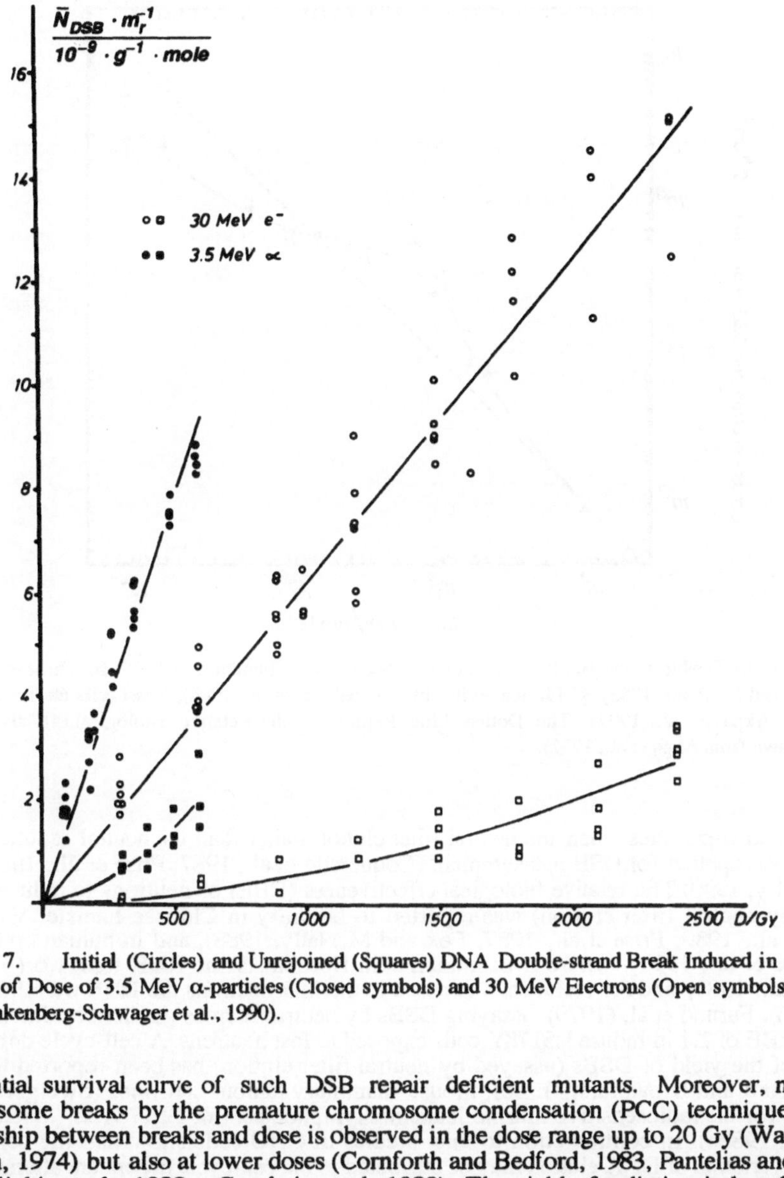

Figure 7. Initial (Circles) and Unrejoined (Squares) DNA Double-strand Break Induced in Yeast as a Function of Dose of 3.5 MeV α-particles (Closed symbols) and 30 MeV Electrons (Open symbols) (Redrawn from Frankenberg-Schwager et al., 1990).

exponential survival curve of such DSB repair deficient mutants. Moreover, measuring chromosome breaks by the premature chromosome condensation (PCC) technique, a linear relationship between breaks and dose is observed in the dose range up to 20 Gy (Waldren and Johnson, 1974) but also at lower doses (Cornforth and Bedford, 1983, Pantelias and Maillie, 1985, Iliakis et al., 1988a, Goodwin et al. 1989). The yield of radiation-induced DSBs is lower by a factor of about 25 than the yield of SSBs based on 1000 SSBs per mammalian cell per Gy (Elkind, 1979) and 40 DSBs per cell per Gy (Blöcher, 1982). In oxic cells the yield of radiation- induced DSBs is about three-fold higher than in anoxic cells (e. g. Lennartz et al., 1975a, Kampf et al., 1977a, Frankenberg-Schwager et al., 1979, Radford, 1985a, Prise et al., 1987). The yield of DSBs (measured by the neutral sedimentation technique) increases with increasing LET up to a maximum at 100 - 200 keV/mm of the radiation applied (fig. 8) (Kampf et al., 1977a, 1977b, Cole et al., 1980, Frankenberg et al., 1981, Kampf and Eichhorn,1983, Blöcher, 1988). In agreement with this, α-particles (LET = 128 keV/μm) and Ne ions (LET= 183 keV/μm) are found to be 2.16 and 1.5 times more effective than X-rays at producing chromatin breakage (assayed by the PCC technique) in noncycling HF19 human diploid fibroblasts and in CHO cells (Bedford and Goodhead, 1989, Goodwin et al., 1989). Recently, these findings were confirmed by Cornforth and Goodwin (1991) in noncycling normal human fibroblasts exposed to 3.5 MeV ^{238}Pu α-particles (LET = 116 keV/μm). In contrast, virtually the same induction frequency for DSBs is observed for sparsely ionizing

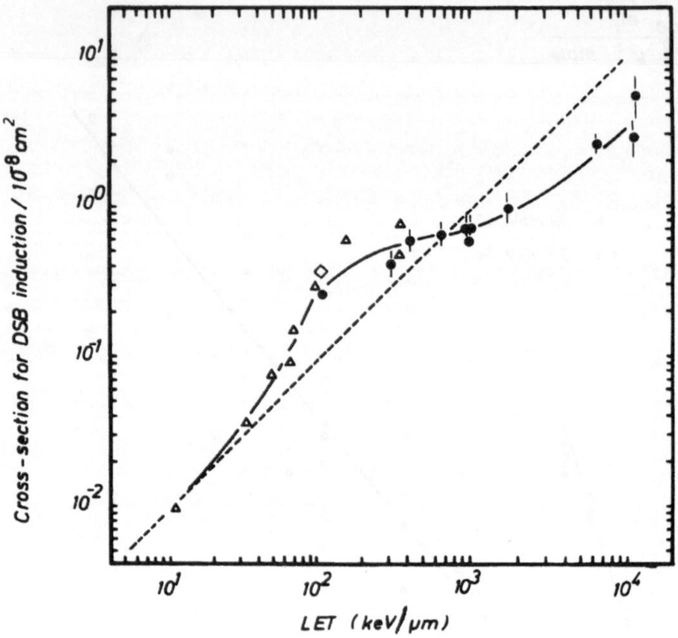

Figure 8. DNA Double-strand Break Induction Cross Section as a Function of LET. (▲) Chinese Hamster V79 (Kampf and Eichhorn, 1983), (◇) Ehrlich ascites tumour cells (Blöcher, 1988), Yeast cells (●) (Frankenberg et al., 1981 , Akpa et al., 1992). The Dotted Line Represents the Relative Biological Effectiveness of Unity (Redrawn from Akpa et al., 1992).

radiations and α-particles when the neutral filter elution rather than the neutral sedimentation technique was applied for DSB measurement (Coquerelle et al., 1987; Prise et al., 1987, Fox and McNally, 1990).The relative biological effectiveness (RBE) of neutrons to induce DSBs (assayed by neutral filter elution) was reported to be unity in Chinese hamster V79 cells (Grdina et al., 1989, Prise et al., 1987, Fox and McNally, 1988), and in human epithelioid cells (Peak et al., 1991), whereas in human skin fibroblasts the RBE was 1.6 (Van der Schans et al., 1983) and for HeLa cells exposed to thermal neutrons the RBE was 2.6 (Maki et al., 1986). Furuno et al. (1979), assaying DSBs by neutral sucrose gradient sedimentation, found an RBE of 1.1 in mouse L5178Y cells exposed to fast neutrons. A cell-cycle dependent variation of the yield of DSBs (assayed by neutral filter elution) has been reported for V79 cells (Radford and Broadhurst, 1986), mouse mammary tumour cell lines (Sweigert et al., 1988) and for the radiosensitive murine leukemic lymphoblast line L5178Y-S (Wlodek and Hittelman, 1988a). In contradiction to this, studies in CHO cells show that the cell-cycle dependent variation of DNA elution may be attributed to the cell-cycle dependent alterations of the physicochemical properties of DNA (especially that associated with DNA replication) rather than to a cell-cycle dependent variation of DSB induction (Okayasu et al.,1988; Iliakis and Okayasu ,1988; Iliakis et al., 1991a; Iliakis et al., 1991b). In agreement with this latter statement is the cell-cycle independent induction of DSBs (assayed by neutral sucrose gradient sedimentation) reported for mouse EAT cells (Blöcher et al. , 1983). An age-dependent decreased induction of DSBs (assayed by neutral filter elution) is reported for unstimulated human peripheral blood lymphocytes exposed to 30 Gy of X-rays (Mayer et al., 1990) which may be related to age-dependent changes in chromatin structure (reviewed by Mayer and Baker (1987)). Preirradiation hyperthermic treatment of rodent cells is reported to yield an increased induction of DSBs (Radford, 1983, 1985,,1986b). In contradiction to this, heat did not affect the induction of DSBs in normal CHO cells (Corry et al.,1977; Iliakis et al., 1990) and its radiosensitive derivative xrs-5 (Iliakis et al., 1990). No difference between radiation-resistant and -sensitive CHO cell lines (Giaccia et al.,1985; Van Ankeren et al. ,1988; Iliakis et al.,1988b), mouse lymphoma cells (Evans et al., 1987) and human cell lines (Van der Schans et al., 1982a, 1983; Radford and Hodgson, 1990) is found for the induction of DSBs and for chromosome breakage (Cornforth and Bedford, 1985). However, in other studies a higher yield of initial DSBs (assayed by neutral filter elution) was observed in radiosensitive cell lines and this was claimed to be the cause of the enhanced sensitivity of these cell lines (Radford,1986b, Kelland et al. ,1988).However, despite the fact that some radiosensitive cell lines show higher levels of induced DSBs than resistent lines, no consistent relationship between radiosensitivity and induction of DSBs was found in human tumour cell lines with widely varying radiosensitivities (Schwartz et al.,1988, McMillan et al., 1990, Schwartz et al., 1991). In agreement with this is the finding that radiation hypersensitive cells isolated from the scid mouse exhibit the same DSB induction frequency as normal mouse BALB/c cells (Biederman et al. ,1991). The effect of glutathione-depletion on the yield of DSBs is relatively small in CHO cells irradiated under oxic or severely hypoxic conditions; only at moderate hypoxia does glutathione depletion sensitize for DSB induction (Van der Schans et al. 1986). In rodent cells, a protective effect of cysteamine against radiation-induced DSBs is observed (Radford , 1986b). In the presence of the radioprotector WR-1065 the yield of radiation-induced DSBs in V79 cells is decreased (Sigdestad et al., 1987).

Rejoining of DNA Double-Strand Breaks

DSBs are rejoined during postirradiation incubation in various types of cells under growth or nongrowth conditions with monophasic (Bradley and Kohn, 1979, Bryant and Blöcher, 1980, Radford, 1987) or biphasic kinetics (Cole et al., 1980, Van der Schans et al., 1982a, 1983, Coquerelle et al., 1987, Radford , 1987, Wlodek and Hittelman, 1987, 1988b, VanAnkeren et al., 1988, Frankenberg-Schwager et al., 1990a, Metzger and Iliakis, 1991, Peak et al.,1991, Frankenberg-Schwager and Frankenberg,1991). The rejoining of radiation-induced DSBs is exponential with time (first order reaction) and independent of dose as shown for mouse cells (Bradley and Kohn , 1979) and Ehrlich ascites tumour cells (Blöcher and Pohlit , 1982). However, there is also evidence for a dose-dependent decrease of the rate of DSB rejoining (Frankenberg-Schwager et al. 1990a, Iliakis et al. 1991c). In a detailed study on the effect of dose on DSB rejoining in yeast (Frankenberg-Schwager et al. 1990a) it was observed that the fraction of the slow component increases with dose, resulting in a dose-dependent increase of the overall $t_{1/2}$ -value for DSB rejoining, although both components of the biphasic kinetics are unsaturated reactions. At very low doses (surviving fraction after PLD repair of about 1) the rapid component is predominant, at high doses (surviving fraction after PLD repair of about 0.5 and less) the slow component is predominant, yielding virtually monophasic kinetics in both these instances, whereas at intermediate doses the kinetics is biphasic. These findings may not only explain why monophasic or biphasic rejoining kinetics are reported in the literature, but may also present an explanation for the dose-dependency of the $t_{1/2}$-value of DSB rejoining in agreement with the experimental evidences that DSB rejoining is an unsatured process. Great variations in the half- time constants $t_{1/2}$ are observed for DSB rejoining under growth conditions, which may depend on the method of DSB measurement: an initial rapid component, with a $t_{1/2}$-value of 3-20 min (Cole et al.,1980,Weibezahn and Coquerelle, 1981, Van der Schans et al. 1983, Radford 1983, 1987, VanAnkeren et al., 1988, Koval and Kazmar, 1988, Metzger and Iliakis, 1991, Iliakis et al. 1991c). is followed by a slow component the $t_{1/2}$ - value of which ranges from 40 min to 4 h (Lehman and Stevens, 1977, Bradley and Kohn, 1979, Bryant and Blöcher, 1980, Radford, 1987, VanAnkeren et al., 1988, Metzger and Iliakis, 1991). DSB rejoining in stationary Ehrlich ascites tumour cells is slower under nongrowth conditions ($t_{1/2}$ = 3h) compared to growth conditions ($t_{1/2}$ =1.8h) (Bryant and Blöcher, 1980). In contrast, the same rate of DSB rejoining was observed in Chinese hamster cells during postirradiation incubation under both conditions (Suzuki et al., 1990). A close correlation exists between DSB rejoining and rejoining of chromosome breaks monitored by the PCC technique. The biphasic, dose-dependent rejoining of chromosome breaks under nongrowth conditions shows an exponential dependence of time with t-values of 2h and 6h for the fast and slow component, respectively (Cornforth and Bedford,1983). The rate of DSB rejoining is independent of the cell cycle phase in which cells are irradiated as shown for V79 cells (Weibezahn and Coquerelle, 1981, Rydberg, 1984), CHO cells (Metzger and Iliakis, 1991), EAT cells (Blöcher et al., 1983) and murine leukemic lymphoblast cells L5178Y (Wlodek and Hittelman,1988b). In contrast, a cell-cycle dependency of DSB rejoining was observed for V79 cells which showed in G1 and S phase a biphasic kinetics ($t_{1/2}$ = 2.7min and $t_{1/2}$ = 27min), whereas mitotic cells exhibited only the slow component of rejoining (Radford,1987).An enhanced capacity of DSB rejoining is reported for CHO cells in late S phase (Resnick and Moore, 1979, Giaccia et al., 1985). The isolation of a CHO mutant with a cell-cycle dependent defect in DSB rejoining suggests the existence of two pathways for DSB rejoining, one operating primarily in late S phase, the other operating either throughout the cell cycle or primarily during G1 and early S phase (Giaccia et al., 1985). An age-related twofold decrease in rejoining of X-ray induced DSBs is reported for human peripheral blood lymphocytes from older (aged 66-78) compared to younger (aged 23-43) donors (Meyer et al., 1989, 1990). The same rate of DSB rejoining is observed for radioresistant and most -sensitive human cell lines of Ataxia telangiectasia patients(Lehman and Stevens ,1977; Van der Schans et al., 1982b, 1983; Coquerelle et al., 1987) but few radiation-sensitive lines do exhibit an impaired rate of DSB rejoining (Coquerelle et al., 1987). Radiation-sensitive cells from Ataxia telangiectasia patients with normal rate of DSB rejoining show, however, a higher fraction of nonrejoined DSBs (Blöcher et al., 1991) and chromosome breaks (Taylor, 1979; Cornforth and Bedforth , 1985).For twelve early passage human tumour cell lines with widely varying radiosensitivities a correlation between the rate of DSB rejoining and radioresistence is reported (Schwartz et al., 1988). In contrast, V79 cells and a 100 times more resistant insect cell line show the same rate of DSB rejoining (Koval and Kazmar, 1988). Some studies show a slower rate of DSB rejoining in radiosensitive rodent cells. For example, radiosensitive murine leukemic lymphoblasts L5178Y-S rejoin DSBs at a slower rate (Wlodek and Hittelman ,1987) and less extensively (Evans et al., 1987) than their radioresistent variant L5178Y-R. Likewise, the radiation hypersensitive cells of the scid mouse show a slower and less extensive rejoining of DSBs compared to the normal mouse BALB/c cells (Biederman et al.,1991). The radiosensitive CHO cell line EM9 was shown to have a slower rate of DSB rejoining (Schwartz et al., 1987), but this was questioned by VanAnkeren et al. (1988). After high LET irradiation the rate of DSB rejoining may (Bryant and Blöcher , 1980; Coquerelle et al., 1987; Blöcher, 1988; Frankenberg-Schwager et al.,1990a, Fox and McNally, 1988, Peak et al. ,1991) or may not be slower (Van der Schans et al., 1983, Maki et al., 1986) compared to low LET irradiation.In contrast, there is one report stating that DSBs induced by ^{238}-Pu α-particles are rejoined faster than those induced by X-rays (Fox and McNally,1990). In the course of DSB rejoining after low and high LET irradiation the linear relationship between induced DSBs and dose is converted into a linear-quadratic or quadratic relationship between unrejoined DSBs and dose (Frankenberg- Schwager et al., 1980a, 1980b, 1982, 1990, Blöcher and Pohlit, 1982, Blöcher, 1988, Kosaka et al. ,1990) and the fraction of unrejoined DSBs is higher after high LET irradiation (fig. 7) (Cole et al., 1980; Coquerelle et al., 1987; Frankenberg-Schwager et al., 1982, Blöcher, 1988, Fox and McNally, 1988, Frankenberg-Schwager et al., 1990a, Peak et al., 1991). Similar findings are reported for interphase chromatin breaks (PCC fragments) after irradiation of noncycling, diploid human fibroblasts with low doses of X-rays: initial PCC fragments depend linearly on dose, whilst residual PCC fragments (present after a 24 h incubation of cells under nongrowth conditions) show a non-linear dependence on dose (Cornforth and Bedford ,1987). Likewise, the fraction of unrejoined chromatin fragments was higher in a human/ hamster hybrid cell-line irradiated with Ne ions compared to X-rays (Goodwin et al., 1989). Hyperthermic treatment reduces the rate and extent of rejoining of radiation-induced DSBs (Corry et al., 1977; Ben-Hur et al., 1978; Dikomey, 1982, Radford,1983, Iliakis et al. ,1990) and enhances the formation of chromosome aberrations (Dewey et al., 1978). No effect of thiol depletion on the rejoining of DSBs is reported for CHO cells irradiated under oxic or hypoxic conditions (Vos et al.,1986). No difference in the time course of DSB rejoining was observed after a 2h incubation in growth medium of primary human skin fibroblasts irradiated in the presence of air, N or N +cysteamine (20 mM) (Van der Schans et al., 1982a). Rejoining of DSBs in irradiated V79 cells is unaffected by the presence of the radioprotective agent WR-1065 (Sigdestad et al. 1987). The mechanism of rejoining of radiation-induced DSBs seems to involve a recombinational process (Resnick, 1976) but other authors suggest that at least part of the DSBs are rejoined by simple ligation (Weibezahn and Coquerelle , 1981). In fact, there is experimental evidence for the occurence of the heteroduplex postulated to be formed during the recombinational repair of DSBs in mammalian cells (Resnick and Moore, 1979, Fonck et al.,1984).

Biological Relevance of DNA Double-Strand Breaks

Evidence that radiation-induced DSBs lead to cell killing are :

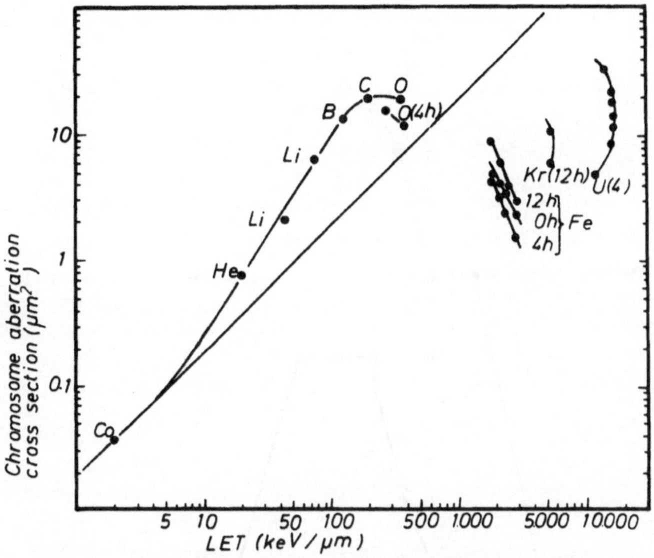

Figure 9. Cross Section of the Induction of Chromosomal Aberrations as a Function of LET. The Data for Lighter Ion Exposure are Obtained 8h after Irradiation (Skarsgard et al., 1967), for the Other Data (Müller, 1985) the Time in Brackets Refers to the Scoring Time after Irradiation (Redrawn from Kraft, 1987).

a) With increasing LET of the radiation both cell killing and yield of unrejoined DSBs increase (Cole et al.,1980, Frankenberg-Schwager et al., 1982, Coquerelle et al., 1987, Blöcher, 1988, Frankenberg-Schwager et al., 1990).

b) Radiation-sensitive human fibroblasts from Ataxia telangiectasia patients show a higher

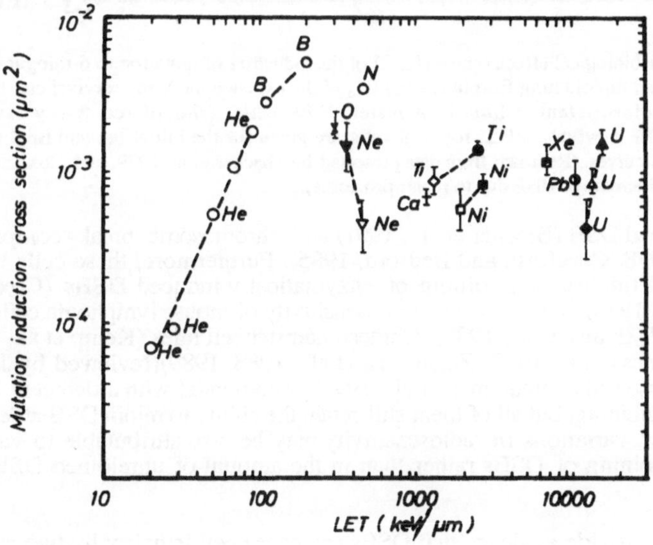

Figure 10. Cross Section of the Induction of Mutation to 6-thioguanine Resistance in Chinese Hamster V79 Fibroblasts as a Function of LET. The Mutation is caused mainly by a Deletion of Parts or all of the Gene Encoding Hypoxanthine-guanine-phosphoribosyl-transferase (HPRT) (Thacker, 1986, Nicklas et al.,1991) (O) data of Thacker et al., (1979), other data from Kranert et al., 1990, (redrawn from Kranert et al., 1990).

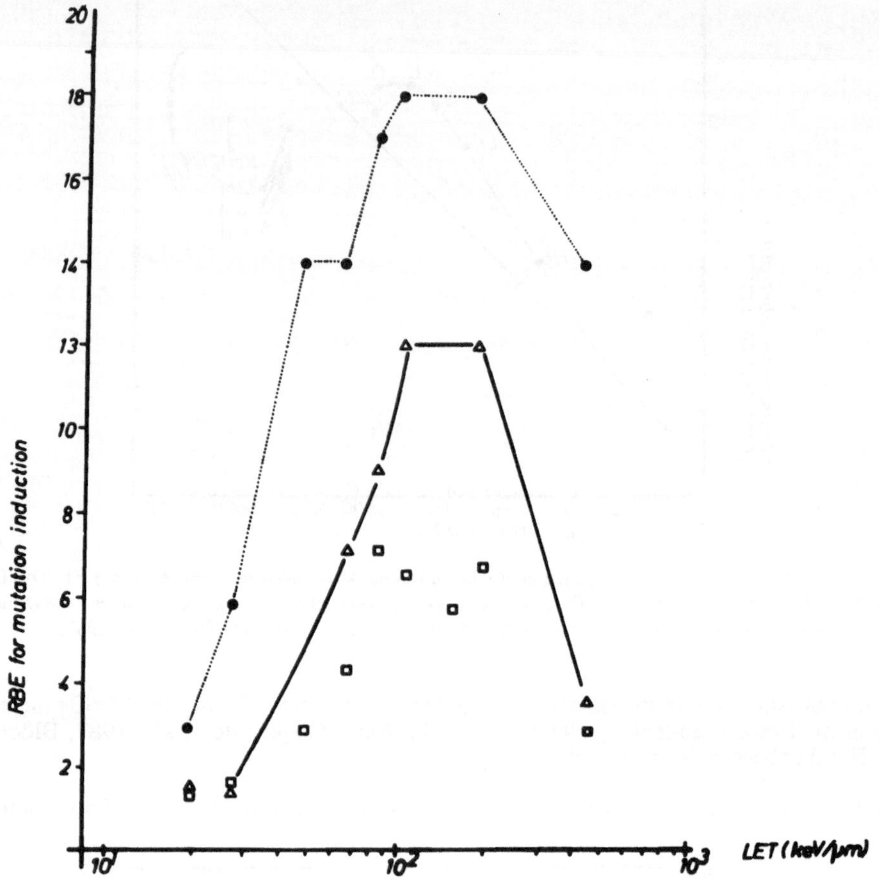

Figure 11. Relative biological effectiveness (RBE) of the induction of mutation to 6-thioguanine resistance in radiosensitive human diploid lung fibroblasts (□) (D_0 of the exponential X-ray survival curve is 1.26 Gy) and in the more radioresistant Chinese hamster V79 cells (shouldered γ-ray survival curve; D = 7.10 Gy at 10% survival). RBEs for V79 cells are given for the initial (●) and final (Δ) slopes of the mutation induction curves. (Redrawn from data presented by Thacker et al.,1979. and Cox and Masson, 1979, to demonstrate the increase of RBE due to repair processes).

level of unrejoined DSB (Blöcher et al. , 1991) and chromosome breaks compared to normal cells (Taylor,1978, Cornforth and Bedford, 1985). Furthermore, these cells were shown to have a reduced fidelity of rejoining of enzymatically induced DSBs (Cox et al. 1986, Debenham et al. 1988). The increased radiosensitivity of mouse lymphoma cells (Wlodek and Hittelman, 1987, Evans et al., 1987), Chinese hamster cell lines (Kemp et al., 1984, Giaccia et al., 1985, Jones et al., 1988, Zdzienicka et al., 1988, 1989)(reviewed by Jeggo (1990)), and of scid mouse cells (Biederman et al., 1991) is correlated with a defect in the rate and/or extent of DSB rejoining, but all of them still retain the ability to rejoin DSB at least to a certain extent. However, variations in radiosensitivity may be also attributable to variations in the extent of misrejoining of DSBs rather than in the amount of unrejoined DSBs (Koval and Kazmar, 1988).

c) Yeast studies provide evidence that DSBs can cause cell lethality by two modes: first, an unrepaired DSB may be lethal by its own and second, two DSBs may interact to form a lethal lesion (binary misrepair) (reviewed by Frankenberg-Schwager and Frankenberg, 1990). Yeast mutants, completely deficient in DSB rejoining, are killed by about one DSB per cell (e.g. Frankenberg et al.,1984a). They do not repair potentially lethal damage (Petin , 1979; Reddy et al., 1982; Frankenberg-Schwager et al., 1985,;1987) and sublethal damage (Reddy et.al., 1982; Frankenberg et al., 1984b; Frankenberg-Schwager et al.,1985). Further-more

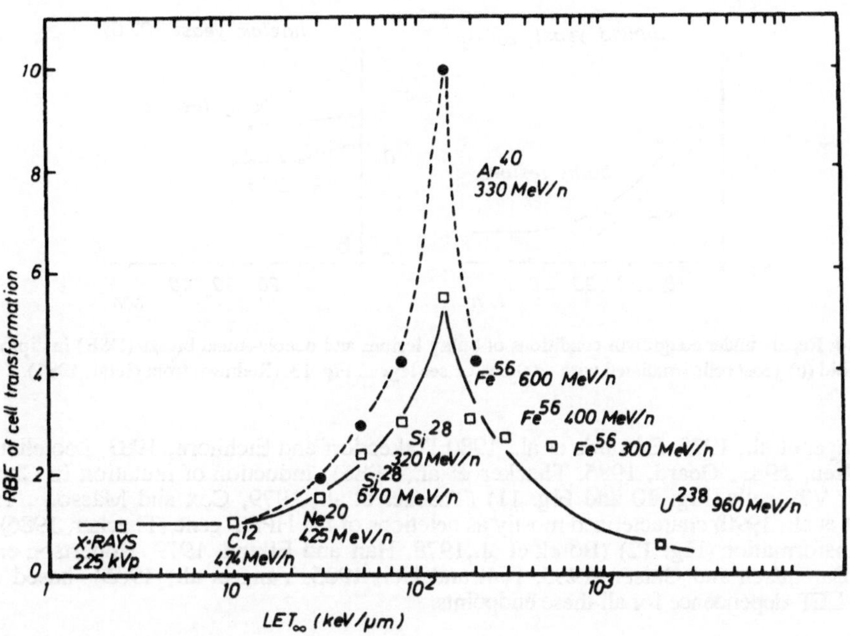

Figure 12. Relative biological effectiveness (RBE) of malignant transformation in confluent mouse fibroblast C3H10T1/2 cells as a function of LET. Open symbols: cells trypsinized and plated immediately after irradiation. Repair incubation at 37 ° C for one day before trypsinization and plating of cells (closed symbols)increases the RBE of malignant cell transformation (Redrawn from Yang et al., 1985).

these mutants do not show a dose-rate effect (Reddy et al., 1982), which is in agreement with the finding that the better survival of normal cells at low dose-rate irradiation is correlated with the rejoining of DSBs during irradiation (Frankenberg-Schwager et al., 1981).

d) DSBs (Fig. 8) (Kampf et al.,1977a, 1977b, Cole et al., 1980, Frankenberg et al., 1981, Van der Schans et al., 1983, Frankenberg et al., 1986) seem to be involved in chromosome aberration production (Fig. 9) (e.g. Neary and Savage, 1966, Skarsgard et al., 1967,

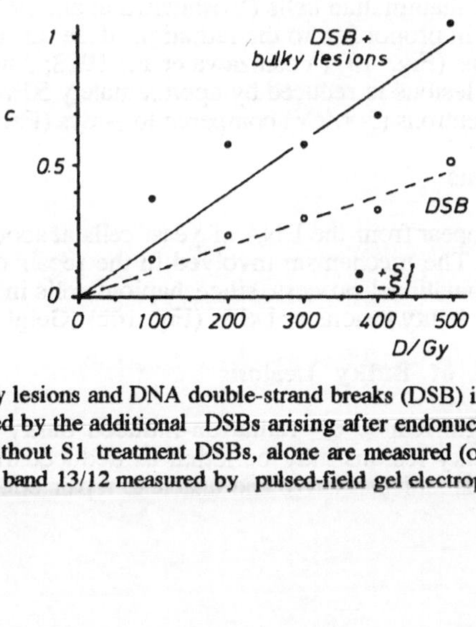

Figure 13. Induction of bulky lesions and DNA double-strand breaks (DSB) in yeast exposed to ^{60}Co γ--rays. Bulky lesions are assayed by the additional DSBs arising after endonuclease S1 treatment of DNA of irradiated cells (●, +S1). Without S1 treatment DSBs, alone are measured (o, -S1). C: Radiation- induced lesions per DNA molecule of band 13/12 measured by pulsed-field gel electrophoresis (Redrawn from Geigl, 1987).

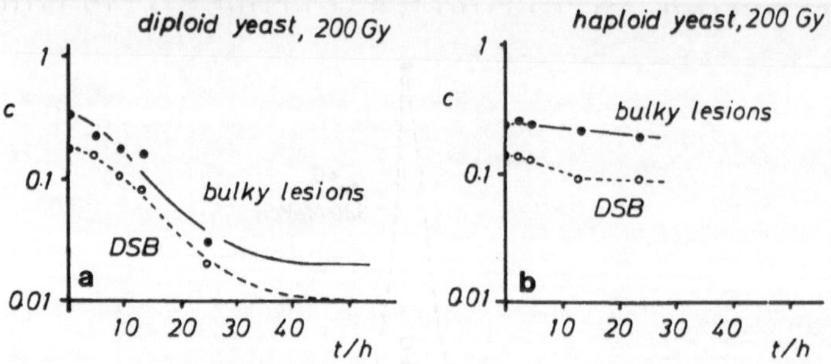

Figure 14. Repair under nongrowth conditions of bulky lesions and double-strand breaks (DSB) in diploid (a) and haploid (b) yeast cells irradiated with 200 Gy. C: see legend Fig. 13. (Redrawn from Geigl, 1987).

Bauchinger et al., 1975, Edwards et al., 1980, Tolkendorf and Eichhorn, 1983, Zootelief and Barendsen, 1983, Geard, 1985, Thacker et al., 1986), induction of mutation in Chinese hamster V79 cells (Fig. 10 and Fig. 11) (Thacker et al., 1979, Cox and Masson , 1979, Kranert et al., 1990) characterized mostly as deletions of the HPRT gene (Thacker, 1986) and cell transformation (Fig. 12) (Borek et al., 1978, Han and Elkind, 1979, Robertson et al., 1983, Barendsen and Gaiser, 1985, Hall and Hei, 1985, Yang et al., 1985), based on a similar LET-dependence for all these endpoints.

Repair processes lead to an enhancement of RBE-values as shown for various endpoints: residual DSBs (Fig. 7) (Frankenberg-Schwager et al., 1982, 1990, Blöcher, 1988), mutation of the non-essential HPRT gene (Fig. 11), and malignant transformation of mouse C3H10T1/2 cells (Fig. 12). There is no evidence for the involvement of DSB in the formation of point mutations based on the fact that DSB repair-proficient and -deficient yeast cells exhibit the same mutation frequency (Magni et al., 1977).

RADIATION-INDUCED BULKY LESIONS

Induction of Bulky Lesions

Bulky lesions are defined as clusters of base damage involving at least three unpaired bases, not necessarily combined with a strand break (Silber and Loeb, 1982, Kohfeldt et al., 1988). These lesions can be detected by the DNA single-strand specific endonuclease S1 resulting in the formation of DSBs. The induction of bulky lesions (S1-sites) by ionizing radiation depends on the structure of DNA: These lesions do not occur in DNA irradiated in vitro, but they are found in the DNA of irradiated yeast (Andrews et al., 1984, Geigl, 1987) and mammalian cells (Yoshizawa et al., 1978, Furuno et al., 1979). Bulky lesions are induced in proportion to the radiation dose at the same or somewhat higher frequency as DSBs (Fig. 13) (Yoshizawa et al., 1978; Andrews et al., 1984, Geigl, 1987). The yield of bulky lesions is reduced by approximately 50% in mouse leukemia cells irradiated with cyclotron neutrons (30MeV) compared to γ-rays (Furuno et al., 1979).

Repair of Bulky Lesions

Bulky lesions disappear from the DNA of yeast cells at about the same rate as DSBs (Fig. 14a) (Geigl, 1987). The mechanism involved in the repair of bulky lesions in yeast seems to require a recombinational process, since haploid cells in stationary phase are not capable of removing either bulky lesions or DSBs (Fig. 14b) (Geigl, 1987).

Biological Significance of Bulky Lesions

The biological significance of radiation-induced bulky lesions remains to be elucidated. Unrepaired bulky lesions may be lethal as deduced from the finding that the diploid yeast mutant rad18 is radiosensisitve and unable to repair bulky lesions, but capable of rejoining radiation-induced DSBs (Geigl and Eckardt-Schupp, 1991). These authors could not, however, exclude the possibility that the increased radiosensitivity of this mutant is due to misrejoining of DSBs.

ACKNOWLEDGEMENT

I want to thank Susanne Beckonert for typing the manuscript and for producing the artwork.

REFERENCES

Ahnström, G., and Edvardsson, K.A., 1974. Radiation-induced single-strand breaks in DNA determined by rate of alkaline strand separation and hydroxylapatite chromatography: an alternative to velocity sedimentation, Int. J. Radiat. Biol. 26: 493-497.

Akpa, T.C., Weber, K.J., Schneider,E., Kiefer, J., Frankenberg-Schwager, M., Harbich, R., and Frankenberg, D.,1992. Heavy ion induced DNA double-strand breaks in yeast, Int. J. Radiat. Biol., accepted for publication.

Andrews, J., Martin-Bertram, H., and Hagen, U., 1984. S 1-nuclease sensitive sites in yeast DNA: an assay for radiation-induced base damage, Int. J. Radiat. Biol., 45: 497-505.

AufderHeide, E., Rink, E., Hieber, L., and Kraft, G., 1987. Heavy ion effects on cellular DNA: Strand break induction and repair in cultured diploid lens epithelial cells, Int.. J. Radiat. Biol., 5: 779-790.

Bacchetti, S., and Benne, R., 1975. Purification and characterization of an endonuclease from calf thymus acting on irradiated DNA, Biochim. Biophys. Acta, 390: 285-297.

Barendsen, G.W., and Gaiser, F.F., 1985. Cell transformation in vitro by fast neutrons of different energies: implications for mechanism, Radiat. Prot. Dosimetry, 13: 145-148 .

Bauchinger, M., Schmid, E., Rimpl, G., and Kühn, H., 1975. Chromosome aberrations in human lymphocytes after irradiation with 15 MeV neutrons in vitro. I. Dose response relation and RBE, Mutat. Res., 27: 103-109.

Bedford, J.S., and Goodhead, D.T., 1989. Breakage of human interphase chromosomes by alpha particles and X-rays, Int.J.Radiat.Biol., 55: 211-216.

Ben-Hur, E., Elkind, M.M., and Riklis, E., 1978. Cancer therapy by hyperthermia and radiation, Urban und Schwarzenberg, Baltimore, München, pp 29-36.

Biedermann, K.A., Sun, J.R., Giaccia, A.J., Tosto, L.M., and Brown, J.M., 1991. Scid mutation in mice confers hypersensitivity to ionizing radiation and a deficiency in DNA double-strand break repair, Proc. Natl. Acad. Sci. USA, 88: 1394 -1398.

Blazek, E.R., Peak, J.G., and Peak, M.J., 1989. Evidence from non-denaturing filter elution that induction of double-strand breaks in the DNA of Chinese hamster V79 cells by γ- radiation is proportional to the square of dose, Radiat. Res., 119: 466-477.

Blöcher, D., 1982. DNA double-strand breaks in Ehrlich ascites tumour cells at low doses of X-rays. Determination of induced breaks by centrifugation at reduced speed, Int. J. Radiat. Biol., 42: 317-328.

Blöcher, D., 1988. DNA double-strand break repair determines the RBE of a- particles, Int. J. Radiat. Biol., 54: 761-771.

Blöcher, D., and Pohlit, W., 1982. DNA double-strand breaks in Ehrlich ascites tumour cells at low doses of X-rays. Can cell death be attributed to double-strand breaks ? Int. J. Radiat. Biol., 42: 329-338 .

Blöcher, D., Nüsse,M., and Bryant, P.E., 1983. Kinetics of double-strand break repair in the DNA of x-irradiated synchronized mammalian cells, Int. J. Radiat. Biol., 43: 579-584.

Blöcher, D., Sigut , D., and Hannan, M.A., 1991. Fibroblasts from ataxia telangiectasia (AT) and AT heterozygotes show an enhanced level of residual DNA double-strand breaks after low dose rate g-irradiation as assayed by pulsed field gel electrophoresis, Int. J. Radiat. Biol., 60: 791-802.

Borek, C., Hall, E.J., and Rossi, H.H., 1978. Malignant transformation in cultured hamster embryo cells produced by X-rays, 430 keV monoenergetic neutrons, and heavy ions, Cancer Res., 38: 2997-3005.

Bowden, G.T., and Casunic, M.D., 1981. Hyperthermic potentiation of the effects of a clinically significant X-ray dose on cell survival, DNA damage, and DNA repair, Bradley, M.O., and Kohn, K.W., 1979, X-ray induced DNA double-strand break production and repair in mammalian cells as measured by neutral filter elution, Nucl. Acids. Res., 7: 793 -804 .

Breimer, L.H.,1988. Ionizing radiation-induced mutagenesis, Br.J.Cancer, 57: 6-18.

Breimer, L.H., and Lindahl, T., 1985. Enzymatic excision of DNA bases damaged by exposure to ionizing radiation or oxidizing agents, Mutat.Res., 150: 85-90.

Brent, T.P., 1976. Purification and characterization of human endonucleases specific for damaged DNA. Analysis of lesions induced by UV or X-radiation, Biochim. Biophys. Acta, 454: 172-183.

Bryant , P.E., and Blöcher, D., 1980. Measurement of the kinetics of DNA double-strand break repair in Ehrlich ascites tumour cells using the unwinding method, Int. J. Radiat. Biol., 38: 335-347.

Bryant , P.E., Warring, R. and Ahnström,G.,1984. DNA repair kinetics after low doses of X-rays. A comparison of results obtained by the unwinding and nucleoid sedimentation methods, Mutat. Res.131: 19 - 26.

Cerutti, P.A., 1974. Effects of ionizing radiation on mammalian cells, Naturwissenschaften, 61: 51-59.

Cerutti, P.A., 1976. Photochemistry and Photobiology of Nucleic Acids, Vol 2. Academic Press, New York, pp 375-401 .

Chiu, S., Oleinick, N.L., Friedman, L.R., and Stambrook, P.J., 1982. Hypersensitivity of DNA in transcriptionally active chromatin to ionizing radiation., Biochim. Biophys. Acta, 699: 15-21 .

Chiu, S., Sokany, N.M., Friedman, L.R., and Oleinick, N.L., 1984. Differential processing of ultraviolet or ionizing radiation-induced DNA-protein crosslinks in Chinese hamster cells, Int. J. Radiat. Biol., 46: 681 -690.

Chiu, S., Friedman, L.R., Xue, L., and Oleinick, N.L., 1986. Modification of DNA damage in transcriptionally active vs. bulk chromatin, Int. J. Radiat. Biol., 12: 1529 - 1532.

Ciejek, E.M., Tsai, M.J., and O'Malley , B.W., 1983. Actively transcribed genes are associated with the nuclear matrix, Nature, 306: 607-609 .

Clark, E.P., and Lett, J.T., 1976. The effect of hyperthermia on the rejoining of DNAstrand breaks in X-irradiated CHO cells, Radiat. Res., 67: 519ff.

Clark, E.P., Dewey, W.C., and Lett, J.T., 1981. Recovery of CHO cells from hyper- thermic potentiation to X-rays : repair of DNA and chromatin, Radiat. Res., 85: 302 -319 .

Cole, A., Meyn, R.E., Chen, R., Corry, P.M., and Hittelman, W., 1980. Radiation Biology in Cancer Research. Raven Press, New York, pp 33-58.

Coquerelle, T., and Weibezahn, K.F., 1981, Rejoining of DNA double-strand breaks in human fibroblasts and its impairment in one Ataxia telangiectasia and two Fanconi strains, J. Supramol. Struct. Cell. Biochem., 17: 369-376 .

Coquerelle, T., Bopp, A., Kessler, B., and Hagen, U., 1973. Strand breaks and 5'-end groups in DNA of irradiated thymocytes, Int. J. Radiat. Biol., 24: 397-404.

Coquerelle, T., Weibezahn, K.F., and Lücke-Huhle, C., 1987. Rejoining of double-strand breaks in normal human and Ataxia telangiectasia fibroblasts after exposure to ^{60}Co γ-rays, 241 Am a-particles or bleomycin, Int. J. Radiat. Biol., 51: 209-218.

Cornforth, M.N., and Bedford, J.S., 1983. X-ray induced breakage and rejoining of human interphase chromosomes, Science, 222: 1141 -1143.

Cornforth, M.N., and Bedford, J.S., 1985. On the nature of a defect in cells from individuals with Ataxia telangiectasia, Science, 227: 1589-1591.

Cornforth, M.N., and Bedford, J.S., 1987. A quantitative comparison of potentially lethal damage repair and the rejoining of interphase chromosome breaks in low passage normal human fibroblasts, Radiat. Res., 111: 385 -405.

Cornforth, M.N., and Goodwin, E.H., 1991. The dose-dependent fragmentation of chromatin in human fibroblasts by 3.5 MeV a-particles from Pu-238: Experimental and theoretical considerations pertaining to single-track events, Radiat.Res., 127 : 64-74.

Corry, P.M., and Cole, A., 1968. Radiation-induced double-strand scissions of the DNA of mammalian metaphase cells, Radiat. Res., 36: 528-543.

Corry, P.M., Robinson, S., and Getz, S., 1977. Hyperthermic effects on DNA repair mechanisms, Radiology,123: 475-482.

Cox, R., and Masson, W.K., 1979. Mutation and inactivation of cultured mammalian cells exposed to beams of accelerated heavy ions. III. Human diploid fibroblasts, Int. J. Radiat. Biol., 36: 149 -160.

Cox, R., Debenham, P.G., Masson, W.K., and Webb, M.B.T., 1986. Ataxia telangiectasia: a human mutation giving high frequency misrepair of DNA double-stranded scissions, Mol.Biol.Med., 3: 229-244.

Cress, A.E., and Bowden, G.T., 1983. Covalent DNA-protein crosslinking occurs after hyperthermia and radiation, Radiat. Res., 95: 610-619.

Cutler, R.G., 1976. Ageing, carcinogenesis, and radiation biology. Plenum Press, New York, London, pp 443-492.

Debenham, P.G., Webb, M.B.T., Stretch, A., and Thacker, J., 1988. Examination of vectors with two dominant selectable genes for DNA repair and mutation studies in mammalian cells, Mutat.Res., 199: 145-158.

Dewey, W.C., Sapareto, A., and Betten, D.A., 1978. Hyperthermic radiosensitization of synchronous Chinese hamster cells: relationship between lethality and chromosomal aberrations, Radiat. Res., 76: 48.

Dewey, W.C., Freeman, M.L., Raaphorst, G.P., Clark, E.P., Wang, R.S.L., Highfield, D.P., Spiro, I.J., Tomasovic, S.P., Denman, P.L., and Coss, R.A., 1980. Cell biology of hyperthermia and radiation, in: Radiation Biology and Cancer Research, edts. R.E.Meyn, H.R.Withers, Raven Press, New York, pp 589-621.

Dikomey, E., 1982. Effect of hyperthermia at 42^{0}C and 45^0 C on repair of radiation- induced DNA strand breaks in CHO cells, Int. J. Radiat. Biol., 41: 603 -614.

Dikomey, E., and Franzke, J., 1986. DNA repair kinetics after exposure to X- irradiation and to internal b-rays in CHO cells, Radiat. Environm. Biophys., 25: 189 -194.

Dizdaroglu, M., Dirksen, M.L., Jiang, H., Robbins, J.H., 1987. Ionizing radiation induced damage in the DNA of cultured human cells. Identification of 8,5'-cyclo-2'-deoxy- guanosine, Biochem. J., 241: 929 -932.

Dooley, D.A., Sacks, P.G., and Miller, M.W., 1984. Production of thymine base damage in ultrasound exposed EMTG mouse mammary sarcoma cells, Radiat. Res., 97: 71 -86.

Edgren, M., Revesz, L., and Larsson, A., 1981. Induction and repair of single-strand DNA breaks after X-irradiation of human fibroblasts deficient in glutathione, Int. J. Radiat. Biol., 40: 355-363 .

Edwards , A.A., Purrott, R.J., Prosser, J.S., and Lloyd, D.C., 1980. The induction of chromosome aberrations in human lymphocytes by a-radiation, Int. J. Radiat. Biol., 38: 83 -91.

Eguchi, K., Inada, T., Yaguchi, M., Sath, S., and Kaneko, I., 1987. Induction and repair of DNA lesions in cultured human melanoma cells exposed to a nitrogen-ion beam, Int. J. Radiat. Biol., 52: 115-123.

Elkind, M.M., 1979. DNA repair and cell repair, are they related? Int. J. Radiat. Oncol. Biol. Phys., 5: 1089 - 1094.

Evans, H.H., Ricanati, M., and Horng, M.F., 1987. Deficiency in DNA repair in mouse lymphoma strain L 5178 Y-S, Proc.Natl.Acad.Sci. 84: 7562 -7566.

Fonck, K., Barthel, R., and Bryant, P.E., 1984. Kinetics of recombinational hybrid formation in X-irradiated mammalian cells: a possible first step in the repair of DNA double-strand breaks, Mutat. Res., 32: 113-118.

Fornace, A.J., 1982. Measurement of M.luteus endonuclease sensitive lesions by alkaline elution, Mutat. Res., 94: 263-276.

Fornace, A.J., and Little, J.B., 1977. DNA crosslinking induced by X-rays and chemical agents, Biochim. Biophys. Acta, 477: 343 -355 .

Fornace, A.J., and Little, J.B., 1980. Normal repair of DNA single-strand breaks in patients with Ataxia telangiectasia, Biochim. Biophys. Acta, 607: 432-437.

Fornace, A.J., Nagasawa, H., and Little, J.B., 1980. Relationship of DNA repair to chromosome aberrations, sister chromatid exchanges and survival during liquid holding recovery in X-irradiated mammalian cells, Mutat. Res., 70: 323-336.

Fornace, A.J., Dobson, P.P., and Kinsella, T.J., Repair of γ-ray induced DNA base damage in Xeroderma pigmentosum cells, Radiat. Res., 106: 73 -88.

Fowler, J.F., 1981. Nuclear particles in cancer treatment. Hilger, Bristol.

Fox, J.C., and Mc Nally, N.J., 1988. Cell survival and DNA double-strand break repair following X-ray or neutron irradiation in V79 cells, Int. J. Radiat. Biol., 54: 1021 - 1030.

Fox, J.C.,and Mc Nally, N.J., 1990. The rejoining of DNA double-strand breaks following irradiation with ^{238}Pu α-particles: evidence for a fast component of repair as measured by neutral filter elution, Int. J. Radiat. Biol., 57: 513-522.

Frankenberg, D., Frankenberg-Schwager, M., Blöcher, D., and Harbich, R., 1981. Evidence for DNA double-strand breaks as the critical lesions in yeast cells irradiated with sparsely and densely ionizing radiation under oxic or anoxic conditions, Radiat. Res., 88: 524-532 .

Frankenberg, D., Frankenberg-Schwager, M., and Harbich, R., 1984a. Interpretation of the shape of survival curves in terms of induction and repair/misrepair of DNA double-strand breaks, Brit. J. Can., 49 ÖSuppl VIÄ: 233-238.

Frankenberg, D., Frankenberg-Schwager, M., and Harbich, R., 1984b. Split-dose recovery is due to the repair of DNA double-strand breaks, Int. J. Radiat. Biol., 46: 541-553.

Frankenberg, D., Goodhead, D.T., Frankenberg-Schwager, M., Harbich, R., Bance, D.A., and Wilkinson, R.E., 1986. Effectiveness of 1.5 keV aluminium K and 0.3 keV carbon K characteristic X-rays at inducing DNA double-strand breaks in yeast cells, Int. J. Radiat. Biol., 50: 727-741.

Frankenberg-Schwager, M., 1989. Review of repair kinetics for DNA damage induced in eukaryotic cells in vitro by ionizing radiation, Radiother. Oncol., 14: 307-320.

Frankenberg-Schwager, M., 1990. Induction, repair and biological relevance of radiation-induced DNA lesions in eukaryotic cells, Radiat. Environm. Biophys., 29: 273-292.

Frankenberg-Schwager, M., Frankenberg, D, Blöcher, D., and Adamczyk, C., 1979. The influence of oxygen on the survival and yield of DNA double-strand breaks in irradiated yeast cells, Int. J. Radiat Biol.,36:261-270.

Frankenberg-Schwager, M., Frankenberg, D., Blöcher, D., and Adamczyk, C., 1980a. Repair of DNA double-strand breaks in irradiated yeast cells under nongrowth conditions, Radiat. Res., 82: 498-510.

Frankenberg-Schwager, M., Frankenberg, D., Blöcher, D., and Adamczyk , C., 1980b. The linear relationship between DNA double-strand breaks and radiation dose (30 MeV electrons) is converted into a quadratic function by cellular repair, Int. J. Radiat. Biol., 37: 207-212.

Frankenberg-Schwager, M., Frankenberg, D., Blöcher, D., and Adamczyk, C., 1981. Effect of dose rate on the induction of DNA double-strand breaks in eukaryotic cells, Radiat. Res., 87: 710-717 .

Frankenberg-Schwager, M., Frankenberg, D., Blöcher, D., Harbich, R., and Adamczyk, C., 1982. Irreparable DNA double-strand breaks induced in eukaryotic cells by sparsely or densely ionizing radiation and their importance for cell killing, Mutat. Res., 96: 132

Frankenberg-Schwager, M., Frankenberg, D., and Harbich, R., 1985. Potentially lethal damage, sublethal damage and DNA double-strand breaks, Radiat. Protect. Dosimetry, 13: 171-174.

Frankenberg-Schwager, M., Frankenberg, D., and Harbich, R., 1987. Potientially lethal damage is due to the difference of DNA double-strand break repair under immediate and delayed plating conditions, Radiat. Res., 111: 192-200.

Frankenberg-Schwager, M., Frankenberg, D., Harbich, R., and Adamczyk, C., 1990a. A comparative study of rejoining of DNA double-strand breaks in yeast irradiated with 3.5 MeV a-particles or with 30 MeV electrons, Int. J. Radiat. Biol., 57: 1151-1168.

Frankenberg-Schwager, M., and Frankenberg, D., 1990b. DNA double-strand breaks: their repair and relationship to cell killing in yeast, Int.J.Radiat.Res., 59: 569-575.

Frankenberg-Schwager, M., and Frankenberg, D., 1991. Rejoining of radiation-induced DNA double-strand breaks in yeast, in: Advances in Mutagenesis Research, Vol.3, edt. G. Obe, Springer Verlag, Berlin, pp. 1-27.

Friedberg, E.C., 1985, DNA Repair, Freeman, New York .

Furuno, I., Yada, T., Matsudaira, H., and Maruyama, T., 1979. Induction and repair of DNA strand breaks in cultured mammalian cells following fast neutron irradiation, Int. J. Radiat. Biol., 36: 639-648.

Geard, C.R., 1985. Charged particle cytogenetics: Effects of LET, fluence and particle separation on chromosome aberrations, Radiat. Res., 104: 112-121 .

Geigll, E.M., 1987. Untersuchung von S1-Nuclease-sensitiven Stellen im Genom der Hefe Saccharomyces cerevisiae nach in vivo Bestrahlung mit Gamma-Strahlen, Thesis, GSF-Report 18/87, GSF München.

Geigl, E.M., and Eckardt-Schupp, F., 1991. Repair of gamma ray-induced S1 nuclease hypersensitive sites in yeast depends on homologous mitotic recombination and a RAD18-dependent function, Curr. Genet., 20: 33-37.

Giaccia, A., Weinstein, R., Hu, J., and Stamato, T.D., 1985. Cell cycle dependent repair of double-strand DNA breaks in a g-ray sensitive Chinese hamster cell, Somatic Cell Mol. Genet., 11 : 485-492.

Goodwin, E., Blakely, E., Ivery, G., and Tobias, C., 1989. Repair and misrepair of heavy-ion-induced chromosomal damage, Adv. Space Res., 9: 1083-1089.

Graubmann, S., and Dikomey, E., 1983. Induction and repair of DNA strand breaks in CHO cells irradiated in various phases of the cell cycle, Int. J. Radiat. Biol., 43: 475-483.

Grdina, D.J., and Nagy, B., 1986. The effect of WR-1065 on radiation-induced DNA damage and repair and cell progression in V79 cells, Br.J.Cancer, 54: 933-937.

Grdina, D.J., Sigdestad, C.P., Dale, P.J.,and Perrin, J.M., 1989. The effect of 2-Ö(aminopropyl) aminoÄethanediol on fission-neutron-induced DNA damage and repair, Br. J. Cancer, 59: 17-21.

Grosovsky, A.J., de Boer, J.G., de Jong, P.J., Drobetsky, E.A., and Glickman, B.W., 1988. Base substitutions, frameshifts, and small deletions constitute ionizing radiation-induced point mutations in mammalian cells, Proc.Natl.Acad.Sci., 85: 185-188.

Hall , E.J., and Hei, T.K., 1985. Oncogenic transformation in vitro by radiations of varying LET, Radiat. Protect. Dosimetry,13: 149-151 .

Han, A., and Elkind, M.M., 1979. Transformation of mouse C3H10T1/2 cells by singular and fractionated doses of X-rays and fission spectrum neutrons, Canc. Res., 39: 123-130.

Hariharan, P.V., Eleczko, S., Smith, B.P., and Paterson, M.C., 1981. Normal rejoining of DNA strand breaks in Ataxia telangiectasia fibroblast lines after low X-ray exposure, Radiat. Res., 86: 589-597.

Hohmann, W.F., Placic, B., and Skarsgard, L.D., 1976. The effect of nitroimidazole and nitroxyl radiosensitizers on the post-irradiation synthesis of DNA, Int. J. Radiat. Biol., 30: 247-261 .

Humphrey, R.M., Steward, D.L., and Sedita, B.A., 1968. DNA- strand breaks and rejoining following exposure of synchronized Chinese hamster ovary cells to ionizing radiation, Mutat. Res.,6: 459-465.

Hutchinson, F., 1985. Chemical changes induced in DNA by ionizing radiation, Progr. Nucl.Acid Res.Mol.Biol., 32: 115-154.

Hutchinson, F., 1989. On the measurement of DNA double-strand breaks by neutral elution, Radiat. Res., 120: 182-186.

Iliakis, G., and Okayasu, R., 1988. The level of induced DNA double-strand breaks does not correlate with cell killing in X-irradiated mitotic and G1-phase CHO cells, Int. J. Radiat. Biol., 53: 395-404 .

Iliakis, G., Pantelias, G.E., and Seaner, R., 1988a. Effect of arabinofuranosyladenine on radiation-induced chromosome damage in plateau phase CHO cells measured by premature chromosome condensation: implications for repair and fixation of alpha PLD, Radiat.Res., 114: 361-378.

Iliakis, G., Okayasu, R., and Seaner, R., 1988b. Radiosensitive xrs-5 and parental CHO cells show identical DNA neutral filter elution dose response : implications for a relationship between cell radiosensitivity and induction of DNA double-strand breaks, Int. J. Radiat. Biol., 54: 55-62.

Iliakis, G., Seaner, R., and Okayasu, R., 1990. Effects of hyperthermia on the repair of radiation-induced DNA single- and double-strand breaks in DNA double-strand break repair-deficient and repair-proficient cell lines, Int. J. Hyperthermia, 6: 813-833.

Iliakis, G., Metzger, L., Denko, N., and Stamato, T.D., 1991a. Detection of DNA double-strand breaks in synchronous cultures of CHO cells by means of asymmetric field inversion gel electrophoresis, Int. J. Radiat. Biol., 59: 321-342.

Iliakis, G., Cicilioni, O., and Metzger, L., 1991b. Measurement of DNA double-strand breaks in CHO cells at various stages of the cell cycle using pulsed field gel electrophoresis: Calibration by means of J-125 decay, Int. J. Radiat. Biol., 59: 343-358.

Iliakis, G., Blöcher, D., Metzger, L., and Pantelias, G., 1991c. Comparison of DNA double-strand break rejoining as measured by pulsed field gel electrophoresis, neutral sucrose gradient centrifugation and non-unwinding filter elution in irradiated plateau-phase CHO cells, Int.J.Radiat.Biol., 59: 927-940.

Jeggo, P. A., 1990. Studies on mammalian mutants defective in rejoining double-strand breaks in DNA, Mutat.Res., 239: 1-16.

Jones, N.J., Cox, R., and Thacker,J., 1988. Six complemention groups for ionizing radiation sensitivity in Chinese hamster cells, Mutat. Res., 193: 139-144.

Jorritsma, J.B.M., and Konings, A.W.T., 1983. Inhibition of repair of radiation-induced strand breaks by hyperthermia, and its relationship to cell survival after hyperthermia alone, Int. J. Radiat. Biol., 43: 505-516.

Kampf, G., and Eichhorn, K., 1983, DNA strand breakage by different radiation qualities and relations to cell killing: Further results after the influence of α- particles and carbon ions, Stud. Biophys., 93: 17-26.

Kampf, G.,Regel, K., Eichhorn, K., and Abel, H., 1977a. Radiation-induced DNA strand breaks in dependence on radiation quality and oxygen and their repair in mammalian cells, Stud. Biophys., 61: 53-60.

Kampf, G., Tolkendorf, E., Regel, K., and Abel, 1977b. Cell inactivation and DNA strand break rates after irradiation with xXrays and fast neutrons, Stud. Biophys., 62: 17-24.

Kelland, L.R., Edwards, S.M., and Steel, G.G., 1988. Induction and rejoining of DNA double-strand breaks in human cervix carcinoma cell lines of differing radiosensitivity, Radiat. Res., 116: 526-538.

Kemp, L.M., Sedgwick, S.G., and Jeggo, P.A., 1984. X-ray sensitive mutants of Chinese hamster ovary cells defective in double-strand break rejoining, Mutat. Res., 132: 189-196.

Kinsella, T.J., Dobson, P.P., Russo, A., Mitchell, J.B., and Fornace, A.J., 1986. Modulation of X-ray DNA damage by SR-2508 ± buthionine sulfoximine, Int. J. Radiat. Oncol. Phys., 12: 1127-1130.

Koch, C.J., and Painter, R.B., 1975. The effects of extreme hypoxia on the repair of DNA single-strand breaks in mammalian cells, Radiat. Res., 64: 256-269 .

Körner, I.J., Günther, K., and Malz, W., 1978. Kinetics of DNA single-strand break rejoining in X- and neutron-irradiated Chinese hamster cells, Stud. Biophys., 70: 157-182.

Kohfeldt, E., Bertram, H., and Hagen, U., 1988. Action of gamma endonuclease on clustered lesions in irradiated DNA, Radiat. Environm. Biophys., 27: 123-132.

Kohn, K.W., Erickson, L.C., Ewing, R.A.G., and Friedman, C.A., 1976. Fractionation of DNA from mammalian cells by alkaline elution, Biochemistry, 15: 4629-4637 .

Kosaka, T., Kaneko, I., and Koide, F., 1990. Correlation between non-repairable DNA lesions and fixation of cell damage by hypertonic solutions in Chinese hamster cells, Int. J. Radiat. Biol., 58: 417- 426.

Koval, T.M., and Kazmar, E.R., 1988. DNA double-strand break repair in eukaryotic cell lines having radically different radiosensitivities, Radiat.Res., 113: 268-277.

Kraft, G., 1987. Radiobiological effects of very heavy ions: Inactivation, induction of chromosome aberrations and strand breaks, Nucl. Sci. Applicat., 3: 1-28.

Kranert, T., Schneider, E., and Kiefer, J., 1990. Mutation induction in V79 Chinese hamster cells by very heavy ions, Int. J. Radiat. Biol., 58: 975-988.

Leadon, S.A., 1990. Production and repair of DNA damage in mammalian cells, Health Physics, 59: 15-22.

Lehman, A.R., and Ormerod, M.G., 1970. Double-strand breaks in the DNA of a mammalian cell after X-irradiation, Biochim. Biophys. Acta, 217: 268-277.

Lehman, A.R., and Stevens, S., 1977. The production and repair of double-strand breaks in cells from normal humans and from patients with Ataxia telangiectasia, Biochim. Biophys. Acta, 474: 49-60.

Lennartz, M., Coquerelle, T., and Hagen, U., 1973. Effect of oxygen on DNA strand breaks in irradiated thymocytes, Int. J. Radiat. Biol., 24: 621-625.

Lennartz, M., Coquerelle, T, Bopp, A., and Hagen, U., 1975a. Oxygen effect on strand breaks and specific endgroups in DNA of irradiated thymocytes, Int. J. Radiat. Biol., 27: 577-587.

Lennartz, M., Coquerelle, T.,and Hagen, U., 1975b. Modification of end-groups in DNA strand breaks of irradiated thymocytes during early repair, Int. J. Radiat. Biol., 28: 181-185.

Lett, J. T., and Sun, C., 1970. The production of strand breaks in mammalian DNA by X-rays: At different stages in the cell cycle, Radiat.Res. 44: 771-787.

Lett, J.T., Bertold, D.S., and Keng, P.C., 1986. Effects of LET... on the fate of DNA damage induced in rabbit sensory cells in situ: fundamental aspects. In: Mechanisms of DNA Damage and Repair, edts. M.G.Simic, L.Grossman, A.C.Upton, Plenum Press, New York, pp. 139-150.

Lindahl, T., 1976. New class of enzymes acting on damaged DNA, Nature, 259: 64-66.

Lunec, J., Hesslewood, I.P., Parker, R., and Leaper, S., 1981. Hyperthermic enhancement of radiation cell killing in HeLa S3 cells and its effect on the production and repair of DNA strand breaks, Radiat. Res., 85: 116-125.

Magni, G.E., Panzeri, L., and Sora, S., 1977. Molecular specificity of X-radiation and its repair in Saccharomyces cerevisiae, Mutat. Res., 42: 223-234 .

Maki, H., Saito, M., Kobayashi, T., Kawai, K., and Akabashi, M., 1986. Cell inactivation and DNA single- and double-strand breaks in cultured mammalian cells irradiated by a thermal neutron beam, Int. J. Radiat. Biol., 50: 795-810.

Matsudaira, H., Furuno, I., Ueno, A.M., Shinohara, K.,and Yashizawa, K., 1977. Induction and repair of strand breaks and 3`-hydroxy terminals in the DNA of mammalian cells in culture following g-ray irradiation, Biochem.Biophys.Acta, 476: 97-101.

Mattern, M.R., and Cerutti, P.A., 1975. Age-dependent excision repair of damaged thymine from gamma-irradiated DNA in isolated nuclei from human fibroblasts, Nature, 254: 450-452.

Mattern, M.R., and Welch, 1979. Production and excision of thymine damage in the cells of mammalian cells exposed to high-LET radiations, Radiat. Res., 80: 474- 483.

Mattern, M.R., Hariharan, P.V., Dunlap, B.E., and Cerutti, P.A., 1973. DNA degradation and excision repair in gamma-irradiated Chinese hamster ovary cells, Nature New Biol., 245: 230-232.

Mattern, M.R., Hariharan, P.V., and Cerutti, P.A., 1975. Selective excision of gamma- ray damaged thymine from the DNA of cultured mammalian cells, Biochim. Biophys. Acta, 395: 48-55.

Mayer, P.J., and Baker, G.T.III., 1987. Chromatin, in: Encyclopedia of Aging, ed. G.L. Maddox, Springer, New York, pp. 112-115.

Mayer, P.J., Lange, C.S., Bradley, M.O., and Nichols, W.W., 1989. Age-dependent decline in rejoining of X-ray-induced DNA double-strand breaks in normal human lymphocytes, Mutat. Res., 219: 95-100.

Mayer, P.J., Lange, C.S., Bradley, M.O., and Nichols, W.W., 1990. Rejoining of X-ray- induced DNA double-strand breaks declines in unstimulated human lymphocytes aging in vivo, Biomed. Adv. in Aging, ed. A.L. Goldstein, Plenum Publishing Corporation, 195-206.

McCready, S.J., Jackson, D.A., and Cook, P.R., 1982. Attachement of intact superhelical DNA to the nuclear cage during replication and transcription, Progr. Mutat. Res., 4: 113-130.

McGhie, J.B., World, E., Pettersen, E.O., and Moan, J., 1983. Combined electron radiation and hyperthermia. Repair of DNA strand breaks in NHIK 3025 cells irradiated and incubated at 37^0C, 42.5^0 C or 45^0 C, Radiat. Res., 96:31-40.

McMillan, T.J., Cassoni, A.M., Edwards, S., Holmes, A., and Peacock, J.H., 1990. The relationship of DNA double-strand break induction to radiosensitivity in human tumour cell lines, Int. J. Radiat. Biol., 58: 427-438.

McWilliams, R.S., Cross, W.G., Kaplan, J.G., and Birnboim, H.C., 1983. Rapid rejoining of DNA strand breaks in resting human lymphocytes after irradiation by low dose of 60-Co-gamma rays and of 16.6 MeV neutrons, Radiat. Res., 94: 499-507.

Metzger, L., and Iliakis, G., 1991. Kinetics of DNA double-strand break repair throughout the cell cycle as assayed by pulsed field gel electrophoresis in CHO cells, Int. J. Radiat. Biol., 59: 1325-1341.

Meyn , R.E., and Jenkins, W.T.,1984. Modification of radiation-induced DNA lesions by oxygen, Radiat. Res., (abstr): 83.

Meyn, R.E., Jenkins, W.T., and Murray, D., 1986. Radiation damage to DNA in various animal tissues: a comparison of yields and repair in vivo and in vitro, In: Mechanism of DNA Damage and Repair, edts. M.G.Simic, L.Grossman, A.C.Upton, Plenum Press, New York London, pp 151-158.

Mills , M.D., and Meyn, R.E., 1981. Effects of hyperthermia on repair of radiation-induced DNA strand breaks, Radiat. Res., 87: 314-328.

Mills , M.D., and Meyn, R.E., 1983. Hyperthermic potentiation of unrejoined DNA strand breaks following irradiation, Radiat. Res., 95: 327-338.

Moran, M.F., and Ebisuzaki, K., 1987. Base excision repair of DNA in γ-irradiated human cells, Carcinogenesis, 8: 607-610.

Müller. W., 1985. Chromosomenaberrationen der chinesischen Hamster Zell-Linie V79 nach Röntgen- und Schwerionenbestrahlung, Thesis, GSI 85-3 Report.

Neary, G.J., and Savage, J.R.R., 1966. Chromosome aberrations and the theory of RBE. II. Evidence from track-segment experiments with protons and alpha particles, Int. J. Radiat. Biol., 11: 209-223 .

Nicklas, J. A., O'Neill, J.P., Hunter, T.C., Falta, M.T., Lippert, M.J., Jakobsen-Kraus, D., Williams, J.R., and Albertini, R.J., 1991. In vivo ionizing irradiations produce deletions in the hprt gene of humans lymphocytes T., Mutat. Res., 250: 383-396.

Okayasu, R., Blöcher, D., and Iliakis, G., 1988. Variation through the cell cycle of DNA neutral filter elution dose response in X-irradiated synchronous CHO cells, Int. J. Radiat. Biol., 53: 729-747.

Okayasu, R., and Iliakis, G., 1989. Linear DNA elution dose reponse curves obtained in CHO cells with non-unwinding filter elution after appropriate selection of the lysis conditions, Int. J. Radiat. Biol., 55: 569-581.

Oleinick, N.L., Chiu, S., and Friedman, L.R., 1985. Gamma radiation as a probe of chromatin structure: damage to and repair of active chromatin in the metaphase chromosome, Radiat. Res., 98: 629-641.

Oleinick, N.L., Chiu, S., Friedman, L.R., Xue, L., and Ramakrishnan, N., 1986. Mechanism of DNA damage and repair, Plenum Press, New York London, pp 181-192.

Oleinick, N.L., Chiu, S., Xue, L., Ramakrishnan, N., and Friedman, L.R., 1990. Properties of DNA crosslinks (DPCs) induced by ionizing radiation, J.Cell. Biochem. (Suppl.) 14A (abstract): UCLA Symposium: 80.

Palcic, B., and Skarsgard, L.D., 1972. The effect of oxygen on DNA single-strand breaks produced by ionizing radiation in mammalian cells, Int. J. Radiat. Biol., 21: 417-433.

Pantelias, G.E., and Maillia, H.D., 1985. Direct analysis of radiation-induced chromosome fragments and rings in unstimulated human peripheral blood lymphocytes by means of the premature chromosome condensation technique, Mutat. Res., 149: 67 .

Paterson, M.C, 1978. Use of purified lesion recognizing enzymes to monitor DNA repair in vivo, Adv. Radiat. Biol., 7: 1-53.

Paterson, M.C., and Smith, P.J., Ataxia telangiectasia: an inherited human disorder involving hypersensitivity to ionizing radiation and related DNA damaging chemicals, Ann. Rev. Genet., 13: 291-318.

Paterson, M.C., Smith, B.P., Lohman, P.H.M., Anderson, A.K., and Fishman, L., 1976. Defective excision repair of gamma-ray damaged DNA in human (Ataxia telangiectasia) fibroblasts, Nature, 260: 444-446.

Patil, M.S., Smith, B.P., and Hariharan, P.V., 1985. Radiation induced thymine base damage and its excision repair in active and inactive chromatin of HeLa cells, Int. J. Radiat. Biol., 48: 691-700.

Peak, M.J., Peak, J.G., Carnes, B.A., Chang Liu, C.M., and Hill, C.K., 1989. DNA damage and repair in rodent and human cells after exposure to JANUS fission spectrum neutrons: A minor fraction of single-strand breaks as revealed by alkaline elution is refractory to repair, Int. J. Radiat. Biol., 55: 761-772.

Peak, M.J., Wang, L., Hill, C.K., and Peak, J.G., 1991. Comparison of repair of DNA double-strand breaks caused by neutron or gamma radiation in cultured human cells, Int. J. Radiat. Biol., 60: 891-898.

Petin, V.G., 1979. Effect of gamma and alpha irradiation on survival of wild-type and sensitive mutants of yeast, Mutat. Res., 60: 43-49.

Prise, K.M., Davies, S., and Michael, B.D., 1987. The relationship between radiation-induced DNA double-strand breaks and cell kill in hamster V79 fibroblasts irradiated with 250 kVp x-rays, 2.3 MeV neutrons or Pu α-particles, Int. J. Radiat. Biol., 52: 893-902.

Radford, I.R., 1983. Effects of hyperthermia on the repair of X-ray induced DNA double-strand breaks in mouse L cells, Int. J. Radiat. Biol., 43: 551-558.

Radford, I.R., 1985. The level of induced DNA double-strand breakage correlates with cell killing after X-irradiation, Int. J. Radiat. Biol., 48: 45-54.

Radford, I.R., 1986a. Effect of radiomodifying agents on the ratios of X-ray-induced lesions in cellular DNA: use in lethal lesions determination, Int. J. Radiat. Biol., 49: 621-637.

Radford, I.R., 1986b. Evidence for a general relationship between the induced level of DNA double-strand breakage and cell killing after X-irradiation of mammalian cells, Int. J. Radiat. Biol., 49: 611-620.

Radford, I.R., and Broadhurst, S., 1986. Enhanced induction by X-irradiation of DNA double-strand breakage in mitotic as compared with S phase V79 cells, Int. J. Radiat. Biol., 49: 909-914.

Radford, I.R., 1987. Effect of cell- cycle position and dose on the kinetics of DNA double-strand breakage repair in X-irradiated Chinese hamster cells, Int. J. Radiat. Biol., 52: 555-563.

Radford, I.R., 1990. Lysis solution composition and non-linear dose-response to ionizing radiation in the non-denaturing DNA filter elution technique, Int. J., Radiat. Biol., 57: 479-483.

Radford, I.R., and Hodgson, G.S., 1990. A comparison of the induction of DNA double-strand breakage and lethal lesions by X-irradiation in Ataxia telangiectasia and normal fibroblasts, Radiat. Res., 124: 334 -335.

Ramakrishnan, N., Chiu, S., and Oleinick, N.L., 1987. Yield of DNA- protein crosslinks in γ-irradiated Chinese hamster cells, Cancer Res., 47: 2032-2035 .

Reddy, N.M.S., Anjaria, K.B., and Subrahmanyam, P., 1982. Absence of a dose rate effect and recovery from sublethal damage in rad52 strain of diploid yeast Saccharomyces cerevisiae exposed to gamma rays, Mutat. Res., 105: 145-148.

Resnick, M.A., 1976. The repair of double-strand breaks in DNA: a model involving recombination, J. Theor. Biol., 59: 97-106 .

Resnick, M.A., and Moore, P.D., 1979. Molecular recombination and the repair of DNA double-strand breaks in CHO cells, Nucl. Acids Res., 6: 3145-3160.

Revesz, L., 1985. The role of endogenous thiols in intrinsic radioprotection, Int. J.Radiat. Biol., 47: 361-368.

Revesz, L., and Edgren, M., 1984. Glutathione dependent yield and repair of single- strand breaks in irradiated cells, Brit. J. Canc. Res., 49 (Suppl VI): 55-60.

Ritter, M.A., Cleaver, J.E., and Tobias, C.A., 1977. High LET radiations induce a large proportion of non-rejoining DNA breaks, Nature, 266: 653-655.

Robertson, J.B., Koehler, A., George, J., and Little, J.B., 1983. Oncogenic transformation of mouse Bulb/3T3 cells by plutonium-238 alpha particles, Radiat. Res., 96: 261-274.

Robinson, S.I., Small, D., Izderda, R., McKnight, A.S., and Vogelstein, B., 1983. The association of transcriptionally active genes with the nuclear matrix of the chicken oviduct, Nucl. Acid Res., 11: 5113-5130.

Roots, R., Yang, T.C., Craise, L., and Blakely, E.A., 1979. Impaired repair capacity of DNA breaks induced in mammalian cellular DNA by accelerated heavy ions, Radiat. Res., 78: 38-49.

Roots, R., Yang, T.C., Craise, L., Blakely, E.A., and Tobias, C.A., 1980. Rejoining capacity of DNA breaks induced by accelerated carbon and neon ions in the spread Bragg peak, Int. J. Radiat. Biol., 38: 203-210.

Roots, R., Holley, W., Chatterjee, A., Irizarry, M., and Kraft, G., 1990. The formation of strand breaks in DNA after high- LET irradiation : a comparison of data from in vitro and cellular systems, Int. J. Radiat. Biol., 58: 55-70.

Roti-Roti, J.L., and Winward, R.T.,1981. The effects of hyperthermia on the protein-to-DNA ratio of isolated HeLa cell chromatin, Radiat. Res., 74: 159-169.

Rydberg, B., 1984. Repair of DNA double-strand breaks in colcemid -arrested mitotic Chinese hamster cells, Int. J. Radiat. Biol., 46: 299-304.

Rydberg, B., 1985. DNA strand-breaks induced by low-energy heavy ions, Int. J. Radiat. Biol., 47: 57-62.

Sakai, K., and Okada, S., 1981. An alkaline separation method for detection of small amount of DNA damage, J. Radiat. Res., 22: 415-424 .

Sakai, K., and Okada, S., 1984. Radiation-induced DNA damage and cellular lethality in cultured mammalian cells, Radiat. Res., 98: 479-490.

Sakai, K., Suzuki,s., Nakamura, N.,and Okada, S., 1987. Induction and subsequent repair of DNA damage by fast neutrons in cultured mammalian cells, Radiat. Res., 110: 311-320.

Sawada, S., and Okada, S, 1970. Rejoining of single-strand breaks of DNA in cultured mammalian cells, Radiat. Res., 41: 145-162.

Schwartz, J.L., Giovanazzi, S., and Weichselbaum, R.R., 1987, Recovery from sublethal and potentially lethal damage in a X-ray sensitive CHO cell, Radiat. Res., 111: 58- 67.

Schwartz, J.L., Rotmensch, J., Giovanazzi, S., Cohen, M.B., and Weichselbaum, R.R., 1988, Faster repair of DNA double-strand breaks in radioresistant human tumor cells, Int. J. Radiat. Oncol. Biol. Phys.,15: 907-912.

Schwartz, J.L., Mustafi, R., Beckett, M.A., Czyzewski, E.A., Farhangi, E., Grdina, D.J., Rotmensch, J., and Weichselbaum, R.R., 1991. Radiation-induced DNA double-strand break frequencies in human squamous cell carcinoma cell lines of different radiation sensitives, Int. J. Radiat. Biol., 59: 1341-1353.

Sigdestad, C.P., Treacy, S.H., Knapp, L.A., and Grdina, D.J., 1987. The effct of 2-((amionopropyl)amino) ethanethiol (WR-1065) on radiation induced DNA double-strand damage and repair in V79 cells, Br.J.Cancer, 55: 477-482.

Silber, J.R., and Loeb, L.L., 1982. S1 nuclease does not cleave DNA at single-base mismatches, Biochim. Biophys. Acta, 656: 256-264.

Skarsgard, L.D., Kihlman, B.A., Parker, L., Pujara, C.M., Richardson, S., 1967. Survival, chromosome abnormalities and recovery in heavy ion and X-irradiated mammalian cells, Radiat. Res. Suppl 7: 208-221.

Skov, K.A., Palcic, B., and Skarsgard, L.D., 1979. Radiosensitization of mammalian cells by misonidazole and oxygen: DNA damage exposed by Micrococcus luteus enzymes, Radiat. Res., 79: 591-600.

Suzuki, S., Eguchihasa, K., Kosaka, T., Watanabe, J., Ohara, H., and Kaneko, J., 1990. Time-lapse microscopy and DNA double-strand breakage of Chinese hamster cells under conditions promoting or preventing PLD repair after irradiation with Co-60 -gamma-rays, Int. J. Radiat. Biol., 58: 769-780.

Sweigert, S.E., Rowley, R., Waters, R.L., and Dethlefsen, L.A., 1988. Cell cycle effect on the induction of DNA double-strand breaks by X-rays, Radiat. Res., 116: 228 -244.

Szumiel, I., 1981, Intrinsic radiosensitivity of proliferating mammalian cells, Adv. Radiat. Biol., 9: 281-321.

Taylor, A.M.R., 1978, Unrepaired DNA strand breaks in irradiated Ataxia telangiectasia lymphocytes suggested from cytogenetic observations, Mutat. Res., 50: 407-418.

Thacker J., Stretch A., and Stephens M.A., Mutation and inactivation of cultured mammalian cells exposed to beams of accelerated heavy ions II. Chinese hamster V79 cells, Int. J. Radiat. Biol., 36: 137-148.

Thacker , J., 1986, The nature of mutants induced by ionizing radiation in cultured hamster cells. III. Molecular characterization of hprt deficient mutants induced by γ-rays or α-particles showing that the majority have deletions of all or part of the hprt gene, Mutat. Res., 160: 267-275.

Thacker, J., Wilkinson, R.E., and Goodhead, D.T., 1986, The induction of chromosome exchange aberrations by carbon ultrasoft X-rays in V79 hamster cells, Int. J. Radiat. Biol., 49 : 645-656.

Thompson, L.H., Brookman, K.W., Dillehay, L.E., Carrano, A.V., Mazrimas, J.A., Mooney, C.L., and Minkler, J.L., 1982. A CHO-cell strain having hypersensitivity to mutagens, a defect in DNA strand-break repair, and an extraordinary baseline frequency of sister chromatid exchange, Mutat. Res., 95: 427.

Tolkendorf,E.,and Eichhorn K., 1983. Effect of ionizing radiation of different linear energy transfer on the induction of cellular death and of chromosome aberrations in cells of the Chinese hamster, Stud. Biophys., 95: 43 -56.

Turner, D.R., Griffith, V.C., and Morley, A.A., 1982. Aging in vivo does not alter the kinetics of DNA strand break repair, Mech. Aging. Dev., 19: 325-331.

Van Ankeren, S.C., Murray, D., and Meyn, R.E., 1988. Induction and rejoining of γ- ray induced DNA single- and double- strand breaks in Chinese hamster AA8 cells and in two radiosensitive clones, Radiat.. Res., 116: 511-525.

Van der Schans, G.P., Centen, H.B., and Lohman, P.H.M., 1979. The induction of gamma-endonuclease-susceptible sites by gamma rays in CHO cells and their cellular repair are not effected by the presence of thiol compounds during irradiation, Mutat. Res., 59: 119-122.

Van der Schans, G.P., Centen, H.B., and Lohman, P.H.M., 1981. Chromosome Damage and Repair, Plenum Press, New York London, pp 355-359 .

Van der Schans, G.P., and Centen, H.B., 1982a. Progress in Mutation Research, Vol 4. Elsevier, Amsterdam, pp 285-299.

Van der Schans, G.P., Centen, H.B., and Lohman, P.H.M., 1982b. A cellular and molecular link between cancer, neuropathology and immune deficiency, Wiley, Chichester New York Brisbane, pp 291-303.

Van der Schans, G.P., Paterson, M.C., and Cross, W.G., 1983. DNA strand breaks and rejoining in cultured human fibroblasts exposed to fast neutrons or gamma rays, Int. J. Radiat. Biol., 44: 75-85.

Van der Schans, G.P., Vos, O., Ros-Verhej, W.S.D., and Lohman, P.H.M., 1986. The influence of oxygen on the induction of radiation damage in DNA in mammalian cells after sensitization by intracellular glutathione depletion, Int. J. Radiat. Biol., 50: 453-465.

Vos, O., Van der Schans, G.P., and Ros-Verhej, W.S.D., 1986. Reduction of intracellular glutathione content and radiosensitivity, Int. J. Radiat. Biol., 50: 155-165.

Waldren, C.A., and Johnson, R.T., 1974. Analysis of interphase chromosome damage by means of premature chromosome condensation after X-and ultraviolet irradiation, Proc. Nas., 71: 1137.

Ward, J.F., Blakely, W.F., and Joner, E.I., 1985. Mammalian cells are not killed by DNA single-strand breaks caused by hydroxyl radicals from hydrogen peroxide, Radiat. Res., 103: 383-392.

Warters, R.L., and Roti Roti, J., 1978. Production and excision of 5',6'-dihydroxy- dihydrothymine type products in the DNA of preheated cells, Int. J. Radiat. Biol., 34: 381-384.

Warters , R.L., and Childers, T.J., 1982. Radiation induced thymine base damage in replicating chromatin, Radiat. Res., 90: 564-574.

Warters, R.L., Lyons, B.W., Axtell-Bartlett, J., 1987. Inhibition of repair of radiation induced DNA damage by thermal shock in Chinese hamster ovary cells, Int. J. Radiat. Biol., 51: 505-518.

Watanabe, M., and Horrikawa, M.,1977. Analyses of differential sensitivities of synchronized HeLa S3 cells to radiation and chemical carcinogen during the cell cycle, part IV. X-rays, Mutat. Res., 44: 413-426.

Weibezahn, K.F., and Coquerelle, T., 1981. Radiation-induced DNA double-strand breaks are rejoined by ligation and recombination processes, Nucl. Acids Res., 9: 3139-3150.

Wheeler, K.T., and Wierowski, J. V., 1983. DNA repair kinetics in irradiated undifferentiated and terminally differentiated cells, Radiat. Environm. Biophys., 22: 3-19.

Wheeler, K.T., and Nelson, G.B., 1987. Saturation of a DNA repair process in dividing and nondividing mammalian cells, Radiat. Res., 109: 109-117.

Wlodeck, D., and Hittelman, W.N., 1987. The repair of double-strand DNA breaks correlates with radiosensitivity of L5178 Y-S and L5178 Y-R cells, Radiat. Res., 112: 146-155.

Wlodeck, D., and Hittelman,W.N., 1988a. The relationship of DNA and chromosome damage to survival of synchronized X-irradiated L5178 Y cells. I. Initial damage, Radiat. Res., 115: 550-566.

Wlodek, D., and Hittelman, W.N., 1988b. The relationship of DNA and chromosome damage to survival of synchronized X-irradiated L5178 Y cells. II. Repair, Radiat. Res., 115: 566-576.

Woods, W.G., Lopez, M., and Kaslvonijan, M., 1982. Normal repair of gamma- irradiation-induced single- and double-strand DNA breaks in retinoblastoma fibroblasts, Biochim. Biophys. Acta, 698: 40-48.

Yang, T.C., Craise, L.M., Mei, M.-T., and Tobias, C.A., 1985. Neoplastic transform ation by heavy charged particles, Radiat.Res., 104: S177-S187.

Yoshizawa, K., Furuno, I., Yada, T., and Matsudaira,H., 1978. Induction and repair of strand breaks and 3'-hydroxy terminals in the DNA of mouse brain following gamma irradiation, Biochim. Biophys. Acta, 521: 144-154.

Zdzienicka, M.Z., Tran, M., Van der Schaans, M.P., and Simons, J.W.I.M., 1988. Characterization of an X-ray hypersensitive mutant of V79 Chinese hamster cells, Mutat. Res., 194: 239- 250.

Zdzienicka, M.Z., Jaspers, N.G.J., Van der Schaans, G.P., Natarajan, A.T., and Simons, J.W.I.M., 1989. Ataxia-telangiectasia-like Chinese hamster V79 cell mutants with radioresistant DNA synthesis, chromosomal instability, and normal DNA strand break repair, Cancer Res., 49: 1481-1485.

Zotelief, J., and Barendsen, G.W., 1983. Dose-effect relationships for induction of cell inactivation and asymmetrical chromosome exchanges in three cell lines by photons and neutrons of different energy, Int. J. Radiat. Biol., 43: 349-362.

ASSESSMENT OF HEAVY ION INDUCED DNA STRAND BREAKS IN MAMMALIAN CELLS

J. Heilmann and H. Rink

GSI Darmstadt, Bonn University (FRG)

INTRODUCTION

Ionizing radiation has both beneficial and hazardous properties.This is true for HZE-particles as well. Their biophysical characteristics make them an ideal tool for an improved radiotherapy. On the other hand, heavy ions contribute substantially to the radiation hazard that will be imposed upon people in long term space missions.

It has been evidenced in a number of studies (for a review see Kraft, 1987) that the biological effects of accelerated particles differ significantly from those induced by x-rays. The study of induced DNA-lesions is an important topic in the evaluation of primary damage. Especially the induction and repair of DNA double-strand breaks are essential for the understanding of cell killing as well as of mutations (Frankenberg-Schwager et al, 1980; Kranert et al, 1990).

The presented study focuses on the intracellular induction of DNA strand breaks. Furthermore, rejoining of strand breaks was monitored as a first step towards more detailed studies of repair mechanisms. Data on cellular recovery will be briefly discussed.

MATERIALS AND METHODS

Studies on DNA strand breaks were performed with either bovine lens epithelial cells or human fibroblasts. Both cell lines were non permanent. For recovery experiments, the V79 chinese hamster cell line was used. Irradiations were carried out with a 300 kVp X-ray unit (Müller) or at the UNILAC accelerator facility at GSI Darmstadt.

For strand break measurements, the alkaline unwinding technique and subsequent hydroxyapatite chromatography (Ahnström and Erixon, 1973) as well as the alkaline and neutral filter elution were applied (Bradley et al., 1976; Bradley and Kohn, 1979). Strand break rejoining was monitored with the alkaline unwinding method. Cellular recovery was studied principally following the protocol given by Hahn and Little (1972). After irradiation, cells were incubated under non-growth conditions for different periods of time and then plated to test their colony forming ability.

RESULTS AND DISCUSSION

The measurements of the induction of DNA strand breaks always yielded purely exponential dose-effect curves independent of the detection system. Cross sections were

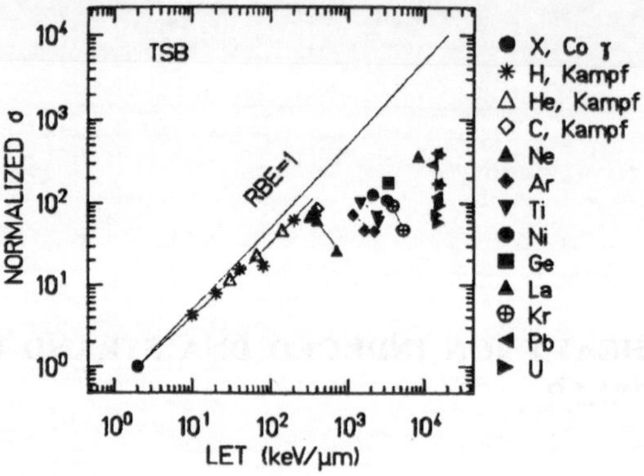

Figure 1. Normalized Cross Sections of DNA Strand Break Induction as Function of LET. Closed Symbols Represent the Authors' Results.

derived from the slopes of least-squares fits. Normalization was achieved by calculating a relative cross section as the ratio of the cross section for the test radiation and the cross section obtained from corresponding x-ray experiments.

Figs. 1 and 2 show normalized cross sections for the induction of total strand breaks and double strand breaks. Data in the LET range up to 300 keV/μm are from Kampf (1983). These measurements were performed using ultracentrifugation in sucrose gradients.

Although data are limited, the hook structures can be seen for data in the high-LET region (Pb, U). The cross sections of the total amount of strand breaks always correspond to RBE values smaller than unity. This is not the case for the induction of double strand breaks, where an RBE greater than one was measured with Ne-ions at LET values around 300keV/μm. The results shown in Fig. 1 quite nicely continue the data from (Kampf,1983) for high LET values, despite the different methods applied. Concerning the cross sections for induction of double strand breaks, neutral elution data are smaller by a factor of two than could be expected from sucrose-sedimentation studies, but the systematics itself remain unaffected.

The separate measurements of total strand breaks and double strand breaks allowed the calculation of the DSB/SSB ratio as a parameter for the severeness of damage. Fig. 3 is a compilation of data from Christensen and Tobias (1972) obtained with ϕX174 DNA, from Taucher-Scholz et al. (1991) with SV40 DNA and our own data (filled symbols). In this

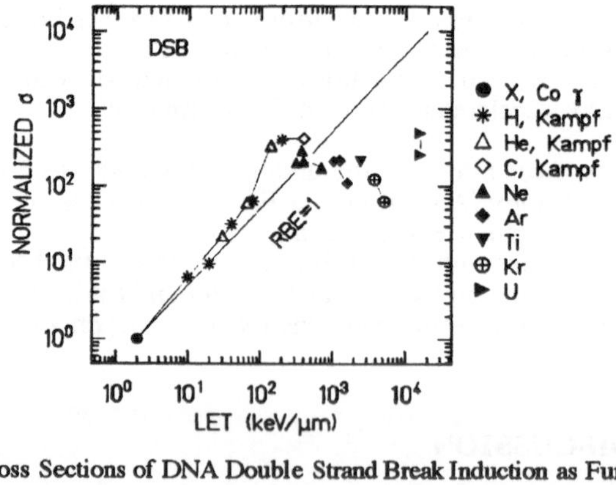

Figure 2. Normalized Cross Sections of DNA Double Strand Break Induction as Function of LET. Choice of Symbols as in Fig.1.

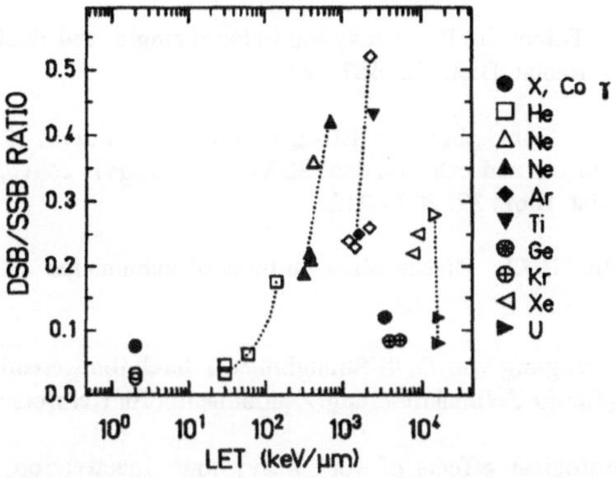

Figure 3. DSB/SSB Ratios as Function of LET. Open Smbols are Data from Kampf (1983); Christensen and Tobias (1972); and Taucher-Scholz et al (1991). Closed Symbols Refer to the Authors' Results.

graph, the DSB/SSB ratio sharply increases for any particular ion with LET. However, this ratio never exceeds a value of about 0.5, demonstrating that even for very high LET values the majority of DNA lesions seems to comprise single strand breaks and alkali labile lesions.

The rejoining of DNA strand breaks is indicative of repair in living cells after exposure to ionizing radiation (Catena and Mattoni, 1985), but it does not prove whether this repair is genetically correct. It was demonstrated that rejoining is slowed down very much after heavy ion irradiation. Furthermore, a high amount of residual damage was always detectable, even after long incubation times. In the LET range beyond 10^4 keV/μm rejoining could not be detected.

In contrast, cellular recovery could not be detected at all after heavy ion exposure. In some experiments, an apparent increase in the surviving fraction was always accompanied by a subsequent loss of cells. Thus, the population tested for survival was not identical with the population initially irradiated.

The observed detachment of cells may be due to membrane damage induced by radiation. For instance the haemolysis of erythrocytes as a parameter of membrane damage was dramatically increased after heavy ion exposure (Schön et al., 1989). Damage in biological membranes may have a strong impact on the fate of cells after heavy ion exposure which is neglected up to now.

In summary, cross sections of the intracellular induction of DNA strand breaks show the same dependency from LET as reported for other biological endpoints. The comparison of DNA strand-break rejoining and cellular recovery has shown that the extent of cellular response after heavy ion irradiation is of very limited significance.

REFERENCES

Ahnström, G. and K. Erixon, 1973. Radiation induced strand breakage in DNA from mammalian cells. Strand separation in alkaline solution. Int. J. Radiat. Biol., 23: 285 - 289.

Bradley, M.O., L.C. Erickson, and K.W. Kohn, 1976. Normal DNA strand rejoining and absence of DNA crosslinking in progeroid and aging human cells. Mutat. Res., 37:279 - 292.

Bradley, M.O. and K.W. Kohn, 1979. X-ray induced double strand break production and repair in mammalian cells as measured by neutral filter elution. Nucl. Acids Res., 7: 793 - 804.

Catena, C. and A. Mattoni, 1985. DNA strand break rejoining in human lymphocytes during the S phase. Int. J. Radiat. Biol., 47: 489 - 496.

Christensen, R.C. and C.A. Tobias, 1972. Heavy-ion-induced single- and double-strand breaks in ϕX-174 replicative form DNA. Int. J. Radiat. Biol., 22: 457 - 477.

Frankenberg-Schwager, M., D. Frankenberg, D. Blöcher, and C. Adamczyk, 1980. The linear relationship between DNA double-strand breaks and radiation dose 30 MeV electrons is converted into a quadratic function by cellular repair. Int. J. Radiat. Biol., 37: 207 - 212 .

Hahn, G.M., and J.B. Little, 1972. Plateau-phase cultures of mammalian cells. Curr. Top. Radiat. Res. Quarterly, 8: 39 - 83.

Kampf, G., 1983. Die Erzeugung von DNS-Strangbrüchen durch ionisierende Strahlen unterschiedlicher Qualität und ihre Bedeutung für die Zellinaktivierung. Zentralinstitut für Kernforschung, Rossendorf, KfR 504.

Kraft, G., 1987. Radiobiological effects of very heavy ions: inactivation, induction of chromosome aberrations and strand breaks. Nucl. Sci. Appl., 3: 1 - 28.

Kranert, T. , E. Schneider, and J. Kiefer, 1990. Mutation induction in V79 chinese hamster cells by very heavy ions. Int. J. Radiat. Biol., 58: 975 - 987.

Schön, W. , H. Gärtner, G. Bartosz, and G. Kraft, 1989. Measurement of haemolysis of human erythrocytes as an indication of radiation induced membrane damage. In GSI Scientific Report, page 181.

Taucher-Scholz, G., M. Schneider, J. A. Stanton, C. Wiese, and G. Kraft. Induktion von Strangbrüchen in viraler und eukaryotischer DNA nach Schwerionenbestrahlung. In 4. Symposium molekulare und zelluläre Mechanismen der biologischen Strahlenwirkung, Programm und Kurzfassungen, Februar 1991, page 22, Essen, 1991.

DOES THE TOPOLOGY OF CLOSED SUPERCOILED DNA AFFECT ITS RADIATION SENSITIVITY?

C.E. Swenberg, J.M. Speicher, and J.H. Miller*

Radiation Biochemistry Department
Armed Forces Radiobiology Research Institute
8901 Wisconsin Ave, Bethesda, MD 20889-5145
*Biology and Chemistry Department
Pacific Northwest Laboratory, Richland, WA 99352

ABSTRACT

Several families of negatively supercoiled topoisomers of the plasmid pIBI30 were ^{60}Co γ-irradiated and assayed for the induction of strand scission by agarose gel electrophoresis. Form-I DNA for all topoisomers decreased exponentially with increasing dose. The radiation sensitivity ($1/D_{37}$) was dependent on the average linking difference (ΔL) associated with a given supercoiled family. For $|\Delta L| < 2.5$ the radiation sensitivity of DNA decreased with increasing $|\Delta L|$, whereas for $|\Delta L| > 2.5$ enhanced radiation sensitivity was observed with increasing $|\Delta L|$. These results are in agreement with data reported by Miller, *et al.* (1991) also for pIBI30 (250kV X-irradiated), but are inconsistent with experiments by Milligan, *et al.* (1992) for ^{137}Cs γ-irradiated pUC18. Our results are suggestive of several mechanisms that could be operative in explaining the dependence of DNA radiation sensitivity on topology.

INTRODUCTION

Plasmid DNA is a useful model system for investigating the initial radiation damage in the absence of repair processes due to the ease of quantifying strand scission by agarose gel electrophoresis. In this paper we investigate the dependence of ^{60}Co γ-radiation sensitivity on the conformational state of DNA. It is almost an axiom of radiation research that smaller target volumes are less sensitive to ionizing radiation. We present experimental evidence that disputes the validity of this axiom in some circumstances. Our data strongly suggest that the organization of DNA in small domains can have pronounced effects on its radiation sensitivity.

It is well known that covalently closed duplex DNA is topologically constrained and can assume a large number of topologically inequivalent conformations. These different DNA conformational states can be classified by their linking number (Cozzarelli, Boles, and White, 1990). A more useful quantity, the linking difference (ΔL) denotes the difference between linking numbers of a given topoisomer and fully relaxed plasmid DNA. Native plasmids all have $\Delta L < 0$, a condition corresponding to more base pairs per turn than predicted by the Watson-Crick model. To address whether standard target theory (Fowler, 1964) might be inapplicable for small domain sizes, *i.e.* whether DNA's topology alters its radiation sensitivity, we have prepared supercoiled families of pIBI30 and have measured their response to ^{60}Co γ-irradiation. Our results strongly suggest that DNA's conformational state affects its radiation response. For DNA in superhelical states an increase in $|\Delta L|$ corresponds to a decrease in target size as is indicated by an increase in electrophoretic mobility. In our studies for DNA characterized by low linking difference ($|\Delta L| < 2.5$) the response was that expected on the basis of standard target theory. However, for smaller, more compact DNA corresponding to larger negative linking differences ($|\Delta L| > 2.5$), the radiation sensitivity (defined as the negative of the slope of the semi-logarithmic plot of the fraction of Form-I DNA as a function of dose) was found to increase with increase in $|\Delta L|$. Several possible competing processes responsible for the enhanced sensitivity of DNA to radiation-induced strand scission at high $|\Delta L|$ values are suggested. Our results compare favorably with data on plasmid pIBI30 reported for 250kV X-rays by Miller and co-workers (1991), but are not in accord with recent results for ^{137}Cs γ-irradiated pUC18 by Milligan, *et al.* (1992).

EXPERIMENTAL METHODS

Protocol for preparation of supercoiled families (topoisomers) of pIBI30

Plasmid pIBI30 was isolated from Escherichia coli by alkaline lysis (Ausubel, *et al.*, 1989). Eukaryotic topoisomerase-I (topo-I) was purchased from Bethesda Research Laboratory (BRL) and was prepared for assay following the recommendations of the supplier. A relaxation protocol was modified from the published procedure of Singleton and Wells (1982). In our method 30 to 50μg of plasmid were relaxed with 30 units of enzyme in the presence of 1.4, 6.4, 9.7, 12, or 16.4 μmol dm^{-3} ethidium bromide at 37° C for four hours in a total volume of 150μl per reaction tube. Native plasmid was subjected to the same manipulations by combining it in a reaction mixture containing all ingredients with the exception of topo-I and ethidium bromide. Since topoisomers were desired in quantity and in order to maximize use of the enzyme, a typical preparation involved twelve reaction tubes (six for each of two topoisomers). The individual tubes were combined to a single tube following incubation and prior to extraction.

Rigorous adherence to recommended standard DNA clean-up protocols (Sambrook, Fritsch, and Maniatis 1989, and Davis, Dibner, and Battey 1986) was found necessary to avoid contamination of sample with chemical substances and to insure removal of residual proteins and introduced enzymes. Fig. 1 depicts an admittedly exaggerated illustration of the absorption spectra of improperly prepared DNA. Topoisomer reaction volumes were extracted once (always with equal volume) with buffer saturated phenol (BRL, catalog number 5513UA) and once with phenol/chloroform/isoamyl alcohol (proportions: 25::24::1) removing the aqueous phase to a fresh tube following each extraction. The resulting volume was subjected to three extractions with water saturated ethyl ether with careful removal of the organic layer to waste. Following each extraction a brief centrifugation (5000 rpm for 1 to 5 min) aided in phase separation. Finally, resulting volumes were incubated for ≈ 10 min at 65° C with constant flow of N_2 gas to purge residual ether from the system.

Ethanol precipitation of topoisomer preparations was adapted from standard protocols. Approximately 200μl of topoisomer solution was combined with 200μl of deionized water (Millipore, Milli-Q water system) to dilute solution salts below the level of co-precipitation with DNA. To this was added 44μl of 3 mol dm^{-3} sodium acetate (pH 5.2) and the volume thoroughly mixed by vortexing before addition of chilled 95% ethanol to a total volume of ≈ 1.5ml. The tubes were again vortexed and chilled on dry ice for ≈ 20 min. Centrifugation in a microcentrifuge at 4° C for 10 min pelleted the DNA. The pellets were carefully rinsed in 0.5ml of chilled 70% ethanol. Centrifugation at 4° C for 4 min prepared the samples for lyophilization (approximately one hour). The DNA was dissolved in a small volume of 50 mmol dm^{-3} potassium phosphate buffer (pH 7.2) and stored at 4° C for a minimum of 24 hours. If the lyophilization step of the topoisomer preparation involved multiple tubes, they were combined to a single tube using a vigorous wash- through method which took advantage of the innate stability (resistance to shearing) of plasmid (supercoiled) DNA. Topoisomer preparations were

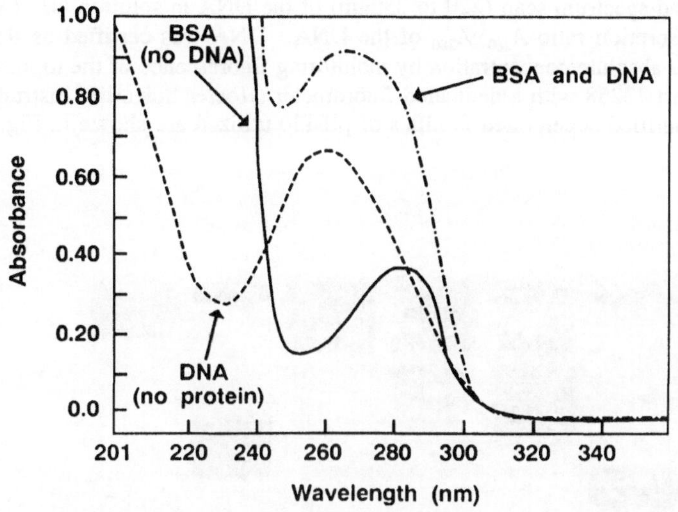

Figure 1. Spectrophotometric scans of solutions of plasmid DNA (pIBI30), protein (Bovine Serum Albumen), and mixed protein/DNA.

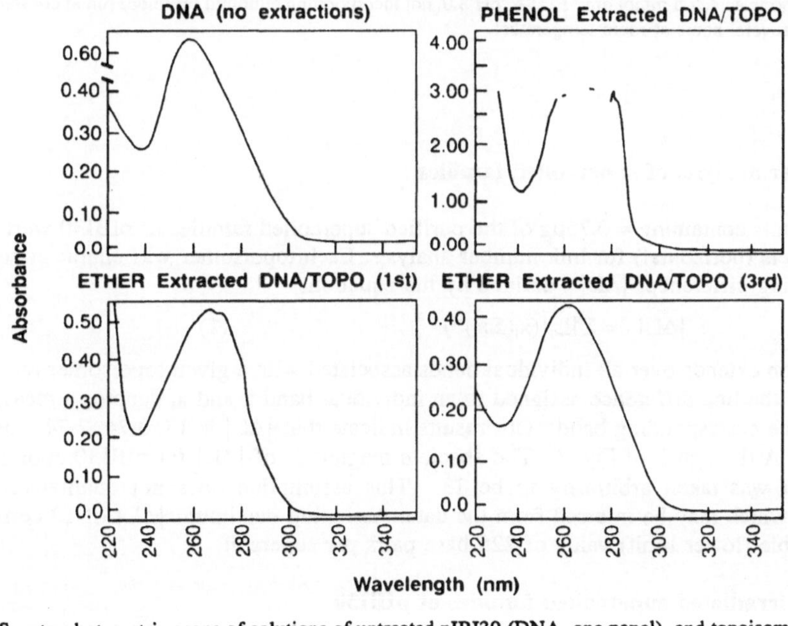

Figure 2. Spectrophotometric scans of solutions of untreated pIBI30 (DNA, one panel), and topoisomerase-I treated pIBI30 (DNA/TOPO) showing decreased levels of contaminants at stages of the purification protocol (see text for details).

brought to a final concentration of ≈ 0.2μg/μl following a careful dilution protocol. DNA concentration was determined by scanning sample solutions at 260nm on a Hewlett Packard 8450A diode array spectrophotometer. DNA purity was determined by monitoring the shape of the curve of a broad-spectrum scan (220 to 350nm) of the DNA in solution (see Fig. 2) and by monitoring the absorption ratio $A_{260}::A_{280}$ of the DNA. DNA was certified as RNA-free and double-checked for absolute concentration by monitoring fluorescence of the topoisomers in the presence of Hoechst 33258 with a dedicated fluorometer (Hoefer Scientific Instruments, model TKO-100). The purified supercoiled families of pIBI30 utilized are shown in Fig. 3.

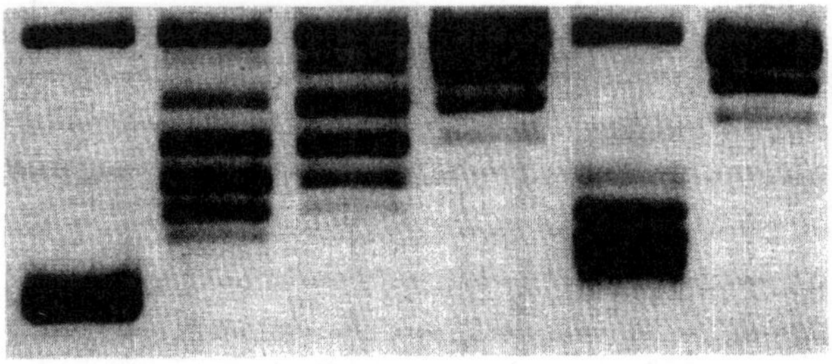

A B C D E F

Figure 3. Electrophoretic *signature* of supercoiled families (topoisomers) of pIBI30. Native plasmid was relaxed with topoisomerase-I in the presence of (A) 0.0, (B) 12.0, (C) 9.7, (D) 6.4, (E) 16.4, and (F) 1.4 μmol dm^{-3} ethidium bromide (see text). Electrophoretic conditions: 7 x 10 cm gel, 1.6% agarose in TBE buffer (89 mmol dm^{-3} Tris, 89 mmol dm^{-3} boric acid, 2.5 mmol dm^{-3} EDTA, pH 8.0, not incorporating ethidium bromide) run at constant voltage (75 Volts) for several hours at room temperature.

Link number analysis of supercoiled families

Aliquots containing ≈ 0.75μg of the purified supercoiled families of pIBI30 were applied to agarose gels (horizontal) for link number analysis. Each topoisomer was characterized by an average linking difference, $|\Delta L|$, defined by the equation:

$$|\Delta L| = \Sigma|L_i|(x_i(\Sigma x_i)^{-1}) \qquad (1)$$

where the sum extends over all individual bands associated with a given topoisomer (A to F, see Fig. 3), L_i is the link difference assigned to an individual band i, and x_i denotes a measurement of area for the corresponding band. Our results indicate that $|\Delta L|$ is 13, 3.74, 2.74, 0.96, 5.92, and 0.75 for A through F of Fig. 3. The absolute magnitude of $|\Delta L|$ for pIBI30 upon isolation from *E. coli* was taken arbitrarily to be 13. This assumption does not qualitatively affect conclusions which may be inferred from the data. Assigning maximum $|\Delta L|$ of 13 corresponds to a reasonable (lower limit) value of 225 base pairs per supercoil.

Analysis of irradiated supercoiled families of pIBI30

Supercoiled families (0.2μg/μl) were exposed to ^{60}Co γ-irradiation at room temperature. The dose rate was ≈ 10 Gy min^{-1}. Dosimetry was performed using a calibrated tissue-equivalent ionization chamber (calibration traceable to N.I.S.T.) and following the AAPM TG21 protocol (American Association of Physicists in Medicine, Task Group 21, Radiation Therapy Committee,

1983). Irradiated topoisomers were stored at 4° C for at least two weeks prior to assay in tightly capped eppendorf tubes which were further protected from desiccation by enclosure within sealed 50ml conical centrifuge tubes. DNA was adversely affected by storage, but the effect stabilized after approximately 14 days (data not shown). Irradiated samples were prepared just prior to analysis by dilution with potassium phosphate buffer (50 mmol dm^{-3}) and bromophenol blue/glycerol tracking dye solution to a concentration of 0.075µg/5µl. 5µl of this preparation were applied to each of several lanes such that the entire spectrum of experimental treatments (radiation doses) including control (no radiation) were represented at least twice on each of two lane-sets on a single gel. Also included on the gel were marker bands composed of native pIBI30 delineating the extremes of the linear response region - separately determined (data not shown).

Agarose gels (1.6%, 15 x 20cm, in TBE buffer {89 mmol dm^{-3} Tris, 89 mmol dm^{-3} boric acid, 2.5 mmol dm^{-3} EDTA}, pH 8.0) incorporating ethidium bromide (0.5µg/ml) were run at constant voltage (75 Volts) for two hours at room temperature. Electrophoresis resolved samples into three distinct bands with the nicked circular band widely separated from the rapidly migrating supercoiled band. The band corresponding to linearized plasmid (Form-III) migrated slower than Form-I, but faster than Form-II. Following electrophoresis, gels were photographed using Polaroid 665 positive/negative film with a Polaroid MP-4 camera. The photographic negatives were analyzed with a microdensitometer (Molecular Dynamics model 300B). Densitometric analysis yielded measurements of band densities. A log/lin plot of the supercoiled (Form-I) band densities (normalized to control) versus dose provided a means for determining the D_{37} dose according to the equation:

$$\text{Form-I}(D) = d(\exp(-D/D_{37})) \qquad (2)$$

where d is the intercept, D is dose, and $1/D_{37}$ is an estimate of the sensitivity of the supercoiled family to strand scission. This relationship is illustrated in Fig. 4 for two supercoiled families of pIBI30 (family A, native plasmid, and family C).

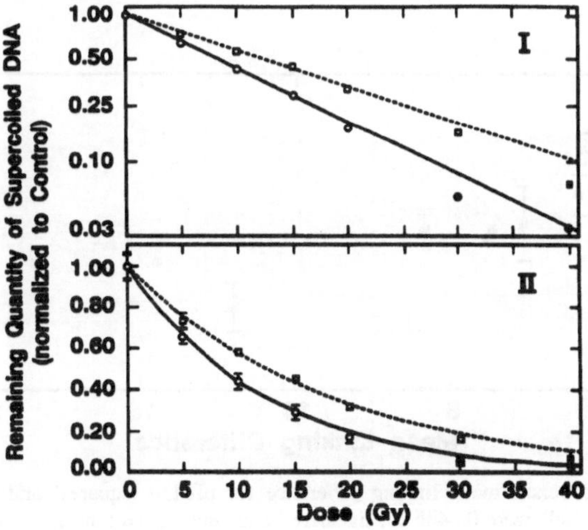

Figure 4. Loss of supercoiled pIBI30 DNA (Form-I) with increasing dose of ^{60}Co γ-irradiation. Figure curves (both panels) show the least squares fit of the data (according to equation (2)). Data represented as closed symbols were near or below the threshold of reliable detection and were not used for least squares curve-fitting. Circles represent native pIBI30 (A from Fig. 3); squares denote supercoiled family C. Panel I shows a log/linear plot of the data. Panel II shows the same data ± one standard deviation in a linear/linear format.

Correction factors for supercoiled families of pIBI30

It is well known that the binding of the intercalator ethidium bromide to DNA depends on the conformational state of DNA (Morgan, *et al.* , 1979). Correction factors for interpreting ethidium bromide/DNA uv-excitation intensities for topoisomers of pUC18 are reported in the literature (Milligan, Arnold, and Ward, 1992). We have also determined correction factors for topoisomers prepared from plasmid pIBI30 in our lab (Fig. 3). Linearized pIBI30 (L-pIBI30) was prepared, purified and brought to concentrations of 0.04 and 0.03μg/5μl. Five supercoiled families and native pIBI30 were also prepared, purified and concentrated similarly. Finally, topoisomer/L-pIBI30 mixtures were prepared which contained 0.08 and 0.06μg/5μl (combined DNA species). Agarose gels were prepared containing 20 lanes to provide four lanes each for topoisomers, topoisomer/L-pIBI30 mixtures, and L-pIBI30 - all of a uniform concentration (0.04, or 0.03μg/5μl), and 3 lanes of a (0.01μg/5μl) reference L-pIBI30 standard; 5μl of sample were applied to a lane. This experimental protocol facilitated the determination of correction factors when supercoiled families and L-pIBI30 were combined and when they were separated but adjacent to one another.

Analysis was by agarose gel electrophoresis and microdensitometry following the protocols already detailed for irradiated samples with the exception that densitometric analysis yielded density values for each of the distinct bands for Form-I, Form-II, and Form-III pIBI30. Correction factors, (CF), corresponding to weaker binding of ethidium to supercoiled DNA relative to the open-circular, or linear conformations, were determined using the formula:

$$CF = (L - NC)/SC \qquad (3)$$

where L, NC, and SC denote the fluorescence intensities of linear, nicked circular, and supercoiled pIBI30 DNA. The average correction factors for the supercoiled families of pIBI30 are shown in Fig. 5. Included, also, are correction factors reported by Milligan, *et al.* (1992) for pUC18 investigated under conditions similar to ours.

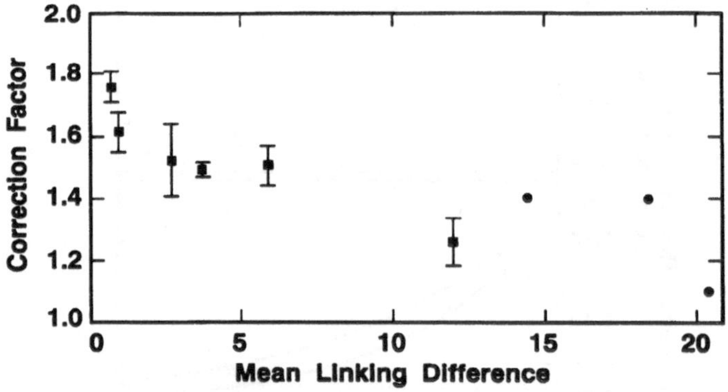

Figure 5. Correction factors versus mean linking difference for pIBI30 (squares) and pUC18 (circles). DNA quantities applied to agarose gels were 0.04μg for linearized and supercoiled families of pIBI30. Supercoiled and linearized pIBI30 were assayed independently in adjacent lanes (this figure) and combined in a single lane (data not shown). Error bars for pIBI30 indicate one standard deviation. pUC18 data are from Milligan, Arnold, and Ward (1992).

When we undertook the investigation of correction factors for topoisomers of pIBI30 we were uncertain of their value in the interpretation of our data. We remain uncertain although we note that, clearly, there are differences for CF's for supercoiled families as shown in Fig. 5. However, we note that the value of CF for a topoisomer would not be expected to remain unchanged over a dose response curve and, in fact, for each dose a separate determination of CF would seem to be required. G(SSB) values reported by Milligan, *et al.* (1992) and sensitivity(Gy^{-1}) values reported in this work and in the work of Miller, *et al.* (1991) are derived from $1/D_{37}$ values which are themselves derived from the slopes of plots like fig. 4 (top). It would seem to us that application of CF values as a data filter would serve only to shift the data framework without altering the slope of the response curve or the value of $1/D_{37}$. Therefore, we decided, albeit arbitrarily, to ignore CF values in the interpretation of our data.

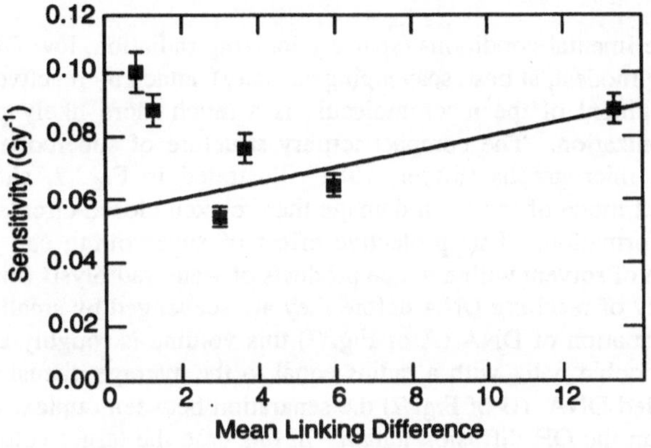

Figure 6. The sensitivity of topoisomers of pIBI30 to the induction of single strand scission (SSB) by ^{60}Co γ-radiation. The data points were obtained by a least squares fit of the survival of Form-I DNA versus dose, *i.e.* from plots like those shown in Fig. 4.

RESULTS

Our data, as evident from Fig. 4, suggest that the degree of superhelicity of topoisomers of pIBI30 influences their sensitivity to ionizing radiation. In Fig. 6 we plot the sensitivity of topoisomers of pIBI30 to the induction of single strand scission (SSB) by ^{60}Co γ-radiation. The data points were obtained by a least squares fit of the survival of Form-I DNA versus dose from plots like those shown in Fig. 4. Results shown in Fig. 6 agree with data reported by Miller and co-workers (1991) for 250 kV X-rays. Both studies report an approximate linear relationship of radiation sensitivity as a function of linking difference (for $|\Delta L| > 2.5$). We attribute the difference between plot intercept values primarily to differences in buffer systems. Miller, *et al.* (1991) used a Tris buffer which is known to have a radioprotective effect, while we employed a potassium-phosphate buffer offering little in the way of protection from ionizing radiation.

For small values of $|\Delta L|$ (*i.e.* values < 2.5) the sensitivity to ionizing radiation appears to decrease with an increase in $|\Delta L|$. It is only this portion of our data that follows standard radiation target theory. To estimate the G-value for single strand scission, G(SSB), we note that at the D_{37} dose there is an average of one SSB per plasmid (*i.e.* the concentration of SSB at D = D_{37} is equal to the concentration of plasmid molecules), hence

$$G(SSB) = [DNA]/(D_{37}Gy^{-1})(\rho/kg\ dm^{-3}) \quad (4)$$

where ρ is the density of the solution (taken to be unity) and [DNA] is DNA concentration/μmol dm^{-3}. Milligan, *et al.* (1992) found G(SSB) to be insensitive to superhelical density for pUC18 (2686 base pairs) with a value of about $4x10^{-4}$ μmol J^{-1}. By contrast our results for pIBI30 (2926 bp) show a dependence of G(SSB) on superhelical density (data derived from Fig. 6, but not shown). For families A and C of Fig. 3 we find G(SSB) to be $8.8x10^{-3}$, and $6.0x10^{-3}$ μmol J^{-1} respectively.

DISCUSSION

Under our experimental conditions (sparsely ionizing radiation, low DNA concentration, and a solvent with only modest, at best, scavenging capacity) attack by reactive species produced in the aqueous environment of the macromolecule is a much more likely pathway for DNA damage than direct ionization. The compact tertiary structure of supercoiled DNA, which is evident from electron micrographs (Stryer, 1981) illustrated in Fig. 7, should make it less sensitive to this indirect mode of radiation damage than relaxed closed-circular DNA which has a more extended conformation. This protective effect of supercoiling can be understood by considering the volume of solvent within which products of water radiolysis (mainly OH radicals) have a finite probability of reaching DNA before they are scavenged by small-molecule solutes. For the relaxed conformation of DNA (A of Fig. 7) this volume is roughly approximated by a tube surrounding the double helix with a radius equal to the average diffusion distance of OH radicals. For supercoiled DNA (B of Fig. 7) the separation between duplexes in an interwound branch may be less than the OH diffusion length. In this case the target volume per nucleotide is smaller than it is in the relaxed state and a greater portion of the absorbed dose is lost in the production of radical species that are scavenged before they can react with DNA.

The tendency of negative supercoiling to produce strand separation may also reduce the sensitivity of supercoiled DNA to radiation-induced strand scission relative to the relaxed state. This protective mechanism follows from the shielded base hypothesis (Ward, 1975, Ward and

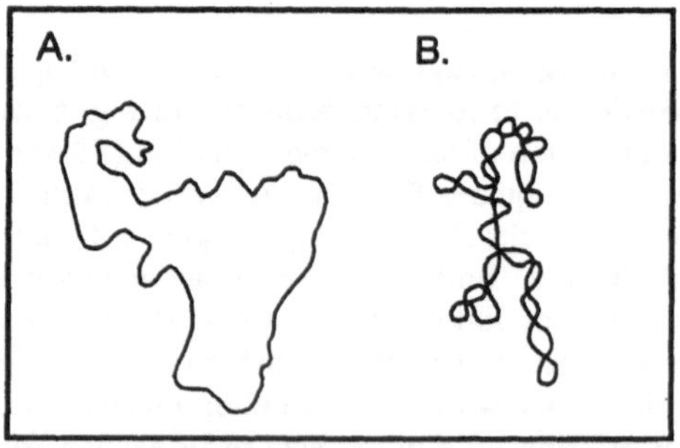

Figure 7. Electron micrograph of mitocondria plasmid DNA illustrating the difference between fully relaxed and supercoiled conformational states.

Kuo, 1978) which asserts that the yield of radiation-induced base damage is limited by the solvent accessibility of sites for OH radical addition in the standard B-DNA double helix. Strand separation increases the accessiblilty of these sites which increases the probability that OH radical attack on DNA leads to modified bases rather than backbone cleavage.

If the above considerations of target volume and solvent accessibility are valid and our observations of an increase in the sensitivity of plasmid DNA to radiation-induced strand breaks with negative superhelicity for both X- and γ-radiation are not due to an experimental artifact, then the physiological levels of negative supercoiling (superhelical densities of the order of -0.05) must affect DNA interactions leading to strand breaks in ways that more than compensate for the protective effects of compactness and single-strandedness. One possible mechanism by which this might occur is through a change in the energetics of the β-phosphate elimination pathway (Beesk, *et al.* , 1979) that follows hydrogen abstraction from the C4' position of sugar moieties. Even though in duplex B-DNA, hydrogens on sugar moieties are about 10 times more accessibile to solvent than base aromatic carbon atoms (Ward, 1985), not more than 20% of the OH radicals reaching DNA react with sugars (Schulte-Frohlinde, 1988). This suggests that OH addition to the double bonds of DNA bases is energetically favored in relaxed DNA over abstraction of hydrogen from ribose. However, in supercoiled DNA hydrogen abstraction from sugar opens a channel (backbone cleavage) to a much lower energy state of the molecule than can be achieved by base modifications resulting from OH addition to double bonds. This may provide a thermodynamic driving force that makes strand scission more competitive with base damage as the elastic energy stored in supercoiled DNA increases.

Exceptions to the general rule that base damage does not lead to strand scission have been found for the single-stranded homopolymers poly(U) and poly(dA) (Adinarayana, *et al.* , 1988). In these systems base radicals resulting from OH addition are able to abstract hydrogen from sugar moieties. Thus far, this process has not been detected in double-stranded DNA, where the more rigid tertiary structure probably makes it more difficult for carbon-centered base radicals to get close enough to sugar moieties to allow hydrogen abstraction. Evidence for the importance of strand flexibility in base-to-sugar radical transfer is seen from the salt dependence of the lifetime of peroxyl radicals in poly(U) (Schulte-Frohlinde, *et al.* , 1986) and the higher yield of strand breaks in poly(dA) relative to poly(A) (Adinarayana, *et al.* , 1988). In the latter case differences in sugar pucker and base stacking make poly(A) a more rigid polymer (Evans and Sarma, 1976). Although the single-stranded characteristics that negative supercoiling confers to duplex DNA probably increase OH addition to bases at the expense of hydrogen abstraction from sugar, a significant fraction of the primary base damage may be converted to strand breaks by base-to-sugar radical transfer. The preference for strand separation in regions of low GC content and the observed decay of base radicals to strand breaks in poly(dA) support this mechanism of sensitization by negative supercoiling.

Our data suggest that DNA response to ionizing radiation can be a strong function of its conformational state in addition to its intrinsic base pair composition. For these reasons the lack of a dependence of pUC18 radiation sensitivity on superhelical density (Milligan, *et al.* , 1992) and our data on plasmid pIBI30 showing just such a relationship, in fact, might have been expected. An understanding of the consequences of other variables (not investigated) including plasmid size and base composition, temperature, DNA and scavenger concentrations, and ionic environment might elucidate the reasons for the apparent discrepancies between our results and those of Milligan and co-workers (1992).

We have recently shown (unpublished data, see also Holwitt, Koda, and Swenberg, 1990) that the radioprotectants WR-1065 [N-(2-mercaptoethyl)-1,3-diaminopropane] and its disulfide, WR-33278, modulate the relaxation of supercoiled pIBI30 DNA by calf thymus topoisomerase-I.

Topoisomerase-I is a ubiquitous eukaryotic enzyme that relaxes supercoiled DNA by single-strand cleavage and re-ligation (Maxwell and Gellert, 1986, D'Arpa and Liu, 1989). Most mechanisms of radioprotection proposed for WR-1065 have focused on its scavenging properties (Philip, 1980, Durand, 1983). Because supercoiled domains exist in both prokaryotic and eukaryotic cells, the results reported here and the observed stimulation of topo-I induced unwinding of supercoiled DNA (Holwitt, Koda, and Swenberg, 1990) suggest that the WR compounds may confer some protection to the genome by decreasing the superhelicity of DNA. The data that we report in this paper leads us to believe that a decrease in DNA supercoiling should result in a decrease in DNA damage. If this is the case, then the critical DNA damage sites would correspond to chromosomal regions where the superhelicity is large. Alternately, a decrease in superhelicity may change the functional properties of DNA. This mechanism suggests changes in metabolic processes, a virtually unexplored field although there have been reports that WR-1065 enhances DNA repair (Lowenstein, *et al.* , 1989, Riklas, *et al.* , 1988, Swenberg, 1989). The actual process responsible for DNA's radiation sensitivity to increase when confined to very small domains, and the molecular mechanism underlying the stimulation of eukaryotic type I topoisomerase by WR-1065 and WR-33278 are not known.

REFERENCES

Adinarayana, M., Bothe, E., and Schulte-Frohlinde, D., 1988, Hydroxyl radical-induced strand break formation in single-stranded polynucleotides and single-stranded DNA in aqueous solutions as measured by light scattering and conductivity, Int. J. Radiat. Biol., 54, 723-737.

American Association of Physicists in Medicine, Task Group 21, Radiation Therapy Committee, 1983. A protocol for the determination of absorbed dose from high-energy photon and electron beams, Med. Phys., 10, 741-771.

Ausubel, F.M., Brent, R., Kingston, R.E., Moore, D.D., Seidman, J.G., Smith, J.A., and Strabl, K., 1989, Current Protocols in Molecular Biology, (Wiley Interscience), 1.71-1.74.

Beesk, F., Dizdaroglu, M., Schulte-Frohlinde, D., and von Sonntag, C., 1979, Radiation-induced DNA strand breaks in deoxygenated aqueous solutions. The formation of altered sugar and end groups, Int. J. Radiat. Biol., 36, 565-576.

Cozzarelli, N.R., Boles, T.C., and White, J.H., 1990, Primer on the topology and geometry of DNA supercoiling, in DNA topology and its biological effects, N.R. Cozzarelli, and J.C. Wang, eds., (Cold Spring Harbor Laboratory Press), 139-184.

Davis, L.G., Dibner, M.D., Battey, J.F., 1986. Basic Methods in Molecular Biology, (Elsevier Science Publishing Co. Inc.).

D'Arpa, P., and Liu, L.F., 1989, Topoisomerase-targeting antitumor drugs, Biochem. Biophys. Acta, 989, 163-177.

Durand, R.E., 1983, Radioprotection by WR-2721 *in vitro* at low oxygen tensions. Implications for its mechanism of action, Br. J. Cancer, 47, 387-392.

Evans, F.E., and Sarma, R.H., 1976, Nucleotide rigidity, Nature, 263, 567-572.

Fowler, J.F., 1964, Differences in survival curve shape for formal multi-target and multi-hit models, Phys. Med. Biol., 9, 177-188.

Holwitt, E.A., Koda, E., and Swenberg, C.E., 1990, Enahncement of topoisomerase-I mediated unwinding of supercoiled DNA by the radioprotector WR-33278, Radiat. Res., 124, 107-109.

Lowenstein, E., Gleeson, J.L., Hecht, E., Factor, R., Goldfischer, C., Cajigas, A., and Steinberg, J.J., 1989, Excision repair is enhanced by WR2721 radioprotection, in Terrestrial Space Radiation and Its Biological Effects, eds., P.D. McCormack, C.E. Swenberg, and H. Bucker, (Plenum Press, New York), 697-714.

Maxwell, A., and Gellert, M., 1986, Mechanistic aspects of DNA topoisomerase, Adv. Protein Chem., 38, 69-86.

Miller, J.H., Nelson, J.M., Ye, M., Swenberg, C.E., Speicher, J.M., and Benham, C.J., 1991, Negative supercoiling increases the sensitivity of plasmid DNA to single-strand break induction by X-rays, Int. J. Radiat. Biol., 59, 941-949.

Milligan, J.R., Arnold, A.D., Ward, J.F., 1992. The effect of superhelical density on single strand break yield for gamma irradiated plasmid DNA, Rad. Res., 132, 69-73.

Morgan, A.R., Lee, J.S., Pulleyblank, D.E., Murray, N.L., and Evans, D.H., 1979, Ethidium fluorescence assays. Part I. Physiochemical studies, Nucleic Acids Res., 7, 547-569.

Phillips, T.L., 1980, Rationale for initial trials and future development of radioprotectants, Cancer Clin. Trials, 3, 165-173.

Riklas, E., Kob, R., Green, M., Prager, R., Marko, R., and Mintsberg, M., 1988, Increased radioprotection attained by DNA repair enhancement, Pharmacol. Ther., 39, 311-322.

Schulte-Frohlinde, D., Behrens, G., Onal, A., 1986, Lifetime of peroxyl radicals in poly(U), poly(A), and single- and double-stranded DNA and their reaction with thiols, Int. J. Radiat. Biol., 50, 103-110.

Schulte-Frohlinde, D., 1988, The effect of oxygen on the OH radical-induced strand break formation in vitro and in vivo, Basic Life Sci., 49, 403-417.

Sambrook, J., Fritsch, E.F., Maniatis, T., 1989, Molecular Cloning: A Laboratory Manual, 2nd edition, (Cold Spring Harbor Laboratory Press).

Singleton, C.K., and Wells, R.D.,1982, The facile generation of covalently closed circular DNAs with defined negative superhelical densities, Anal. Biochem., 122, 253-257.

Stryler, L., 1981, Biochemistry, 2nd edition, (W.H. Freeman, San Francisco), 574.

Swenberg, C.E., 1989, DNA and radioprotection, in Terrestrial Space Radiation and Its Biological Effects, eds., P.D. McCormack, C.E. Swenberg, and H. Bucker, (Plenum Press, New York), 675-695.

Ward, J.F., 1975, Molecular mechanisms of radiation-induced damage to nucleic acids, Advances in Radiation Biology, J.T. Lett and H. Adler, eds., (Academic Press, New York), vol. 5, 181-239.

Ward, J.F., and Kuo, I., 1978, Radiation damage to DNA in aqueous solution: a comparison of the response of the single-stranded form with that of the double-stranded form, Radiat. Res., 75, 278-285.

Ward, J.F., 1985, Biochemistry of DNA lesions, Radiat. Res., 104, S103-S111.

PLASMIDS AS TESTSYSTEM FOR THE DETECTION OF DNA STRAND BREAKS

J. Wehner, G. Horneck and H. Bücker

DLR, Institute for Aerospace Medicine
Biophysics Division
Linder Höhe, 5000 Köln 90, FRG

ABSTRACT

The E. coli plasmid pUC19 was used to analyse the formation of strand breaks in the DNA. Strand breaks were induced by endonuclease DNAase I in the presence of Mg^{2+} or Mn^{2+} ions. Single as well as double strand breaks were analysed by gel electrophoresis. This testsystem will be used for detecting damage at the molecular level of DNA induced by radiation, especially by ionizing or vacuum-ultraviolet (V-UV) radiation.

INTRODUCTION

Studies on biological effects of V-UV radiation suggest that the induction of strand breaks (sb) in DNA may be one of the major reasons for inactivation by this highly energetic electromagnetic radiation (Johnson-Thompson et al. 1984, Dose, 1986, Hieda,et al. 1986, Munakata et al. 1991). In addition, a variety of other kinds of DNA damage which occur after V-UV irradiation, e.g. base modifications, apurinic and apyrimidinic (AP) sites (Maezawa, et al. 1984). The chain of events, from induction of DNA damage by V-UV radiation to processing of these damages by the cellular repair machinery is not well understood. The understanding of these effects of the radiation, e.g. inactivation and mutation, requires analysis of induced DNA damages at the molecular level (Seidmann, et al. 1985).

The use of plasmids makes it possible to analyse radiation damage at the molecular level of DNA without any interference from cellular repair. The target gene in the plasmids serves as a surrogate for endogenous genes. Using these plasmids, the target gene can be treated with the V-UV radiation *in vivo* or *in vitro*. Due to their high transparency for V-UV light and their ability to tolerate desiccation periods, these plasmids are ideal candidates to study the biological effects of V-UV (Coohill, 1986). A method will be introduced that allows one to determine the efficiency of the induction of single and double strand breaks in the DNA of pUC 19 by V-UV radiation.

Biological Effects and Physics of Solar and Galactic Cosmic Radiation, Part A, Edited by C.E. Swenberg *et al.*, Plenum Press, New York, 1993

MATERIALS AND METHODS

Test System

We used the E. coli high copy plasmid pUC 19 (Sambrook et al. 1989) with the *lac Z* gene serving as target (Figure 1). This gene encloses 363 nucleotides (Yanish-Perron,et al. 1985). The relative small size of the mutagenesis target gene decreases the probability of recovering spontaneous deletion mutations. On the other hand the small size makes it possible to sequence the target gene in one step with two primers.

Induction of DNA Strand Breaks

2 μg plasmids in aqueous solution were treated with 10 ng DNAase I (Serva) and 50 μg BSA in the presence of 10 mM $MgCl_2$ or 0,66 mM $MnCl_2$ at 37° C. After incubation for certain time intervals up to 60 min, depending on the used ion, aliquots were taken for analysis.

Analysis of DNA Strand Breaks

DNA strand breaks were identified by gel electrophoresis. The treated plasmids were separated in a 0.8 % agarose gel with the running conditions 35 V for 20 h. After electrophoresis the gels were stained with ethidium bromide, photographed with Polaroid film and the negatives were scanned and the fraction of DNA in the different bands was calculated.

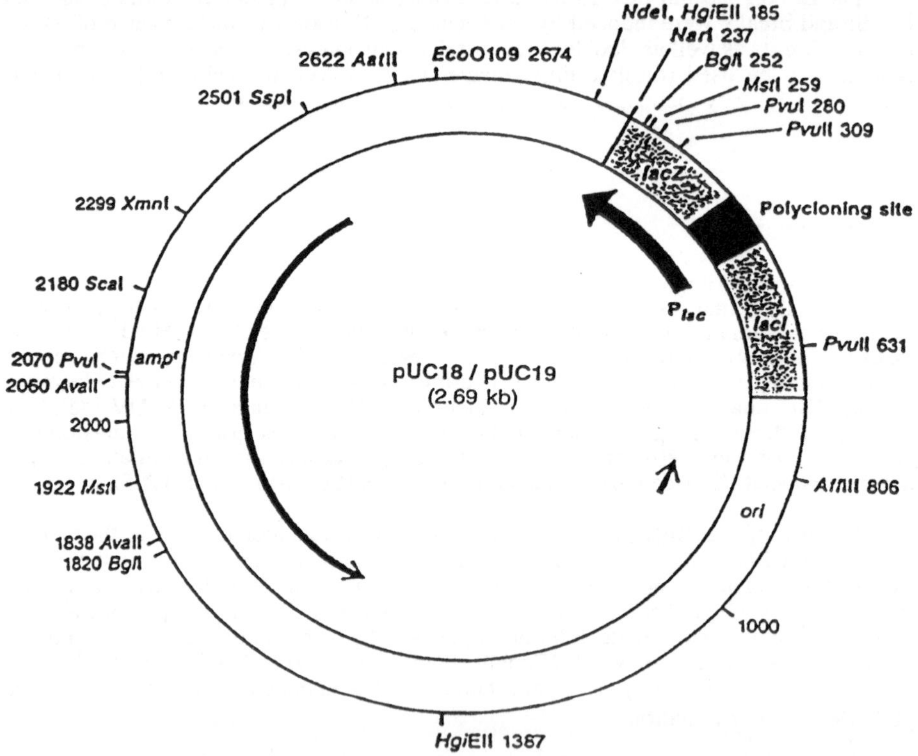

Figure 1. E. coli Plasmid pUC 19 with Target Gene (lac Z = Polycloning Site) (Sambrook et al. 1989)

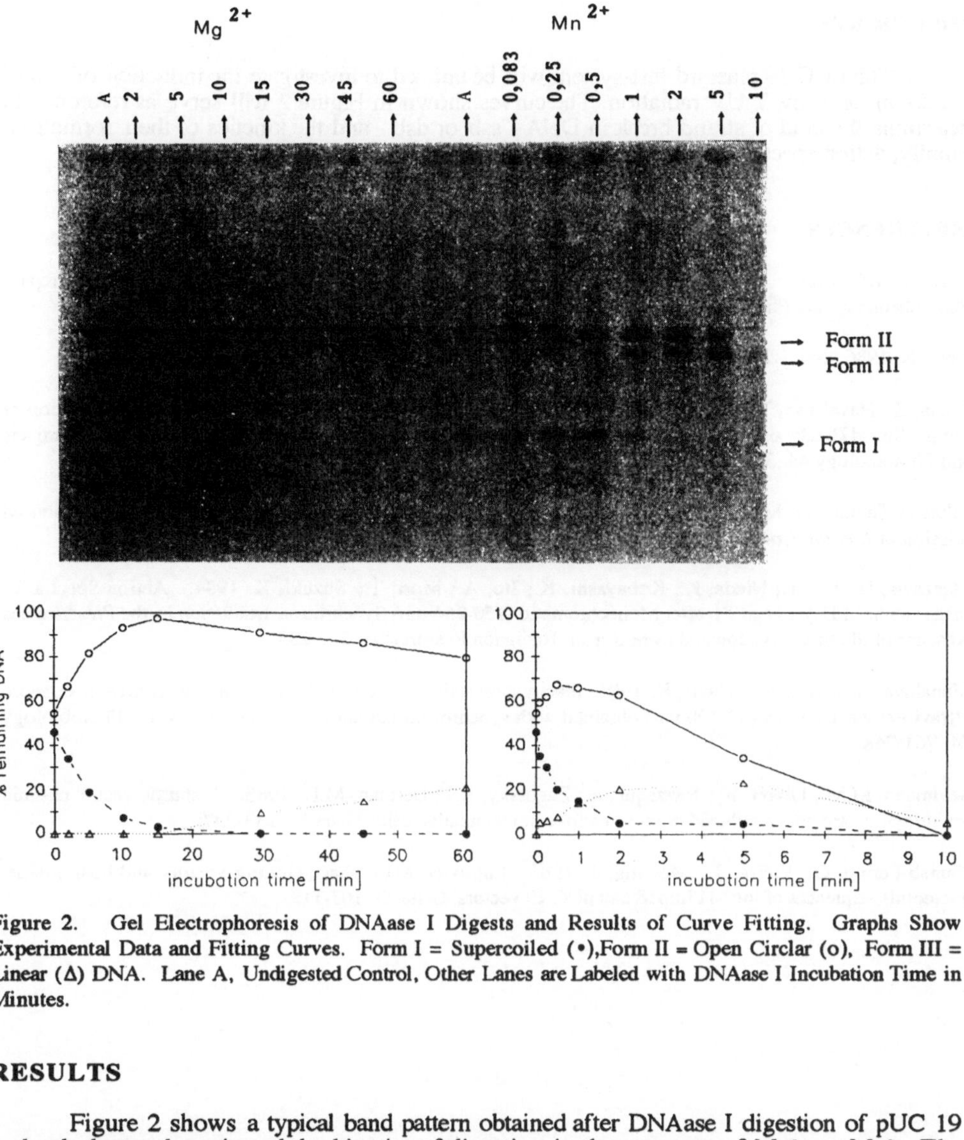

Figure 2. Gel Electrophoresis of DNAase I Digests and Results of Curve Fitting. Graphs Show Experimental Data and Fitting Curves. Form I = Supercoiled (•),Form II = Open Circlar (o), Form III = Linear (Δ) DNA. Lane A, Undigested Control, Other Lanes are Labeled with DNAase I Incubation Time in Minutes.

RESULTS

Figure 2 shows a typical band pattern obtained after DNAase I digestion of pUC 19 and gel electrophoresis and the kinetics of digestion in the presence of Mg^{2+} or Mn^{2+}. The bands represent different three dimensional structures of the plasmid. Form I, indicates supercoiled DNA, Form II open circular DNA and Form III linear DNA. In E. coli cells, the transition between the different three dimensional structures is catalysed by topoisomerases. It can also be induced by radiation, V-UV or ionizing radiation. One single strand break (ssb) of DNA transfers its structure from supercoiled to an open circular one, a double strand break (dsb) leads to a linear molecule. During DNAase I digestion, ssb and dsb are induced in the DNA. The relation between ssb and dsb and the kinetics of formation depend on the kind of ion present during DNAase I treatment. In the presence of Mg^{2+} ions, preferably ssb's are produced leading to a sharp decrease in open circular DNA (Figure 2, left side). Linear DNA, resulting from dsb, appears only after more than 15 min of incubation. In the presence of Mn^{2+}, dsb are predominantly induced. This is indicated by the appearance of linear DNA directly after starting of incubation and a drop in open circular DNA with incubation time (Figure 2, left side). Hence, this method allows for differentiation between the induction of ssb and dsb by different agents.

Discussion

The pUC 19 plasmid test-system will be utilized to investigate the induction of strand breaks in DNA by V-UV radiation. The curves shown in Figure 2 will serve as reference to determine the kind of strand break in DNA - ssb or dsb - and the kinetics of their formation. Finally, action spectra for induction of ssb and dsb will be obtained.

References

Coohill, T.P. 1986. Virus-cell interactions as probes for Vacuum-ultraviolet radiation damage and repair. Photochemistry and Photobiology 44, pp. 359-363.

Dose, K. 1986. Survival under space vacuum. Biochemical aspects. Adv. Space Res. 6 (12): 307-312.

Hieda, K.; Hayakawa, Y.; Ito, A.; Kobayashi, K.; Ito, T. 1986. Wavelength dependence of the formation of Single Strand Breaks and Base Changes in DNA by the Ultraviolet Radiation above 150 nm. Photochemistry and Photobiology 44: 379-383.

Johnson-Thompson, M.; Halpern, J.B.; Jackson, W.M.; Georges, J. 1984. Vacuum UV LASER induced scission of *Simian Virus 40* DNA. Photochemistry and Photobiology 39: 17-24.

Maezawa, H.; Ito, T.; Hieda, K.; Kobayashi, K.; Ito, A.; Mori, T.; Suzuki, K. 1984. Action Spectra for Inactivation of Dry Phage T1 after Monochromatic (150-254 nm) Synchrotron Irradiation in the Presence and Absence of Photoreactivation and Dark Repair. Radiation Research 98: 227-233.

Munakata, N.; Saito, M.; Hieda, K. 1991. Inactivation action spectra of *Bacillus subtilis* Spores in extended ultraviolet wavelengths (50-300 nm) obtained with synchrotron radiation. Photochemistry ans Photobiology 54: 761-768.

Seidmann, M.M.; Dixon, K.; Razzaque, A.; Zagursky, R.J.; Berman, M.L. 1985. A shuttle vector plasmid for studying carcinogen induced point mutations in mammalian cells. Gene 38: 233-237.

Yanish-Perron, C.; Vieira, J.; Messing, J. 1985. Improved M13 phage cloning vectors and host strains: nucleotide sequences of the M13mp18 and pUC 19 vectors. Gene 33: 103-119.

HIGH RESOLUTION GEL ELECTROPHORESIS METHODS FOR STUDYING SEQUENCE-DEPENDENCE OF RADIATION DAMAGE AND EFFECTS OF RADIOPROTECTANTS IN DEOXYOLIGONUCLEOTIDES

B. Mao*, C.E. Swenberg**, Y. Vaishnav** and N.E. Geacintov*

*Chemistry Department, New York University
New York, N.Y., 10003, U.S.A.

**Radiation Biochemistry Department, Armed Forces Radiobiology Research Institute, Bethesda, MD 20889, U.S.A.

INTRODUCTION

Ionizing radiation is known to produce both single and double-strand breaks in DNA (von Sonntag et al., 1981; Hutchinson, 1985; Ward, 1988). In aqueous DNA solutions, radiolysis of water produces OH· radicals which react with DNA bases and sugar residues; hydrogen atom abstraction from the sugar moieties produces radicals which ultimately give rise to strand scission. In the case of low LET radiation (e.g. γ-irradiation), DNA damage caused by OH· radicals generated in the bulk of the aqueous solution is called the "indirect effect" (see for example, Skov, 1984; Achey and Durea, 1974; van Rijn et al., 1985; Roots et al., 1985; Siddiqi and Bothe, 1987; Schulte-Frohlinde, 1989). In the case of high LET radiation, e.g. charged heavy particles, a direct deposition of energy within the DNA molecules is dominant in producing DNA damage by the "direct effect" (see Holley et al., 1990, for example). The formation of DNA strand breaks, in particular double-strand breaks, are rather strongly correlated with ionizing radiation-induced cell killing (Coquerelle, 1978; Elkind, 1985). Other harmful effects of ionizing radiation (Beebe, 1982), including mutations (Waters et al., 1991; Jaberaboansari et al., 1991; Raha and Hutchinson, 1991; Geacintov and Swenberg, 1992) have been well documented.

High resolution electrophoretic gels (Maxam and Gilbert, 1980) are ideal for monitoring the occurrence of DNA strand breaks (Tullius, 1987). However, there are only a few reports on the use of these techniques in studies of radiation damage in DNA (Henner et al., 1982, 1983). In this work we have further explored the utility of these electrophoresis methods to study the formation of strand breaks induced in short deoxyoligonucleotides (11-base pairs) in the single and double-stranded forms in the presence and absence of the known radioprotectant molecule cysteamine (van der Shans, 1970; Roots and Okada, 1972; Bird, 1980; Ward, 1983; Held et al., 1984; Smoluk et al., 1988; Zheng et al., 1988).

EXPERIMENTAL METHODS

The deoxyribooligonucleotides 5'-d(CACATGTACAC) (**X**) and its complement 5'-d(GTGTACATGTG) (**Y**)were synthesized by the phosphoramidite method using a Biosearch Cyclone automated DNA synthesizer (Milligen-Biosearch Corp., San Rafael, CA). The oligonucleotides were purified by oligonucleotide purification columns (Applied Biosystems,

Biological Effects and Physics of Solar and Galactic Cosmic Radiation, Part A, Edited by C.E. Swenberg *et al.*, Plenum Press, New York, 1993

Foster City, CA), and by HPLC using a Rainin Dynamax C_4 column (Rainin Instrument Co., Woburn, MA) using 0.1 M triethylamine-acetate/acetonitrile solvent mixtures. The oligonucleotide **X** was labeled at the 5'-end with [γ^{32}P]ATP purchased from New England Nuclear Corporation (Boston, MA) and employing a T_4 polynucleotide kinase 5'-terminus labeling system (Bethesda Research Laboratories, Gaithersburg, MD). Prior to irradiation, the oligonucleotide was further purified using a 20% polyacrylamide denaturing (7 M urea) gel electrophoresis system , which was also used in all subsequent experiments with DNA exposed to ionizing radiation.

The duplexes **X:Y** (d(CACATGTACAC)· d(GTGTACATGTG)) were formed by annealing the two individual oligonucleotides following standard procedures which involved heating a stoichiometric solution (20 mM sodium phosphate buffer, pH 7) of **X** and **Y** to 85 °C and slow (overnight) cooling to 4 °C. The formation of **X:Y** duplexes was ascertained by electrophoresis on native 20 % polyacrylamide gels (without urea) at 4 °C; the duplex samples were characterized by bands which migrated significantly slower than the single-stranded oligonucleotides.

All irradiations were carried out at 4 °C in air-saturated 20 mM sodium phosphate buffer solutions at pH 7.0. The single- or double-strand concentrations were 0.25 and 0.5 μM (expressed in terms of strand molarities), respectively. The samples were irradiated with either ^{60}Co γ-radiation or neutrons at the Armed Forces Radiobiology Research Institute at Bethesda, MD as described elsewhere in these proceedings (Swenberg et al., 1992).

RESULTS

γ-Irradiation

Typical densitometer tracings of Maxam-Gilbert gels are shown in Fig. 1 for the single-strand **X** (2.5 μM nucleotide concentration) for the case of the unirradiated control, and **X** exposed to 100 and to 200 Gy. Analogous results for the double-stranded **X:Y** oligonucleotides (5 μM) are shown in Fig. 2. Characteristic double-maxima due to frank strand breaks are observed in both the single- and double-stranded oligonucleotides, but are best defined in the case of the irradiated duplex **X:Y**, particularly at the higher dosage of 200 Gy. Besides the intense band due to the original 11-mer (lowest mobility broad band on the

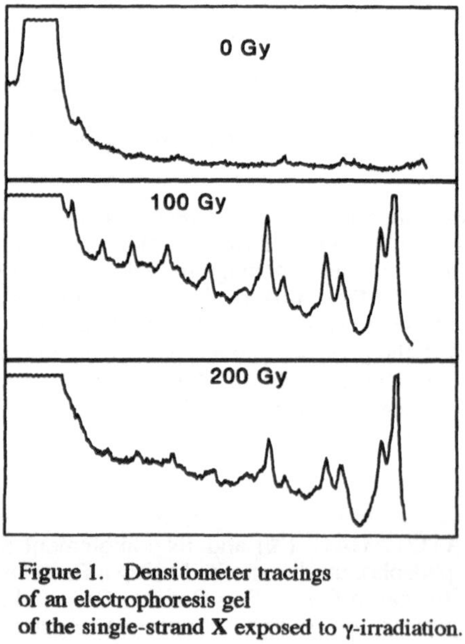

Figure 1. Densitometer tracings of an electrophoresis gel of the single-strand **X** exposed to γ-irradiation.

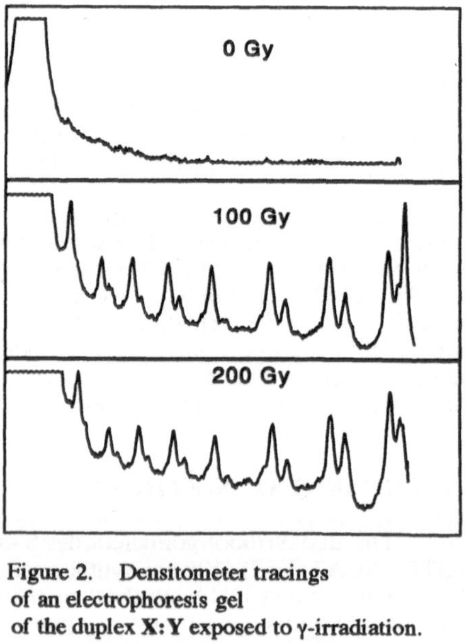

Figure 2. Densitometer tracings of an electrophoresis gel of the duplex **X:Y** exposed to γ-irradiation.

left), eight other types of fragments, corresponding to 10-, 9-, 8-, 7-, 6-, 5-, 4-, and 3-mers are distinguishable. The bands due to the nucleotide monomer and dimer fragments are missing, because the mobilities of these fragments are so high that they are located beyond the scale shown in Figs. 1-6.

Each type of fragment (the different fragments differ from one another by the number of nucleotides) is characterized by double bands in the electrophoretic gels. The occurrence of double-bands for each type of 5'-end-labeled fragment has been previously observed by Henner et al. (1983) and attributed to the occurrence of two types of 3'-termini; the slower migrating fragment contains a 3'-phosphoryl group, while the faster moving fragment contains a glycolate moiety attached to the 3'-phosphoryl group via the 2"-OH group of glycolic acid.

In the case of the irradiated single-stranded oligonucleotide **X**, the maxima due to the shorter fragments are superimposed on a background whose amplitude increases with increasing size of the electrophoresed fragments. In the case of the γ-irradiated double-stranded **X:Y** this background appears to be less pronounced, and the amplitudes of the bands due to strand breaks appear to be greater (Fig. 2) than in the single-stranded case (Fig. 1).

Neutron-irradiation

In the case of the single-strand oligonucleotide **X**, double-bands can be recognized only in the case of the 3-mer, 4-mer, and 5-mer (Fig. 3). At the higher dosage, the higher molecular weight bands tend to disappear. In the case of the duplex (Fig. 4), bands due to all eight fragments, together with the faster moving phosphate glycolate bands, are recognizable.

Estimation of fractions of damaged oligonucleotides

The fraction of damaged oligonucleotides was estimated by comparing the radioactivities in the 11-mer band, and in all of the shorter fragment bands (including the monomer and dimer bands not shown in Figs. 1-6). This was accomplished by integrating the areas under the densitometer tracings and comparing the areas under the 11-mer band, and summing the areas under the tracings corresponding to all lower mobility fragments (including the background). The area under the intense 11-mer band was estimated by a serial

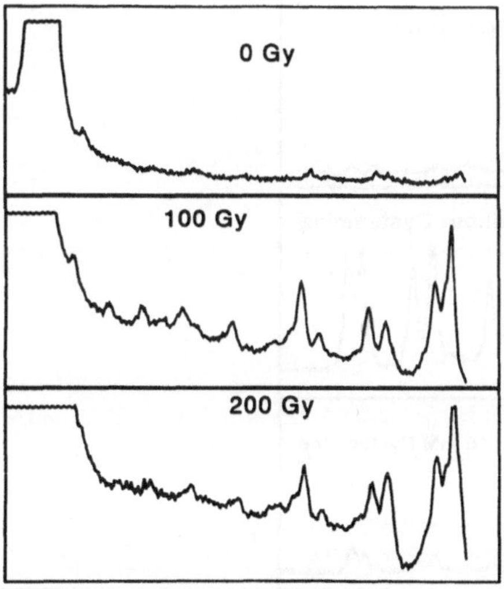

Figure 3. Densitometer tracings of an electrophoresis gel of the single-strand **X** exposed to fission neutron-irradiation.

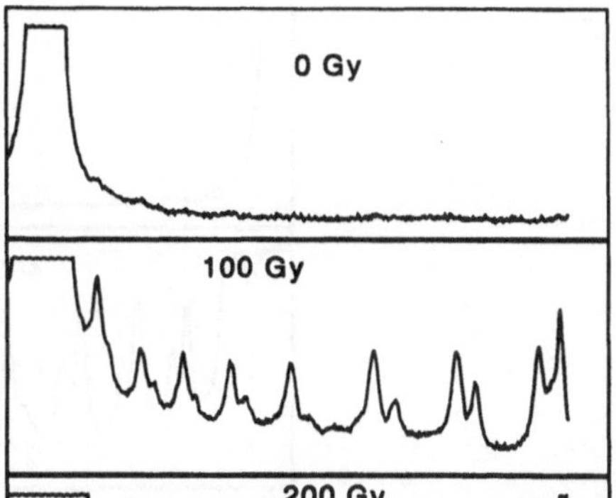

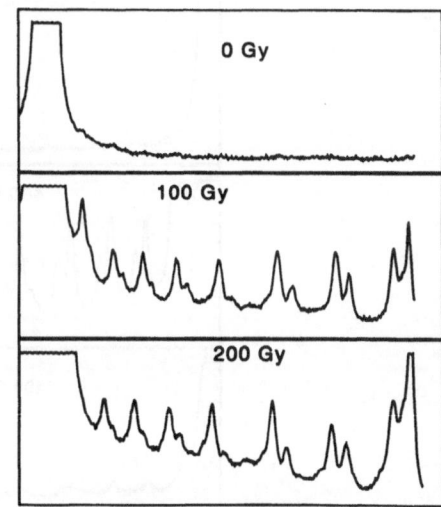

Figure 4. Densitometer tracings of an electrophoresis gel of the duplex **X:Y** exposed to fission neutron-irradiation.

dilution procedure, thus producing bands of sufficiently low amplitudes for densitometric analysis. In all cases, the fraction of damaged oligonucleotides estimated in this manner varied from 4 - 5 %.

Effects of the radioprotectant cysteamine

In the presence of 10 mM cysteamine, there is a large reduction in strand break formation in the γ-irradiated duplex **X:Y** (Fig. 5) even at the relatively high dosage of 400 Gy. In the case of neutron irradiation (also at a dosage of 400 Gy), a protective effect is also observed, although it appears to be less pronounced than in the case of γ-irradiation (Fig. 6).

The protection factors (PF) = $(PF_0)/(PF_{cyst})$, defined here as the ratio of areas under the densitometric traces due to damaged fragments (including the background) in the absence (PF_0) and presence of cysteamine (PF_{cyst}), were evaluated for both types of irradiation. For γ-irradiation, $(PF)_\gamma \approx 16$, while in the case of neutron irradiation this factor was significantly smaller with $(PF)_n \approx 2.3$.

DISCUSSION

Single strand breaks and base/sugar damage

Two kinds of radiation damage can be observed and distinguished with the high resolution denaturing polyacrylamide gels: (1) strand breaks which are characterized by rather sharp bands due to fragments with different numbers of nucleotides, and (2) unspecified, probably multiply damaged oligonucleotide fragments which give rise to pronounced backgrounds in the autoradiograms, and the continuous background levels in the densitometer tracings. We presume that this background, whose contribution rises with increasing dosage, is due to modification of the bases and to damaged sugar residues at the 3'-ends of the oligonucleotides. The formation of multiply damaged DNA fragments at the dosages employed in this work is consistent with quantitative estimates made previously by Ward and Kuo (1978). Multiple damage per oligonucleotide fragment is expected to give rise to a heterogeneity of molecular weights and thus to a broadening of each individual fragment band; ultimately, as the level of damage is increased, the broadening should lead to an

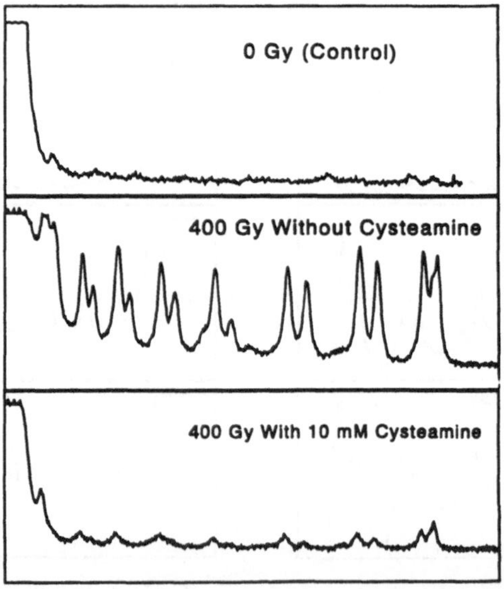

Figure 5. Densitometer tracings of an electrophoresis gel of the duplex **X:Y** exposed to γ-irradiation with and without cysteamine (10 mM).

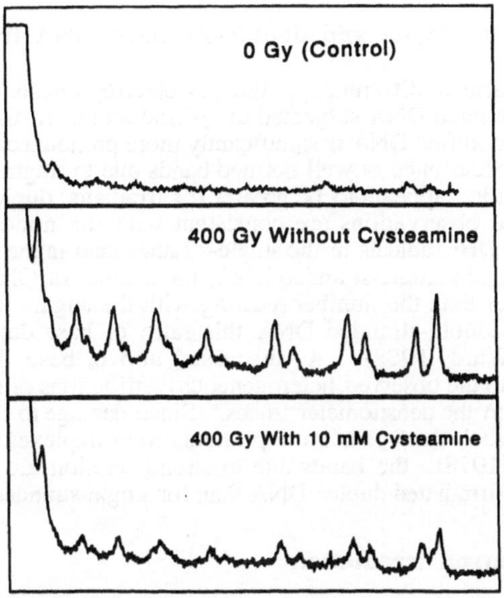

Figure 6. Densitometer tracings of an electrophoresis gel of the duplex X:Y exposed to fission neutron-irradiation with and without cysteamine (10 mM).

overlapping of the individual bands and eventually their ultimate disappearance into the background.

It is evident from Figs. 1-4 that the amplitudes of the shorter molecular weight fragments (e.g. the 3-mer, 4-mer, and 5-mer), after subtraction of the background, are larger than those of the higher molecular weight fragments. It thus appears that there are more frank strand breaks closer to the 5'-end of **X** than to the 3'-end. However, Henner et al. (1982) reported that strand scission events in various DNA fragments 12-150 nucleotides long occur uniformly at all nucleotide sites. The apparent preference for strand scission near the 5'-end observed here may involve factors other than intrinsic differences in the efficiencies of strand scission, as noted below.

The mechanisms of DNA strand scission induced by ionizing radiation involves mainly hydrogen abstraction from the sugar residues by OH· and other radicals, especially in the case of γ-irradiation where the indirect effect dominates (Achey and Durea, 1974; von Sonntag et al., 1981; Skov, 1984; van Rijn et al., 1985; Siddiqi and Bothe, 1987; Ward, 1988; Schulte-Frohlinde, 1989). Since the irradiations were performed in the presence of atmospheric oxygen, hydrogen abstraction from the sugar moieties, followed by the formation and decomposition of peroxyl radicals, results in strand breaks (von Sonntag, 1981; Liphard et al., 1990). Each fragment formed may have suffered earlier base damage, or may undergo further modification under the action of ionizing radiation. At the relatively high dosages of 100-200 Gy, the higher the molecular weight of a fragment, the greater the probability of multiply damaged sites on a given fragment. Thus, the number and distribution of damaged sites per fragment should be greater the larger the fragment. In other words, the heterogeneity of multiply damaged fragments should increase with increasing molecular weight. Therefore, the gel bands associated with higher molecular weight fragments (e.g. the 6-mer, 7-mer, 8-mer, 9-mer and 10-mer bands in Fig. 1 and Fig. 3) tend to be less well defined than the three smaller 3-, 4-, and 5-mer fragments. Also, with increasing dose and in the case of the single stranded **X**, the bands due to the higher molecular weight fragments appear to merge into the background, an effect which is consistent with the model proposed here.

Patterns of damage in single- and double-stranded DNA fragments

There is a significant difference in the gel electrophoresis patterns of the single-stranded and double-stranded DNA subjected to γ-irradiation. In the single-stranded case, the background due to modified DNA is significantly more pronounced than in the case of the duplex. However, the occurrence of well defined bands due to fragments differing from one another by one nucleotide, appears to be greater for irradiated duplexes than for irradiated single strands. These observations are consistent with the notion that bases are more susceptible to attack by OH· radicals in the single- rather than in the double-stranded forms (Ward and Kuo, 1978). In single-stranded DNA, the number of OH· radicals reacting with the bases is 7-10 greater than the number reacting with the sugars (as measured by strand scission), whereas in double-stranded DNA this ratio of base damage/sugar damage is between 2.8 and 4 (Ward, 1988). As discussed above, base damage is expected to significantly contribute to the observed heterogeneous distributions of molecular weights, and thus to the background in the densitometer traces. Since damage to the sugar moieties leads to strand scission, and such damage is more pronounced in duplexes than in single-stranded DNA (Ward and Kuo, 1978), the bands due to strand scission are more prominent in the electrophoretic gels of γ-irradiated duplex DNA than for single-stranded DNA.

Effects of γ- and neutron irradiation

For low LET γ-irradiation, DNA damage occurs primarily via the indirect effect. In the case of fission neutrons, damage primarily results from high energy recoil protons (Watt, 1988). Strand scission in the indirect mechanism is more likely in duplex DNA than in single-stranded DNA, whereas for the direct mechanism, strand scission in single-stranded and double-stranded DNA is expected to be comparable. Recently, Spotheim-Morizot et al. (1990) have suggested that single-strand breaks induced in supercoiled DNA irradiated with fission neutrons are caused primarily by OH· radicals. Our results are consistent with this view as there are no discernible differences in the electrophoresis patterns of either single- or double-stranded DNA exposed to either γ- or neutron irradiation (Figs. 1-4) and the fractions of damaged oligonucleotides (4 - 5%) are similar in both cases.

Effects of the radioprotective agent cysteamine

Cysteamine, has an electrical charge of +1 at pH 7.0, and binds to the negatively charged DNA polyion, thus providing a significant degree of protection against damage produced by OH· radicals and strand scission (Roots and Okada, 1972; Smoluk et al., 1988; Zheng et al., 1988; Spotheim-Morizot et al., 1991). The diffusion length of OH· radicals is about 60 Å (Roots and Okada, 1975) and thus radical scavenging by thiols in the bulk solution constitutes an important radioprotection mechanism (Swenberg, 1988); in addition, thiols can interact directly with radicals on the DNA molecules by proton transfer, thus further reducing DNA damage. In aerated solutions, the protective effect of the endogeneous thiol, glutathione (GSH), results predominantly from the scavenging of radicals; strand breaks are caused by the decay of DNA peroxyl radicals and these apparently do not significantly react with GSH (Liphard et al., 1990).

Cysteamine causes a dramatic reduction of strand scission in the duplex **X:Y** exposed to γ-irradiation (Fig. 5); at a 10 mM cysteamine concentration, the protection factor $(PF)_\gamma \approx$ 16. Similar effects were observed in the case of plasmid DNA by Roots and Okada (1972) and by Spotheim-Morizot et al. (1991). These results are in accord with the dominance of the indirect effect and the role of OH· radicals in causing DNA damage by the low LET γ-irradiation. For fission neutron irradiation, there is also a marked reduction in DNA damage in the presence of cysteamine (Fig. 6), although the protection factor $(PF)_n$ is only 2.3, considerably smaller than in the case of γ-irradiation. These observations suggest that a fraction of strand breaks induced by neutron irradiation are produced by mechanisms different from those operative for low LET irradiation. These mechanisms probably involve the direct formation of DNA radicals by recoil protons, α-particles, etc. (Spotheim-Morizot et al., 1991).

SUMMARY AND CONCLUSION

The polyacrylamide high resolution gel system is suitable for quantitatively determining the occurrence of strand breaks in oligonucleotides. The apparent base-sequence effects observed here in oligonucleotides exposed to γ- and fission neutron irradiation seems to be related to the lengths of the fragments generated by strand scission, rather than to different probabilities of breaks occurring at different sites within the oligonucleotide. The heterogeneous distributions of molecular weight of the degradation products (fragments of reduced chain length) is attributed to multiple base-damage. High resolution gel methods allow for quantitative estimates of radioprotection factors. These techniques can also be used for evaluating the sequence-dependence of strand-breaks in DNA with unusual tertiary structures, or with DNA complexed with proteins.

ACKNOWLEDGEMENTS

The portion of the work carried out at New York University was supported by the Office of Health and Environmental Research of the U.S. Department of Energy, Grant DE-FGO2-86ER60405.

REFERENCES

Achey, P. and Duryea, H., 1974. Production of DNA strand breaks by hydroxyl radicals. Int. J. Radiat. Biol. 25: 595-601.

Beebe, G.W., 1982. Ionizing radiation and health. Am. Scientist 70: 35-44.

Bird, R.P., 1980. Cysteamine as a protective agent with high LET radiations. Radiat. Res. 82: 290-296.

Coquerelle, T, Hagen, U., Köhnlein, W. and Crump, W. , 1978. Radiation effects on the biological function of DNA, in "Effects of Ionizing Radiation on DNA", Bertinchamps, A.J., Hütterman, J., Köhnlein, W. and Téoule, R., Eds., Springer-Verlag, Berlin, New York, pp. 261-302.

Elkind, M.M., 1985. DNA damage and cell killing, Cancer 56: 2351-2363.

Geacintov, N.E. and Swenberg, C.E., 1992. Chemical, molecular biology, and genetic techniques for correlating DNA base damage induced by ionizing radiation with biological end-points, in: "Physical and chemical Mechanisms in Molecular Radiation Biology", Varma, M.N. and Glass, W.A., Eds., Plenum Press, New York, pp. 452-474.

Held, K.D., Harrop, H.A. and Michael, B.D. , 1984. Effects of oxygen and sulphydril-containing compounds on irradiated transforming DNA. II. Glutathione, cysteine and cysteamine. Int. J. Radiat. Biol. 45: 615-626.

Henner, W.D., Grunberg, S.M. and Haseltine, W.A., 1982. Sites and structure of γ radiation-induced DNA strand breaks. J. Biol. Chem. 257: 11750-11754.

Henner, W.D., Rodriguez, L.O., Hecht, S.M. and Haseltine, W.A., 1983. γ Ray induced deoxyribonucleic acid strand breaks. J. Biol. Chem. 258: 711-713.

Holley, W.R., Chatterjee, A. and Magee, J.L., 1990. Production of DNA strand breaks by direct effects of heavy charged particles. Radiat. Res. 121: 161-168.

Hutchinson, F., 1985. Chemical changes induced in DNA by ionizing radiation. Progr. Nucleic Acid Res. Mol. Biol. 32: 115-154.

Jaberaboansari, A., Dunn, W.C., Preston, R.J., Mitra, S. and Waters, L.C., 1991. Mutations induced by ionizing radiation in plasmid replicated in human cells. II. Sequence analysis of α-particle-induced mutations. Radiat. Res. 127: 202-210.

Liphard, M., Bothe, E. and Schulte-Frohlinde, D. , 1990. The influence of glutathione on single-strand breakage in single-stranded DNA irradiated in aqueous solution in the absence and presence of oxygen. Int. J. Radiat. Biol. 58: 589-602.

Maxam, A.M. and Gilbert, W., 1980. Sequencing end-labeled DNA with base-specific chemical cleavage. Methods. Enzymol. 65: 499-560.

Raha, M. and Hutchinson, F., 1991. Deletions induced by gamma rays in the genome of Escherichia coli. J. Mol. Biol. 220: 193-198.

Roots, R. and Okada, S., 1972. Protection of DNA molecules of cultured mammalian cells from radiation-induced single-strand scissions by various alcohols and SH compounds. Int. J. Radiat. Biol. 21: 329-342.

Roots, R. and Okada, S., 1975. Estimation of lifetimes and diffusion distances of radicals involved in X-ray-induced DNA strand breaks or killing of mammalian cells. Radiat. Res. 64: 306-320.

Roots, R., Chatterjee, A., Chang, P., Lommel, L. and Blakely, E.A., 1985. Characterization of hydroxyl radical-induced damage after sparsely and densely ionizing radiation, Int. J. Radiat. Biol. 47: 157-166.

Schulte-Frohlinde, D., 1989. Studies of radiation effects on DNA in aqueous solution. The L.H. Gray Lecture. ICRU News, December issue, 4-15.

Siddiqi, M.A. and Bothe, E., 1987. Single- and double-strand break formation in DNA irradiated in aqueous solution: dependence on dose and OH radical scavenger concentration. Radiat. Res. 112: 449-463.

Skov, K.A., 1984. The contribution of hydroxyl radicals to radiosensitization: a study of DNA damage. Radiat. Res. 99: 502-510.

Smoluk, G.D., Fahey, R.C. and Ward, J.F., 1988. Interaction of glutathione and other low-molecular weight thiols with DNA: evidence for counterion condensation and coion depletion near DNA. Radiat. Res. 114: 3-10.

Spotheim-Morizot, M., Charlier, M. and Sabattier, R., 1990. DNA radiolysis by fast neutrons. Int. J. Radiat. Biol. 57: 301-313.

Spotheim-Morizot, M., Franchet, J., Sabattier, R. and Charlier, M., 1991. DNA radiolysis by fast neutrons. II. Oxygen, thiols and ionic strength effects, Int. J. Radiat. Biol. 59: 131-1324.

Swenberg, C.E., 1988. DNA and radioprotection, in: "Terrestrial Space Radiation and its Biological Effects", NATO ASI Series A: Life Sciences Vol. 154, McCormack, P.D., Swenberg, C.E. and Bücker, eds., Plenum Press, New York, pp. 675-695.

Swenberg, C.E., Speicher, J.M. and Miller, J.H., 1992. Does the topology of closed supercoiled DNA effect its radiation sensitivity? These Proceedings.

Tullius, T.D., 1987. Chemical "snapshots" of DNA: using the hydroxyl radical to study the structure of DNA and DNA-protein complexes. Trends Biochem. Sci. 12: 297-301.

van der Schans, G.P. and Blok, J., 1970. The influence of oxygen and sulphhydryl compounds on the production of breaks in bacteriophage DNA by gamma-rays. Int. J. Radiat. Biol. 17: 25-38.

van Rijn, K., Mayer, T., Blok, J., Verberne, J.B. and Loman, H., 1985. Reaction rate of OH radicals with ϕX174 DNA: influence of salt and scavenger. Int. J. Radiat. Biol. 47: 309-317.

von Sonntag, C., Hagen, U., Schön-Bopp and Schulte-Frohlinde, D., 1981. Radiation-induced strand breaks in DNA: chemical and enzymatic analysis of end groups and mechanistic aspects. Adv. Radiat. Biol. 9: 109-142.

Ward, J.F. and Kuo, I., 1978. Radiation damage to DNA in aqueous solution: a comparison of the response of the single-stranded form with that of the double-stranded form. Radiat. Res. 75: 278-285.

Ward, J.F., 1983. Chemical aspects of DNA radioprotection, in: "Radioprotectors and Anticarcinogens", Nygaard, O.F. and Simic, M.G., eds., Academic Press, New York, pp. 73-85.

Ward, J.F., 1988. DNA damage produced by ionizing radiation in mammalian cells: identities, mechanisms of formation, and repairability. Progr. Nucleic Acid Res. Mol. Biol. 35: 95-125.

Waters, L.C., Skipi, M.O., Julian Preston, R., Mitra, S. and Jaberaboansari , 1991. Mutations induced by ionizing radiation in plasmid replicated in human cells. I. Similar, nonrandom distribution of mutations in unirradiated and X-irradiated DNA. Radiat. Res. 127: 190-201.

Watt, D.E., 1988. Absolute biological effectiveness of neutrons and photons. Radiation Protection Dosimetry 23: 63-67.

Zheng, S., Newton, G.L., Gonick, G., Fahey, R.C. and Ward, J.F., 1988. Radioprotection of DNA by thiols: relationship between the net charge on a thiol and its ability to protect DNA. Radiat. Res. 114: 11-27.

THYMINE DIMER FORMATION MEDIATED BY THE PHOTOSENSITIZING PROPERTIES OF PHARMACEUTICAL CONSTITUENTS

L.F. Salter, B.S. Martincigh, K. Bolton, S.R. Aliwell and S.J. Clemmett

Department of Chemistry and Applied Chemistry, University of Natal, King George V Avenue, Durban, 4001, South Africa

ABSTRACT

Thymine dimer formation is a major photochemical lesion of UV-irradiated DNA and has been implicated as a precursor in skin cancer. Certain pharmaceutical constituents, such as the sunscreen absorbers para-aminobenzoic acid, Uvinul DS49, Eusolex 232 and the tranquillizer chlorpromazine, have the potential to photosensitise thymine dimerisation in thymine-containing systems. The yields of thymine dimer, obtained from the UV-irradiation of thymine substrates in the presence of the photosensitisers, were determined by reverse phase HPLC. Computer simulations of the experimental results were used to establish the kinetics and mechanisms of photosensitised thymine dimerisation in the various systems investigated. Initial studies involved the photosensitisation of free thymine base in aqueous medium. In this system, sunscreen agents were found to be effective photosensitisers of thymine dimerisation. The PABA photosensitisation of thymine dimer formation in the more biologically relevant system, pUC19 plasmid DNA, was investigated. The kinetic mechanism for the photosensitised dimerisation of contiguous thymines in pUC19 plasmid DNA at pH 7 is reported.

INTRODUCTION

Exposure of DNA to either ultraviolet (UV) radiation alone or UV radiation in conjunction with exogenous chemicals results in the formation of photo-induced lesions (Sutherland, 1977; Sutherland and Griffin, 1981). The primary result of these lesions is the introduction of mutations which interfere with cellular processes (Holbrook et al., 1988) and are instrumental in sunlight induced carcinogenesis (Brash, 1988). A major UV-induced photoproduct is the cis-syn cyclobutane pyrimidine dimer which occurs via the photochemical cycloaddition of contiguous pyrimidine bases in a DNA strand (Kittler and Löber, 1977; Raghunathan et al., 1990). The most significant of these dimers found in biological systems are those formed between adjacent thymine bases (Franklin and Haseltine, 1986; Holbrook et al., 1988; Ananthaswamy and Pierceall, 1990).

Direct thymine dimerisation can be caused by light of wavelengths less than 290 nm. However, due to the absorption of these wavelengths by the stratospheric ozone layer, minimal direct thymine dimerisation occurs from the radiation incident on the earth's surface. Photosensitised dimerisation can occur as a result of a sensitiser molecule absorbing the incident radiation ($\lambda > 300$ nm), to form a long-lived triplet excited state molecule which then transfers its energy, by a collisional encounter, to a thymine base in DNA. This excited base then undergoes a photocycloaddition reaction with a vicinal thymine residue.

Sunscreen agents, which are designed to absorb harmful UV rays (λ < 300 nm) that cause skin erythema, have the potential to act as photosensitisers. Para-aminobenzoic acid (PABA) and its derivatives have been shown to possess mutagenic and carcinogenic potential (Hodges et al., 1977; Sutherland, 1982; Sutherland and Griffin, 1984; Rutherford et al., 1990). It is also possible that other sunscreen-absorbers related to benzophenone compounds, e.g. Uvinul DS49 and Eusolex 232, may also induce dimerisation; benzophenone has been shown to increase thymine dimer yields in systems of the free base (Kilfoil and Salter, 1988). In addition, it was speculated that the phenothiazine tranquillizer chlorpromazine (CPZ), which can cause enhanced sensitivity of the skin to sunlight (Bruinsma, 1972), could photosensitise thymine dimer formation. It is therefore relevant to understand the kinetics and mechanisms of thymine dimerisation in the presence of sunscreen and other pharmaceutical constituents.

The work presented here investigates the ability of the sunscreen constituents PABA, Uvinul DS49 and Eusolex 232, as well as the drug CPZ, to photosensitise thymine dimer formation in DNA and related systems. Initial studies involved the photosensitisation of free thymine base in aqueous medium. This work was then extended to the more complex system involving pUC19 plasmid DNA and the photosensitiser PABA. For each system investigated, a kinetic mechanism was proposed for the photosensitised dimerisation of thymine. The rate constants for the elementary reactions were obtained from experiment, values cited in literature and Stern-Volmer plots. The mechanisms were verified by comparison of the experimental data with those calculated from the proposed mechanism and the associated rate constants. The computer program CAKE (Computer Analysis of Kinetic Equations) was used to obtain the calculated data. A good agreement between the simulated and experimental data for the various experimental conditions investigated, validates the mechanism and rate constants proposed.

EXPERIMENTAL METHODS

The photosensitisers, Uvinul DS49 and Eusolex 232, were purified before use; PABA and CPZ were used without further purification. The pUC19 plasmid DNA was prepared by culture in Escherichia coli host cells followed by extraction and purification using standard microbiological techniques.

Irradiations were carried out in a 1 mm quartz cuvette using an Osram HBO 500W high pressure mercury lamp. Solutions were degassed with N_2 before irradiation. Light intensities were measured with a chemical actinometer or a spectroradiometer. Aqueous thymine solutions were irradiated, together with the respective photosensitisers, using a 324 nm filter. Aqueous solutions of the pUC19 DNA were irradiated in the presence of PABA using a pyrex filter (λ > 300 nm). After irradiation, the DNA samples were acid hydrolysed to excise the dimers.

Dimer yields were determined by reverse-phase HPLC using either a 5 ODS 2 or 5 ODS(30) C_{18} column, depending on the thymine-containing system being analysed. Solvent systems utilised were 10% MeOH:H_2O, pH 3 and H_2O, pH 3. The dimer yield was quantitated from a cis-syn dimer calibration curve.

The experiments were performed for a range of irradiation times, thymine-containing substrate concentrations, photosensitiser concentrations and pHs in order to study the various factors affecting dimer yield. When one of these parameters was varied, the others were kept constant.

Kinetic mechanisms were proposed and assessed for their ability to correctly predict dimer yields for the various experimental conditions investigated.

RESULTS AND DISCUSSION

A general mechanism for the photosensitised reaction of thymine in solution is summarised below (adapted from Charlier and Hélène, 1972)

1.	S	→	^{1}S	photoexcitation of S
2.	^{1}S	→	S	internal conversion
3.	^{1}S	→	^{3}S	intersystems crossing
4.	^{3}S	→	S	internal conversion
5.	^{3}S + T	→	^{3}T + S	energy transfer
6.	^{3}S + T	→	S< >T	non-dimer photoproduct formation
7.	^{3}S + S	→	2S	sensitiser self-quenching
8.	^{3}T	→	T	internal conversion
9.	^{3}T + T	→	T< >T	dimer formation
10.	^{3}T + T	→	2T	thymine self-quenching
11.	^{3}S + T	→	S + T	quenching of ^{3}S by T
12.	^{3}T + S	→	S + T	quenching of ^{3}T by S

where S is the photosensitiser, T is thymine, ^{1}S is singlet photosensitiser, ^{3}S is triplet photosensitiser, ^{3}T is triplet thymine, S< >T are non-dimer photoproducts and T< >T is cis-syn thymine dimer.

The proposed reaction mechanisms for the photosensitised reaction of aqueous thymine base with Uvinul DS49, Eusolex 232, CPZ and PABA were based on this general mechanism. Rate constants for the elementary reactions in the mechanism were determined

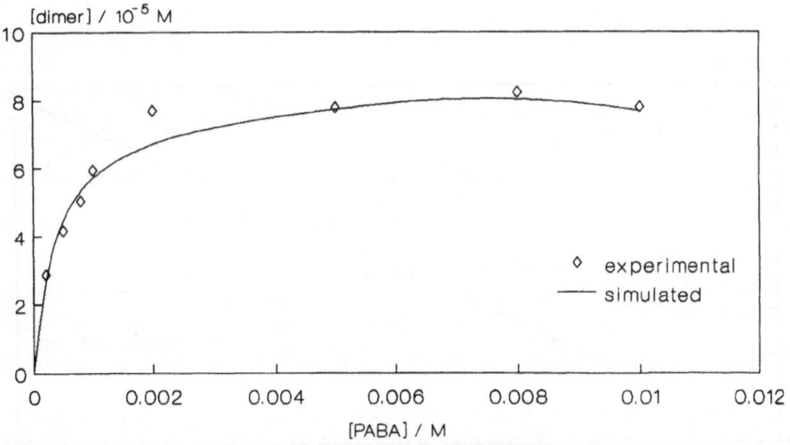

Figure 1. Thymine Dimer Yield as a Function of Photosensitiser Concentration for the PABA Photosensitisation of Aqueous Thymine Base at pH 7. The Plateau that occurs at High [PABA] is a Result of the Quenching of Triplet PABA which Limits Dimer Formation.

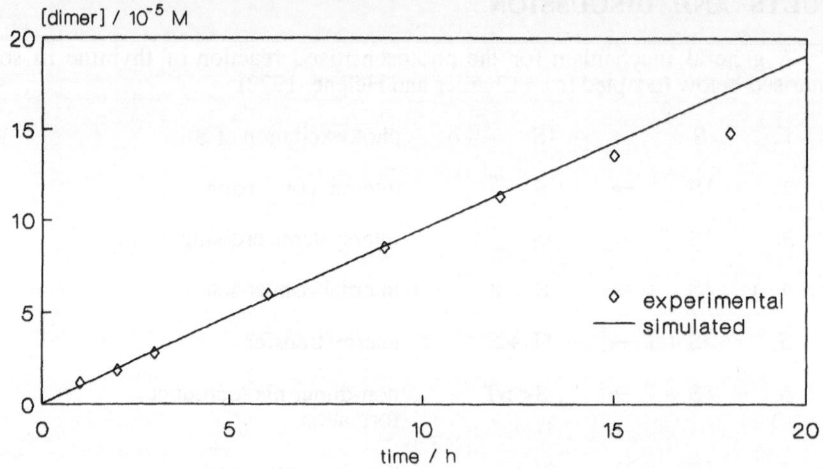

Figure 2. **Thymine Dimer Yield as a Function of Irradiation Time for the PABA Photosensitisation of Aqueous Thymine Base at pH 7.**

from the light absorbed, literature values and calculated values. Reactions and their rate constants were incorporated in or deleted from the general mechanism according to the photosensitised system investigated.

Experimental dimer yields were plotted as a function of initial photosensitiser concentration or irradiation time or initial thymine concentration. Examples of such plots are shown in Figures 1, 2 and 3 for the case where free thymine base was irradiated in the presence of the photosensitiser, PABA. All three graphs show that the dimer yield increases with an increase in the parameters that were varied, i.e. PABA concentration, irradiation time and thymine concentration.

Similar results were obtained for the photoreaction of Uvinul DS49 and Eusolex 232 with aqueous thymine base (Bolton, 1991). No dimerisation was observed for the CPZ photosensitisation of thymine base at 324 nm due to the weak absorbance of CPZ at this wavelength. At other wavelengths, thymine dimer formation was not detected due to the

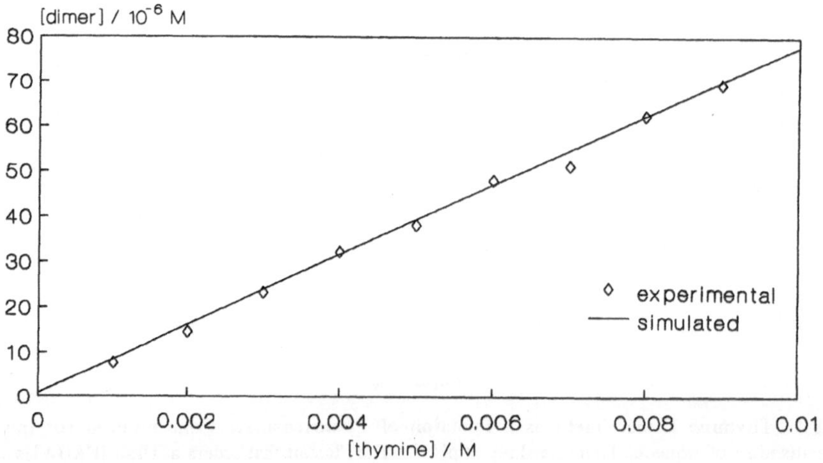

Figure 3. **Thymine Dimer Yield as a Function of Thymine Concentration for the PABA Photosensitisation of Aqueous Thymine Base at pH 7.**

efficient quenching of excited state thymine by CPZ (viz. reaction 12 in the above mechanism). However, various non-dimer thymine photoproducts were observed in the photoreaction of CPZ with thymine (Bolton, 1991).

The sunscreen agents (PABA, Uvinul DS49 and Eusolex 232) were thus shown to photosensitise dimerisation in free thymine base and the kinetic mechanisms proposed accounted for the observed thymine dimer yields.

The mechanism proposed for the PABA photosensitised reaction of free thymine base (Aliwell, 1991) was adapted to account for thymine dimerisation in a more complex and biologically relevant system - the in vitro irradiation of pUC19 plasmid DNA at pH 7. This mechanism, together with the assigned rate constants, is shown below

P	→	^1P	$k_1 = 4.8 \times 10^{-5}\ s^{-1}$
^1P	→	P	$k_2 = 1.3 \times 10^{9}\ s^{-1}$
^1P	→	P	$k_3 = 1.5 \times 10^{10}\ s^{-1}$
^1P	→	^3P	$k_4 = 1.2 \times 10^{9}\ s^{-1}$
^3P	→	P	$k_5 = 1.8 \times 10^{5}\ s^{-1}$
^3P + P	→	2P	$k_6 = 2.9 \times 10^{9}\ M^{-1}\ s^{-1}$
^3P + X	→	P + X	$k_7 = 2.9 \times 10^{9}\ M^{-1}\ s^{-1}$
^3P + TcT	→	P + ^{3}TcT	$k_8 = 3.7 \times 10^{9}\ M^{-1}\ s^{-1}$
^{3}TcT	→	TcT	$k_9 = 8.1 \times 10^{7}\ s^{-1}$
^{3}TcT	→	T<>T	$k_{10} = 1.4 \times 10^{8}\ s^{-1}$

where P is ground state PABA, ^1P is singlet state PABA, ^3P is triplet state PABA, TcT is a ground state contiguous thymine pair, ^{3}TcT is a triplet state contiguous thymine pair, T< >T is cis-syn thymine dimer and X is all non-contiguous thymine bases in DNA. The rate constants were assigned as follows: k_1 was determined from the measured intensity of the light absorbed by PABA. k_2, k_3 and k_4 were calculated from the data given for the

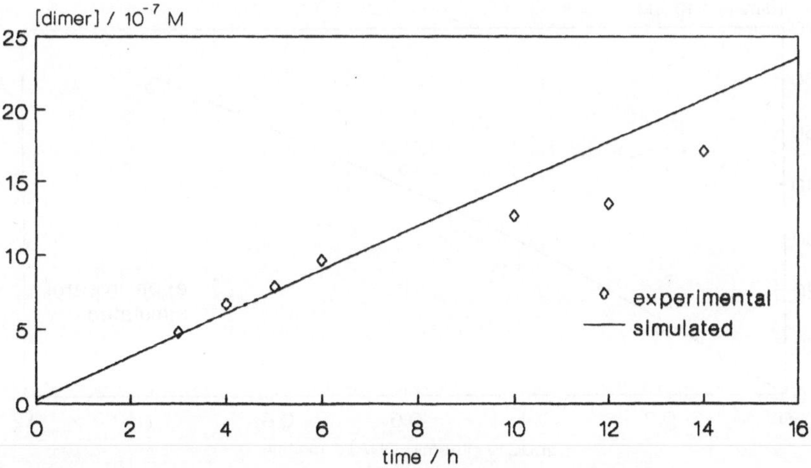

Figure 4. Thymine Dimer Yield as a Function of Irradiation Time for the PABA Photosensitisation of pUC19 Plasmid DNA at pH 7.

fluorescence of excited PABA (Aliwell, 1991) and the quantum yield of intersystems crossing from PABA singlet to triplet (Rutherford et al., 1990). k_5, k_7, k_9 and k_{10} were determined using Stern-Volmer steady state analysis of the kinetic mechanism. k_6 was obtained by determining the maximum rate constant for the diffusion-controlled reaction of ions in solution under Smoluchowski boundary conditions (Lin et al., 1975). k_8 was determined by calculating the rate constant for the diffusion-controlled triplet transfer reaction of triplet PABA with a ground state contiguous thymine pair.

The rate constants k_2, k_3, k_4, k_5 and k_6 are identical to those used in the mechanism of PABA photosensitisation of free thymine base at pH 7. These rate constants are applicable to both PABA photosensitised systems since the corresponding reactions only involve PABA photochemistry and thus the rate constants are unaffected by the thymine-containing substrate involved.

Figures 4 and 5 show that there is fair agreement between predicted dimer yields (from CAKE) and those obtained experimentally, thus supporting the above kinetic mechanism proposed for the PABA photosensitised thymine dimerisation in pUC19 plasmid DNA. The deviation that occurs between the observed and simulated dimer yields in Figure 4 is due to the formation of various PABA photoproducts which only become significant at the longer irradiation times. These photoproducts are not accounted for in the mechanism since insufficient quantities were isolated to enable identification of these.

CONCLUSION

The photosensitised dimerisation of thymine mediated by certain pharmaceutical agents - p-aminobenzoic acid, Uvinul DS49, Eusolex 232 and chlorpromazine - has been investigated.

The sunscreen agents, PABA, Uvinul DS49 and Eusolex 232, were shown to induce dimerisation of thymine in aqueous solution and thus are potential carcinogens. Kinetic mechanisms for the photosensitised dimerisation of free thymine base in aqueous medium were proposed. Rate constants were assigned for the individual reactions. The good correlation between the experimentally determined thymine dimer yields and computer simulated dimer yields verified the mechanisms and rate constants proposed for the three systems investigated.

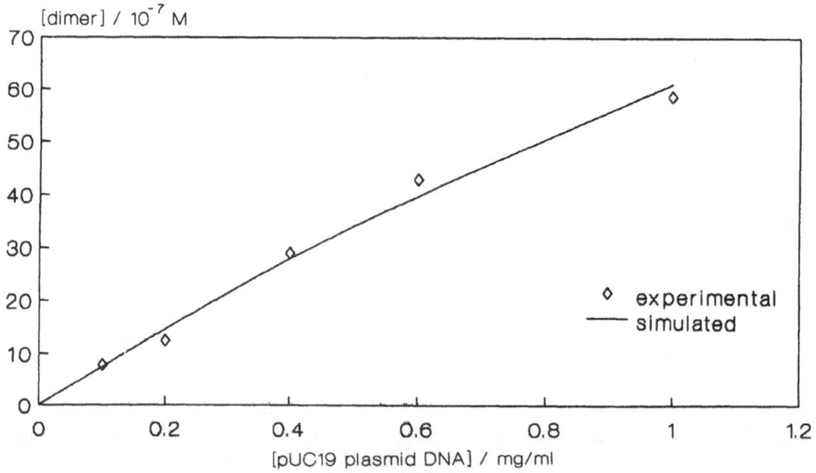

Figure 5. Thymine Dimer Yield as a Function of DNA Concentration for the PABA Photosensitisation of pUC19 Plasmid DNA at pH 7.

Chlorpromazine did not cause dimerisation of thymine in aqueous solution. However, the photoreaction of CPZ with thymine resulted in the formation of non-dimer thymine photoproducts (also found with PABA) thus indicating the mutagenic and carcinogenic potential of CPZ.

Photosensitised thymine dimerisation in a more biologically relevant system (pUC19 plasmid DNA) was investigated. The mechanism proposed for the PABA photosensitised dimerisation of contiguous thymine residues in pUC19 plasmid DNA at pH 7, together with the rate constants assigned from experimentally determined, literature and calculated values, is reported. The fair agreement between experimental and simulated dimer yields supports the proposed kinetic mechanism. The validity of the mechanism is further confirmed by the applicability of the rate constants for PABA photochemistry to the various systems investigated.

These studies are being extended, via immunological assays of irradiated *E. coli* systems, to elucidate the mechanisms of in vivo UV-induced damage in DNA. In principle, similar mechanisms should apply. Analogous mechanisms could also be used to undertake the computer simulation of other radiation-induced (X- ray, γ- ray) base damage in DNA.

REFERENCES

Aliwell, S.R., 1991, Para-aminobenzoic acid photosensitised dimerisation of thymine, M.Sc. Thesis, University of Natal, Durban, South Africa.

Ananthaswamy, H.N. and Pierceall, W.E., 1990, Molecular mechanisms of ultraviolet radiation carcinogenisis, Photochem. Photobiol., 52:1119.

Bolton, K., 1991, Studies of the photochemical reactions of thymine with selected sensitisers, M.Sc. Thesis, University of Natal, Durban, South Africa.

Brash, D.E., 1988, UV mutagenic photoproducts in *Escherichia coli* and human cells: a molecular genetics perspective on human skin cancer, Photochem. Photobiol., 48:59.

Bruinsma, W., 1972, Adverse reaction profiles of drug eruptions, Dermatologica, 145:377.

Charlier, M. and Hélène, C., 1972, Photochemical reactions of aromatic ketones with nucleic acids and their components I - Purine and pyrimidine bases and nucleosides, Photochem. Photobiol., 15:71.

Franklin, W.A. and Haseltine, W.A., 1986, The role of the (6-4) photoproduct in ultraviolet light-induced transition mutations in *E. coli*, Mutat. Res., 165:1.

Hodges, N.D.M., Moss, S.H. and Davies, D.J.G., 1977, The sensitising effect of a sunscreening agent, p-2-*p*-aminobenzoic acid, on near UV induced damage in a repair deficient strain of *Escherichia coli*, Photochem. Photobiol., 26:493.

Holbrook, S.R., Pearlman, D.A. and Kim, S.-H., 1988, Molecular Models of Photodamaged DNA, Rev. Chem. Intermed., 10:71.

Kilfoil, V.J. and Salter, L.F., 1988, The kinetics of photosensitised thymine dimerisation, Int. J. Chem. Kin., 20:645.

Kittler, L. and Löber, G., 1977, Photochemistry of the nucleic acids, Photobiol. Rev., 2:39.

Lin, S.H., Li, K.P. and Eyring, H., 1975, in: 'Physical Chemistry: An Advanced Treatise', Vol VII, Eyring, H., Henderson, D. and Jost, W., (eds.), Academic Press, New York.

Raghunathan, G., Kieber-Emmons, T., Reiri, R. and Alderfer, J., 1990, Conformational features of DNA containing a cis-syn photodimer, J. Molc. Struc. Dyn., 7:899.

Rutherford, C.E., Salter, L.F. and Thomas, R.C., 1990, pH Effects on p-Aminobenzoic Acid Photosensitised Dimer formation from the free Thymine Base, J. Photochem. Photobiol. A, 52:337.

Sutherland, B.M., 1982, p-Aminobenzoic acid - sunlamp sensitisation of pyrimidine dimer formation and transformation in human cells, Photochem. Photobiol., 36:95.

Sutherland, J.C., 1977, Photophysics and photochemistry of photoreactivation, Photochem. Photobiol., 25:435.

Sutherland, J.C. and Griffin, K.P., 1981, Absorption spectrum of DNA for wavelengths greater than 300 nm, Radiat. Res., 86:399.

Sutherland, J.C. and Griffin, K.P., 1984, p-Aminobenzoic acid can sensitise the formation of pyrimidine dimers in DNA: direct chemical evidence, Photochem. Photobiol., 40:391.

THE ASSAY OF 2' DEOXYURIDINE-3'-MONOPHOSPHATE (dUMP) BY 3' PHOSPHORYLATION AND TWO DIMENSIONAL THIN LAYER CHROMATOGRAPHY: A POTENTIAL RADIATION MARKER OF DNA DAMAGE

Antonio Cajigas and J.J. Steinberg

Departments of Pathology, Radiation Biology and Oncology
Albert Einstein College of Medicine
F-538, 1300 Morris Park Avenue
Bronx, New York 10461

ABSTRACT

Deoxyuridine is a stable pyrimidine adduct in DNA, and can be detected via ^{32}P labeling cold DNA. Deoxyuridine (dU; monophosphate - MP) forms by the deamination of deoxycytidine, remains in DNA via deficient uracil glycosylase excision repair, or diminished scavenging by thymidylate synthetase, dUTPase and/or dihydropyrimidine dehydrogenase activities. The assessment of dUMP adduct formation and normal base composition is carried out by "shot-gun" 5'-phosphorylation of representative deoxyribonucleotides with all four standard ^{32}P-alpha- triphosphates (dNMP's as monophosphate) through nick translation. Subsequent 3'-monophosphate digest (micrococcal nuclease and spleen phosphodiesterase) "sister exchanges" a radioactive $^{32}PO_4$= to the nearest neighbor cold nucleotide. Separation in two dimensional PEI-cellulose is carried out in acetic acid (1.0 M, pH 3.5 with NH_4OH), and 5.6M $(NH4)_2SO_4$, 0.12M Na_2EDTA, 0.035M $(NH_4)HSO_4$ to pH 4. The technique was applied to human placental DNA, control and altered calf thymus DNA with cold stoichiometric replacement of dUMP; and adult Drosophila DNA, which lacks dUTPase and is deficient in uracil glycosylase, is heavily methylated, and surprisingly is devoid of dUMP. Scintillation detection of dUMP is more sensitive than densitometry, but densitometry more accurately measures adducts close to other dNMP's. dUMP's ubiquity in DNA is not completely explained by chemical deamination, since this forms slowly in vitro. Nor does enzymatic defense completely preclude dUMP's presence. This technique easily and quickly quantifies the low molecular weight adduct dUMP in DNA both in vivo and in vitro.

Key words: DNA adducts, postlabel, nick-translation, two dimensional thin layer chromatography, deoxyuridine, uracil, deoxyuridine-3-monophosphate, antiviral, radiation therapy, anti-neoplastic

INTRODUCTION

The cell maintains a number of strategies to exclude the incorporation of uracil into genomic DNA. These include: deoxyuridine triphosphate nucleotidohydrolase (dUTPase) (Williams,1984), dihydropyrimidine dehydrogenase (Tuchman et al., 1984), and uracil glycosylase activities. Regardless, a number of natural and unnatural (iatrogrenic) chemical, biochemical, and enzymatic pathways exist to overwhelm protective enzyme systems, and insert uracil either in place of thymine by chemical cytosine deamination (Kochectov and Budovskii, 1972), or enzymatic dCMP deaminase. Native uracil exists in: the phages of Bacillus subtilis PBS1, PBS2, dut- and ung-mutants of T4 phage (Warner and Duncan, 1978), Escherichia coli forming short Okazaki fragments (sof mutants; 5), protozoan dinoflagellates of Crypthecodinium cohnii (as hydroxymethyluracil;) (Rae and Steele, 1978), Drosophila (Green and Deutsch, 1984), and in humans with Bloom's syndrome (Yamamoto and Fujiwara, 1986), B12, folate deficiency, and/or antifolate methotrexate therapy (Richards et al., 1984; Goulian et al., 1980).

The present technique (Steinberg, Cajigas, and Brownlee, 1992) which enhances the ability to detect dUMP derives from a history of TLC separation techniques carried out to separate tRNA (Kuchino et al., 1987; Ro-Choi and Busch, 1974), and mRNA polynucleotides (Iserentant and Fiers, 1979). Also, the technique derives impetus from ^{32}P-postlabeling techniques, which have been critical in the separation of high-molecular weight carcinogen DNA adduct formation, and other "indigenous" ("I" spots) adducts (Randerath, 1964, 1981; Beland et al., 1990; Harris, 1989; Shields and Harris, 1990; Wilson et al., 1988, 1988; Manchester etal., 1988, 1990; Park et al., 1988; Haseltine et al., 1983). The present technique, an alternate technique to the others described, labels representative fractions of all four deoxynucleotides in DNA, in situ, affording better adduct detection. The TLC separation retains normal deoxynucleotides retention values -which allows rapid visual assessment of dUMP content in DNA, and DNA quality.

MATERIALS AND METHODS

Nucleic acids: Calf thymus DNA (Type I; Highly polymerized), 2'-deoxyuridine 5'-triphosphate (dUTP), 2'- deoxy-5'-nucleotides (dAMP, dCMP, dGMP, dTMP, and dUMP) as cold monophosphate controls were purchased from Sigma Chemicals, Inc., (St. Louis). pBR322 DNA was obtained from Boehringer-Mannheim (Germany). Alpha-^{32}P-radiolabeled (3000 μCi/mmol; 8-16 μCi per "nick") dATP, dCTP, dGTP, and dTTP were purchased from Amersham, Inc., or New England Nuclear (Dupont, Inc.; Cambridge, MA). Transfer ribonucleic acid (tRNA) from Type XXI/Escherichia coli, strain W, was purchased from Sigma.

Enzymes: Micrococcal nuclease (EC 3.1.31.1; activity: 100-200 μmolar units per mg of protein), and spleen phosphodiesterase II (EC 3.1.16.1; activity: 13.5 units per mg of protein) were purchased from Sigma Chemicals, Inc., (St. Louis). DNase I, and Escherichia coli DNA-polymerase I were purchased from Boehringer-Mannheim (Germany).

Chromatographic solvents and plates: First dimension: 1 M acetic acid, pH adjusted to 3.5 with NaOH was purchased from J.T. Baker (New Jersey). Second dimension: 74 g of $(NH4)_2SO_4$, 0.4 g of $(NH_4)HSO_4$ (Aldrich Chemical Company, Inc.) and 4 g of Na_2EDTA in 100 ml of dH_20. TLC plates, polyester polyethyleneimine cellulose were purchased from Machery-Nagel-Merck, Inc., (Germany). Reagents: 30.7 μM dimethyl sulfate (DMS; EM Science, NewJersey) for methylation with dimethyl sulfate reaction mix:200 mM sodium cacodylate (Sigma), 130 mM sodium perchlorate (Sigma) and 1 mM EDTA at pH 7.0; for depurination: 5.0 M sodium chloride, 0.5 M sodium acetate, 4.0M ammonium acetate (Fisher Scientific Company); precipitants and buffers: 100% ethanol, 70% ethanol, TE buffer at pH 8.0, PBS buffer; DNA digest buffer: 20 mM succinate, and 8 mM calcium chloride at pH of 6.0.

Equipment: Laser densitometry is by Beckmann (New York); and TLC scintillation counting is with a computerized Ambis (Boston, MA). Kodak XAR-5 film (20.3 x 25.4 cm) by Eastman Kodak Company (Rochester, NY). Kits: DNA extraction ASAP columns and nick translation kits were purchased from Boehringer-Mannheim.

Substrate DNA, ^{32}P-labeling and nick-translationmethods for TLC analysis of formed-DNA adducts: Calf thymus DNA is incubated at 0.5 μg/μl concentration (PBS buffer). This technique of detection is sensitive from one adduct per 10^5

to 10^8 nucleotides (14-24). ^{32}P has been incorporated into DNA constituents mononucleotides by [^{32}P]-alpha-dNTP nick-translation (Rigby et al., 1977). [Nick-translation will remove a small percent of adducts. Nevertheless, equal amounts of adducts are labeled, in situ in DNA, and some may be uniquely labeled as opposed to post-labeling after enzymatic digest, which may not label all adducts]. At the end of nick translation native DNA is coprecipitated with 2 μl (10 μg) of tRNA (Ausubel et al., 1990). Un-incorporated counts are meticulously removed by three cold ethanol washings, and buffer (TE at pH 8.0) re-suspension of pellet. Subsequent digest is carried out by 40 μl of spleen phosphodiesterase II (activity: 0.03375 units per ul) and 10 ul of micrococcal nuclease (activity: 0.2-0.4 μmolar units per ul) in 40 μl of 20 mM sodium succinate, 8 mM $CaCl_2$, pH 6.0, at 37°C for 16 hours. Acetone precipitation (0.5 ml), and collection of supernatant, precedes evaporation to dryness (vacuum dessicator). Resuspension in 10 μl distilled water, and spotting of 20,000 -100,000 DPM (typically 1-10μl) occurs next. The monophosphate separation is easily carried out by two dimensional PEI-cellulose TLC employing solvents in the first and second dimensions (Beland et al., 1990; Manchester et al., 1988; Park and Ames, 1988). Next, Ambis quantification of ^{32}P, with control DPM to account for quenching, is carried out. This is then followed by autoradiography at 24 and 72 hour at -70° C with dual screens. Laser densitometry follows.

Partial replacement of dTMP by dUMP: Two μl of dUTP (1.8 x 10^{-9} moles/μl) was added to standard nick-translation of calf thymus DNA.

DNA: Includes human placental DNA, calf-thymus DNA, and adult Drosophila melanogaster DNA (the kind gift of Dr. Scott Hawley)

Ambis computer assisted TLC scintillation counting: This is carried out directly after drying the second phase. Typical readings require fifteen to thirty minutes, and chromatograms are faithfully reproduced. Controls are added to the TLC plate to establish quenching, which is significant below 500 DPM (1:10). A linear regression to control for quenching generated the equation of $Y = 0.51X - 945$ ($R^2 = 0.999$), where "Y" is CPM from Ambis, and "X" is true DPM based on standard ^{32}P controls.

Beckmann laser densitometry computer assisted analysis of the TLC XAR-5 autoradiograms: This has been almost as sensitive as scintillation counting and is helpful in determining adduct to nucleotide ratios, retention factor, quantification of proximate adducts to known dNMP's, area of spot exposure, and spot density (on gray scale) which reflects quantity of each labeled phosphate. The ability to superimpose graphically each film with control spots ultimately eases the analyses of product.

RESULTS

The results of the following TLC's are presented and discussed to include: 1. freshly prepared calf thymus and human placental DNA; 2. cold stoichiometric replacement of dUMP for dTMP; 3. pBR322; 4. Calf thymus DNA in distilled water or PBS at 4°C; 5. adult Drosophila melanogaster DNA; 6. densitometry vs. scintillation counting.

Control calf thymus or human placental DNA

Findings for calf-thymus DNA digest include (Figure 1A & 1B; 24 hour autoradiogram): 1. Normal retention times of all radio-labeled monophosphates to cold UV-markers. Autoradiogram demonstrates, in clockwise position, the major bases of DNA as their deoxyribonucleotides (Table 1A): adenine (Retention factor and coordinates {Rf} X,Y -cm = 2.7, 11.2), thymine (Rf = 7.5, 10.0), guanine (Rf = 3.6, 5.5), and cytosine (Rf = 12.8, 14.9). Our percentages of dNMP label of calf thymus DNA are (Tables 1 & 2): dTMP - 44.6%, dAMP - 22.3%, dCMP - 10.6%, dGMP -18.7%, dUMP - 3.8%. One additional major spot (6.2%) exists after 24 hours autoradiography which represents deaminated dAMP to dIMP via contaminating adenosine deaminase in the spleen enzyme preparations. After 72 hours autoradiography, four additional minor spots are more clearly visible: 5-me-dAMP, two spots close to dCMP representing 5- me-dCMP, and dUMP. Human placental DNA is essentially devoid of methylated bases and dUMP -not surprising in rapidly dividing cells.

Over-labeling of dTMP is consistent, and unexplained in nick-translation (Table 1B). Further, some variation in CPM exists due to differences in quenching at lower counts (10:1 at 200 DPM) vs. higher counts (2:1 over 10,000 DPM). Lastly, densitometry integrates area, and not gray scale density, and these numbers appear in Table 1A.

Cold stoichiometric replacement of dUMP for dTMP (Figure 2)

We have been able to routinely detect dUMP on our TLC's. Naturally occurring dUMP coincides with cold dUMP, and incorporated, radiolabeled dUMP (densitometry Rf -

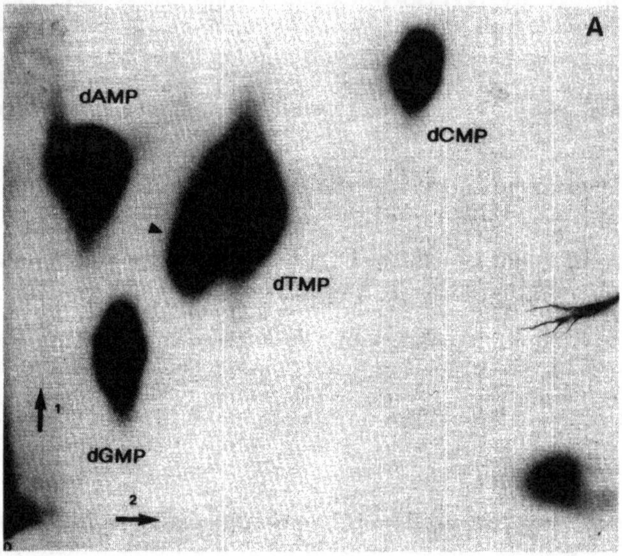

Figure 1A. Control calf thymus DNA (24 hour autoradiogram): Autoradiogram demonstrates, in clockwise position, the major bases of DNA as their nucleotides: adenine (Rf = 2.7, 11.2), thymine (Rf = 7.5, 10.0), guanine (Rf = 3.6, 5.5), and cytosine (Rf = 12.8, 14.9). Arrowhead represents dIMP; arrows demonstrate direction of first and second dimensions in chromatography.
Figure 1B. Control calf thymus DNA - 3-D Ambis: 3-D Ambis representation of scintillation counts for control DNA.

Table 1A. Densitometry Results.

A **DENSITOMETRY**

	CT DNA	dUTP	Drosophila	pBR322
Rx values				
dAMP	2.7/11.2	2.9/11.6	3.1/11.0	1.4/8.6
dAMP'	--------	--------	3.8/11.8	1.6/9.1
dCMP	12.8/14.9	12.7/15.3	12.8/14.8	10.6/11.9
dCMP'1	--------	---------	---------	--------
CMP'2	--------	---------	---------	--------
dGMP	3.6/5.5	3.6/5.7	3.6/5.7	3.7/3.0
dIMP	5.8/8.5	5.8/8.8	6.1/8.9	6.5/4.8
dTMP	7.5/10.0	7.4/10.3	7.5/10.4	8.3/5.8
dUMP	--------	10.5/9.1	--------	11.5/6.1
Area (mm^2)				
dAMP	644	631	579	263
dAMP'	---	---	46	31
dCMP	336	271	343	384
dCMP'1	---		---	---
dCMP'2	---		---	---
dGMP	387	417	438	386
dIMP	291	318	275	397
dTMP	967	407	1155	636
dUMP	---	875	---	49
Percent Volume ($AUmm^2$)				
dAMP	21.35	19.61	21.95	11.2
dAMP'	-----	-----	.15	.7
dCMP	11.05	8.48	7.95	15.3
dCMP'1	-----		-----	---
dCMP'2	-----		-----	---
dGMP	12.31	16.17	9.47	21.5
dIMP	6.22	7.44	5.88	13.8
dTMP	47.38	8.66	48.98	37.3
dUMP	-----	39.42	-----	.1

Table 1B. Ambis Scintillation Results.

B **AMBIS**

NUCLEOTIDES	Control DNA	dUTP	Dros. DNA	pBR322
dAMP	12863/22.32%	15426/26.50%	12207/23.24%	3337/14.80%
dCMP	6121/10.62%	6244/10.73%	5677/10.81%	3529/15.65%
dGMP	7338/12.73%	8298/14.25%	6404/12.19%	3563/15.80%
dIMP	3444/5.98%	4394/7.55%	3900/7.43%	3086/13.68%
dTMP	25686/44.57%	5689/9.77%	22492/42.83%	6634/29.42%
dUMP	2178/3.78%	18161/31.20%	------------	2402/10.65%

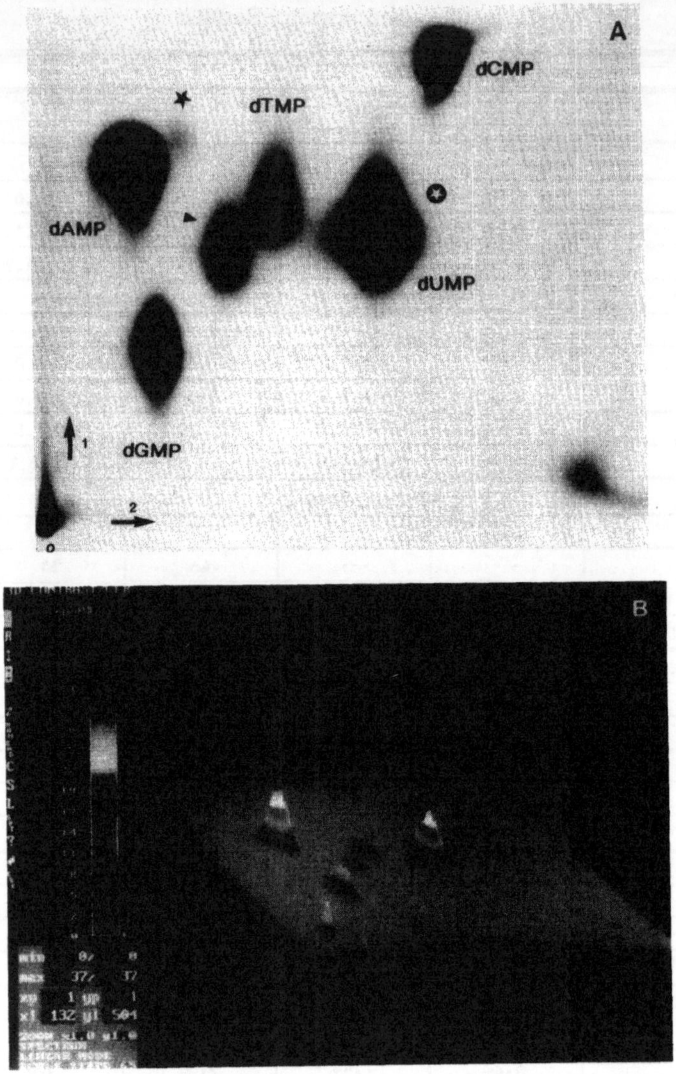

Figure 2A. Cold stoichiometric replacement of dUMP for dTMP: Autoradiogram demonstrates radiolabeled dUMP (black-bordered star; densitometry Rf -10.5, 9.1; densitometry percentage 39% at highest mM substitution). Black star represents methylated-dAMP; and arrowhead is dIMP.

Figure 2B. Cold stoichiometric replacement of dUMP for dTMP - 3-D Ambis: Figure demonstrates three dimensional representation of Ambis beta-scintillation counts for dUMP and control dNMP's.

10.5, 9.1; densitometry percentage 39% at highest mM substitution). Figures 2A & 2B demonstrates our ability to easily detect these abnormal cold nucleotides from normal. The concentration of dUTP replacement began at 4 mM (0.1 nm added). dUMP's incorporation in DNA produces one unique spot in each case. The addition of 4 mM dUTP, after incorporation as dUMP, significantly competes with dTMP labeling and it's un-determined neighbor. Further, the addition of dUMP may enhance our ability to detect methyl-dAMP, methyl-dCMP, and possibly hydroxymethyl-dCMP.

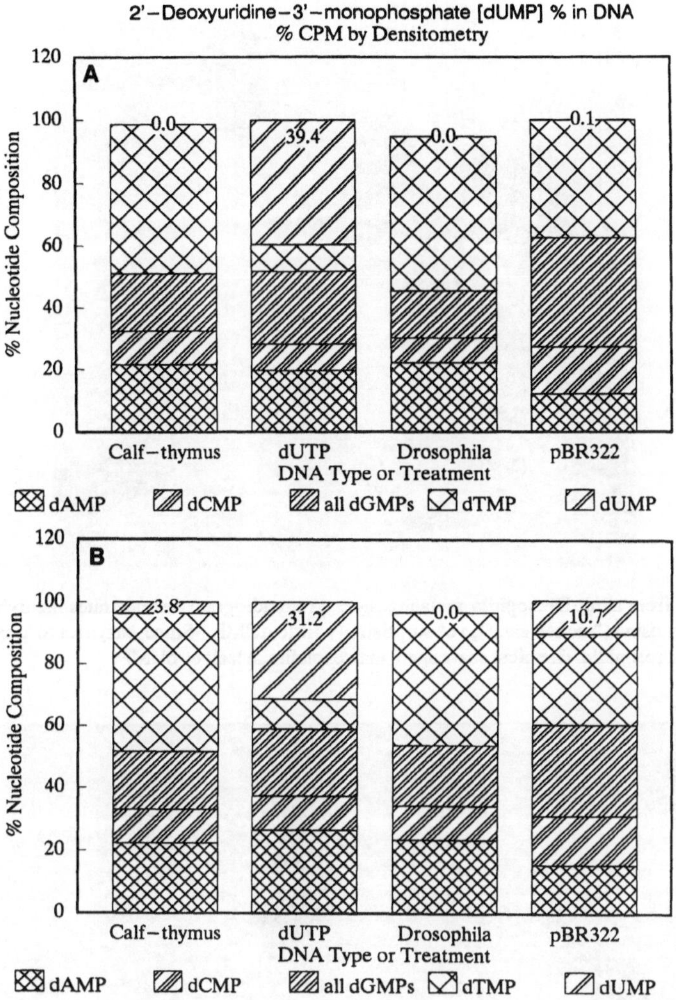

Figure 3. Figure presented as histograms of DNA dNMP's and adduction: Densitometry (A) vs. scintillation counts (B).

Calf thymus DNA in distilled water or PBS at pH 7.8:

This was carried out at fixed intervals for up to 60 days (Table 1 & Figure 3). Findings of these naturally occurring dNMP's in DNA include the persistence of dUMP - likely from the deamination of dCMP - which has only increased approximately 14% above initial assay.

DNA from adult Drosophila melanogaster (Figure 4):

This DNA is from an elderly fly. Surprisingly, little detectable dUMP is evident, given low amounts of uracil glycosylase, and no dUTPase activity. Autoradiogram demonstrates methylation at adenine and cytosine, yet surprisingly little dUMP. Since enzymes to preclude dUMP are barely present in Drosophila, chemical scavengers may explain the lack of dUMP (and may also account for insects' relative resistance to oxidant stress by radiation injury).

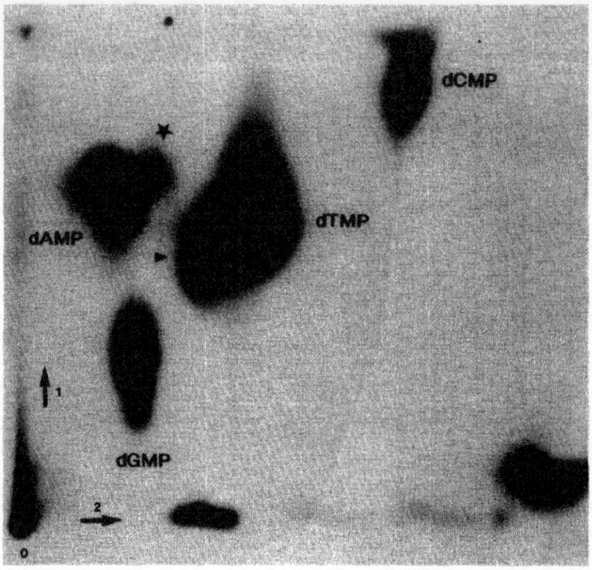

Figure 4. DNA from adult Drosophila melanogaster: Autoradiogram demonstrates methylation at adenine (black star) and cytosine ("mouse ears"), yet surprisingly little dUMP. Since enzymes to preclude dUMP are barely present in Drosophila, chemical scavengers may explain the lack of dUMP.

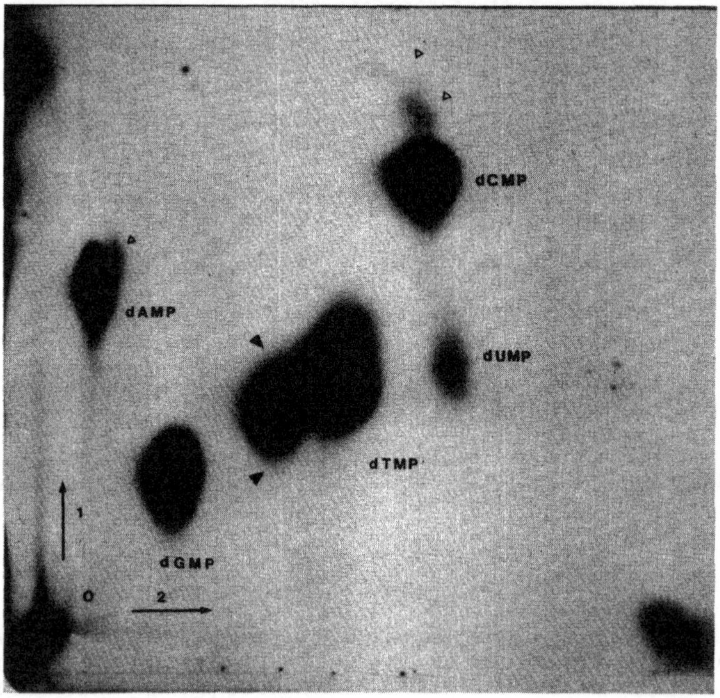

Figure 5. pBR322: Results in pBR322 support the findings in calf thymus DNA: the same retention times for all dNMP's; the presence of dUMP; some formation or suggestion of other adducts near A and C. Open triangles are methylated-dNMP's. The surprising increment in dUMP may be related to lower initial counts plated, and altered percentages especially after linear regression correction. The black arrowheads outline dIMP.

pBR322 (Figure 5)

This was employed to both control for possible artefact present in genomic DNA, and validate the technique for shorter DNA fragments (2 kilobases). We therefore extended the technique to pBR322. Results in pBR322 support the findings in calf thymus DNA (Table 1 & Figure 3): the same retention times for all dNMP's; the presence of dUMP; some formation or suggestion of other adducts near A and C. Open triangles are methylated-dNMP's. The surprising increment in dUMP may be related to lower initial counts plated, and altered percentages especially after linear regression correction.

Densitometry vs. scintillation counting (Table 1 & Figure 3)

Ambis correlated closely with densitometry, but exceeded densitometry percent by 6.5% overall, with an average range of -2.7% (dGMP) to +36% (dCMP). Ambis is more successful in detecting dUMP, and other less discrete dNMP's than densitometry. Yet, densitometry better dissects away borders between migration patterns of close dNMP's, e.g. especially methylated dNMP's.

Other phosphates within the field may represent normal minor base nucleotides or adducts. Autoradiography not only demonstrates four primary spots, but is accompanied by a specific "fingerprint".

DISCUSSION

Overview

The ability to assay modified pyrimidines naturally shifts attention to the surprising substantial occurrence of uracil in DNA. Uracil effects on DNA occur at the base level, within a sequence, in three dimensional conformation, and in DNA's interaction with regulatory proteins. Cytosine deamination may produce uracil in DNA, and small amounts of uracil have been reported in DNA (Richards et al., 1984). There is a well described excision repair enzyme (uracil glycosylase) present to remove uracil, or another enzyme scavenges uracil prior to its incorporation into DNA (dUTPase). Uracil glycosylase can also present a mechanism for single and double stranded breaks in DNA via removal of uracil opposite a single strand break which forms an irreparable double strand break. This can also occur by removing uracil from opposing strands of heavily substituted DNA. Further, uracil glycosylase activity may lead to error prone repair due to misincorporation of bases (Hutchinson, 1985; Hanawalt et al., 1979). Modification in purine bases also leads to the diminution of uracil removal from DNA (Duker et al., 1982). Some of the ways the nitrogenous bases in DNA may incur damage is in the form of tautomeric shifts, depyrimidination (1,300 events/cell/day), deamination (cytosine, 350 events/cell/day), and alkylation (Friedberg, 1985,1987). Deamination occurs in the three bases consisting of exocyclic amino groups (cytosine, adenine and guanine) and results in theconversion of these bases into uracil, hypoxanthine and xanthine, respectively. Deamination can result in base mispairing during replication, resulting in transitions (G:C- ->A:T), abnormalities in protein (histone) binding, and abnormalities in transcription, and hence, translation (Richards et al., 1984; Goulian et al., 1980; Duker et al., 1982), which ultimately effect errors in gene regulation. It hasbeen demonstrated that pyrimidine bases determine the formation of bulge-out structures. The poor stacking qualities of uracil appear to be responsible for this. The absence of bulge-out structures in DNA purine-pyrimidine- purine sequences is related to the relatively strong stacking proclivity of dT residues compared to that of dU residues (vanden Hoogen et al., 1990). Lastly, dU may account for transition of B-DNA to Z-DNA (Richards et al., 1984).

dUMP as a marker of DNA damage

The basis for a choice of dUMP in DNA as a marker of chronic xenobiotic or endogenous damage is based on (Steinberg et al., 1989,1990; Gil et al., 1990): 1. DNA is a molecule significantly at risk for the ill effects of many drug therapies, a number of disease states, toxic environmental exposure, e.g. mercury as an inhibitor of dUTPase, long-term vitamin deficiency (B12 and folate), teratogenic agents, or sequelae of aging; 2. the formation of adducts is a chemical mechanism and may therefore be stoichiometric with dose; 3. many of these DNA adducts are stable, and therefore measurable; 4. an interindividual balance of DNA repair enzymes exist (Randerath, 1964, 1981; Beland et al., 1990; Harris, 1989; Shields and Harris, 1990; Wilson et al., 1988, 1988; Manchester etal., 1988, 1990; Park et al., 1988; Haseltine et al., 1983) and may or may not adequately protect DNA from the long-term effects of these adducts, therefore, fluctuating adduct content may be measurable. The analysis of stable DNA adducts can act as markers for DNA damage, the ability of DNA enzymes to repair, and provide long term sequential information on tissue damage.

A Biochemical Application

Drosophila: Drosophila was historically felt to contain an excess of dUMP. Indeed, an activity that inhibits dUTPase has been reported in Drosophila melanogaster. This suggests that a negative regulatory mechanism is exerted on Drosophila dUTPase that could allow a greater opportunity for the incorporation of uracil into DNA (Nation et al., 1989). Drosophila lacks most uracil-DNA glycosylase with the exception of third instar larvae.. The amount of uracil-DNA glycosylase correlates well with the amount of DNA in actively replicating cells (Morgan and Chlebek, 1988). One finding in the present study is the essential absence of uracil in quiescent Drosophila DNA which is heavily methylated, and lacks the ability to prevent dUTP formation or uracil excision. It is possible that non- enzymatic chemical means may be available to prevent uracil's formation in Drosophila.

Thin layer chromatography

The benefits of this chromatographic technique from others are:

a) The improved ability to radiolabel nucleotide adducts versus standard post-labeling techniques: Post-labeling after digestion may prevent the enzymatic labeling of some classes of adducts. Yet, within DNA the t ransfer of the 5'-phosphate to the 3'-position is easily carried out.

b) Higher resolution on TLC than HPLC: One employs a new TLC plate with eachTLC. The ability to resolve nucleotides on HPLC diminishes after each run. Further, greater visible separation is more easily affected with TLC.

c) Ease of technique: The technique employs "off-the-shelf" items, without long preparations of TLC plates, or multiple dimensional TLC analyses, as most post-labeling techniques suggest.

d) Quantification of low molecular weight adducts: Standard post-labeling technology has been geared to large chemotherapeutic or carcinogenic adduct formation with high molecular weight nucleotide adducts (greater than 400 mw).

Table 2. Human Diseases of Accelerated DNA Damage and DUMP Content.

Anti-folate, methotrexate, antineoplastic, or antiviral therapies*
Alcoholism*
Ataxia telangiectasia+
Bloom's syndrome+
Dihydorpyrimidine dehydrogenase deficiency+ (uraciluria)
Down's syndrome+ (diminished uracil glycosylase)
Environmental & radiation exposure*
Fanconi's anemia+
Mercury toxicity*+

* Diseases of accelerated DNA damage
\+ Diseases of diminished DNA repair

Our conclusions from these TLC studies are: We believe our results reflect consistent findings related to deoxynucleotide composition in DNA. We believe we can determine dUMP base composition in DNA or in kilobase size fragments or motifs. The technique is able to measure DNA damage in tissue samples. The strength of this technique is based on its ability to 32P label even significantly altered DNA, and DNA to which significant chemical alteration has occurred. It offers a rapid test of DNA integrity prior to more extensive DNA manipulations.

SUMMARY

The assay serves as a mode to quickly analyze dUMP content in DNA. The possibility of the application of this assay as a marker of dUMP from the metabolic consequences of disease or chemical injury to tissue samples and biopsy specimens is feasible.

ACKNOWLEDGEMENTS

In honor of Dr. Charles Swenberg; Drs. Betsy & Arthur Upton's retirement. Our thanks to Dr.Sidney Goldfischer whose support is appreciated. Supported in part by the American Diabetes Association, and the American Federation for Aging Research. This study was further funded by a grant from the Jewish Communal Fund by the Family of Lauren & J. Ezra Merkin. A.C. is a Fellow of the New York State Health Research Council.

REFERENCES

Ausubel, F.M., Brent, F., Kingston, R.E., Moore, D.D., Seidman, J.G., Smith, J.A., and Struhl, K. (1990), in Current Protocols in Molecular biology (CPMB; Volume I,II), Greene publishing associates and Wiley-Interscience.

Beland, F.A., Fullerton, N.F., Marques, M.M., Melchior, W.B., Smith, B.A., Poirier, M.C. (1990). DNA adduct formation in relation to tumorigenesis in mice continuously administered 4-aminobiphenyl. Mutation and Environment, Part A, 81-90, Wiley-Liss, Mendelsohn, N.L., & Albertini, R.J., eds.

Duker, N.J., Jensen, D.E., Hart, D.M., Fishbein, D.E., 1982, Preturbations of enzymatic uracil excision due to purine damage in DNA. Proc. Natl. Acad. Sci. U.S.A., 79: 4878-82.

Friedberg, E.C., ed., 1985, in DNA Repair, W.H. Freeman, NY.

Friedberg, E.C., Backendorf, C., Burke, J., et al., 1987. Molecular aspects of DNA repair. Mutat Res., 184: 67-86.

Gil, D., Brownlee, M., Steinberg, J.J., 1990. Abstract, Proc. Geront. Soc. Am. 30: 216A.

Goulian, M., Bleile, B., Tseng, B.Y., 1980. The effect of methotrexate on levels of dUTP in animal cells. J. Biol. Chem., 255: 10630-37.

Green, D.A., Deutsch, W.A., 1984. Direct determination of uracil in [32P,uracil-3H]Poly(dA*dT) and bisulfite- treated phage PM2 DNA. Anal. Biochem. 142: 497-503.

Hanawalt, P.C., Cooper, P.K., Ganesan, A.K., Smith, C.A., 1979. DNA repair in bacteria and mammalian cells. Ann. Rev. Biochem., 48: 783-836.

Harris, C.C., 1989. Interindividual variation among humans in carciongen metabolism, DNA adduct formation and DNA repair. Carcinogenesis, 10: 1563-6.

Haseltine, W.A., Franklin, W., Lippske, J.A., 1983, Enviro. Hlth. Pers., 48: 29-41.

Hutchinson, F., 1985. Chemical changes induced in DNA by ionizing radiation. Prog. Nucleic Acid Res. Mol. Biol., 32: 115-54.

Iserentant, D., Fiers, W., 1979. Secondary structure of the 5' end of bacteriophage MS2 RNA. Eur. J. Biochem. 102: 595-604.

Kochectov, N.K., and Budovskii, E.I., 1972, in Organic chemistry of nucleic acids. Plenum Press, NY.

Kuchino, Y., Hanyu, N., Nishimura, S., 1987. Analysis of modified nucleosides and nucleotide sequence of tRNA. Methods in Enzymology, 155: 379-397.

Manchester, D.K., Weston, A., Choi, J.S., Trivers, G. E., Fennessey, P.V., Quintana, E., Farmer, P.B., Mann, D.L., Harris, C.C., 1988. Detection of benzo[a]pyrene diol epoxide-DNA adducts in human placenta. Proc. Natl. Acad. Sci. U.S.A. 85: 9243-7.

Manchester, D.K., Wilson, V.L., Hsu, I.C., Choi, J.S., Parker, N.B., Mann, D.L., Weston, A., Harris C.C., 1990. Synchronous fluorescence spectroscopic, immunoaffinity chromatographic and 32P-postlabeling analysis of human placental DNA known to contain benzo[a]pyrene diol epoxide adducts. Carcinogenesis, 11: 553-9.

Morgan, A.R., Chlebek, J., 1988. Quantitative assays for uracil-DNA glycosylase of high sensitivity. Biochem Cell Biol. 66: 157-60.

Nation, M.D., Guzder S.N., Giroir, L.E., Deutsch, W.A., 1989. Control of drosophila deoxyuridine triphosphatase. Biochem J. 259: 593-6.

Park, J.W., Ames, B.N., 1988. 7-Methylguanine adducts in DNA are normally present at high levels and increase on aging: Analysis by HPLC with electrochemical detection. Proc. Natl. Acad. Sci. U.S.A., 85: 7467-7470.

Rae,P.M.M., Steele, R.S., 1978. Modified bases in the DNAs of unicellular eukaryotes: An examination of distributions and possible roles, with emphasis on hydroxymethyluracil in dinoflagellates. Biosystems 10: 37-53.

Randerath, K., Randerath, E., 1964. Ion-exchange chromatography of nucleotides on poly-(ethyleneimine)-cellulose thin layers. J. Chromatog. 16: 111-125.

Randerath, K., Reddy, M.V., Gupta, R.C., 1981. ^{32}P-labeling test for DNA damage. Proc. Natl. Acad. Sci. U.S.A., 78: 6126-6129.

Richards, R.G., Sowers, L.C., Laszlo, J., Sedwick, W.D., 1984. The occurrence and consequences of deoxyuridine in DNA. Adv. Enz. Regul., 22: 157-85.

Rigby, P.W.J., Dieckmann, M., Rhodes, C., and Berg, P., 1977. Labeling deoxyribonucleic acid to high specific activity in Vitro by nick translation with DNA polymerase I J. Mol. Biol., 113: 237-251.

Ro-Choi, T.S., Busch, H., 1974. Low-molecular-weight nuclear RNA's. (Chapt. 5) in The Cell Nucleus, Vol. III, H. Busch, ed.

Shields, P.G., Harris, C.C., 1990. Environmental causes of cancer. Med. Clin. North Am., 74: 263-77.

Steinberg, J.J., Cajigas, A., Brownlee, M., 1992. Enzymatic 5' phosphorylation and 3' sister phosphate exchange (ESPSPE): A two dimensional TLC technique to measure DNA nucleotide damage, J. Chromatography,574: 41-55.

Steinberg, J.J., Gleeson, J.L., Passman, R., Brownlee, M., 1989. Life shortening: A metabolic consequence of DNA damge by advanced glycosylation products. Gerontologist, 29: 186-187A.

Steinberg, J.J., Gleeson, G.L., Gil, D., 1990. Arch. Enviro. Med. 45: 80-87.

Steinberg, J.J., et al., 1990. Alteration of DNA repair in neurodegenerative disease of aging. in UCLA Symposium: Molecular Biology of Aging, C. Finch & T. Johnson, eds., A. Liss, NY.

Tuchman, M., Stoeckeler, J.S., Kiang, D.T., et al., 1984. Familial pyrimidinemia and pyrimidinuria associated with severe fluorouracil toxicity. N. Engl. J. Med. 313: 245-249.

Tye,H.K., Chien, J., Lehman, I.R., Duncan, B.K., Warner, H.R., 1978. Uracil incorporation: A source of pulse- labled DNA fragments in the replication of the Escherichia coli chromosome. Proc. Natl. Acad. Sci. USA, 75: 233-237.

van den Hoogen, Y.T., Erkelens, C., de Vroom, E., van der Marel, G.A., van Boom, J.H., Altona, C., 1990. Influence of uracil on the conformational behaviour of RNA oligonucleotides in solution. Eur. J. Biochem., 173: 295-303.

Warner, H., Duncan, B.K., 1978. In vivo synthesis and properties of uracil-containing DNA. Nature 272: 32-37.

Williams, M.V., 1984. Deoxyuridine triphosphate nucleotidohydrolase induced by Herpes Simplex Virus Type I. Journal of Biological Chemistry 259: 10080-10084.

Wilson, V.L., Basu, A.K., Essigmann, J.M., Smith, R.A., Harris, C.C., 1988. O6-Alkyldeoxyguanosine detection by 32P-postlabeling and nucleotide chromatographic analysis. Cancer Res. 48: 2156-61.

Wilson, V.L., Masui, T., Smith, R.A., Harris, C.C., 1988. Genomic 5-methyldeoxycytidine decreases associated with the induction of squamouns differentiation in cultured normal human bronchial epithelial cells. Carcinogenesis, 9: 2155-9.

Yamamoto, Y., Fujiwara, Y., 1986. Abnormal regulation of uracil-DNA glycosylase induction during cell cycle and cell passage in Bloom's syndrome fibroblasts. Carcinogenesis 7: 305-310 .

Tarnawski, A.J., Stoeckeler, J.S., Kiang, D.T., et al. (1984). Familial polyhidramnios and polyadenoma associated with severe Hypomagnesemia... N. Engl. J. Med. 313: 244-249.

Tye, B.K., Chien, J., Lehman, I.R., Duncan, B.K., Warner, H.R., 1978. Uracil incorporation: A source of pulse-labeled DNA fragments in the replication of the Escherichia coli chromosome. Proc. Natl. Acad. Sci. USA. 75: 233-237.

van den Hoogen, Y.T., Erkelens, C., de Vroom, E., van der Marel, G.A., van Boom, J.H., Altona, C., 1990. Influence of uracil on the conformational behaviour of RNA oligonucleotides in solution. Eur. J. Biochem. 173: 295-303.

Warner, H., Duncan, B.K., 1978. In vivo synthesis and properties of uracil-containing DNA. Nature 272: 32-34.

[illegible]
Type I. Journal of Biological Chemistry 259: 10989-10994.

Wilson, V.L., Basu, A.K., Essigmann, J.M., Smith, R.A., Harris, C.C., 1988. O6-alkyldeoxyguanosine detection by 32P-postlabeling and nucleotide chromatographic analysis. Cancer Res. 48: 2156-2161.

Wilson, V.L., Masui, T., Smith, R.A., Harris, C.C., 1988. Genomic 5-methyldeoxycytidine decreases associated with the induction of squamous differentiation in cultured normal human bronchial epithelial cells. Carcinogenesis 9: 2155-2159.

Yamamoto, T., Fujiwara, J., 1986. Abnormal regulation of uracil-DNA glycosylase induction during cell cycle and cell passage in Bloom's syndrome fibroblasts. Carcinogenesis 7: 305-310.

FREE-RADICAL YIELDS IN PROTON IRRADIATION OF ORIENTED DNA: RELATIONSHIP TO ENERGY TRANSFER ALONG DNA CHAINS

J.H. Miller*, D.L. Frasco*, M. Ye*, C.E. Swenberg**, L.S. Myers, Jr., and A. Rupprecht***

*Pacific Northwest Laboratory, Richland, WA 99352
**Armed Forces Radiobiology Research Institute, Bethesda, MD 20814
***University of Stockholm, S-106 91 Stockholm, Sweden

ABSTRACT

Spatial patterns of energy deposition on the nanometer scale are currently believed to be a major factor in determining the biological effectiveness of ionizing radiation. If the most common precursors of biologically significant lesions are clusters of ionization in or near DNA, then intramolecular energy and charge transfer along DNA chains could be very important in lesion development. This paper describes investigations of these phenomena through model calculations and measurements of radical yields in oriented DNA exposed to proton irradiation.

INTRODUCTION

The idea that the critical lesions in radiation biology result from clustering of ionizations on a nanometer scale can be traced to the early work of Lea (1947), Howard-Flanders (1958), and Barendsen (1964). More recent support for this model of the biological effects of ionizing radiation comes from the analysis of cell killing by radiations with different linear-energy-transfer (LET) based on computer simulations of track structure in water (Goodhead, 1987). Although these biophysical models ignore the complexity of the cellular medium and the macromolecular structures that regulate its function, their basic conclusion can be rationalized by the high scavenging capacity of the chemical environment of DNA (Roots and Okada, 1975) and the efficiency of repair of minor perturbations of DNA structure (Doetsch and Cunningham, 1990).

From arguments of this type, one concludes that the most common precursor of cytotoxic and mutagenic effects from radiation exposure is a cluster of ionization in or, at least, very near to the DNA molecule. The involvement of macromolecules in the early stages of lesion production opens the possibility of intramolecular energy and charge transfer following excitation or ionization by the radiation field. These processes acting in the presence of traps for energy and charge provide a mechanism for concentrating energy deposited in macromolecular systems that is independent of stochastic processes in the slowing down of charged particles. Conversely, energy and charge transfer along DNA chains may dissipate clusters of excitation and ionization before biologically significant lesions are formed. In general, the existence of energy and charge migration in macromolecules tends to decouple lesion production from the stochastics of energy deposition

Biological Effects and Physics of Solar and Galactic Cosmic Radiation, Part A, Edited by C.E. Swenberg *et al.*, Plenum Press, New York, 1993

just as ordinary diffusion tends to decouple radiation chemistry on a long time scale from track effects in the radiolysis of homogeneous solutions of small molecules.

Recent findings at several laboratories have raised questions about the assumption that radiation-induced DNA damage remains localized on the nanometer scale during lesion formation. Observations by Arroyo et al. (1986) that the yield of neutron-induced free radicals in oriented DNA fibers was dependent on the orientation of the sample relative to the neutron flux were attributed to energy transfer between stacked DNA bases. Al-Kazwini et al. (1990) presented evidence that electrons can move along DNA chains for distances up to about 100 base pairs. Data obtained by van Lith et al. (1986) on microwave conductivity in pulsed radiolysis suggest that excess electrons in frozen DNA solutions can migrate for distances of the order of 100 nm in the structured water layers around macromolecular chains.

Simultaneous with these experimental results, there has been a renewed interest in nonlinear modes of vibrational excitation in DNA called solitons (Baverstock and Cundall, 1988). Several models for solitary waves in DNA have been proposed (Englander et al., 1980; Yomosa, 1984; Takeno and Homma, 1987; Zhang, 1987, Muto et al., 1988). In Yomosa's model, about 0.4 eV of vibrational energy is transported along DNA at a rate of about 100 nm/ns as a disruption of hydrogen bonds between complementary bases. The broad spectrum of excitation associated with the absorption of energy from ionizing radiation in biological systems (Bednár, 1985) should provide ample opportunity to overcome the energy threshold required to initiate nonlinear phenomena like solitons (van Zandt, 1989).

The first section of this paper discusses models for the effects of energy and charge transfer on free-radical yields in oriented DNA samples exposed to direct proton-beam irradiation. Experiments designed to detect an orientation dependence of radical yields from proton irradiation of oriented DNA are described in the second section. Our conclusions are presented in the final section.

MODELING RADICAL YIELDS IN PROTON IRRADIATION OF ORIENTED DNA

We have investigated (Miller et al., 1988; Miller and Swenberg, 1990) mechanisms for the observation by Arroyo et al. (1986) that DNA base radicals induced by neutrons are dependent on the orientation of DNA fibers relative to the neutron flux. Monte Carlo codes developed by Wilson and Paretzke (1981) and scoring algorithms contributed by Charlton (1985) were used to model energy deposition in oriented DNA under direct proton-beam irradiation. Table I summarizes our results for a 1 MeV proton flux incident on an oriented DNA sample either parallel or perpendicular to the fiber direction. Although the average amount of energy deposited in the parallel case is 5 times greater than in the perpendicular case, the average separation between the excitations or ionizations in the same DNA chain is about 100 times greater in the parallel case. The pattern of deposition in the parallel case is also more sensitive to uncertainty in the fiber orientation relative to the proton beam.

If the pattern of energy deposition events along a DNA chain in the parallel case is as diffuse as these model calculations suggests, then only long-range modes of intramolecular energy or charge transfer could couple the excitations and ionization in the molecule in ways

Table 1. Energy Deposition in Oriented DNA by 1 MeV Protons

Orientation	Energy Deposited (eV)	Event Separation (nm)
0°	293	224
0° ± 10°	150	47
90°	62	2
90° ± 10°	62	2

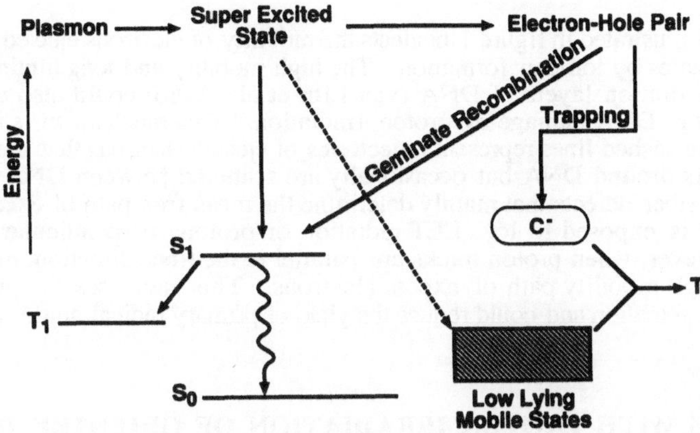

Figure 1. Decay Modes of Energy Absorbed in Solid DNA Samples

that would make the yield of free radicals orientation dependent. For example, recent work by Georghiou (1990) indicates that singlet excitations of bases in calf thymus DNA at room temperature move only 1 or 2 base pairs before they are irreversible trapped. Clearly, this type of excitation has a low probability of interacting with other energy deposition events in same DNA chain and should have equivalent effects in both irradiation geometries.

Energy absorbed from ionizing radiation in oriented DNA fibers should produce modes of excitation that are considerably more mobile than singlet excitons (Al-Kazwini et al., 1990; van Lith et al., 1986; Yomosa, 1984). Figure 1 illustrates a mechanism by which solitons might influence free-radical yields when DNA at 77° K is exposed to a proton beam parallel to the molecular orientation. The interaction of DNA with protons and secondary electrons produces highly excited electronic states that decay primarily through formation of electron-hole pairs. Usually this decay is accompanied by some conversion of electronic to vibrational energy. Sufficiently large vibrational excitations in DNA can be self cohering (van Zandt, 1989). Recent work by Bernhard (1989) suggests that excess electrons are trapped on cytosine by reversible proton transfer from guanine. If many super-excited states are produced in the same DNA chain by a proton flux that is parallel to the molecular orientation, then the probability of interaction between trapped electrons and solitons increases. This interaction may induce electron transfer to thymine where TH· is formed by irreversible protonation at C6.

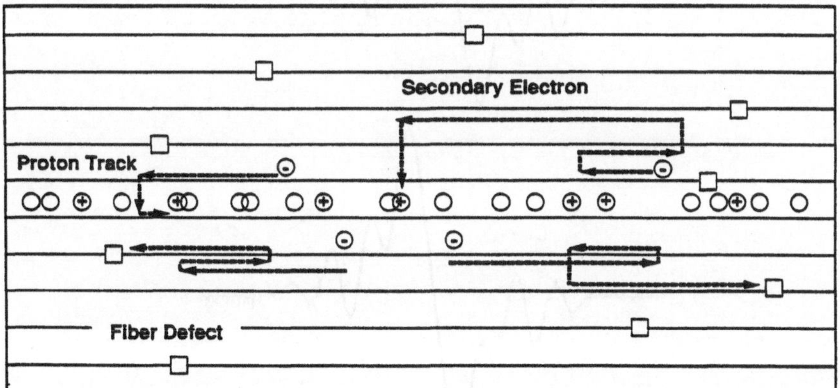

Figure 2. Schematic of Electron-hole Recombination in Oriented DNA Exposed to a Proton Flux that is Parallel to the DNA Fibers.

The model illustrated in figure 1 neglects the mobility of electrons ejected in the decay of super-excited states by ion-pair formation. The high mobility and long lifetime of excess electrons in the hydration layers of DNA (van Lith et al., 1986) could also contribute to orientation effects in DNA damage by proton irradiation. This mechanism is illustrated in figure 2, where the dashed lines represent trajectories of ejected electrons that move primarily in hydration layers around DNA but occasionally are scattered between DNA chains. The squares represent fiber defects that mainly determine the mean free path of excess electrons when the sample is exposed to low LET radiation or protons perpendicular to the fiber orientation. However, when proton tracks are parallel to the fiber direction, many positive ions lie in the high-mobility path of excess electrons. This increases the probability of electron-hole recombination and could reduce the yield of primary radical anions and cation in the parallel case.

EXPERIMENTS WITH PROTON IRRADIATION OF ORIENTED DNA

Preparation of oriented DNA samples (Rupprecht, 1966) to look for effects like those illustrated in figures 1 and 2 involves wet spinning of high molecular weight DNA into fibers that are wound on a spool to form a thin sheet of oriented DNA. Samples for perpendicular

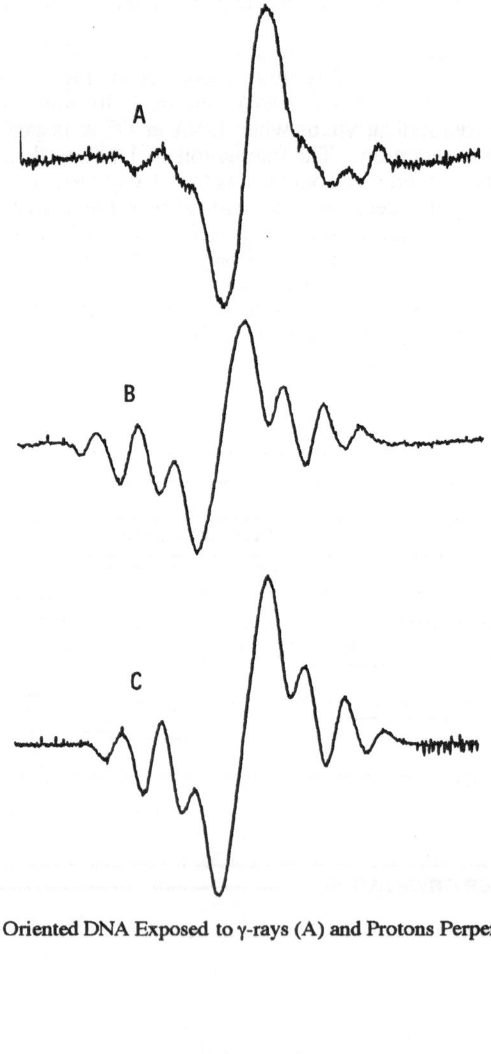

Figure 3. EPR Spectra of Oriented DNA Exposed to γ-rays (A) and Protons Perpendicular (B) or Parallel (C) to the DNA Fibers.

irradiation were made by pressing together a sufficient number of sheets to give a thickness greater than the range of a 4 MeV proton, which is about 0.5 mm. Samples for parallel irradiation were sliced from a block of oriented DNA that had a thickness slightly less than the 3 mm inside diameter of quartz tubes used to transfer irradiated samples to the electron paramagnetic resonance (EPR) spectrometer. All samples were approximately 1 cm long and weighed about 15 mg.

For irradiation, the samples were placed on a copper block in contact with a reservoir of liquid nitrogen and held in place by a thin polyester film. After cooling to 77° K, the sample was placed in a vacuum chamber attached to the beam line of the accelerator. Samples were exposed to graded doses of 4 MeV protons in the range of 20 to 60 kGy. The dose rate of 2.5 kGy/min was less than the value of 10 kGy/min recommended by Henriksen and Snipes (1970) to avoid sample heating. After irradiation, the samples were removed from the vacuum chamber and transferred to a precooled EPR tube. During this transfer, the sample lost contact with liquid-nitrogen cooled surfaces for less than one second as it fell through a funnel into the EPR tube. Several samples were exposed to γ-rays for comparison with published data (Gräslund et al., 1971) and the results of proton irradiation. In this case the sample could be sealed into an EPR tube before irradiation due to the penetrating power of the radiation.

The EPR spectra shown in figures 3A-3C were observed after 4 kGy of γ-rays, 56 kGy of protons incident perpendicular to the fiber orientation, and 48 kGy of protons parallel to the DNA chains, respectively. All three spectra are composed of a central line that we associate with primary radical anion and cation species with varying amounts of modulation in the wings resulting from TH· production. The greater amount of TH· observed with proton irradiation is not likely to be due to sample warming during the transfer from the proton beam line to the EPR spectrometer because the irradiated samples were kept in contact with liquid nitrogen cooled surfaces and the magnitude of the central line relative to the structure in the wings did not change as a function of proton dose. The greater proportion of TH· with proton irradiation may be due to the higher dose required to detect radicals since the total radical yield per unit of dose was more than an order of magnitude lower for protons than for γ-rays; however, the shape of the EPR spectrum did not change significantly with proton dose in the range investigated. The recommendation of Henriksen and Snipes (1970) regarding the dose rate was based on their experience with 6.5 MeV electrons which may not apply to proton irradiation; hence, the higher yield of TH· that we observed with protons may have resulted from sample heating due to a dose rate that was too large.

Unlike the results reported for neutrons (Arroyo et al., 1986), EPR spectra of radicals produced by direct proton irradiation of oriented DNA in parallel and perpendicular geometries were not significantly different. Figure 4 shows that, within experimental error, total radical yields were also independent of the orientation of DNA fibers relative to the proton flux. To obtain these results, differential EPR spectra were recorded digitally and

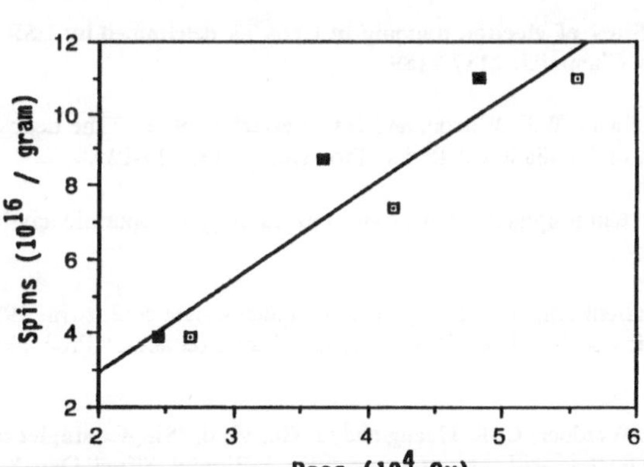

Figure 4. Total Radical Yields in Parallel (■) and Perpendicular (□) Proton Irradiation of Oriented DNA at 77° K.

double integrated to give the area under the absorption lines. This area was converted to number of spins by comparison with a standard sample of 2,2 diphenyl-1-picrylhydrazyl (DPPH) dissolved in paraffin.

CONCLUSIONS

We did not find any evidence for long-range energy or charge transfer in DNA from experiments in which oriented DNA was exposed to direct proton-beam irradiation. This may be due to the high doses and dose rates used in our experiments. The dose and dose rate could be reduced by redesigning the sample holder and transfer system to avoid the limitations on sample size imposed by the present system. Experiments with larger samples, higher proton energies for greater penetration, and improved EPR detection sensitivity might reveal orientation effects that are not present in our data due to sample heating or other processes that destroy free radicals at high exposure levels (Bernhard, 1981).

ACKNOWLEDGEMENTS

The authors gratefully acknowledge support for this work by the Office of Health and Environmental Research of the U.S. Department of Energy under contract DE-AC06-76RLO 1830, the Armed Forces Radiobiology Research Institute, and the Swedish Medical Science Research Council.

REFERENCES

Arroyo, C.M., A.J. Carmichael, C.E. Swenberg, and L.S. Myers, Jr., 1986, "Neutron-induced free radicals in oriented DNA." Int. J. Radiat. Biol. 50: 789-793.

Al-Kazwini, A.T., P. O'Neill, G.E. Adams, and E.M. Fielden, 1990, "Radiation-induced energy migration within solid DNA: The role of misonidazole as an electron trap." Radiat. Res. 121: 149-153.

Barendsen, G.W., 1964, "Impairment of the proliferative capacity of human cells in culture by α-particles with differing linear-energy transfer." Int. J. Radiat. Biol. 8: 453-466.

Baverstock, K.F. and R.B. Cundall, 1988, "Long range energy transfer in DNA." Radiat. Phys. Chem. 32: 553-556.

Bednár, J., 1985, "Electronic excitations in condensed biological matter." .Int. J. Radiat. Biol. 48: 147-166.

Bernhard, W.A., 1981, "Solid-state radiation chemistry of DNA: The bases." Adv. Radiat. Biol. 9: 199-280.

Bernhard, W.A., 1989, "Sites of electron trapping in DNA as determined by ESR of one-electron-reduced oligonucleotides." J. Phys. Chem. 93: 2187-2189.

Charlton, D.E., D.T. Goodhead, W.E. Wilson, and H.G. Paretzke, 1985, "The deposition of energy in small cylindrical targets by high LET radiations." Radiat. Prot. Dosim. 13: 123-125.

Doetsch, P.W. and R.P. Cunningham, 1990, "The enzymology of apurinic/apyrimidinic endonucleases." Mutat. Res. 236: 173-201.

Englander, S.W., N.R. Kallenbach, A.J. Heeger, J.A. Krumhansl, and S. Litwin, 1980, "Nature of the open state in long polynucleotide double helices: Possibility of soliton excitation." Proc. Natl. Acad. Sci. 77: 7222-7226.

Georghiou, S., S. Zhu, R. Weidner, C.-R. Huang and G. Ge, 1990, "Singlet-singlet energy transfer along the helix of a double-stranded nucleic acid at room temperature" J. Biomol. Struct. Dyn. 8: 657-674.

Goodhead, D.T., 1987, "Physical basis for biological effect." Nuclear and Atomic Data for Radiotherapy and Related Radiobiology (International Atomic Energy Agency, Vienna), pp. 37-53.

Gräslund, A., A. Ehrenberg, A. Rupprecht, and G. Ström, 1971, "Ionic base radicals in γ-irradiated DNA." Biochim. Biophys. Acta 254: 172-186.

Henriksen, T. and W. Snipes, 1970, "Radiation-induced radicals in thymine: ESR studies of single crystals." Radiat. Res. 42: 255-269.

Howard-Flanders, P., 1958, "Physical and chemical mechanisms in the injury of cells by ionizing radiation." Adv. Biol. Med. Phys. 6: 553-603.

Lea, D.E., 1947, "The induction of chromosome structural changes by radiation: detailed quantitative interpretation." British Journal of Radiology, Supplement 1: 75-83.

Miller, J.H., W.E. Wilson, C.E. Swenberg, L.S. Myers, Jr. and D.E. Charlton, 1988, "Stochastic model of free radical yields in oriented DNA exposed to densely ionizing radiation at 77K." Int. J. Radiat. Biol. 53: 901-907.

Miller, J.H. and C.E. Swenberg, 1990, "Free-radical yields in DNA exposed to ionizing radiation: Role of energy and charge transfer." Can. J. Phys. 68: 962-966.

Muto, V., J. Halding, P.L. Christiansen and A.C. Scott, 1988, "Solitons in DNA." J. Biomol. Struct. Dyn. 4: 873-894.

Roots, R. and S. Okada, 1975, "Estimation of life times and diffusion distances of radicals in x-ray-induced DNA strand breaks or killing of mammalian cells." Radiat. Res. 64: 306-320.

Rupprecht, A., 1966, "Preparation of oriented DNA by wet spinning." Acta Chem. Scand. 20: 494-504.

Takeno, S. and S. Homma, 1987, "Kinks and breathers associated with collective sugar puckering in DNA." Progress of Theoretical Physics 77: 548-562.

van Lith, D., J.M. Warman, M.P. de Haas, and A. Hummel, 1986, "Electron migration in hydrated DNA and collagen at low temperatures." J. Chem. Soc., Faraday Trans. 1 82: 2933-2943.

van Zandt, L.L., 1989, "DNA solitons with realistic parameter values." Phys. Rev. A 40: 6134-6137.

Wilson, W.E. and H.G. Paretzke, 1981, "Calculations of distribution of energy imparted and ionization by fast protons in nanometer sites." Radiat. Res. 87: 521-537.

Yomosa S., 1984, "Solitary excitations in deoxyribonucleic acid (DNA) double helices." Phys. Rev. A 30: 474-480.

Zhang, C.-T., 1987, "Soliton excitations in deoxyribonucleic acid (DNA) double helices." Phys. Rev. A 35: 886-891.

FREE RADICAL FORMATION IN AMINO ACIDS INDUCED BY HEAVY IONS

T. Heck [a], J. Hüttermann [a], and G. Kraft [b]

a) Fachrichtung Biophysik, Univ. 'd. Saarl.
6650 Homburg, Germany

b) Gesellschaft für Schwerionenforschung
6100 Darmstadt, Germany

INTRODUCTION

The use of heavy ions in the realm of physical tools probing radical formation mechanisms in biological objects is of considerable importance. Not only can the influence of parameters like dose or LET on the efficiency be measured; by using different particles and velocities it is also possible to achieve the same LET values but with different track diameters. For elucidating radiation chemical pathways the study of free radicals by Electron Spin Resonance (ESR)- spectroscopy is an appropriate approach which allows study of both the structural parameters and the concentration of radical intermediates in suitable systems. In this report we present first results of a systematic study of both quantitative and structural aspects of free radical formation in amino acids after low- and high LET irradiation. The structural assignments are derived from comparison with results obtained from specific radical formation in different aqueous glasses and by spectra simulation using single crystal data.

MATERIALS AND METHODS

The substances studied were glycine, alanine and arginine. They were bought in the highest availaible purity (Sigma, Merck) and used without further treatment. Heavy-ion bombardments were performed at the Unilac (GSI, Darmstadt) using Argon-40 ions of 7 and 15 MeV/u. Low-LET irradiations were carried out with 100 kV X-rays. Specific radical formation by hydrogen atoms was performed using a sulphuric glass (H_2SO_4/H_2O). Electron reactions were studied in an alkaline glass ($NaOH/H_2O$). The experimental procedures for sample preparation, data acquisition and analyses are reported elsewhere. (Ayscough et al., 1965; Hüttermann et al.,1991).

RESULTS AND DISCUSSIONS

Quantitative Aspects

Fig. 1 shows the dose yield curves for alanine and glycine after irradiation at 300 K with argon ions. For comparison, the corresponding curves for X-irradiation are included. The dose effect curves for heavy charged particle bombardment show the same characteristics as those measured for low-LET-irradiation. All curves consist of a linear part at lower dose

Biological Effects and Physics of Solar and Galactic Cosmic Radiation, Part A, Edited by C.E. Swenberg *et al.*, Plenum Press, New York, 1993

and level off to saturation at higher doses. This behaviour can be described mathematically by an exponential function (Müller-Broich, 1964):

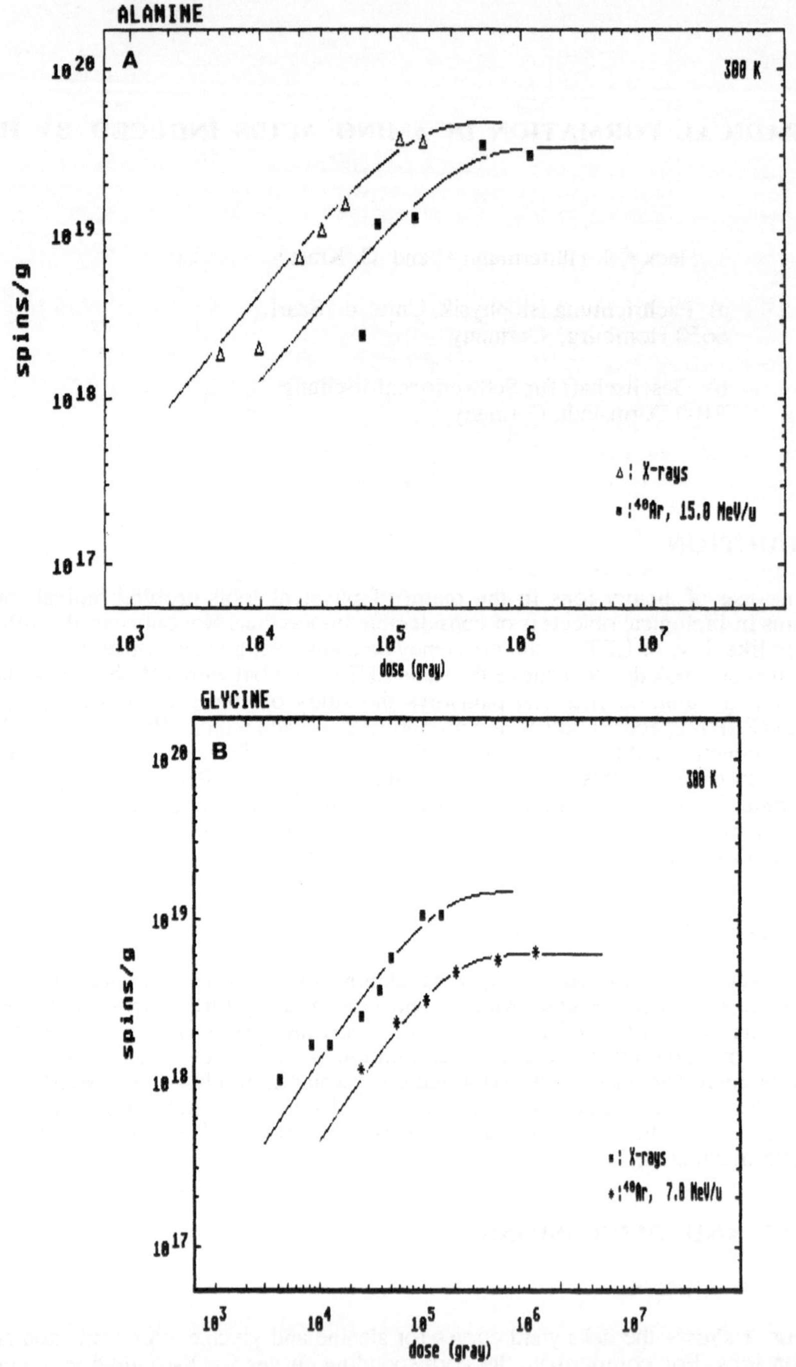

Figure 1. Dose-effect Curves of Alanine and Glycine after Irradiation at 300 K with X -rays and Argon 40 Ions.

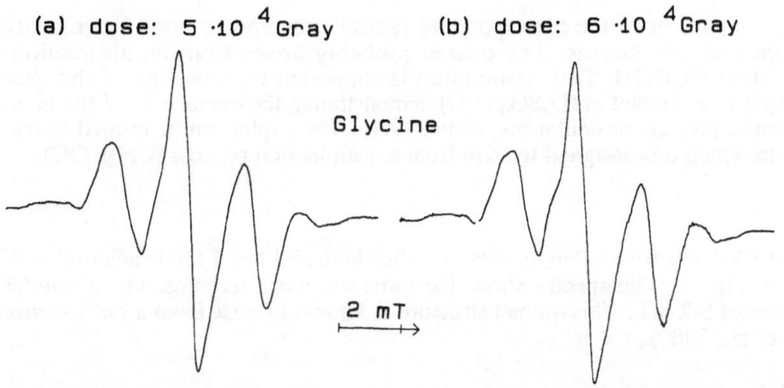

Figure 2. ESR- Spectra of Glycine after High-and Low-LET Irradiation at 300 K Irradiation with (a) Argon-40 Ions, 7 MeV/u and (b) X-Rays.

$$C_D = C_\infty \; (1 - e^{-D/D_{37}})$$

C_∞ is the saturation concentration and D_{37}, the characteristic dose, is the dose at which 63 % of the saturation concentration are obtained. For both substances the curve for heavy ion bombardment is shifted to higher doses and the saturation level is lower compared to the curve for X-irradiation. The relative effectiveness (Dertinger-Jung, 1969) calculated from these dose effect curves are 0.40 for alanine and 0.76 for glycine, respectively.

STRUCTURAL ASPECTS

Glycine

In Fig. 2 the ESR-spectrum of powdered glycine after heavy ion bombardment at 300° K is shown in comparison with the spectrum measured after X-irradiation. Both spectra reveal structural features which are independent of dose as well as of radiation quality.

The spectra indicate the presence of two components: the pre-dominant pattern is a triplet with a spectral extent of 4.5 mT whereas a minor feature has outer lines with an overall splitting of 7.2 mT (Fig. 3).

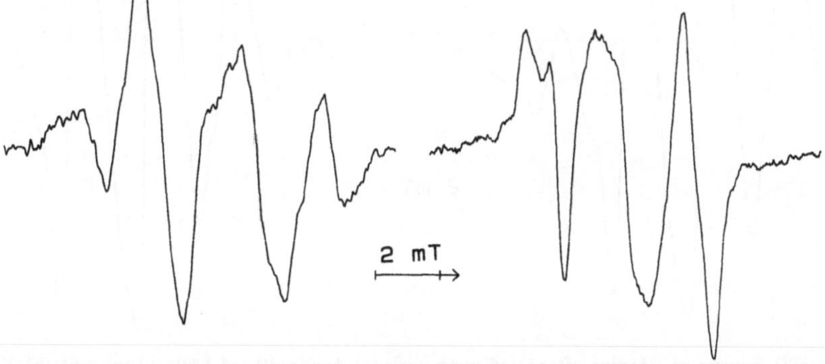

Figure 3. Isolated Components of the Room Temperature Spectrum of Glycine: (a) Abstraction - and (b) Deamination Radical.

An assignment of the corresponding radical structures is aided by comparison with results from aqueous glasses. The quartet probably arises from an abstraction process, yielding NH_3+CHCOOH. This assumption is supported by a change of the quartet from (H_2SO_4/H_2O) to a doublet in (D_2SO_4/D_2O) demonstrating the interaction of the unpaired spin with the amino protons or deuterons, respectively. The triplet can be related to results from 8N NaOD in which it is assigned to arise from a deamination process (CH_2COO).

Alanine

The ESR-spectra of alanine obtained after high and low LET-irradiation at 300°K are presented in Fig. 4. The spectra show the same structural features, i.e. a quintet with an overall extent of 9.2 mT. This quintet structure is known to arise from a deamination process (Samskog et al., 1980).

Arginine

The spectra of arginine measured after low-and high-LET irradiation are presented in Fig. 5. The spectra obtained after heavy ion bombardment are dose-independent (Fig. 5a,b), whereas X-irradiation of arginine yields different spectra for low and high values of dose (Fig. 5c,d). For low doses the same pattern is found independent of radiation quality, that is a complex spectrum with an overall extent of 8.0 mT (Fig. 5a,c). With increasing X-ray dose this pattern changes into a triplet structure of 2.4 mT (Fig. 5d).

Using computer analysis, the complex structure could be extracted to be composed of two patterns: a quintet (Fig. 6a) of 8.0 mT and the triplet of 2.4mT.

The dose dependence found after X-irradiation can be explained by a decrease of the relative radical concentration of the quintet pattern. In comparison with results from H_2SO_4/H_2O glasses the quintet can be identified as side chain radical of the form -$CHCH_2$- (Fig. 6b). The origin of the triplet pattern is not fully understood at the moment. We assume tentatively a radical formation process in the guanidium group.

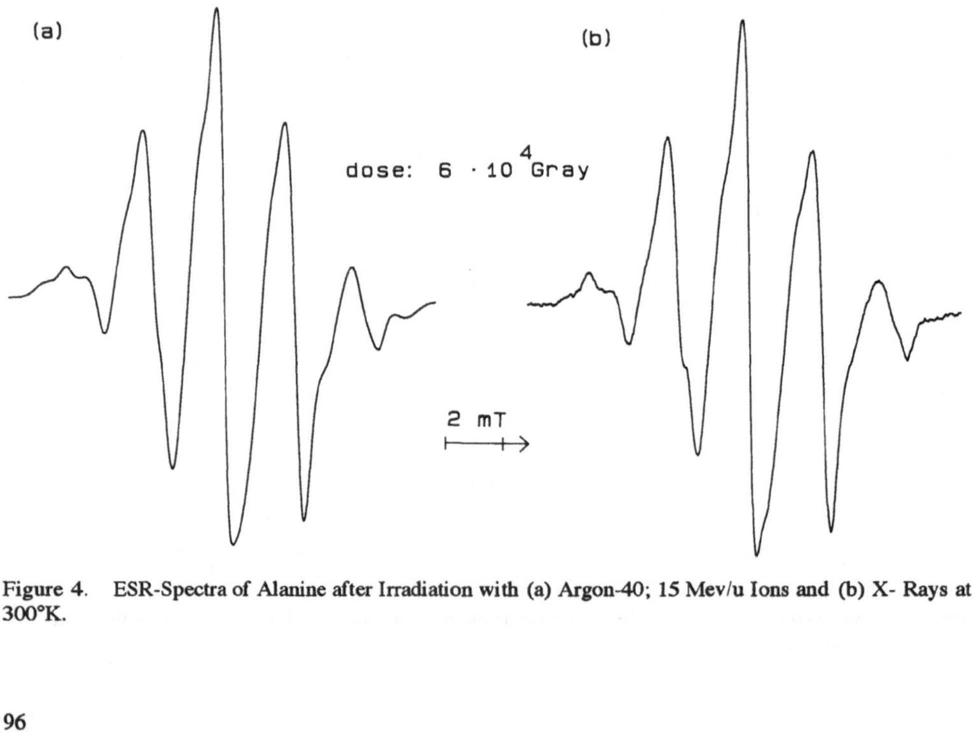

Figure 4. ESR-Spectra of Alanine after Irradiation with (a) Argon-40; 15 Mev/u Ions and (b) X- Rays at 300°K.

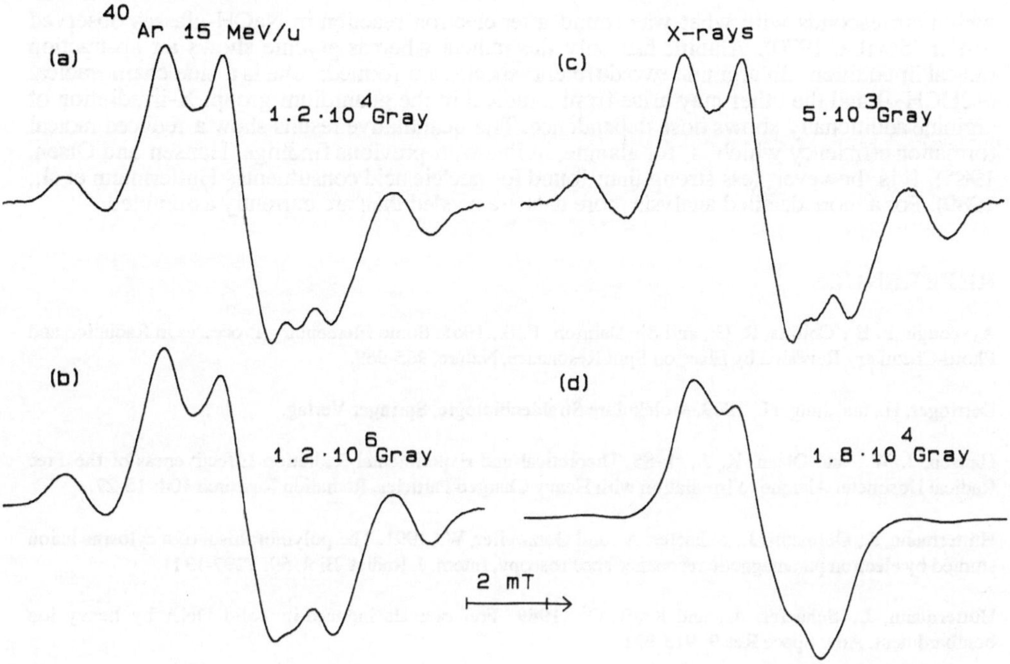

Fig. 5. ESR- Spectra of Arginine Obtained after Bombardement with Argon-40 Ions (a,b) and X-rays (c,d).

SUMMARY

The formation of radicals by low and high LET-radiation in the amino acids studied so far is found to be independent of radiation quality. For glycine a deamination radical is found

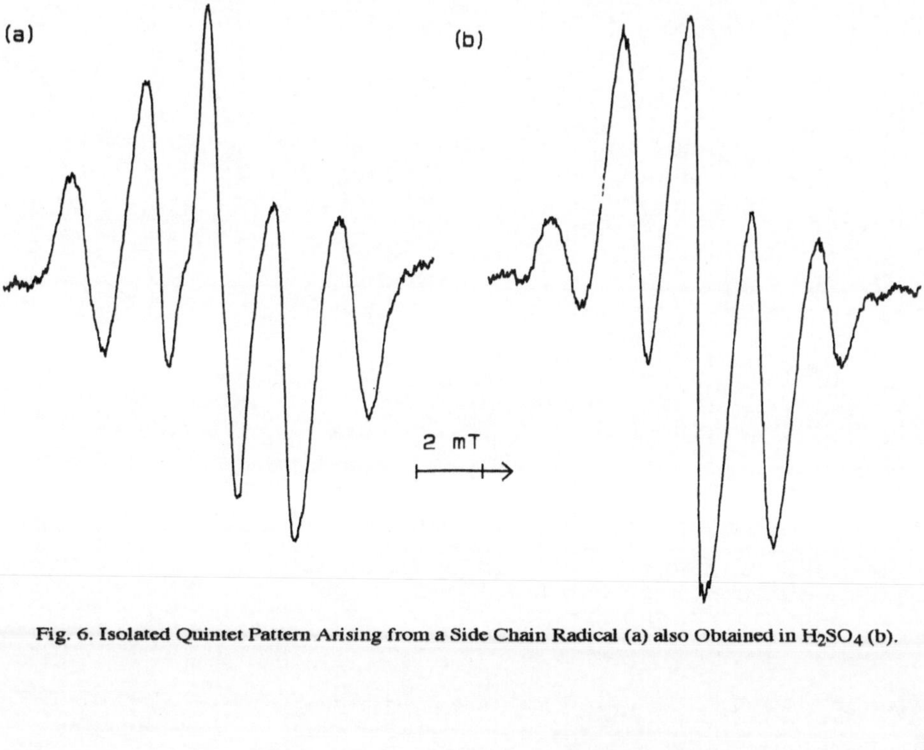

Fig. 6. Isolated Quintet Pattern Arising from a Side Chain Radical (a) also Obtained in H_2SO_4 (b).

which corresponds with what was found after electron reaction in NaOH-glasses observed earlier (Sevilla, 1970). Alanine has only this radical whereas glycine shows an abstraction radical in addition. In arginine two different species are formed: one is a side chain radical ($-CHCH_2-$) and the other may arise from a radical in the guanidium group. X-irradiation of arginine additionally shows dose dependence. The quantitative results show a reduced radical formation efficiency which is, for alanine, in line with previous findings (Hansen and Olsen, 1985). It is, however, less strong than found for nucleic acid constituents(Hüttermann et al., 1989). For a more detailed analysis more data are needed than are currently assembled.

REFERENCES

Ayscough, P. B., Collins, R. G., and Sir Dainton, F. S., 1965. Some Elementary Processes in Radiation and Photo-Chemistry Revealed by Electron Spin Resonance, Nature, 965-969.

Dertinger, H., and Jung, H., 1969, Molekulare Strahlenbiologie, Springer Verlag.

Hansen, J. W., and Olsen, K. J., 1985. Theoretical and Experimental Radiation Effectiveness of the Free Radical Dosimeter Alanine to Irradiation with Heavy Charged Particles, Radiation Research 104: 15-27.

Hüttermann, J., Ohlmann, J., Schaefer, A., and Gatzweiler, W., 1991. The polymorphism of a cytosine anion studied by electron paramagnetic resonance spectroscopy, Intern. J. Radiat. Biol. 59: 1297-1311.

Hüttermann, J., Schaefer, A., and Kraft, G., 1989. Free radicals induced in solid DNA by heavy ion bombardment. Adv. Space Res.9: 915-921.

Müller-Broich, A., 1964, Spektrogr. Untersuchungen mittels EPR über die Wirkung ionisierender Strahlen auf elementare biolog. Objekte, Akademie der Wissenschaften u. d. Literatur, Abhandlung d. math.-nat. Klasse, 5.

Samskog, P. O., Nilsson, G., Lund, A., and Gillbro, T., 1980, Primary Reactions in irradiated Deuterio a-Amino Acids Studied by Pulse Radiolysis and ESR Spectroscopy, Journal of Phys. Chem. 84: 2819-2823.

Sevilla, M. D., 1970, Radicals formed by Reaction of Electrons with Amino acids in an Alkaline Glass, Journal of Phys. Chem. 74: 2096-2102.

THE BIOSTACK CONCEPT AND ITS APPLICATION IN SPACE AND AT ACCELERATORS: STUDIES ON BACILLUS SUBTILIS SPORES

G. Horneck

DLR Institut für Flugmedizin
Abt. Biophysik
Linder Höhe, 5000 Köln 90, Germany

INTRODUCTION

It is now well documented that exposure of living systems to space ionizing radiation may have deleterious biological effects (reviewed by Bücker 1977, Bücker and Facius 1980, Akoev et al. 1981, Horneck 1988, 1992, Hagen 1989). Among the large variety of radiations in space which cover a wide spectrum of mass and energy, it is likely that HZE particles are the most effective species. As a working definition, HZE particles are considered those that have a $Z > 2$ and that are able to penetrate at least 1 mm of spacecraft or spacesuit shielding (Grahn 1973). Depending on their charge, HZE particles will have a minimum initial energy of 10-35 MeV/u and, near the end of their range, they will deposit energy at a rate in excess of 100 keV/μm. HZE particles amount to about 1% of the nuclear component of cosmic radiation.

The heterogeneous composition of cosmic radiation and the low flux of HZE particles (Simpson 1983) demand for an experimental methodology to localize the path of each HZE particle relative to the biological target and to correlate physical with biological events. Such effect particle correlations were accomplished in spaceflight experiments in two ways:

1) by use of visual track detectors that were sandwiched between layers of either biological objects in resting state, like viruses, bacterial spores, plant seeds or shrimp cysts, or of embryonic systems, like insect eggs (Biostack concept) (Bücker et al. 1973, Bücker 1975; Bücker et al. 1984), or
2) by use of nuclear track detectors that were in fixed orientation to biological targets of interest, like implantations beneath the scalp of animals (Haymaker et al. 1975) or helmet devices for astronauts (Osborne et al. 1975).

This paper concentrates on the Biostack concept (Bücker and Horneck 1975). As biological test-system spores of *Bacillus subtilis* have been used in space experiments as well as at accelerators. DNA injury, inactivation, mutation induction and the efficiency of repair processes will be discussed as well as the impact parameter dependence of the inactivation. This report summarizes studies that have been performed for about 20 years in the laboratory of Prof. H. Bücker at the DLR in Köln.

THE BIOSTACK CONCEPT

Single particle effect correlation methods, on the basis of the Biostack concept, were first applied in spaceflight experiments (recently reviewed by Horneck 1992). In these

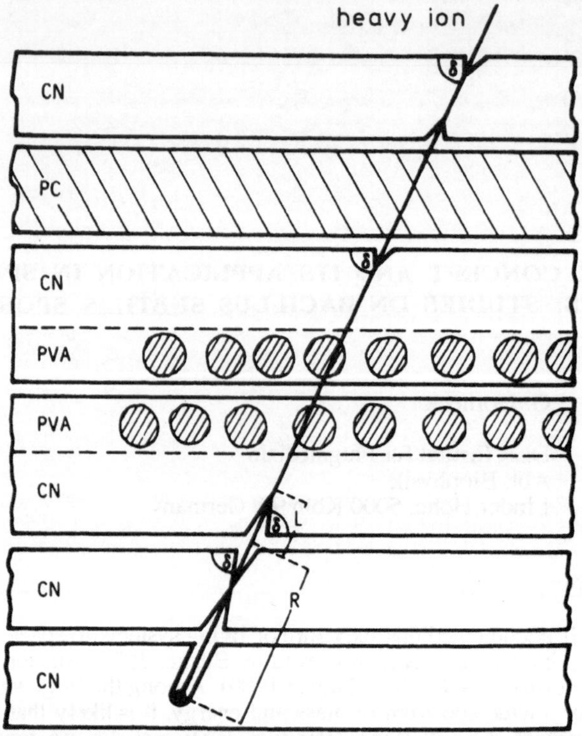

Figure 1. **The Biostack Concept to Localize Biological Effects Produced by Single HZE Particles of Cosmic Radiation (Modified from Enge et al. 1974). Biostack Experiments were Flown on Board Apollo 16, 17, ASTP, Spacelab 1, LDEF, and Cosmos 1887 and 2004.**

Biostack and Biobloc experiments, monolayers of selected biological objects, fixed in position, were sandwiched between visual nuclear track detectors (Figure 1). Postflight analysis comprised (1) localization of each HZE particle's trajectory in relation to the biological specimens; (2) separate investigation of the response of each biological individual hit, in regard to radiation effects; (3) determination of the impact parameter (i.e. the distance between particle track and sensitive target); (4) determination of the physical parameters (Z, E, LET) of the relevant HZE particles, and (5) correlation of the biological effect with the HZE particle parameters. High precision methods for localization of HZE particle tracks within the biological layer have been developed by Schäfer et al. (1977) and Facius et al. (1990). Sandwiches of this type of combination of biological layers and nuclear track detectors were flown on several US or USSR space missions as well irradiated at accelerators. In a variety of test systems, injuries were traced back to the traversal of a single HZE particle, such as somatic mutations in plant seeds, development disturbances and malformations in insect and salt shrimp embryos, or inactivation in bacterial spores (Bücker and Horneck 1975).

BACILLUS SUBTILIS SPORE AS MODEL SYSTEM IN RADIATION BIOLOGY

Bacterial spores offer the opportunity to test a variety of radiobiological endpoints at the cellular and subcellular level, both in space experiments and at accelerators. As a dormant system, they are capable of a long shelf life, and they are resistant to heating and to desiccation. Their repair processes do not operate during treatment and can be switched on during subsequent incubation. Bacterial spores having a cytoplasmic core with a geometrical cross section of 0.2-0.3 μm^2 (Figure 2) are suitable test organisms in heavy ion research,especially for sounding the biological effectiveness in dependence of the impact parameter. We have followed the radiobiological chain of events, from the chemical damage

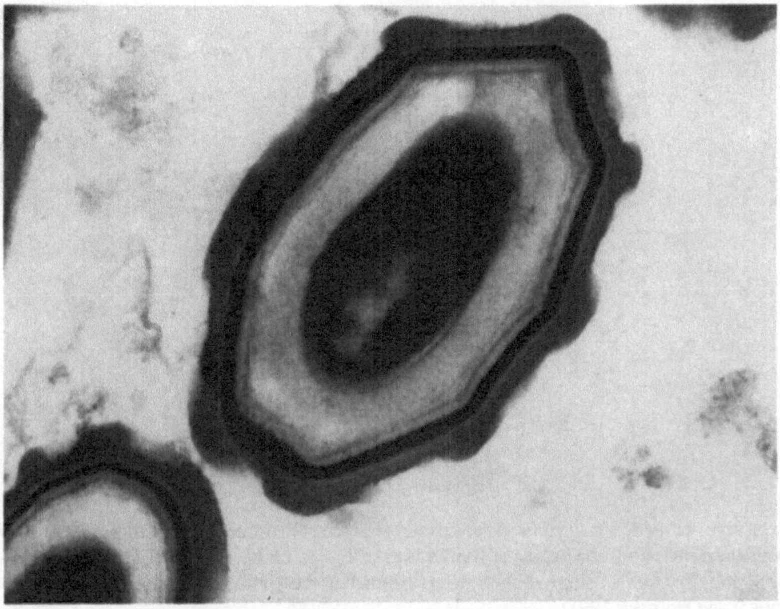

Figure 2. Electronmicrograph of a Spore of *B. subtilis* Showing the Cytoplasmic Area (Spore Core) and the Surrounding Layers, the Membrane, the Cortex and the Spore Wall (Photo Kindly Provided by S. Pankratz). The Long Axis of the Whole Spore is 1.2 mm, the Core Area 0.25 mm^2.

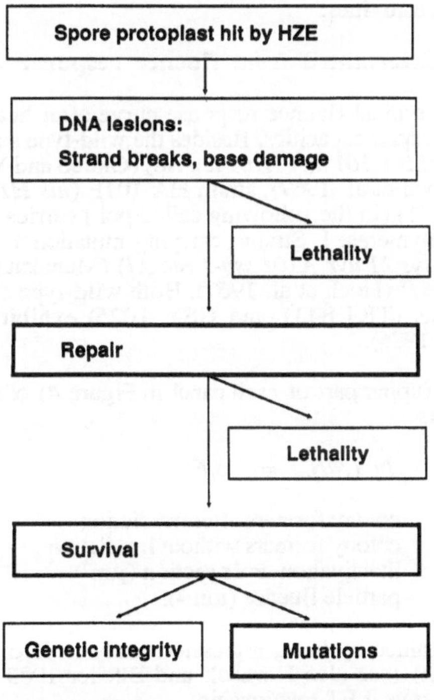

Figure 3. Scheme of the Chain of Events after an HZE Particle Affects a Bacterial Spore. DNA Lesions Will be Investigated by Biochemical Analysis, Inactivation Will be Determined as Loss of Colony Formers, Repair Will be Studied by Comparing Strains of Different Repair Capacity, and Mutagenesis by Scoring Histidine Revertants.

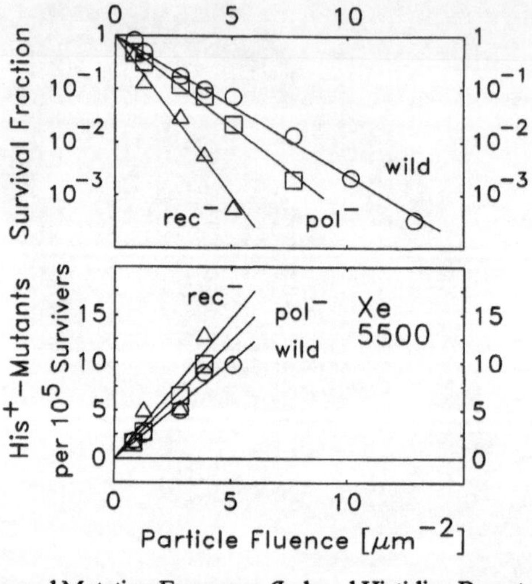

Figure 4. Surviving Fraction and Mutation Frequency (Induced Histidine Revertants per 10^5 survivors) of *B. subtilis* Spores as a Function of Fluence of Xenon Ions of E_n = 11.8 MeV/u and LET = 5500 keV/μm. (O = Wild Type, r = pol-, Δ = rec-). The Irradiation was Performed at the UNILAC of the GSI, Darmstadt (Modified from Baltschukat and Horneck 1991).

to the DNA up to the genetic manifestation of the radiation injury (Figure 3), using this proper test system of a bacterial spore; several isogenic strains are available that only differ in their repair capacity. This allows to compare the responses to heavy ions of spores of the same size but of different radiation sensitivities.

HZE particle effects, determined from fluence response curves

Figure 4 shows a typical fluence response curve after heavy ion irradiation of *B. subtilis* spores of different repair capacities. Besides the wild-type strains with regard to DNA repair capability, HA 101 (*his H101 met B101 leu A8*) (Okubo and Yanagida 1968) and GSY 1026 (*trp C2 met B4*) (Hoch et al. 1967), strain HA 101F (*his H101 met B101 leu A8 pol A1*) (K.G. Gass et al. 1971) (in the following called pol-) carries the mutation *pol A1* that causes a defect in DNA polymerase I. Strains, carrying mutation *rec A1* (called rec-) are TKJ 8431 (*his H101 met B101 lys 21 uvr A101 ssp-1 rec A1*) (Munakata, pers. comm.) and GSY 1025 (*trp C4 met B4 rec A1*) (Hoch et al. 1967). Both wild-type strains (HA101 and GSY 1026) or both rec- strains (TKJ 8431 and GSY 1025) exhibit the same sensitivity to X-irradiation (Baltschukat 1986).

All survival curves (upper part of each panel in Figure 4) of all strains are purely first order. The relation used is

$$ln\ (N/N_o) = -\sigma_\iota F \qquad (1)$$

with
N = colony formers after irradiation
N_o = colony formers without irradiation
σ_ι = inactivation cross section (μm^2)
F = particle fluence (μm^{-2}).

The constant σ_ι was determined by linear regression analysis. If σ_ι is plotted as function of LET (Figure 5, upper part) (see also Horneck and Bücker 1983, Baltschukat 1986), the curves show the following σ_ι vs. LET relationship:

1.) The cross section σ_ι increases with increasing LET up to about 100 keV/μm.

2.) At LET values from 200-1000 keV/μm, the σ_ι vs. LET curve saturates.

3.) At LET values above 1000 keV/μm, the σ_ι vs. LET dependence is described by a separate function for each ion. At very high LET values, specific for each ion, σ_ι decreases, resulting in the so-called "σ_ι-LET hooks".

A similar dependence on LET has been observed for a variety of biological endpoints studied so far in different biological systems, such as inactivation, mutation induction, chromosomal aberrations, or induction of DNA strand breaks (reviewed by Todd and Tobias 1974, Leith et al. 1983, Blakely et al. 1984, Kiefer 1985, Kraft 1987). In our experiments, the so-called saturation cross section coincides with the average projected area of the spore core, which has been determined from electron micrographs of spores of *B. subtilis* wild-type (for example Figure 2) as 0.25 μm^2. At higher LET values (> 4000 keV/μm), when separate σ_ι vs. LET curves are obtained for each ion, the σ_ι values exceed the geometrical cross section of the spore core by up to a factor of 5 (wild-type), 6 (pol$^-$), or 11 (rec$^-$), respectively. With increasing LET, they diminish forming the so-called σ_ι-LET hooks.

In the lower part of Figure 5, the σ_ι values are plotted versus energy per nucleon, E_n, separately for each strain. With respect to the wild type strain, the main findings are:

1.) All curves, extrapolated back to $E_n = 0$, - within experimental error - intersect the ordinate at approximately 0.25 μm^2, irrespective of the atomic number of the ion. This value coincides with the "saturation cross-section" of the σ_ι vs. LET curves, and is equivalent to the geometrical cross-section of the spore core.

2.) For the group of lighter ions (boron and carbon), σ_ι decreases with increasing E_n.

3.) For neon, σ_ι is nearly independent of E_n.

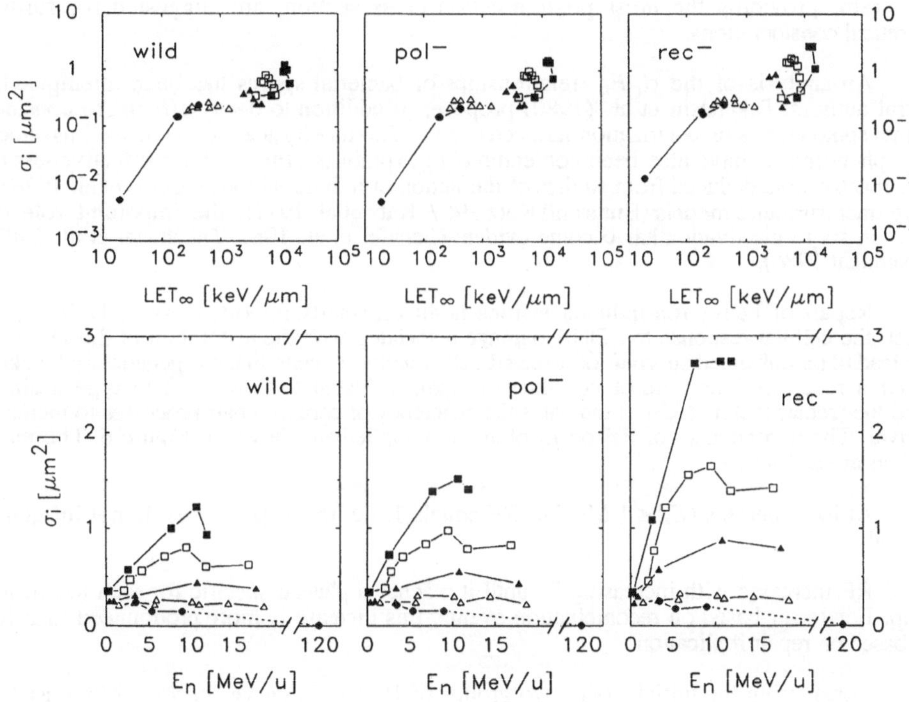

Figure 5. Inactivation Cross Section, σ_i, of *B. subtilis* Spores as a Function of LET (Upper Part). The Curves Demonstrate an Increase of σ_i with Increasing LET at LET ≤ 100 keV/μm, a Saturation at LET of 200-1000 keV/μm and the so-called σ–LET Hooks at LET > 2000 keV/μm. The lower part shows σ_i as a Function of Energy per Nucleon, E_n. All Curves of All Strains Converge at $E_n = 0$ roughly at 0.25 μm^2, the Suggested Projected Area of the Core of the Spore. (o = B, C; Δ = Ne, Δ = Fe, Ni, Kr; = Xe, = Pb, U) (from Baltschukat and Horneck 1991).

4.) For heavier ions, σ_ι increases with increasing E_n, up to a maximum (or saturation) at about $E_n = 10$ MeV/u; this increase becomes steeper with increasingly heavier ions. Similar observations have been made by Schäfer et al. (1981) and Takahashi et al. (1983).

Spores of pol- and rec- strains show σ_ι vs. E_n relationships similar to those of the wild type, in line with findings (1) through (4). Concordant with increasing sensitivity against X-rays from wild type through pol- to rec- spores (Baltschukat and Horneck 1991), for heavier ions the slopes and the maximum values of the σ_ι vs. E_n curves increase in the same direction: wild < pol- < rec-. It is extremely noteworthy that also for these repair deficient strains, all σ_ι vs. E_n curves, extrapolated back to $E_n = 0$, intersect with the ordinate at about 0.25 μm^2.

Models relating the increase of σ_ι with E_n to track diameter characteristics predict that extrapolation of the σ_ι vs. E_n curves back to $E_n = 0$ might yield the size of the critical target structure (Kraft 1987). Since the extrapolated cross section coincides with the geometrical area of the spore core, this innermost part of the spore, containing the dehydrated cytoplasma and DNA (Gould 1978), most likely is the sensitive site of the spore, as already pointed out by Powers et al. (1968) for lighter ions.

In nuclear emulsions, the phenomenon of "thindown" is observed in the track of a heavy ion approaching the end of its range: as the ion slows down, the track first increases in width to a maximum value and then thins down like a sharpened pencil. According to Katz (1986), in this track regime, energetic δ-rays play a dominant part in the production of the action cross section. This may also be valid for the inactivation of spores, since "thindown" is reflected by the σ_ι vs. E_n curves (Figure 5). For each ion, this "thindown" effect, i.e. the maximum value and the slope, increases from wild type through pol- to rec- strain. Onsetting repair might blur the initial extent and spectrum of damage. Therefore, the responses of the rec- strain, providing the most pristine action cross section, are suggested for further theoretical considerations.

An analysis of the σ_ι-E_n relationships of bacterial spores has been attempted by several authors. Takahashi et al. (1980) propose, in addition to δ–ray effects, to consider thermal spike effects or polarization-induced events. Thermophysical events as well as shock wave phenomena, have also been conjectured to explain the far-reaching effectiveness of spore inactivation, deduced from studies of the action of individual ions (Facius et al. 1978a). Using track structure models (Butts and Katz 1967, Katz et al. 1971), the important role of δ-rays in spore inactivation has become evident (Schäfer et al. 1981, Takahashi et al. 1983, Baltschukat 1986).

Repair of heavy ion induced lesions is an especially important issue. DNA repair reflects the cellular response to a DNA damage associated with the restoration of the DNA. It may lead to an enhanced survival of affected cells relative to cells that are genetically blocked in such a response. The ratio of σ_i values of repair deficient to those of wild type strain is called the "repair factor" (RF). It indicates the efficiency of certain repair processes to increase survival. The dependences of RF on E_n of the ion applied are shown in Figure 6. The main findings are as follows:

1.) At low energies ($E_n \leq 1$ MeV/u) RF equals 1, i.e. repair processes do not influence survival.

2.) RF increases with increasing E_n until it reaches a plateau specific for each ion or ion group. For repair based on recombination events, this increase is more pronounced than for that based on repair replication.

3.) Considering the initial slope, two groups of RF vs. E_n curves become obvious: For lighter ions ($Z \leq 10$) the curves are relatively flat, for heavier ions ($Z \geq 26$), a steep slope is obtained. Note that in the experiment with carbon at 120 MeV/u, RF for repair requiring recombination is as high as that after X-ray treatment.

This shows that, in most cases, DNA repair enhances survival after heavy ion irradiation.

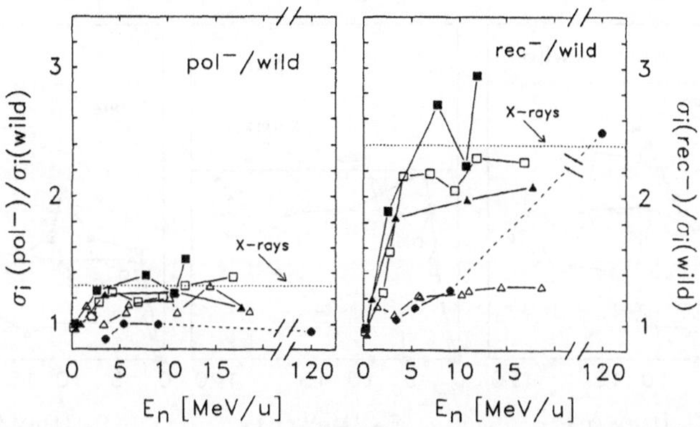

Figure 6. Repair Factor RF, i.e. Ratio of σ_i (Repair Deficient) to σ_i (Wild-type) as a Function of E_n. (o = B, C; D = Ne, Δ = Fe, Ni, Kr; = Xe, = Pb, U), (Dotted line: RF after X-irradiation) (From Baltschukat and Horneck 1991).

Repair processes involving recombination events are more effective than those requiring repair replication.

Mutagenesis is another endpoint of high biological relevance, since it represents another form of DNA damage in irradiated cells. The lower part of Figure 4 shows the frequency of reversion to histidine prototrophy versus fluence. The mutation induction curves are linear. They are described by

$$M = \sigma_m F \tag{2}$$

with M = mutation frequency
σ_m = mutation induction cross section (μm²)
F = particle fluence (μm⁻²).

Figure 7 shows the mutation induction cross section, σ_m, versus E_n, separately for each strain. Separate curves are obtained for each ion. The main observations are as follows:

1.) At low energies per nucleon ($E_n \leq 1$ MeV/u) no mutations are induced.

2.) Few, if any, mutants are induced by light ions ($Z \leq 10$).

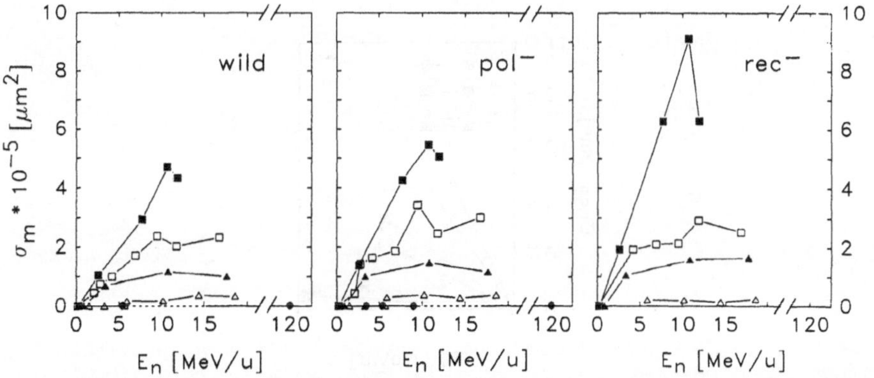

Figure 7. Mutation Induction Cross Section, σ_m to Histidine Prototrophy of *B. subtilis* Spores as a Function of E_n. (o = B, C; Δ = Ne, Δ = Fe, Ni, Kr; = Xe, = Pb, U) (From Baltschukat and Horneck 1991).

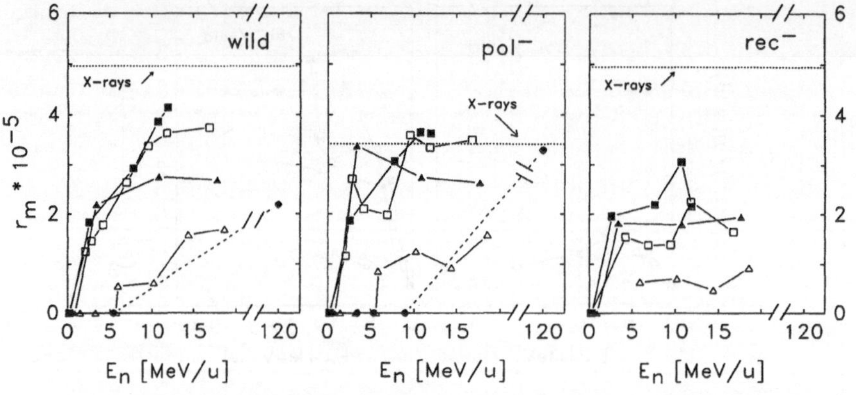

Figure 8. Mutagenic Efficiency per LethalEvent, r_m, as a Function of E_n. (o = B, C; Δ = Ne, Δ = Fe, Ni, Kr; = Xe, = Pb, U) (Dotted Line: r_m after X-irradiation) (from Baltschukat and Horneck 1991).

3.) For heavier ions (Z ≥ 26) σ_m increases with E_n. The increment becomes steeper with increasing atomic number of the ion. At $E_n > 10$ MeV/u, σ_m levels off or slightly decreases.

Hence, mutagenesis also shows the "thindown" phenomenon similar to inactivation (Figure 5). However, the differences between strains of different repair capacity are smaller than those for inactivation. Only after irradiation with uranium ions at $E_n > 7$ MeV/u, are considerably higher σ_m values seen in rec- than seen in wild type or pol- strain.

The r_m values indicating the mutagenic efficiency per lethal event are plotted as a function of E_n in Figure 8. The maximum r_m value reached for pol- is equal to that seen after X-ray treatment, whereas for wild type and rec- it is less than that obtained by X-rays. In this plot, the mutagenic action of carbon at 120 MeV/u becomes evident.

DNA is considered to be the most important and sensitive target in biological systems. Beside base damage, DNA strand breaks are considered as the major lesion in the genome of irradiated biological systems. DNA double strand breaks are especially conjectured to be lethal events in cellular systems. Figure 9 shows the cross section for the induction of double strand breaks in vegetative cells of *B. subtilis* as a function of E_n. Again, a similar pattern is observed as for inactivation, mutation induction and efficiency of repair.

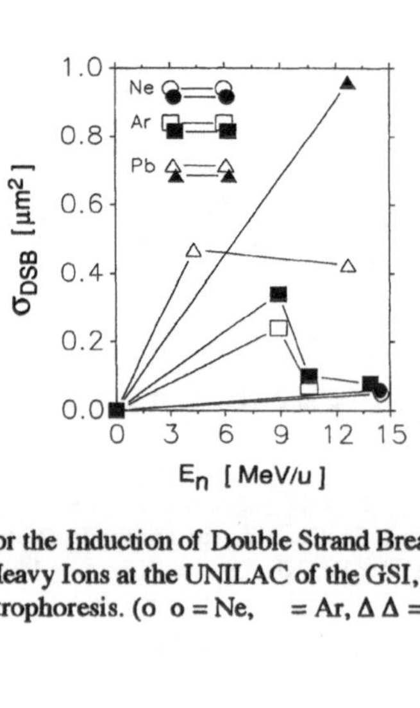

Figure 9. Cross Section, σ_{DSB} for the Induction of Double Strand Breaks in the DNA of Vegetative Cells of *B. subtilis* after Irradiation with Heavy Ions at the UNILAC of the GSI, Darmstadt. The Strand Breaks were Determined by Pulse Field Gel Electrophoresis. (o o = Ne, = Ar, Δ Δ = Pb) (Micke et al. in Preparation).

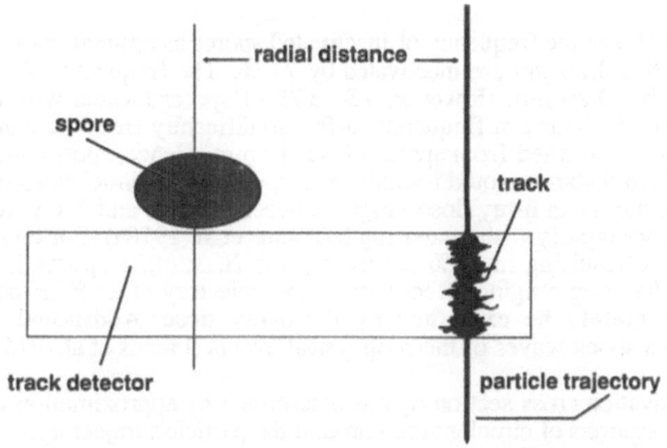

Figure 10. Determination of the Impact Parameter, b, as the Radial Distance of the Center of the Spore from the Trajectory of the HZE Particle.

With these studies, we have demonstrated that DNA strand breaks, repair and mutagenesis being well-established biological processes for sparsely ionizing radiation also occur after irradiation with HZE particles. The magnitude of the effects increases with E_n and Z of the respective ion. It is of special interest that after heavy ion irradiation, the maximum values - RF for repair and r_m for mutagenesis in pol- - roughly coincide with the respective values obtained after X-irradiation. Therefore, one might conclude that especially in those circumstances where responses are similar to those to X-rays the biological effect of heavy ions may be ascribed mainly to the action of the sparsely ionizing secondary electrons. This especially applies to very heavy ions ($Z \geq 54$ and $E_n \geq 10$ MeV/u) as well as to carbon ($Z = 6$, $E_n = 120$ MeV/u).

Inactivation cross section determined from single particle analysis

We have also investigated the response of *B. subtilis* spores to the HZE component of cosmic radiation by applying the Biostack concept (Figure 1 and 10). The method allows one to determine the inactivation probability as a function of the radial distance to each particle's trajectory. For that purpose, spores in suspension of 0.01 % aqueous polyvinyl alcohol solution (PVA) were spread onto cellulose nitrate (CN) track detector sheets. After drying, a fixed contact of the spore monolayer with the track detector was achieved. These sheets, carrying the biological layer, were stacked between pure plastic detector sheets in the typical arrangement of the biological units of Biostack (Figure 1) (Horneck et al. 1974). The complete Biostack experiment package accommodated in a hermetically sealed aluminum container was stowed inside of the spacecraft (Apollo or Spacelab 1) and exposed to cosmic radiation. Total absorption by the spacecraft wall was approximately 2.4 g/cm^2 (Bücker 1975). The flux of HZE particles with LET $\geq$ 130 keV/μm was 0.3 - 0.7 particles/cm^2 d.

After retrieval and disassembly, CN foils carrying a spore layer were etched (6 N NaOH, 30°C) only on the CN-side resulting, along the trajectory of an HZE particle, in an etch cone at the side opposite to the spore layer. This procedure was followed by individual etching of each etch cone under microscopical control (pin-point etching), until the tip of the etch cone approaches the spore layer up to a few microns. By extrapolation of the etch cone, the target area in the spore layer was determined. The target area was defined as an area of a radius of 5 μm off the point of intersection of an HZE particle with the spore layer. Spores of the target area were individually transferred onto nutrient agar by micromanipulation; germination, outgrowth and growth were followed until a microcolony was formed. The original position of each spore relative to the HZE particle's trajectory, i.e. the impact parameter (Figure 10), was obtained from microphotographs and after break-through-etching of the etch cone (Schäfer et al. 1977). The accuracy in determining the impact parameter - depending on the dip angle of the HZE particle - was up to $\pm$ 0.2 μm, which is well below the size of a bacterial spore. The physical parameters of the respective HZE particle, such as charge and energy loss, were obtained from the adjacent physical detector foils.

Figure 11 shows the frequency of inactivated spores as a function of impact parameter b. Spores within b ≤ 0.25 μm are inactivated by 73 %. The frequency of inactivated spores drops abruptly at b > 0.25 μm. However, 15 - 30% of spores located within 0.25 < b < 3.8 μm are still inactivated. This μm frequency differs significantly from the control value of (12 ±3)% dead spores, determined from spores of b > 10 mm. Hence, spores are inactivated well beyond 1 μm, which distance would roughly correspond to the dimensions of a spore. At the distance of 1 μm, the mean δ-ray dose ranges between 0.1 Gy and 1 Gy, depending on the particle, and declines rapidly with increasing b (Facius et al. 1978a). For comparison, the D_{37} for electrons (dose resulting in 37% survivors) for *B. subtilis* spores is about 800 Gy. Therefore, the radial long-ranging-effect around the trajectory of an HZE particle (up to b = 3.8 μm) cannot merely be explained by the δ-ray dose. Additional mechanisms are conjectured, such a shock waves or thermophysical events (Facius et al. 1978a).

The inactivation cross section σ_ι was determined by approximation as the sum of the inactivation cross sections of circular zones around the particle's trajectory:

$$\sigma_\iota = 2\,\pi \Sigma P_n\, R_n\, (\Delta R) \qquad (3)$$

with P_n = inactivation probability in the circular zone n
R_n = distance of the center of the circular zone width from the particle trajectory
ΔR = width of the circular zone

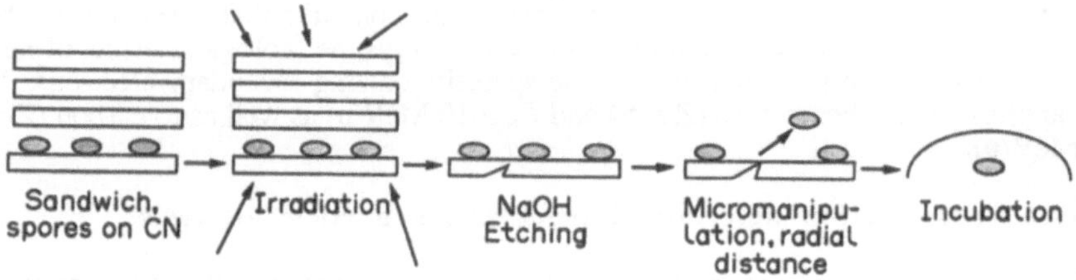

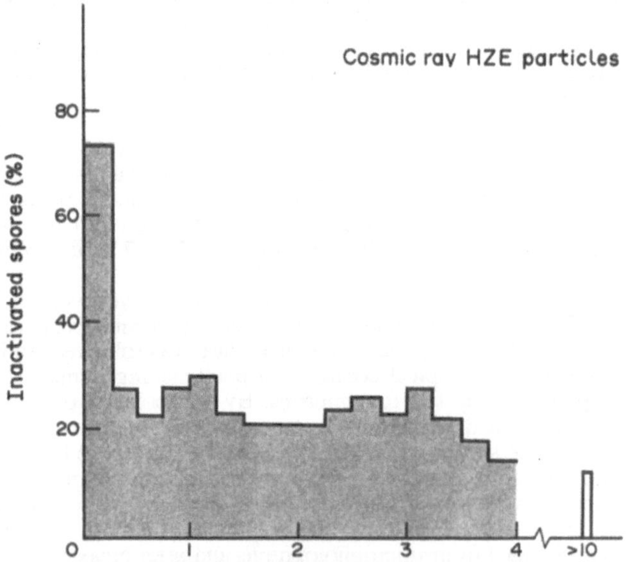

Figure 11. Inactivation (%) of *B. subtilis* Spores by Single HZE Particles (Z ≥ 12, LET > 100 keV/μm, E_n = 10-300 MeV/u) of Cosmic Radiation as a Function of the Impact Parameter. Results of the Biostack Experiment Flown onboard of ASTP (From Horneck et al. 1989).

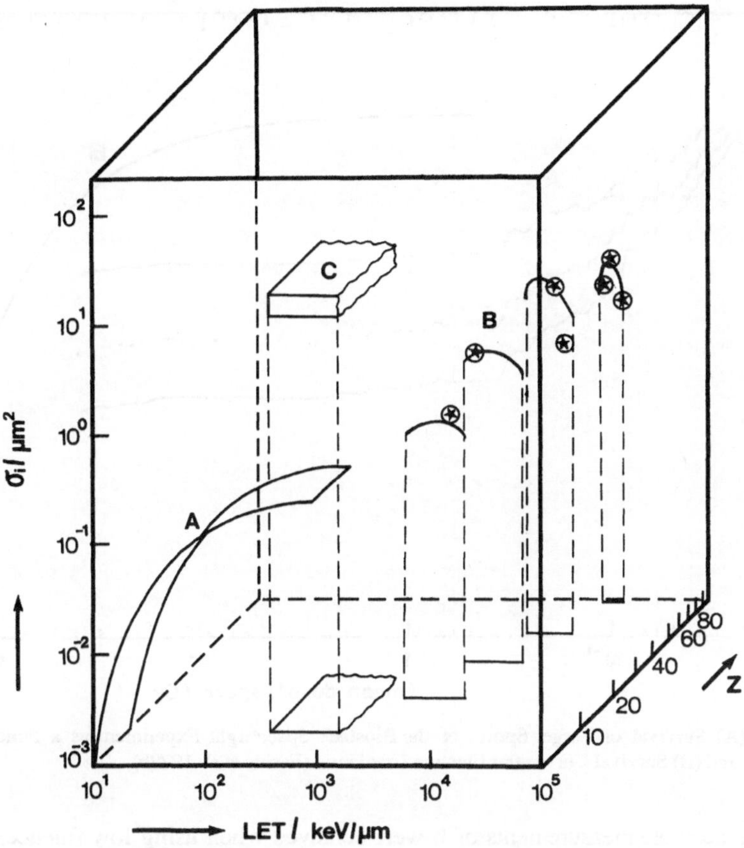

Figure 12. Inactivation Cross Section, σ_i, as Function of LET and Atomic Number Z, Determined from Fluence Effect Curves (A: LBL data, B: GSI data) and from Impact Parameter Dependence in Single-particle Effect Correlations (C: SpaceData, H : GSI Data).

Taking into account an inactivating efficiency up to b = 3.8 μm for the spaceflight experiments, σ_ι amounts to 5,24 μm^2. The responsible HZE particles are composed of a mixed field in atomic number - with iron being the most frequent ion recorded by this method and in LET values above 100 keV/μm. In Figure 12, σ_ι from the spaceflight experiments is compared to the σ_ι values obtained from fluence effect curves. It is roughly 20 times higher than it would be expected from the fluence effect data. This value is also about 20 times higher than the geometrical cross section of the spore core (see Figure 2).

In order to compare the spaceflight results with data from irradiations with X-rays or electrons, the mean dose was calculated for each spore of an impact parameter b $\leq$ 4.5 μm by integrating the radial δ-ray dose function over a sphere with a radius of 0.6 mm - representing the size of a spore (Facius et al. 1978b). With these data a survival curve was constructed which is shown in Figure 13 together with a survival curve after electron irradiation. Taking the 96 % survival level, a RBE value in the range of 600 is obtained.

Attempts to correlate the biological response, i.e. loss of colony forming ability, with the physical data of the respective HZE particles showed that ions of Z $\geq$ 22 are more lethal than lighter ions (Facius et al. 1979). Concerning the dependence of the biological response on energy and LET of the HZE particles, no monotonic trend was clearly observable. However, it turned out that for particles with very high energies (about 300 MeV/u) or in the high LET region (500-700 keV/μm) survival was significantly reduced (Facius et al. 1979).

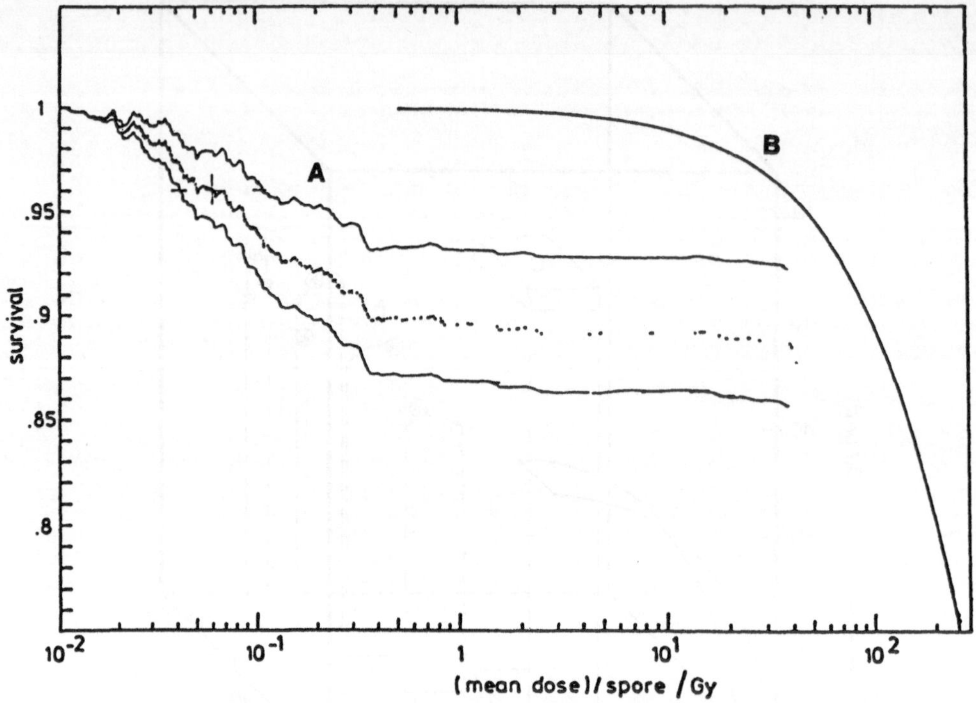

Figure 13. **(A) Survival of Target Spores of the Biostack Spaceflight Experiment as a Function of the Specific Dose and (B) Survival Curve after Electron Irradiation (Facius et al. 1978b).**

More accurate measurements of b were achieved when using low fluences of heavy ions at the UNILAC of the GSI in Darmstadt, since in this case all ions impinge vertically to the spore layer. Three different approaches of the Biostack concept were applied as follows.

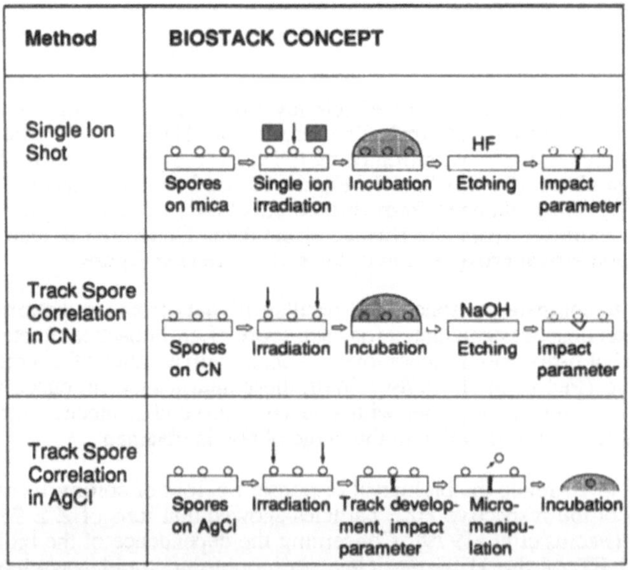
Method
BIOSTACK CONCEPT
Single Ion Shot
HF
Spores on mica
Single ion irradiation
Incubation
Etching
Impact parameter
Track Spore Correlation in CN
NaOH
Spores on CN
Irradiation
Incubation
Etching
Impact parameter
Track Spore Correlation in AgCl
Spores on AgCl
Irradiation
Track develop-ment, impact parameter
Micro-manipu-lation
Incubation

Figure 14. **Methods of Single-particle Effect Correlation Using the Biostack Concept, as Applied at the UNILAC of the GSI, Darmstadt on *B. subtilis* Spores.**

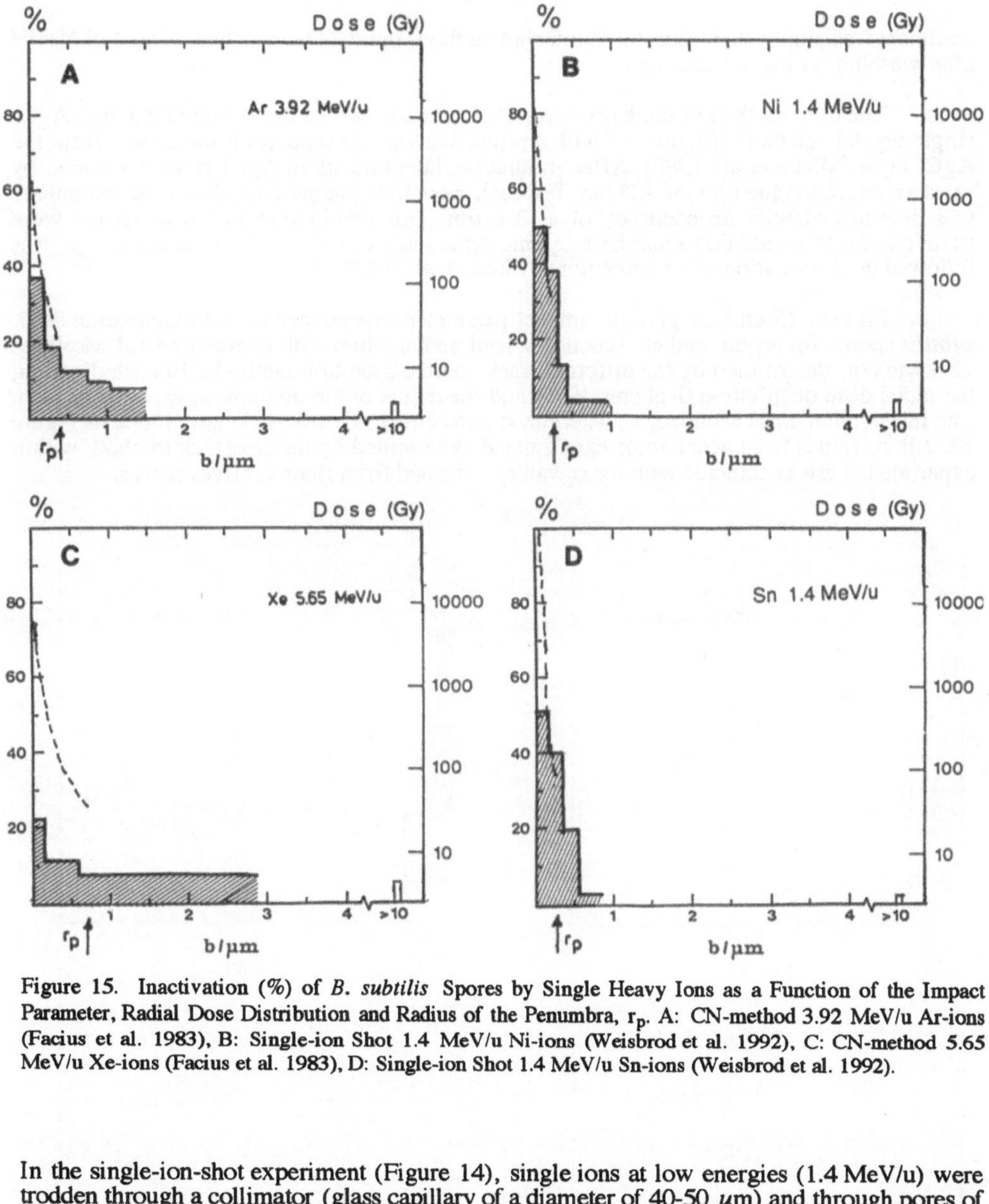

Figure 15. Inactivation (%) of *B. subtilis* Spores by Single Heavy Ions as a Function of the Impact Parameter, Radial Dose Distribution and Radius of the Penumbra, r_p. A: CN-method 3.92 MeV/u Ar-ions (Facius et al. 1983), B: Single-ion Shot 1.4 MeV/u Ni-ions (Weisbrod et al. 1992), C: CN-method 5.65 MeV/u Xe-ions (Facius et al. 1983), D: Single-ion Shot 1.4 MeV/u Sn-ions (Weisbrod et al. 1992).

In the single-ion-shot experiment (Figure 14), single ions at low energies (1.4 MeV/u) were trodden through a collimator (glass capillary of a diameter of 40-50 μm) and through pores of 1 μm in diameter of a nuclear track filter (mica foil) (Weisbrod et al. 1992). The bacterial spores, mounted on a mica detector foil, could be selected and positioned - one by one - in front of the collimator by use of a microscope with an accuracy of 1 μm. After irradiation of each selected spore with 1 ion, the spore layer was covered with nutrient agar to allow germination and outgrowth in special incubation chambers. Microscopical control confirmed that the spores had not changed position by this procedure (scratch marks). After incubation, the biological layer was washed off and the latent tracks in the mica were etched with 40% HF. The position of the etch cones relative to the spores and thereby the impact parameter was obtained with an accuracy of $\pm$ 0.1 μm by superimposition of the different micrographs (reference marks!) (Weisbrod et al. 1992).

For track spore correlation by means of CN detectors (see Figure 14), spores were mounted in a monolayer onto a CN foil similar to the space experiments. They were irradiated with low fluences (approximately 10^6 ions/cm^2) (Facius et al. 1983). Following irradiation, treatment was similar as in the single-ion-shot method, but tCN was etched using 2 N NaOH after washing off the nutrient agar.

The third method of track spore correlation made use of spores mounted onto AgCl single-crystal detectors (Figure 14) with a protective varnish separating the spores from the AgCl layer (Micke et al. 1988). After irradiation, latent tracks in AgCl were developed by short exposure to blue light of 405 nm. For each spore in the target area, the impact parameter was determined with an accuracy of ± 0.1 μm. For viability tests, these spores were transferred into incubation chambers by micromanipulation. Growth on nutrient agar was followed until formation of a microcolony (Micke et al. 1988).

Figures 15 and 16 give the impact parameter dependence of the inactivation of *B. subtilis* spores for argon, nickel, xenon, tin, lead and uranium with energies of 1.4 MeV/u to 12.9 MeV/u, determined by the different track spore correlation methods. In each diagram, the radial dose distribution (Kellerer 1977) and the radius of the penumbra, r_p, are indicated. The inactivation cross sections, σ_ι determined according to equation (3) are plotted in Figure 12. All σ_ι values from accelerator experiments, determined by the Biostack method, within experimental errors coincide with the σ_ι values, obtained from fluence effect curves.

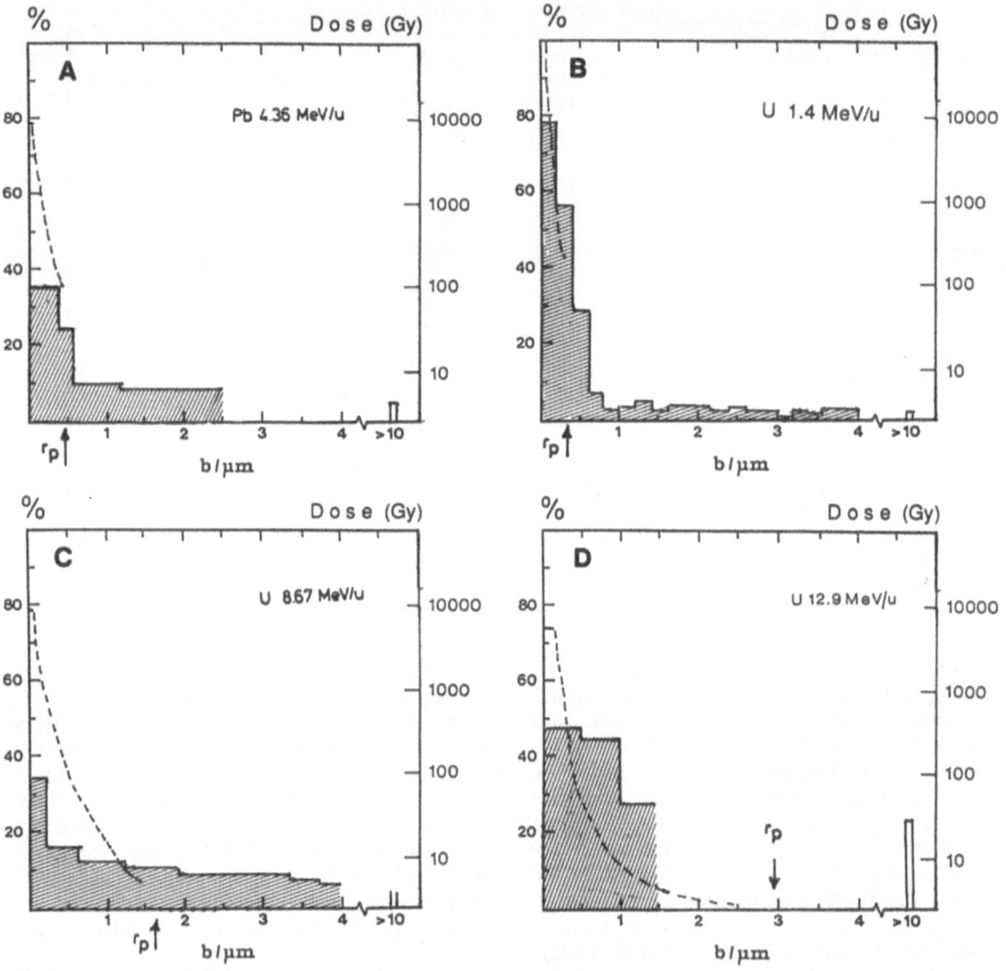

Figure 16. Inactivation (%) of *B. subtilis* Spores by Single Heavy ions as a Function of the Impact Parameter, Radial Dose Distribution and Radius of the Penumbra, r_p. A: CN-method 4.36 MeV/u Pb-ions (Facius et al. 1983), B: Single-ion Shot 1.4 MeV/u U-ions (Weisbrod et al. 1992), C: CN-method 8.67 MeV/u U-ions (Facius et al. 1983), D: AgCl-method 12.9 MeV/u U-ions (Micke et al. 1988).

Comparing the results of these different approaches of single-particle-effect-analysis in space and at accelerators, we can recognize the following:

1) The inactivation probability for spores, centrally hit, is always substantially less than 1.

2) The effective range of inactivation extends far beyond the range of impact parameter where inactivation of spores by δ-rays can be expected. This far-reaching effect is less pronounced for ions of low energies (1.4 MeV/u), a phenomenon which might reflect the "thindown effect" at the end of an ion's path (Katz, 1986). However, in none of the accelerator experiments so far performed, the lethal effect extends as far as it was observed in the space experiments.

3) The inactivation cross section from single particle analysis at accelerators roughly coincides with the one obtained from fluence effect curves and is far below the value found in the space experiment.

The dependence of inactivated spores from impact parameter points to a superimposition of two different inactivation mechanisms: A short ranged component reaching up to 1 μm that may be traced back to the δ-ray dose and a long ranged one that extends at least to somewhere between 4-5 μm off the particle's trajectory, for which additional mechanisms are conjectured, such as shock waves, UV radiation, or thermophysical events (Facius et al. 1978a).

REFERENCES

Akoev, I.G., Yurov, S.S. and Akoev, B.I., 1981, A review of comparative analysis of the biological damage induced during spaceflight by HZE particles and space hadrons. Adv. Space Res. 1 No. 14: 75-81.

Baltschukat, K. and Horneck, G., 1991, Responses to accelerated heavy ions of spores of *Bacillus subtilis* of different repair capacity. Radiat. Environ. Biophys, 30: 87-103.

Baltschukat, K., 1986, Wirkung sehr schwerer Ionen auf Sporen von *Bacillus subtilis*: Inaktivierung, Reparatur von Strahlenschäden und Mutationsauslösung, Dissertation, University of Frankfurt.

Blakely, E.A., Ngo, F.Q.H., Curtis, S.B. and Tobias, C.A., 1984, Heavy-ion radiobiology: cellular studies. Adv. Radiat. Biol. 11: 295-390.

Bücker, H., 1975, Biostack, a study of the biological effects of HZE galactic cosmic radiation. In: Biomedical Results of Apollo (eds Johnston R.S., Dietlein L.F. and Berry C.A.), NASA SP-368, pp. 343-354.

Bücker, H., 1977, Results of radiobiological spaceflight experiments, In: Proc. European Symposium on Life Sciences Research in Space, ESA SP-130, pp. 263-270.

Bücker, H. and Facius, R., 1980, Recent radiobiological findings from spaceflight and ground-based studies - An overview. Life Sci. Space Res. 18: 125-130.

Bücker, H. and Horneck, G., 1975, Studies on the effects of cosmic HZE particles on different biological systems in the Biostack expriments I and II flown onboard of Apollo 16 and 17. In: Radiation Research (eds Nygaard O.F., Adler H.I. and Sinclair W.K.), Academic Press, New York, pp. 1138-1151.

Bücker, H. Horneck, G., Allkofer, O.C., Bartholomä, K.P., Beaujean, R., Cüer, P., Enge, W., Facius, R., Francois, H., Graul, E.H., Henig, G., Heinrich, W., Kaiser, R., Kühn, H., Massué, J.P., Planel, H., Portal, G., Reinholz, E., Rüther, W., Scheuermann, W., Schmitt, R., Schopper, E., Schott, J.U., Soleilhavoup, J.P. and Wollenhaupt, H., 1973, The Biostack experiment on Apollo 16. Life Sci. Space Res. 11: 295-305.

Bücker, H., Baltschukat, K., Beaujean, R., Bonting, S.J., Delpoux, M., Enge, W., Facius, R., Francois, H., Graul, E.H., Heinrich, W., Horneck, G., Kranz, A.R., Pfohl, R., Planel, H., Portal, G., Reitz, G., Rüther, W., Schäfer, M., Schopper, E. and Schott, J.U., 1984, Advanced Biostack, Expriment 1 ES 027 on Spacelab-1, Adv. Space Res. 4 No. 10: 83-90.

Butts, J.J. and Katz, R., 1967, Theory of RBE for heavy ion bombardement of dry enzymes and viruses. Radiat. Res., 30: 855-871.

Enge, W., Beaujean, R., Bartholomä, K.-P. and Fukui, K., 1974, The charge spectrum of heavy cosmic ray nuclei measured in the Biostack experiment aboard Apollo 16 using plastic detectors. Life Sci. Space Res. 12: 52-56.

Facius, R., Bücker, H., Hildebrand, D., Horneck, G., Höltz, G., Reitz, G., Schäfer, M. and Toth ,B., 1978a, Radiobiological results from the *Bacillus subtilis* Biostack experiments within the Apollo and ASTP spaceflights. Life Sci. Space Res. 16: 151-156.

Facius, R., Bücker, H., Reitz, G. and Schäfer, M., 1978b, Radial dependence of biological response of spores of *Bacillus subtilis* around tracks of heavy ions. In: Proc. 6th Symp. on Microdosimetry (eds Booz J. and Ebert H.G.) EUR 6064 DE-EN-FR, Harwood Academic Publishers Ltd., London, pp. 977-986.

Facius, R., Bücker, H., Horneck, G., Reitz, G. and Schäfer, M., 1979, Dosimetric and biological results from the *Bacillus subtilis* Biostack experiment within the Apollo-Soyuz Test Project. Life Sci. Space Res. 17: 123-128.

Facius, R., Reitz, G., Bücker, H., Nevzgodina, L.V., Maximova, E.N., Kaminskaya, E.V., Vikrov, A.I., Marenny, A.M. and Akatov, Yu.A., 1990, Reliability of trajectory identification for cosmic heavy ions and cytogenetic effects of their passage through plant seeds. Nucl. Tracks Radiat. Meas. 17: 121-132.

Facius, R., Schäfer, M., Baltschukat, K. and Bücker, H., 1983, Inactivation probability of heavy ion-irradiated *Bacillus subtilis* spores as a function of the radial distance to the particle's trajectory. Adv. Space Res. 3: No. 8, 79-84.

Gass, K.G., Hill, T.C., Goulian, M., Strauss, B.S. and Cozzarelli, N.R., 1971, Altered desoxyribonucleic acid polymerase activity in a methyl methane-sulfonate-sensitive mutant of *Bacillus subtilis*. J. Bacteriol. 108: 364-374.

Gould, G.W., 1978, Practical implications of compartimentalization and osmotic control of water distributions in spores. in: Spores VII (eds. Chambliss G. and Vary J.C.) American Society of Microbiology, Washington D.C., pp. 21-26.

Grahn, D., Ed., 1973, *HZE Particle Effects in Manned Spaceflight*. Radiobiological Advisory Panel, Committee on Space Medicine, National Academy of Sciences, National Academy Press, Washington D.C..

Hagen, U., 1989, Radiation biology in space: a critical review. Adv. Space Res. 9: (10)3-(10)8.

Haymaker, W., Look, B.C., Benton, E.V. and Simmonds, R.S., 1975, The Apollo 17 pocket mouse experiment (BIOCORE) In: Biomedical Results of Apollo (eds Johnston R.S., Dietlein L.F. and Berry C.A.), NASA SP-368, pp. 381-403.

Hoch, J.A., Barat, M. and Anagnostopoulos, G., 1967, Transformation and transduction in recombination defective mutants of *Bacillus subtilis*. J Bacteriol. 89: 1925-1937.

Horneck, G., 1988, Cosmic ray HZE particle effects in biological systems: results from experiments in space. In: Terrestrial Space Radiation and Its Biological Effects (eds. McCormack P.C., Swenberg C.E. and Bücker H.) Plenum Press, New York, pp.129-152.

Horneck, G., 1992, Radiobiological experiments in space: a review. Nucl. Tracks Radiat. Meas. 20: 185-205.

Horneck, G., Facius, R., Enge, W., Beaujean, R. and Bartholomä, K.P., 1974, Microbial studies in the Biostack experiment of the Apollo 16 mission: germination and outgrowth of single *Bacillus subtilis* spores hit by cosmic HZE particles. Life Sci. Space Res. 12: 75-83.

Horneck, G. and Bücker, H., 1983, Inactivation, mutation induction and repair in *Bacillus subtilis* spores irradiated with heavy ions. Adv. Space Res. 3(No. 8): 79-84.

Horneck, G., Schäfer, M., Baltschukat, K., Weisbrod, U., Micke, U., Facius, R. and Bücker, H., 1989, Cell inactivation, repair and mutation induction in bacteria after heavy ion exposure: results from experiments at accelerators and in space. Adv. Space Res. 9: (10)105-(10)116.

Katz, R., 1986, Biological effects of heavy ions from the standpoint of target theory. Adv. Space Res. 6: (11)191-(11)198.

Katz, R., Ackerson, B., Hamayoonfaar, M. and Sharma, S.C., 1971, Inactivation of cells by heavy ion bombardements. Radiat. Res. 47: 402-425.

Kellerer, A.M., 1977, Microdosimetric concepts relevant to HZE-particles. In: Life Sciences Research in Space, ESA SP-130, pp. 271-277.

Kiefer, J., 1985, Review: Cellular and subcellular effects of very heavy ions. Int. J. Radiat. Biol. 48: 873-892.

Kraft, G., 1987, Radiobiological effects of very heavy ions: Inactivation, induction of chromosome aberrations and strand breaks. Nuclear Science Applications 3: 1-28.

Leith, J.T., Ainsworth, E.J. and Alpen, E.L., 1983, Heavy-ion radiobiology: Normal tissue studies. Adv. Radiat. Biol. 10: 191-236.

Micke, U., Schott, J.-U., Horneck, G. and Bücker, H., 1988, Heavy ion radiation effects on single spores of *Bacillus subtilis*. In: Terrestrial Space Radiation and its Biological Effects (eds. McCormack P.C., Swenberg C.E. and Bücker H.) Plenum Press, New York, pp. 193-196.

Okubo, S. and Yanagida, T., 1968, Isolation of a suppressor mutant in *Bacillus subtilis*. J. Bacteriol. 95: 1187-1188.

Osborne, W.Z., Pinsky, L.S. and Bailey, J.V., 1975, Apollo light flash investigations In: Biomedical Results of Apollo (eds. Johnston R.S., Dietlein L.F. and Berry C.A.), NASA SP-368, pp. 355-365.

Powers, E.L., Lyman, J.T. and Tobias, C.A., 1968, Some effects of accelerated charged particles on bacterial spores. Int. J. Radiat. Biol. 14: 313-330.

Schäfer, M., Bücker, H., Facius, R. and Hildebrand, D., 1977, High precision localization method for HZE particles. Rad. Effects 34: 129-130.

Schäfer, M., Facius, R., Baltschukat, K. and Bücker, H., 1981, Contribution of ion-kill and of d-electrons to inactivation cross sections of *Bacillus subtilis* spores irradiated with very heavy ions. In: Proc. 7th Symp. Microdosimetry, Vol. II (eds. Booz J., Ebert H.G. and Hartfiel H.D.) Harwood Academic Publ., London, pp. 1331-1340.

Simpson, J.A., 1983, Elemental and isotopic composition of the galactic cosmic rays. *Ann.* Rev. Nucl. Part. Sci. 33: 323-381.

Takahashi, T., Yatagai, F. and Kitayama, S., 1983, Effect of heavy ions on bacterial spores. Adv. Space Res. 3(No. 8): 95-104 .

Takahashi, T., Yatagai, F. and Matsuyama, A., 1980, Possible long range effects in the inactivation of bacterial cells by heavy ions. Scientific Papers of the Institute of Physical and Chemical Research 74: 51-58.

Todd, P. and Tobias, C.A., 1974, Cellular radiation biology. In: Space radiation biology and related topics (eds. Tobias C.A. and Todd P.) Academic Press, New York, pp. 142-187.

Weisbrod, U., Bücker, H., Horneck, G. and Kraft, G., 1992, Heavy ion effects on bacterial spores: the impact parameter dependence of the inactivation. Rad. Res. (in press).

BIOLOGICAL ACTION OF SINGLE HEAVY IONS ON INDIVIDUAL YEAST CELLS

Michael Kost and Jürgen Kiefer

Strahlenzentrum der Justus-Liebig-Universität
6300 Gießen, Germany

ABSTRACT

Diploid yeast cells placed on nuclear track detector foils were exposed to heavy ions of different atomic number and energy. To assess the action of one single heavy ion on an individual yeast cell, track parameters and the respective colony forming capabilities were determined with the help of computer aided image analysis. The results show that the biological action depends on atomic number and specific energy of the impinging heavy ions. A comparison with the results of a model calculation is presented.

INTRODUCTION

Four years ago, at a similar ASI-meeting held at Korfu, Greece, we presented the essentials of our experimental method, which is in principle based on the known Biostack-concept (Bückel and Facius, 1981), and could show very first results (Kost and Kiefer,1988). In the meantime we improved our method of analysis and in this papaer we will focus on the presentation of some more results and on an attempt to explain the experimental data on the basis of a recent track structure model. There is to stress once more the question: Is there something special concerning the energy deposition pattern of heavy ions? There is some evidence that not only the amount of energy deposited along the particle path, commonly given by the LET, is of importance but also how this energy is deposited. This "how" is described by the track structure and here the penumbra plays a very special role. In fact it is rather impossible to investigate track structure effects in detail with whole cell populations and (globally applied) high particle fluences, therefore the method of analysing single-particle-single-cell effects was chosen. This method is, on the other hand, practically relevant for the estimation of the potential hazards in manned space missions, caused by galactic heavy particles. Because of their very small number this method is the only possible way to go in space experiments.

MATERIALS AND METHODS

Cells and culture conditions

Diploid wild type yeast cells, Saccharomyces cerevisiae, are placed as monolayers on nuclear track detectors of 3.5cm diameter (CN (Kodak) or LEXAN, depending on ion type and energy), embedded in a thin layer of nonnutrient agarose gel, completed by D-Trehalose, providing a suitable protection against damage by the subsequent drying procedure. The cells

Biological Effects and Physics of Solar and Galactic Cosmic Radiation, Part A, Edited by C.E. Swenberg *et al.*, Plenum Press, New York, 1993

were then treated by a soft drying procedure, consisting of a predrying step under a temperature of 30°C for 0.5 hour and a subsequent 6-8 hours lasting main drying step under influence of silicagel and in Argon-atmosphere at 0°C.

After irradiation, the test for colony forming ability was performed by overlaying another layer of nutrient agar, once again completed by Trehalose, on top of the first one. After 6 hours of incubation under growth conditions, surviving cells should have formed a microcolony consisting of at least 4 cells, otherwise the origin-cell was regarded to be inactivated.

Irradiation and dosimetry

All ion exposures were performed at the UNILAC heavy ion linear accelerator of the "Gesellschaft für Schwerionenforschung" GSI, Darmstadt, Germany. The particle fluence was determined by means of a secondary electron monitor, calibrated by counting tracks in suitable detectors (glass for argon and heavier ions, CR39 for neon-ions). Specifically for the "single ion - single cell-method", with track detectors serving as cell carriers, the exact fluence could be determined independantly for each sample.

Method of analysis

The method of analysis is based on a computer-aided image analysing equipment (Fig. 1). A microscope-coupled CCD-camera (512x512 pixels resolution, 256 grey shades) provides the image data for a real-time frame grabber. Storing of pictures and analysing steps are managed by an IBM-compatible PC. The biological samples, consisting of a track detector with a biological layer as described already in the previous section, were irradiated, then an additional layer of nutrient agar was put on the area of interest, providing on the one hand side a suitable growth substrate, on the other side the quality of microscopic images could be improved.

Colony forming ability was tested by incubating under growth conditions for 6 hours, based on the assumption, that after a lag time of about 2 hours, the cells can divide two times forming a colony of 4 cells. Cells which formed such a colony after the above mentioned time were accepted as survivors. After removal of the biological layer, track-etching in 6N NaOH (60°C), with etching time depending on particle type and energy as well as on detector material, revealed the latent particle tracks.

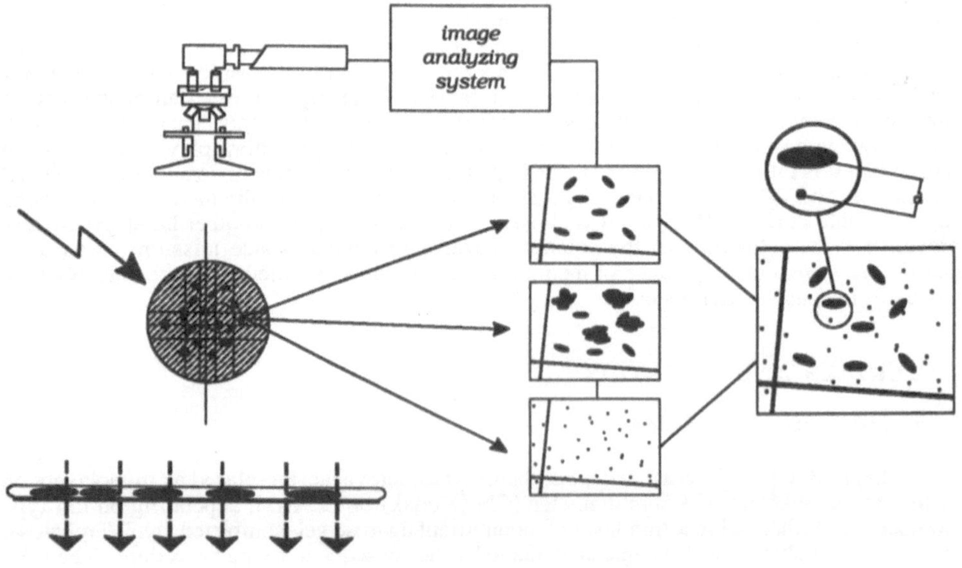

Figure 1. The Principle of the Experimental Method.

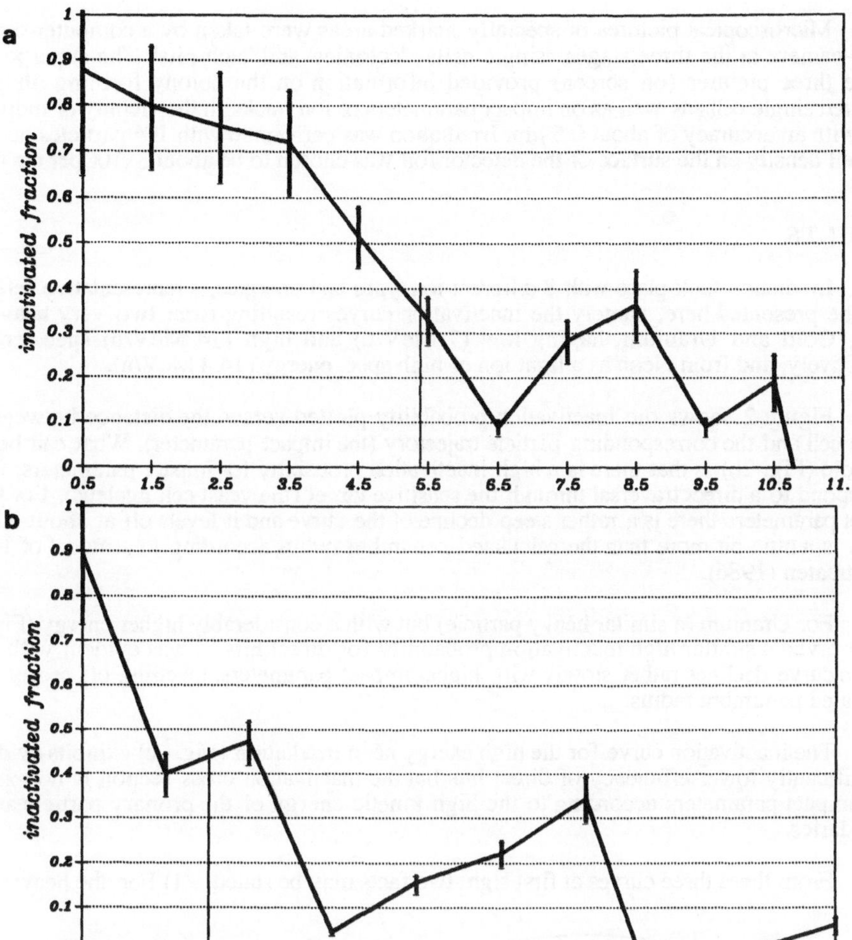

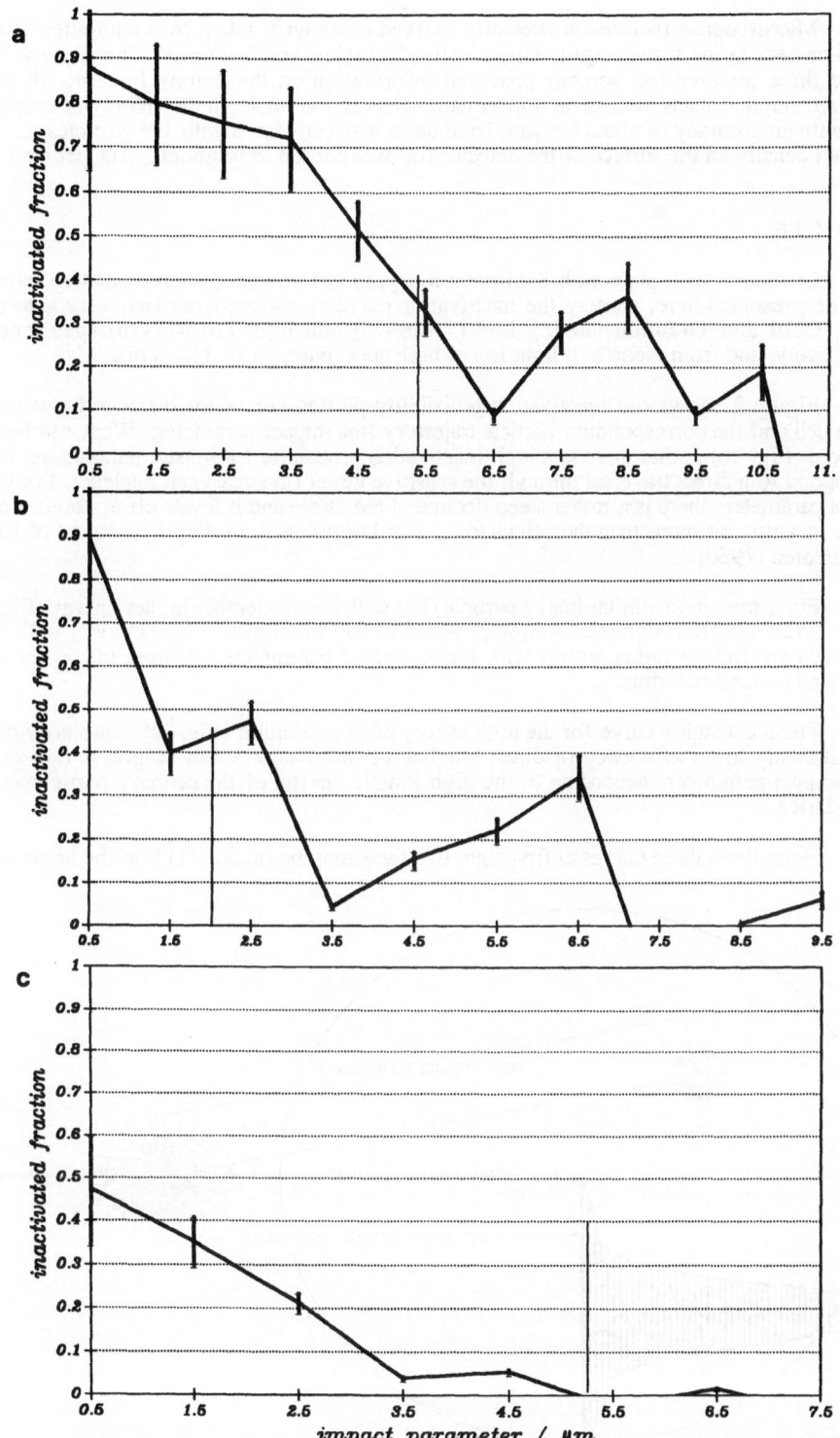

Fig.2. Normalized inactivated fraction of yeast cells for individual impact parameters after irradiation with a) 14 MeV/u Uranium-, b) 7 MeV/u Gold- and c) 14.4 MeV/u Neon-ions.

Microscopical pictures of specially marked areas were taken by a computer-coupled video-camera at the three stages: 'single cells', 'colonies' and 'etch-pits'. The superposition of the three pictures (on screen) provided information on the colony forming ability of specified single cells as well as on impact parameters of ion tracks in the vicinity of individual cells with an accuracy of about 0.5 μm. Irradiation was performed with 10^6 particles per cm^2. The cell density on the surface of the detector foil was chosen to be about $3x10^6$ per cm^2.

RESULTS

Irradiation took place with 8 different ion types and energies, a representative selection shall be presented here, namely the inactivation curves resulting from two very heavy ion types, Gold and Uranium, having low (7 MeV/u) and high (14 MeV/u) spec. energy, respectively, and from Neon as a light ion of high spec. energy (14.4 MeV/u).

Figure 2 shows the inactivation probability plotted versus the distance between one single cell and the corresponding particle trajectory (the impact parameter). What can be seen for Gold (Fig. 2b) is that there is a high inactivation probablity for impact parameters, which correspond to a direct traversal through the sensitive target (the yeast cell nucleus). For larger impact parameters there is a rather steep decline of the curve and it levels off at about 3.5μm, which is a little bit more than the calculated penumbra radius according to a model of Kiefer and Straaten (1986).

For Uranium (a similar heavy particle) but with a considerably higher energy (Fig. 2a) we observed a similar high inactivation probability for direct hits as was evident with gold, but the curve declines rather slowly with higher impact parameters, levelling off nearly at the calculated penumbra radius.

The inactivation curve for the high energy neon irradiation (Fig. 2c) exhibits evidently a significantly lower efficiency of direct hits but the inactivation cross section is non-zero at high impact parameters according to the high kinetic energy of the primary particle and its secondaries.

From these three curves at first sight two facts may be stated: 1) For the heavier ions

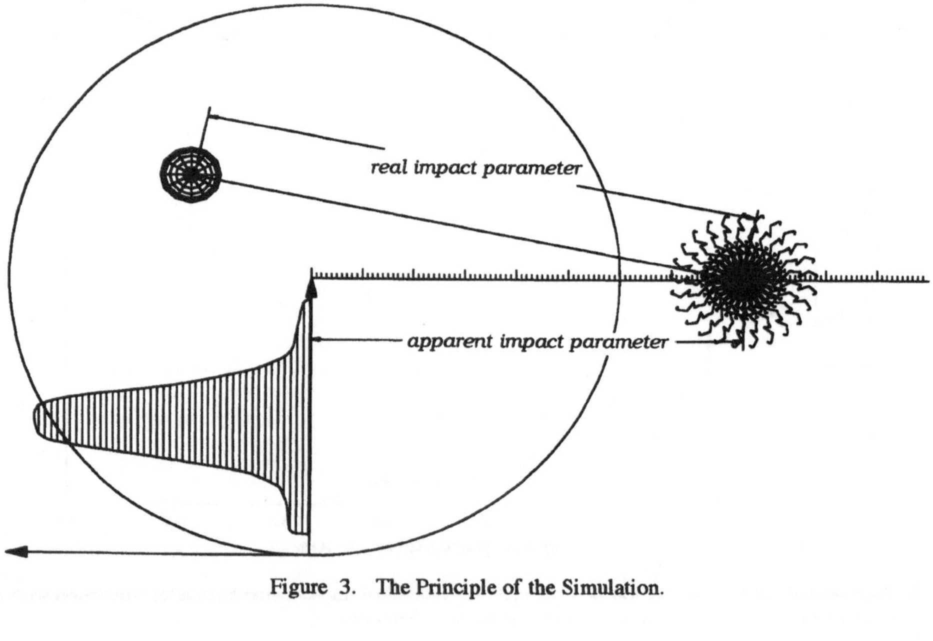

Figure 3. The Principle of the Simulation.

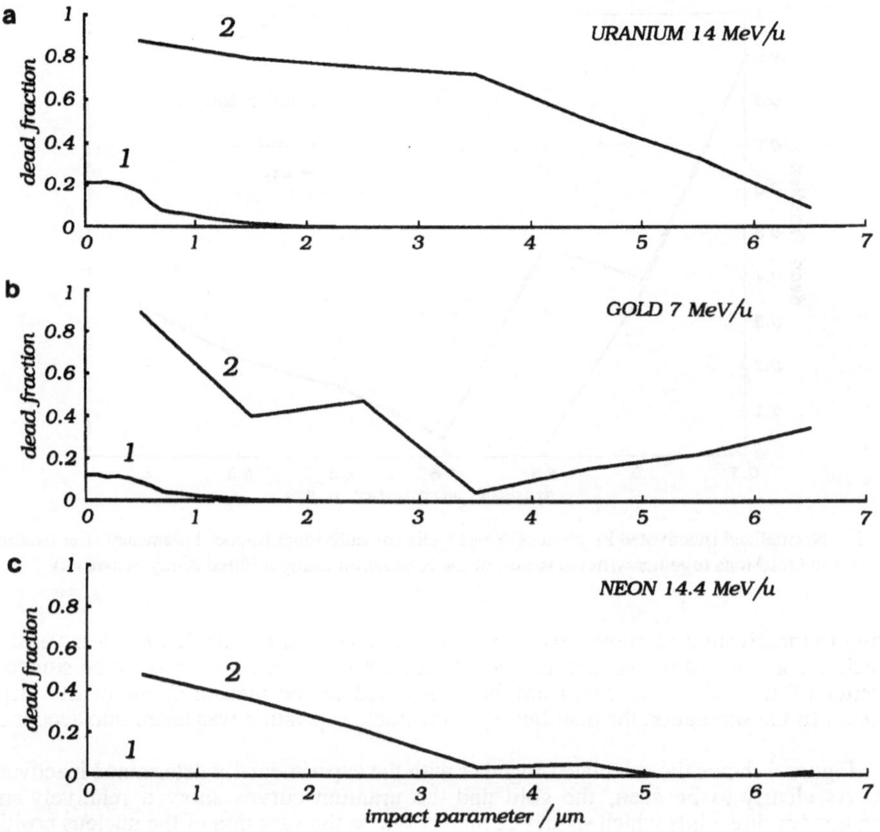

Fig. 4. Normalized Inactivated Fraction of Yeast Cells for Individual Impact Parameters after Irradiation with 14 MeV/u Uranium-, 7 MeV/u Gold- and 14.4 MeV/u Neon-ions (2), together with the Results of the Simulation (1).

there is a high inactivation for small impact parameters which correspond to a direct hit of the sensitive target (but substantially smaller than one), and a low inactivation capacity for the light heavy ion in this impact parameter range; 2) The radius of the sensitive target is 0.6μm. Obviously there is an inactivation possible, even when the particle itself doesn't hit the cell nucleus. That means there is a far reaching action of the ion via the penumbra electrons. Its maximum range seems to be correlated with the particle kinetic energy.

CONCLUSIONS

The question now arises whether it is possible to reproduce these experimental results with the help of an actual track structure model. For that reason a simulation of the action of a single particle on a single yeast cell in its dependence of the impact parameter by means of a rather simple computer program was performed (Fig. 3).

For this simulation several parameters are needed. The amount of energy which is deposited in the sensitive site by the particle calculated according to the model of Kiefer and Straaten. In addition one needs to calculate the response of the biological object to this amount of energy, which gives the degree of inactivation has to be provided. This should be determined by the sensitivity of the yeast strain against X-irradiation because, if the damage

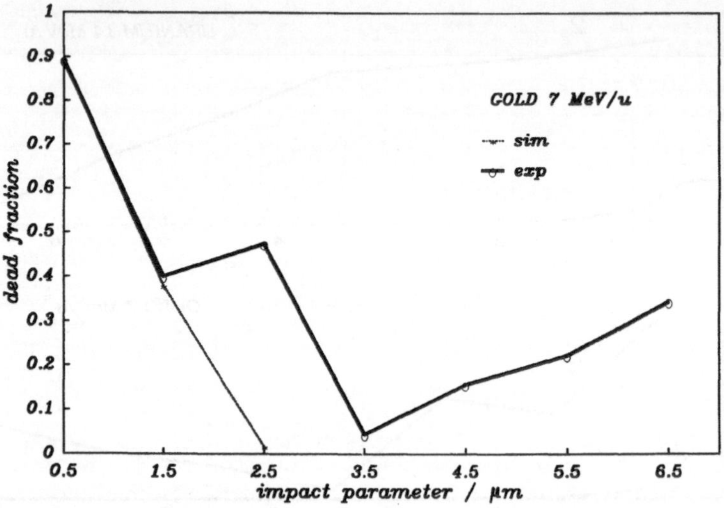

Figure 5. Normalized Inactivated Fraction of Yeast Cells for Individual Impact Parameters after Irradiation with 7 MeV/u Gold-ions together with the Result of the Simulation Using a Fitted X-ray Sensitivity.

resulting in inactivation is caused exclusively by electrons, as is with X-rays. For yeast this approach is complicated by the fact that the cell nucleus may not be assumed to be situated in the center of the cell. This could not be considered in the measurement of the impact parameter. In the simulation the distribution of the nucleus position was taken into account.

Figure 4 shows the calculated together with the experimentally determined inactivation curve. As clearly to be seen, the gold and the uranium curves show a relatively small inactivation for direct hits which should be mainly due to the variation of the nucleus position. The range of inactivation is also small and nearly independent of the calculated penumbra radius, but that could be expected from the model description of the functional dependence of the energy density on the distance from the track center. With neon no significant inactivation can be detected.

There may be several reasons for these large discrepancies: One of the most promising explanations is the assumption that using the X-ray sensitivity in the simulation of the cell response to this special energy deposition pattern may not be justified.

Suppose that the biological effectiveness of heavy-ion-generated electrons is larger than of those electrons produced by X- or gamma-rays then we may assume a higher sensitivity of the cells to this special kind of energy deposition.

If this is taken into account in the calculation then an accordance for small impact parameters can be gained; this is apparent with the gold-example (see Fig. 5). With larger impact parameters the assumption of a higher biological effectivity is not justified because here the origin of the remaining few electrons need not be of importance. This means that the high effectivity with larger impact parameters can not be explained.

Our conclusions can be stated as follows: a) The question, whether direct hits with very heavy ions must definitely lead to 100% inactivation could not be clearly answered; b) The large distance action of the heavy ion via the penumbra electrons could be experimentally confirmed; and c) Simulation of the experimental data by means of a model calculation implies that the expected biological effect cannot be described by a model which attributes the biological effectiveness of a radiation type exclusively to the energy density in the sensitive target deposited by delta electrons.

ACKNOWLEDGEMENTS

This work was supported by the "Gesellschaft für Schwerionenforschung" (GSI), Darmstadt, Germany and by the "Bundesministerium für Forschung und Technologie", Germany. We thank Günther Lenz from GSI for for his help at the accelerator facility.

REFERENCES

Bücker H, and Facius R., The role of HZE , 1981, Particles in space flight: results from spaceflight and ground-based experiments. Acta Astronautica , 8: 1099.

Kost M., and Kiefer J., Biological action of heavy ion irradiation on individual yeast cells. In: Terrestrial space radiation and its biological effects (P.D. McCormack, C.E. Swenberg, H. Bücker, eds.) Plenum Press (1988).

Kiefer J., and Straaten H., 1986, A model of ion track structure based on classical collision dynamics. Phys Med Biol 31: 1201.

REPAIR OF RADIATION INDUCED GENETIC DAMAGE UNDER MICROGRAVITY: AN EXPERIMENTAL PROPOSAL

H.-D. Pross, M. Kost and J. Kiefer

Strahlenzentrum der Justus-Liebig-Universität
Leihgesternerweg 217, 6300 Giessen, Germany

ABSTRACT

The aim of the suggested experiment is to study the influence of microgravity on the repair of radiation induced genetic damage. A mutant of yeast Saccharomyces cerevisiae (rad 54-3), which is temperature-conditional for repair, is exposed to radiation and microgravity and incubated in space at the permissive and restrictive temperature. Thus it is possible to assess the influence of microgravity on cellular repair processes. Ground experiments have already shown the applicability of this concept.

INTRODUCTION

During space flight astronauts are subjected to a multitude of stimuli of various kinds and intensities; radiation and microgravity being two of them. These influences cannot be considered separately. With regard to possible interactions, missions like COSMOS 368, 690 and 782, Soyus/Saljut, Gemini, Biosatelite II and the D1-mission showed inconsistent results with tendencies to synergistic effects (Horneck, 1987).

Unfortunately different kinds of biological systems were used (e.g. microorganisms, air-dried plant seed, beetles, wasps and rats) and the results with respect to repair-mechanisms have been fairly inconclusive due to the complexity of the systems and ill-defined conditions.

The importance of the topic "radiation and microgravity" becomes evident by considering the dosimetry data from U.S. and Soviet missions where doses of more than 70 mGy were received by the astronauts (Benton and Parnell, 1987). An impairment of repair processes by conditions during space flight could thus have severe implications on human health.

In view of indications that repair mechanisms may be delayed or inhibited under microgravity (Horneck, 1987), an experiment is proposed, which addresses answers to these questions in a clear-cut fashion although only for a simple eucaryotic organism (Kiefer, 1987; Kost and Kiefer, 1987).

EXPERIMENTAL APPROACH

Cell line

The cell-line utilised is the diploid yeast Saccharomyces cervisiae, (rad 54-3). Cells of this strain repair DNA double-strand breaks when incubated at 23°C but fail to do so when grown at 36°C. This allows one to assess easily the repair ability under reproducible conditions (Budd and Mortimer, 1982).

Exposure conditions

Stationary cells are collected on membrane filters (0.6 μm pore size) at a density of 5.6 x 10^6 cells/cm^2. During transport and exposure, filters are kept on nutrient-free agar. At the end of the mission cells are recovered from the filters by suspending them in growth medium (M. Frankenberg-Schwager et al.,1980). This growth medium contains succinic acid sodium(Na_2)-salt instead of glucose, providing a more pronounced slope difference between the two temperature curves (D.Frankenberg et al., 1984).

Cells are irradiated (at doses up to 140 Gy) with X-rays on ground, then put into containers and held at +4°C until arriving in orbit. There they are incubated at the two temperatures, so that repair is only operative under microgravity. For landing and transport back to the laboratory the temperature is again decreased to +4°C.

Survival assay

Cells are held eight days on non-nutrient agar; conditions that allow assessment of delayed-plating recovery (DPR) (Frankenberg-Schwager M. et al., 1980). The amount of repair can be evaluated by comparing the survival (colony forming ability) of cells kept under either restrictive or permissive temperature.

Temperature schedule

During transport to the launch site and back to the laboratory, where survival is analyzed the experimental containers were stored at +4°C in order to inhibit cell activity. Under microgravity conditions cells are incubated at either 23°C or 36°C in order to allow or inhibit the repair of the DNA damage.

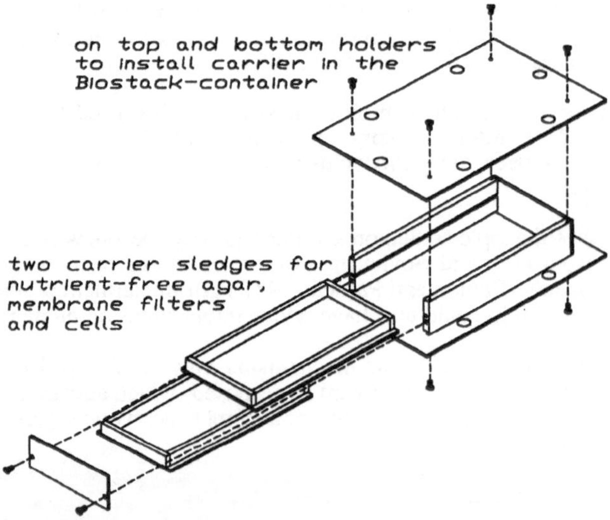

Figure 1. Main Sample Container. Inside Each Sledge, a Net of Polyemid is Installed and Moulded up with Nutrient-free Agar; on the Agar Surface are Membrane Filters with the Irradiated Cells.

Experimental container

A containment is needed that is capable of providing survival conditions during the entire mission and in addition allows for changes of the inside temperature in relative short time periods. It has to be made of material that is non-problematic with respect to crew security, the shuttle itself and the cells. Therefore a special alloy ($AlMgSi_1$) was used along with the plastic Makrolon and screws of special steel (V_4A). The container is schematically described in figure 1.

In order to keep agar, filters and cells in their position in the container a Polyemid net is installed in the sledges and moulded up with the nutrient free agar. Futhermore it is possible to increase the agar viscosity by adding polymeric molecules. The dimensions are chosen in such a way that the container fits into a type II/O - BIORACK-container.

PRELIMINARY RESULTS

The method described above was tested in ground based laboratory experiments. Cells were exposed to graded doses of X-rays and then kept in closed containers at different temperatures for up to 8 days. Results obtained are shown in figure 2 which demonstrate clearly the feasability of the proposed approach.

DISCUSSION

With the concept outlined above it is possible to study the repair of radiation induced genetic damage in a clear-cut way at the cellular level of a simple but well investigated cellular system . One should be able to examine exactly one specific repair-mechanism (for example: double strand breaks). The influence of microgravity - if existent - is then expressed as a difference between ground-control and space samples in the slopes of the 23°C survival curves. Our preliminary ground experiments already demonstrate the feasibility of this approach.

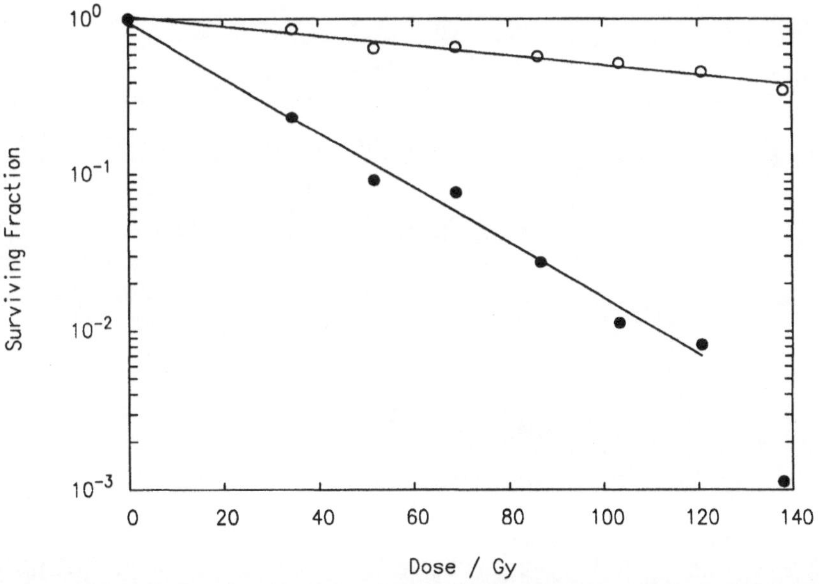

Figure 2. Survival of Saccharomyces Cerevisiae Rad54-3, Incubated for 8 Days at the Permissive (23°C, Open Circles) and Restrictive (36°C, Closed Circles) Temperature under Non-nutrient Conditions and Afterwards Plated on Nutrient Agar.

REFERENCES

Benton, E.V. und T.A. Parnell, 1987. Space radiation dosimetry on U.S. and Soviet manned mission.In: Terrestrial Space Radiation and Its Biological Effekts, edited by P.D. McCormack, C.E. Swenberg, H. Bücker (New York: Plenum Press), 729-794.

Budd, M., and R.K Mortimer, 1982. Repair of double-strand breaks in a temperature conditional radio-sensitive mutant of S.c.Mutat. Res. 103: 19-24.

Frankenberg D., M. Frankenberg-Schwagerand and R. Harbich, 1984. Interpretation of the shape of survival curves in terms of induction and repair/misrepair of DNA double-strand breaks. Br. J. Cancer 49: 233-238.

Frankenberg-Schwager, M., D. Frankenberg , D. Blöcher and C. Adamczyk , 1980. Repair of DNA double-strand breaks in irradiated yeast cells under nongrowth conditions.Radiat. Res. 82: 498-510.

Horneck, G., 1987. Impact of Space Flight Environment on Radiation Response. In: Terrestrial Space Radiation and Its Biological Effekts, edited by P.D. McCormack, C.E. Swenberg, H. Bücker (New York: Plenum Press), 707-714.

Kiefer, J., 1987. UV response of the temperature-conditional rad 54 mutant of the yeast Saccharomyces cerevisiae. Mutation Res., 191: 9-12.

Kiefer J., 1987. Heavy Ion Effects on Cells.In: Terrestrial Space Radiation and Its Biological Effekts, edited by P.D. McCormack, C.E. Swenberg, H.Bücker (New York: Plenum Press), 117-128.

Kost , M. and Kiefer J., 1987. Biological Action of Heavy Ion Irradiation on Individual Yeast Cells.In: Terrestrial Space Radiation and Its Biological Effekts, edited by P.D. McCormack, C.E. Swenberg, H. Bücker (New York: Plenum Press), 197-204.

DNA DAMAGE REPAIR AND MUTAGENESIS IN MUTANTS OF *ESCHERICHIA COLI*

S. I. Ahmad

Nottingham Polytechnic
Dept. of Life Sciences
Nottingham NG11 8NS, England

INTRODUCTION

The DNA of most living organisms is persistently exposed to a variety of deleterious agents. The agent can be physical (e.g. UV light and ionizing radiations) chemical (e.g. ethyl methane sufonate, nitrosourea and mitomycin C; they are too many to be listed) or a combination of the two (e.g. UVA + hydrogen peroxide and UVA + 8-methoxypsoralen). Also certain physiological conditions have been known to induce DNA damage e.g. thymine starvation of thymine auxotrophs and heat shock (Friedberg, 1985; Smith and Burton, 1965; Anathaswamy and Eisenstark, 1976; Smith, 1988; Smith, 1966; Lindahl, 1982; Johansen, et al., 1971).

Different agents damage DNA differently. Damage to DNA can broadly be classified in 5 major types: a monoadduct (involving a single base), a diadduct (involving two bases on the same strand of DNA), DNA strand breaks (single and double strand breaks), inter-strand DNA cross-links (e.g. binding of 8-methoxypsoralen or mitomycin C to DNA) and DNA to protein cross-links.

The mechanisms of induction of DNA damage by different agents vary. The damage can be specific such as binding of 8-methoxypsoralen (PS) in the presence of UVA (UV of 310-400 nm) or mitomycin C (MC) molecules to DNA (Smith, 1988), or non-specific for example UVC light (UV of below 290 nm) and ionizing radiations can damage DNA in a variety of ways including induction of a variety of changes and single and double-strand DNA breaks (Smith, 1966).

DNA is so vital a cellular constituent its integrity is considered to be of prime importance. For the last 3 decades DNA damage and repair have been the subject of exhaustive study and a vast wealth of data have been collected from a number of living organisms. With regards to DNA damage repair and mutagenesis *Escherichia coli* is the best understood organism and many good reviews are available on this subject (Friedberg, 1985; Volkert, 1988; Hanawalt et al., 1979; Eisenstark, 1989; Walker, 1984; Webb, 1977). Due to the existence of a variety of agents inducing a variety of damage *E. coli* and perhaps most living organisms have evolved a number of repair systems to overcome the problem.

Certain DNA repair systems are considered to be constitutive while others are inducible. The SOS system in *E. coli* is known to be inducible. It is likely that this system comes into operation only when a cell encounters heavy DNA damage. Using this system the

Biological Effects and Physics of Solar and Galactic Cosmic Radiation, Part A, Edited by C.E. Swenberg *et al.*, Plenum Press, New York, 1993

cell can overcome the problem more efficiently but it has to pay the price; the rate of mutation among survivors is increased.

Most studies on DNA damage repair and mutagenesis have made use of mutant organisms unable to repair specific types of DNA damage. Phenotypically such mutants are sensitive to one or more DNA damaging agents. Until recently over 65 different mutants of *E. coli* have been isolated which are traceable directly or indirectly to changes in DNA repair (Bachmann, 1990).

In our studies on DNA damage and repair we have isolated and analysed a set of novel mutants of *E. coli* to explore these phenomena. The mutants were orginally isolated as hyper-resistant to UVC (SA236), UVA plus PS (SA270), UVA (SA330) and caffein (SA331). In this report genetic and biochemical analyses of some of these mutants are presented.

MUTANT OF *E. COLI* HYPER-RESISTANT TO UVC

Analysis of the mutant bacteria, SA236, showed that it has simultaneously gained hyper-resistance ability to a number of other damaging agents including UVA, X-rays, UVA + H_2O_2, nalidixic acid, novobiocin, fluorouracil, MC, high concentrations of H_2O_2 and thymineless death (Ahmad, et al., 1980). On the other hand it is more sensitive to ethyl methane sulfonate (EMS), methyl methane sulfonate (MMS) and nitrosoguanidine (NMMG) than its wild type parent strain. Measurement of its growth rate in minimal synthetic medium and microscopic examination showed that the mutant strain does not differ from its wild type parent strain. However, when grown on nutrient agar plates the mutant produces smaller colonies compared to its wild type parent strain and when it is plated on Luria agar larger colonies appear which normally lose the mutational phenotypes.

Further studies showed that SA236 is hyper-mutagenic, both spontaneously and in response to UVC, EMS, MMS and NMMG. Also recombinogenecity of the mutant bacteria is increased (our unpublished data).

Biochemical analysis of the mutant bacteria has shown that certain enzymes of DNA repair are present in higher concentrations than in its wild type parent strain. These are DNA polymerase-I, endonuclease-I and exonuclease-III whose activities are significantly increased while the levels of exonuclease-I, exonuclease-II and DNA ligase are normal as in the wild type parent strain (Ahmad et al., 1980).

It was conceivable that the increased activity of these enzymes in the mutant strain was caused by conformational changes in the proteins. Alternatively the level of expression of various operons was increased. To understand this we studied the *polA* operon in the mutant strain SA236. Recombinant plasmids carrying various sections of the *polA* regulatory regions linked to a *galK* gene (Joyce et al., 1982) were introduced into the mutant strain. Analysis of galactokinase activity showed that it is over-expression of the *polA* operon that leads to enhanced activity of the enzyme (Ahmad and van Sluis, 1987).

Based on this study a model for the regulation of the synthesis of DNA polymerase-I in *E. coli* is proposed (Ahmad and van Sluis, 1987). It is suggested that the *polA* operon in *E. coli* is negatively controlled, analogous to the control mechanism observed for the *Lac* operon. The regulatory region of *polA* contains two putative promoters, P1 and P2 and two putative operators, 01 and 02. These operators and promoters are affected by a represser protein synthesised by the *uvh* (UV hyper-resistant) allele located near the *argH* locus (details of *uvh* are provided in the next paragraph). In the wild type strain the *uvh* product binds to both operators, 01 and 02 and a basal level synthesis of DNA polymerase-I ensues. It is, furthermore, proposed that between 01 and P2 is located a site for the binding of *DnaA* protein (Fuller, et al., 1984). Thus in the wild type *E. coli* the expression of the *polA* operon is controlled by two kinds of represser molecules. One is the *uvh* product and the other is produced from *DnaA*. Over-expression of *polA* in the mutant bacteria is proposed to be due to the loss of *uvh* gene product resulting in a reduced repression of the operon.

Genetic analysis of SA236 revealed that there are at least two mutations located on the chromosome associated with the UV damage. One is *uvh* located at 89 min on the chromosome map of *E. coli* and confers hyper-resistance to UVC. The other mutation is located at 90 min conferring sensitivity to UVC (our unpublished data). The UV sensitive phenotype of the mutant bacterium is expressed only when this mutation is transferred to a wild type strain. When the two mutations are present in the same strain (e.g. in SA236) the UV hyper-resistant mutation is expressed i.e. *uvh* is dominant over the UV-sensitive mutation. The significance of the the uv-sensitive mutation in strain SA236 is not clear.

Introduction of a typical *recA* - mutation in SA236 leads the cell to become as sensitive to UV as a typical *recA*- mutant of *E. coli.* The result implies that the *uvh* allele might be a part of the SOS inducible system. The SOS inducible system in *E. coli* is controlled by two genes, *recA* and *lexA*. A class of *lexA* mutant of *E. coli* (known as *lexA(Def)*) has been isolated showing constitutive expression of the SOS enzymes (Mount, 1977). This mutant, however, has not been reported to show enhanced resistance to UV. One likely reason for this may be that induction of the SOS regulons may not be sufficient to promote efficient DNA repair.

Based on these and our other unpublished results it is proposed that in *E. coli* there are at least two inducible repair systems for the UV induced hyper-repair. One is the SOS system and the other a putative UVH system. In the UVH system the *uvh* gene product is involved in the synthesis of a represser protein. This protein on one hand acts as a co-represser of the *lexA* product; thereby repressing all the SOS regulons. On the other hand this represser also represses certain other operons, not known to belong to the SOS regulons, without the help of *lexA* product (such as those having *polA* and *xth*). Mutation in the *uvh* gene leads to a loss of the repression of the SOS regulons and other non-SOS operons. Constitutive expression of the relevant enzymes results in hyper-repair of DNA and consequently hyper-resistance to UVC and other relevant DNA damaging agents.

For the UV induced hyper-mutagenecity in SA236 it is proposed that the constitutive SOS function includes the enhanced synthesis of *umuCD* and DNA polymerase-I. These two proteins may be acting jointly to introduce more mis-matched bases during repair than normally occur in a wild type strain of *E. coli.* This assumption is based on the finding that a modified DNA polymerase may be responsible for UV induced mutagenesis in *E. coli.* (Lackey et al., 1985; Rabkin et al., 1983). Furthermore, the modification of this enzyme is likely to occur by *umuCD* gene products. A likely reason for the induction of mutation in a wild type *E. coli* is transient reduction in the modified polymerase fidelity leading to enhanced incorporation of mis-matched bases in the synthesising strand and subsequent enhanced mutation frequency. It is possible that in strain SA236 the constitutive synthesis of DNA polymerase-I together with the SOS regulons including *umuCD* leads to increased amounts of modified DNA polymerase in the cell. The modified polymerase, on one hand is responsible for enhanced DNA repair and on the other introduces larger numbers of mis-matched bases in the DNA; thus increasing the mutation frequency.

For enhanced sensitivity and increased mutagenesis of SA236 by alkylating agents it is proposed that this strain carries yet another mutation(s), responsible for removing alkyl group from the DNA. This assumption is based on the fact that SA236 is sensitive to alkylating agents. Mutations in *ada, aidBC, alkAB* and *inm* have been associated with DNA damage by alkylating agents (Bachmann, 1990). When SA236 is treated with such agents the alkylated bases in DNA are retained for a longer period than in wild type cell and during DNA synthesis or repair the modified DNA polymerase finds an additional reason to introduce mis-matched bases in the synthesising strand of DNA. Further experimental data, however, are required to support the hypothesis.

MUTANT OF E. COLI HYPER-RESISTANT TO 8-METHOXYPSORALEN PLUS UVA (PUVA)

The mutant of *E. coli* showing enhanced resistance to PUVA (SA270) (Ahmad and Holland, 1985) was found to be also hyper-resistant to two other cross-linking agents, MC and nitrogen mustard (NM). For MNNG, EMS and UVC it remains as sensitive as its wild type parent strain (our unpublished data). From this it can be postulated that the repair process affecting PS induced DNA damage is different from the repair affecting UVC induced DNA damage. However, when a *recA* - mutation was introduced into SA270 the mutant derivative became as sensitive to PUVA as a typical *recA* - mutant of *E. coli* (Ahmad and Holland, 1985). The result suggests that PUVA induced DNA damage is repaired via the SOS inducible repair system. Also van Houten et al., (1986) and Cole et al., (1976) have proposed that the PUVA induced DNA damage is repaired via recombinational repair process involving *uvrABC* gene products.

SDS-PAGE analysis of total cell extract from SA270 showed that it synthesises a protein of 55-kdal at higher concentrations than its wild type parent strain. This protein was purified using FPLC and its biochemical properties analysed. Our preliminary results suggest that it is a nuclease specific for recognizing and incising PS molecules attached to DNA. Further experiments are required to determine the exact role of the protein in DNA repair.

Genetic analysis of SA270 showed that a mutation, *puvR*, is located at 57.2 min on the linkage map of *E. coli*. From this, by using a selective technique, a mutant of *E. coli* (JH110) sensitive to PUVA was isolated. Genetic analysis of this mutant showed that the PUVA-sensitive mutation, *puvA*, is also located near 57 min and is distinct from the *puvR* locus. Complementation data obtained with a diploid strain *puvR* + /*puvR* - showed that *puvR* + is trans dominant over *puv R* -. Induction studies of the 55-kdal protein showed that the enzyme is inducible by MC (Holland et al., 1991).

Analysis of a number of mutants of *E. coli* deficient in various DNA repair enzyme activities showed that the *recA* -, *rec B* - *C* -, *recF* -, *recN* -, *uvrA* - mutant and a triple mutant *recB* - *C* - *sbcB* - *recF* - were all sensitive to PUVA (Holland et al., 1991). These results lends further weight to the proposal that in *E. coli* the repair of DNA cross-links (such as induced by PS or MC) is mediated by a recombination repair system perhaps via the *RecF* pathway (Horii and Clark, 1973).

Based on the data a model for the enhanced synthesis of the 55-kdal protein and a reason for the hyper-resistance of *E. coli* mutants to PUVA is proposed (Holland et al., 1991). According to this model the PUV operon contains at least three loci, *puvR*, a putative *puvO* and *puvA* and is controlled analogous to the *Lac* operon. *PuvA* is inducible and *puvR* is responsible for the synthesis of a represser molecule that binds to *puvO*. This limits the expression of *puvA* and basal level synthesis of the 55-kdal protein ensues. In the presence of an inducer such as MC or in the mutant bacteria, SA270, the activity of this represser is reduced or eliminated and, as a result, the 55-kdal protein is synthesised in higher concentrations. For the *E. coli* mutant becoming hyper-resistant to PUVA it is proposed that the 55-kdal protein plays an important role for this phenotype. Its basal level synthesis in the wild type cell limits the cellular ability to incise and subsequently repair cross-links from the DNA. Once this limitation is eliminated the cell's efficiency to repair the damage is improved and it becomes hyper-resistant to PUVA.

From these studies we may draw the conclusion that when wild type cells of *E. coli* are challenged by DNA damaging agents a variety of hyper-resistant mutants can be isolated. These mutants usually have mutations in their regulatory genes and as a result the synthesis of relevant DNA repair enzymes is increased. The increased synthesis leads to enhanced DNA repair ability and resulting hyper-resistance. However, the cell may have to pay the price; it can become hyper-mutagenic and its growth may be affected.

This report anticipates that for most DNA damaging agents there are two types of repair systems. A low level repair system to combat low level DNA damage and a high level inducible repair system which comes into effect when there is massive DNA damage. At present we are aware of two inducible repair systems; the SOS and the adaptive repair (McCarthy, et al., 1984; Witkin, 1976). It is likely that there are several inducible systems in *E. coli* for DNA repair. These are required to be explored. Our line of experiments, involving selection of hyper-resistant mutants for given DNA damaging agents and their studies, may be a valuable tool for this purpose.

ACKNOWLEDGEMENTS

The author wishes to thank the Medical Research Council of Great Britain for providing a research grant during which the experiments described in this text were carried out. Also Dr Sandra Kirk's valuable comments on the manuscript are highly appreciated.

REFERENCES

Ahmad, S.I., Atkinson, A., Eisenstark, A. (1980). Isolation and characterization of a mutant of *Escherichia coli K12* synthesising DNA polymerase-I and endonuclease-I constitutively, J. Gen. Microbiol., 117, 419-422.

Ahmad, S.I., Holland, I.B. (1985). Isolation and analysis of a mutant of *Escherichia coli* hyper-resistant to near ultraviolet light and 8-methoxypsoralen, Mut. Res., 151, 43-47.

Ahmad, S.I., van Sluis, C.A. (1987). Inducible DNA polymerase-I synthesis in a UV hyper-resistant mutant of *Escherichia coli*, Mut. Res., 190, 77-81.

Ananthaswamy, H.N., Eisenstark, A. (1976). Near UV induced breaks in phage DNA: sensitization by hydrogen peroxide (a tryptophane product), Photochem. Photobiol. 24, 439-442.

Bachmann, B.J. (1990). Linkage map of *Escherichia coli K12*. Edn. 8, Microbiol. Rev., 54, 130-197.

Cole, R.S., Levitan, D., Sinden, R.R. (1976). Removal of psoralen inter-strand cross-links from DNA of *Escherichia coli*, mechanism and genetic control, J. Mol. Biol., 103, 39-59.

Eisenstark, A. (1989). Bacterial genes involved in response to near ultraviolet radiations, Adv. in Genetics, 26, 99-147.

Friedberg, E.C. (1985). DNA Repair, New York: W.H. Freeman and Company.

Fuller, R.S., Funnell, B.E., Kornberg, A. (1984). The *dnaA* protein complex with the *Escherichia coli* chromosomal replication origin (*oriC*) and other DNA sites, Cell, 38, 889-900.

Hanawalt, P.C., Cooper, P.K., Ganesan, A.K., Smith, C.A. (1979). DNA repair in bacteria and mammalian cells, Ann. Rev. Biochem., 48, 783-836.

Holland, J., Holland, I.B., Ahmad, S.I. (1991). DNA damage by 8-methoxypsoralen plus near ultraviolet light (PUVA) and its repair in *Escherichia coli*: genetic analysis, Mut. Res., DNA Repair, 254, 289-298.

Horii, Z.I., Clark, A.J. (1973). Genetic analysis of the *recF* pathway to genetic recombination in *Escherichia coli K12*: isolation and characterization of mutants, J. Mol. Biol., 80, 327-344.

Johansen, I., Gurvin, I., Rupp, W.D. (1971). The formation of single-strand breaks in intracellular DNA by X-rays. Rad. Res. 48, 599-612.

Joyce, C.M., Kelly, S., Grindley, D.F. (1982). Nucleotide sequence of *Escherichia coli polA* gene and primary structure of DNA polymerase-I, J. Biol. Chem., 257, 1958-1964.

Lackey, D., Krauss, S.W., Linn, S. (1985). Characterization of DNA polymerase I: a form of DNA polymerase I found in Escherichia coli expressing SOS function, J. Biol. Chem., 260, 3178-3184.

Lindahl, T. (1982). DNA repair enzymes. Ann. Rev. Biochem. 51, 61-87.

McCarthy, R.V., Kassan, P., Lindahl, T. (1984). Inducible repair of 0-alkylated DNA pyrimidins in *Escherichia coli*. EMBO J. 3, 545-550.

Mount, D.W. (1977). A mutant of *Escherichia coli* showing constitutive expression of the lysogenic induction and error-prone DNA repair pathway. Proc. Natl. Acid. Sci (USA) 74, 300-304.

Rabin, S.D., Moore, P.D., Strauss, B.S. (1983). *In vitro* by pass of UV-induced lesions by *Escherichia coli* DNA polyperase I: Specificity of nucleotide incorporation. Proc. Natl. Acad. Sci. USA., 80, 1541-1545.

Smith, B.J. and Burton, K. (1965). The integrity of deoxyribonucleic acid from *Escherichia coli 15T* - after thymineless death, Biochem. J., 97, 240-246.

Smith, C.A. (1988). Repair of DNA containing furocoumarin adducts, in: F.P. Gasparro (Ed.), Psoralen DNA photobiology, Vol II, CRC Press, Boca Raton, FL, PP 87-116.

Smith, K.C. (1966). Physical and chemical changes induced in nucleic acids by ultraviolet light. Rad. Res. Suppl. 6, 54-79.

Van Houten, B., Gamper, H., Holbrook, S.R., Hearst, J.E., Sancar, A. (1986). Action mechanism of *uvrABC* excision nuclease on a DNA substrate containing a psoralen crosslink at a defined position, Proc. Nat. Acad. Sci., (USA), 83, 8077-8081.

Volkert, M.R. (1988). Adaptive response of *Escherichia coli* to alkylation damage, Env. and Molec, Mutagensis, 11, 241-255.

Walker, G. (1984). Mutagenic and inducible responses to decoxyribonucleic acid damage in *Escherichia coli*, Microbiol. Rev., 48, 60-93.

Webb, R.B. (1977). Lethal and mutagenic effects of near ultraviolet radiations. Photochem. Photobiol. Rev. (Ed.) K.C. Smith, 2, 169-261.

Witkin, E.M. (1976). Ultraviolet mutagenesis and inducible DNA repair in *Escherichia coli*. Bact. Rev. 40,869-907.

NEOPLASTIC TRANSFORMATION INDUCED BY HEAVY IONS: STUDIES ON TRANSFORMATION EFFICIENCIES AND MOLECULAR MECHANISMS

Ludwig Hieber, Jan Smida*, and Albrecht M. Kellerer

Institute for Radiation Biology, GSF, 8042 Neuherberg
*Institute for Radiation Biology, University of Munich
8000 München, Germany

INTRODUCTION

Numerical estimates of the risk of cancer, caused by the exposure to ionizing radiations, are a topic of considerable current interest, especially in the view of the revision of the atomic bomb dosimetry and the ensuing reanalysis of the epidemiological data for the survivors of the atomic bomb explosions. As far as sparsely ionizing radiations are concerned, the epidemiological data are a good basis for risk estimation. For densely ionizing radiations, however, the linkage to epidemiological data had to be abandoned, when the new dosimetry for the atomic bomb survivors showed that the contribution of neutrons is minimal or, at least, uncertain. The carcinogenic potential of fast neutrons is, however, of importance in some areas of radiation protection. High flying aircraft and manned space activities, especially in the context of a long duration deep space flight, such as manned mission to Mars, cause critical questions with regard to the risks due to densely ionizing radiation.

In view of the lack of epidemiological data, the recent recommendations of revised quality factors for densely ionizing radiations by ICRP had to be derived from radiobiological investigations, and they will have to be subjected to further scrutiny as new experimental information develops. In view of the difficulties, to perform extensive animal experiments, the in vitro system of morphological transformation of mammalian cells by ionizing radiation has taken added importance and it has become one of the main tools for the derivation of quantitative information on the relative biological effectiveness of different types of ionizing radiation.

For sparsely ionizing radiation a series of studies of neoplastic transformation have been performed in different laboratories. This information supplements the experience from the epidemiological follow-up of the atomic bomb survivors and from other collectives that have been exposed occupationally, exposed by accident, or exposed for medical reasons. For densely ionizing radiation, especially for particles of high charge and high atomic number, there are, as stated, only few epidemiological data and - except for radon and its decay products - no epidemiological data are available that may be used for the estimation of cancer risk. Experimental data from transformation studies are, therefore, of special importance.

The aim of the research that is reported here is to increase the knowledge about the inactivation and transformation efficiency of various heavy ions of different ionization density and of different energy. Heavy ions - produced at the UNILAC of the GSI in Darmstadt, Germany - with energies between 4 and 20 MeV/u and with LET values between 170 and 16,000 keV/μm were studied.

Beyond the direct determination of the transformation efficiency of different ionizing radiations, the studies are aimed at further insight into the process of neoplastic transformation of mammalian cells by an assessment of the cellular and molecular mechanisms that accompany the change of cells from 'normal' to the transformed phenotype or to malignancy. A combination of dose-effect relations for cell transformation and for radiation-induced tumors in animals, of data from epidemiologic studies, and of mechanistic studies in molecular biology will be required to elucidate the processes of radiation-induced cell transformation and radiation-caused carcinogenesis in man. The feasibility of manned space activities will critically depend not only on technical countermeasures such as shielding but also on the results of risk estimation for radiation-induced late effects.

METHODOLOGY

Cell Systems and Irradiation Procedures

C3H 10T1/2 mouse embryo fibroblasts developed by Reznikoff et al.(1973) are most widely used for the determination of transformation frequencies, although they have the somewhat undesirable feature that they are already immortalized and show certain changes in chromosome numbers. The cells were cultured in Eagle's basal medium (BRL, Karlsruhe, FRG) supplemented with 10% heat-inactivated fetal bovine serum (Boehringer, Mannheim, FRG or Roth, Karlsruhe, FRG), 50 u/ml penicillin, and 50 μg/ml streptomycin (BRL) at 37°C in 95% air and 5% CO_2. Cells of passage 12 were thawed from stock cultures frozen in liquid nitrogen, subcultured and used for the experiments in passage 14.

Primary Syrian Hamster Embryo (SHE) cells were isolated from 12-day-old embryos, stored in passage two or three, and were cultured in IBR medium (BRL) with 20% fetal bovine serum and antibiotics at 37°C in 90% air and 10% CO_2.

The cells were always exposed during exponential growth. The irradiations with different heavy ions were performed in 3.5 cm petri dishes (Greiner, Frickenhausen, FRG) at the linear accelerator UNILAC of the Gesellschaft für Schwerionenforschung (GSI, Darmstadt, FRG). The technical procedures for the heavy ion exposures have been described previously (Hieber et al., 1989).

Survival and Transformation Assays with C3H 10T1/2 Cells

For the determination of survival and transformation rates after different doses of heavy ion exposure, C3H 10T1/2 cells were trypsinized, counted with a Coulter Counter, and replated in 25 cm^2 flasks or 9 cm petri dishes at cell densities of about 80 and 200 to 300 viable cells per culture vessel for survival and transformation assays, respectively. Surviving fractions were determined from the number of colonies with more than 50 cells per colony after 10 to 14 days of incubation. Transformation frequencies were estimated from the number of foci of type 2 and 3 per surviving cell according to the criteria of Reznikoff et al. (1973), as described previously (Hieber et al., 1987). Transformation frequencies and surviving fractions were estimated per unit of particle fluence or per unit absorbed dose.

Heavy Ion Induced Transformation of SHE Cells

After exposure to 8 MeV/u carbon ions, SHE cells were repeatedly subcultured with $5x10^5$ cells per 75 cm^2 flask. The primary cells aged within a few passages, and immortalized or transformed cell clones overgrew the nonproliferating aged cells. The cells of transformed phenotype were then tested for their ability to grow in semi-solid medium and for tumor induction in athymic nude mice. Animal tumors that appear within 3 to 6 weeks after subcutaneous injection into the mice were then used to establish tumor cell lines. Transformed cell lines and tumor cell lines were analysed with respect to cellular and molecular changes, and the results were compared with those of cell lines that were spontanously transformed or transformed by other types of ionizing radiations.

RESULTS AND DISCUSSION

Inactivation and Transformation of C3H 10T1/2 Cells by Various Heavy Ions

C3H 10T1/2 cells were exposed to various heavy ions of energies between 4.6 and 18.6 MeV/u, e.g. oxygen, carbon, argon, iron, krypton, lead, and uranium ions. Beam parameters, such as energy and LET have been described previously (Hieber et al., 1989). Heavy ion exposures lead always to exponential survival curves. The inactivation cross sections increased with increasing atomic number of the ions; the highest inactivation cross sections (80 to 90 μm^2) were found for lead und uranium ions at low energies. Even with these very heavy ions with highest LET values at least 3 particles have to pass a cell nucleus on average to inactivate a 10T1/2 cell; the geometrical cross section of the nuclei of these cells are about 250 μm^2. The cross sections were also dependent on particle energy, i.e. they decreased with increasing energy. Heavy ions of low atomic number were more effective in cell killing per unit dose than sparsely ionizing γ-rays. At higher LET, however, the relative biological effectiveness (RBE) for cell inactivation decreases; at about 900 to 1000 keV/μm the RBE vs.γ-rays begins to be smaller than unity.

Transformation frequencies per surviving cell were investigated for various heavy ions with different energies and LET values. As shown in Fig.1, the transformation rate decreases with increasing mass of the ions at specified velocity; the data points represent transformation frequencies at the same particle fluence of 1.5×10^6 per cm^2. The transformation efficiencies appear to increase with increasing energy, i.e. decreasing LET, as shown for krypton and uranium ions. A similar result was obtained for argon ions at different fluences (data not shown). The LET dependencies both for survival and cell transformation are in substantial agreement with data of Yang et al. (1985) for ions of relativistic energies and of LET greater than 200 keV/μm.

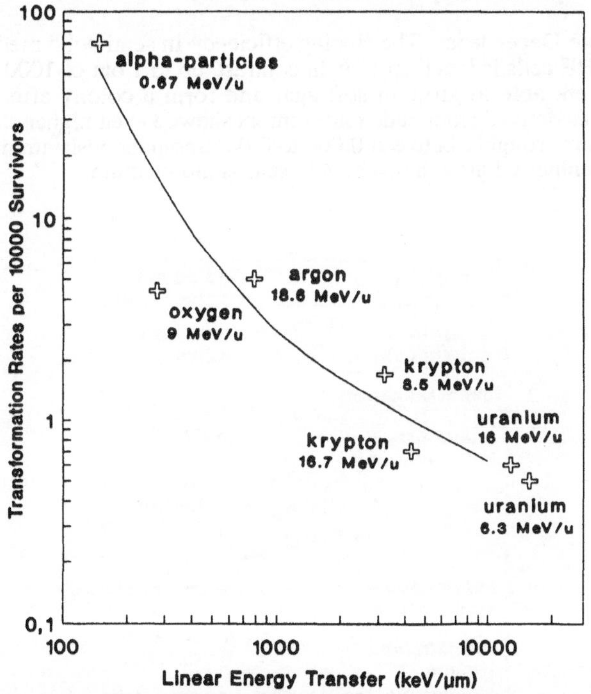

Figure 1. Transformation frequency of C3H 10T1/2 cells versus LET: The cells were exposed to different heavy ions at a particle fluence of 1.5×10^6 per cm^2. The line is given merely to demonstrate the decrease of the transformation rate with increasing LET.

Analysis of Syrian Hamster Embryo Cells Transformed by 8 MeV/u Carbon Ions

In order to understand the process(es) of radiation induced cell transformation and ultimately those of radiation carcinogenesis in man, one needs to study the cellular and molecular changes that take place when a 'normal' cell is neoplastically transformed; in addition, one has to determine whether any such changes are the cause of transformation or a merely concomitant effect.

The scheme in Fig.2 gives an overview of cellular and molecular parameters that have been investigated or are currently under investigation in our laboratory with radiation transformed Syrian Hamster embryo cells. The present data for cellular, cytogenetic, and molecular changes are brought together in Table 1.

Cellular Effects

a.Cell Growth. Cell growth was determined from growth curves and from colony assays. Untransformed, primary SHE cells double their number during exponential growth in 20 to 24 hours. The population doubling time of immortalized and transformed cells was about half this value, i.e. 10 to 12 hours. The plating efficiency, i.e. the fraction of cells that form a colony of more than 50 cells within 10 to 12 days, is about twice as high for immortalized and transformed cells than for primary cells. Hence, immortalized, transformed, and tumor cells have a pronounced growth advantage; they overgrow untransformed cells rapidly.

b. Immortalization. Primary cells age after a certain number of cell doublings, i.e. they exhibit reduced growth rate and finally die. Immortal and transformed cells, in contrast, can be cultured indefinitely over tens of subcultures. This has been found for all radiation transformed and for spontaneously immortalized cell lines.

c. Anchorage Dependency. The cloning efficiency in semi-solid medium, such as soft agar, of primary SHE cells is less than 10^{-6}. In contrast, up to 1 out of 1000 transformed cells and tumor cells were able to grow in soft agar and form a colony after 3 to 4 weeks of incubation. Cell lines derived from nude mice tumors showed even higher cloning efficiencies in semi-solid medium, roughly between 0.004 to 0.04. Spontaneously immortalized cell lines had only a weakly enhanced growth potential in semi-solid medium.

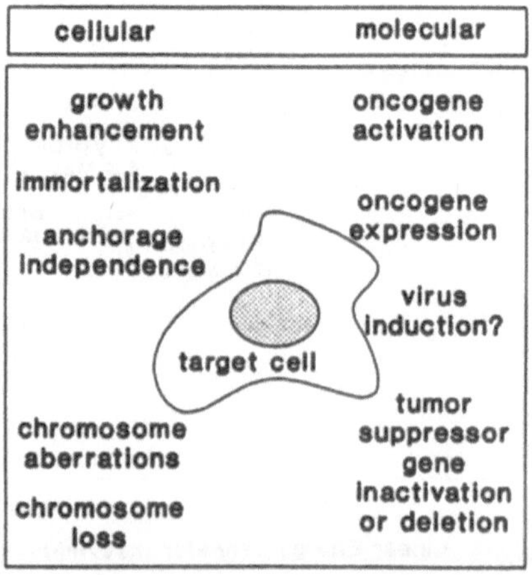

Figure 2. The scheme summarizes the cellular and molecular parameters investigated in radiation-transformed Syrian Hamster embryo cells.

Table 1. Cellular,Cytogenetic, And Molecular Changes in Spontaneously Immortalized And Radiation-transformed Syrian Hamster Embryo Cells.

CELLULAR EFFECTS

CELL TYPE	doubling time	plating efficiency	anchorage dependence	tumori-genicity
normal	20-24 hrs	0.2-0.3	-	-
immortal	12-14 hrs	0.5	+-	-
transformed	12-14 hrs	0.5	++ (+++++)	+

CYTOGENETIC EFFECTS

CELL TYPE	chromosome number	translocations	deletions
normal	44	-	-
immortal	40-70	+-	+
transformed	41-43	+	+

MOLECULAR EFFECTS

CELL TYPE	oncogene activation by point mutation Ha-ras	oncogene expression ras / myc	suppressor gene inactivation
normal	no	low	no
immortal	no	2-3 / 2-4	???
transformed	no	2-3 / 2-4	???

d. Tumorigenicity. All radiation transformed cell lines have been tested for tumor induction in nude mice. After subcutaneous injection of about 10^6 cells of transformed phenotype per mouse all animals developed tumors which can be classified as fibrosarcomas. Spontaneously immortalized cells became at later passages transformed even without further treatment. Cells subcultured more than 40 times became tumorigenic too.

Cytogenetic Effects

Chromosome Number and Chromosome Aberrations. The diploid set of chromosomes of normal Syrian Hamster cells consists of 44 chromosomes. Immortalized, non-tumorigenic cells, as taken from different cell stocks with various passage numbers, exhibit a highly variable number of chromosomes per cell; the numbers varied between 40 and 70. Radiation transformed cell lines and tumor cells, too, exhibit varying distributions of chromosome number; however, the numbers tend to be reduced to 41-43 chromosomes per cell. All cell lines contain various fractions of highly uneuploid cells with chromosome numbers up to 85. These near tetraploid cells reappeared even in subcloned cell populations, after a few subcultures. We are presently investigating whether the chromosome loss in transformed cells is random or specific chromosomes tend to be lost.

In addition there is evidence in the cells of stable chromosome aberrations, such as translocations and deletions in specified chromosomes (S.Endo, personal communication). This poses the question whether such chromosomal changes are related to the transformed phenotype of the cells, an objective of current investigations.

Molecular Effects

a. Oncogene Expression and Oncogene Activation. The expression of two cellular protooncogenes, Ha-ras and myc, has been studied by Northern blot analysis in normal, immortalized, and radiation transformed SHE cells. Both genes are transcribed in untransformed cells at a low level. All transformed cell lines, and 6 out of 8 tumor cell lines have levels of Ha-ras and c-myc transcripts that are enhanced by factors of 2-3 and 2-4, respectively. The enhancement is observed even in cell lines that have just overcome the growth crisis. The enhanced expression of these genes appears to be related to an enhanced growth potential of the cells. It is known that the expression of both genes is cycle dependent; proliferating cells exhibit higher expression than stationary cells (Shuin et al., 1986, Billings et al., 1987).

A Southern blot analysis of the DNA after it had been cut by restriction endonucleases such as EcoRI, BamHI, and HindIII did not provide any evidence for restriction fragment length polymorphisms of these genes. There was also no indication of an amplification of either of both the oncogenes. Furthermore, a sequence analysis of the Ha-ras gene from 15 different tumor cell lines did not show any point mutation in the coding region of the gene. Thus the Ha-ras oncogene is not activated by point mutation in these cells, in contrast to observations for a number of human and animal tumor cells (Guerrero and Pellicer, 1987).

b. Tumor Suppressor Gene Inactivation or Deletion. Neoplastic transformation could be due to an activation of certain oncogenes or an enhanced expression of oncogenes, it could also be caused or could be accompanied by repression or deletion of so-called tumor suppressor genes or anti-oncogenes. There is some evidence for chromosome loss and/or chromosomal deletions in transformed SHE cells and tumor cell lines, and we have, therefore, initiated a study dealing with this subject. For this purpose we have constructed λgt11 cDNA libraries of normal, transformed, and tumor cells. In plaque hybridization tests with the library for normal cells we have, up to now, found 30 clones which carried DNA inserts that did not hybridize with cDNA from the transformed cells, and 40 clones from the library of transformed cells that did not hybridize with the cDNA of normal cells. The DNA inserts are currently under investigation, i.e. they will be identified by sequencing, to determine if these genes are transformation-specific. The extension of these experiments will be a major objective of forthcoming studies.

ACKNOWLEDGMENTS

We are grateful to Dr.Gerhard Kraft and his collegues for extensive support of the experiments at GSI. Miss Sabine Fenn made valuable contributions to the experimental work. The research was financially supported by the European Commission Radiation Protection Programme (Contract No.BI7-0043), by the Deutsche Forschungsgesellschaft (SFB 172, Project No.C-1), and by the Gesellschaft für Schwerionenforschung (Contract: WÜ KE B).

REFERENCES

Billings, P.C., Shuin, T., Lillehaug, J., Miura, T., Roy-Burman, P., and Landolph, J.R., 1987, Enhanced expression and state of the c-myc oncogene in chemically and x-ray transformed C3H 10T1/2 Cl 8 mouse embryo fibroblasts, Cancer Res., 47:3643.

Guerrero, I., and Pellicer, A., 1987, Mutational activation of oncogenes in animal model systems of carcinogenesis, Mutat.Res., 185:293.

Hieber, L., Ponsel, G., Trutschler, K., Fenn, S., Fromke, E., and Kellerer, A.M., 1987, Absence of a dose-rate effect in the transformation of C3H 10T1/2 cells by α-particles, Int.J.Radiat.Biol., 52:859.

Hieber, L., Ponsel, G., Trutschler, K., Fenn, S., and Kellerer, A.M., 1989, Neoplastic transformation of mouse C3H 10T1/2 and Syrian Hamster embryo cells by heavy ions, Adv.Space Res., 9:(10)141.

Reznikoff, C.A., Bertram, J.S., Brankow, D.W., and Heidelberger, C., 1973, Quantitative and qualitative studies of chemical transformation of cloned C3H mouse embryo cells sensitive to postconfluence inhibition of cell division, Cancer Res., 33:3239.

Shuin, T., Billings, P.C., Lillehaug, J.R., Patierno, S.R., Roy-Burman, P., and Landolph, J.R., 1986, Enhanced expression of c-myc and decreased expression of c-fos protooncogenes in chemically and radiation-transformed C3H 10T1/2 Cl 8 mouse embryo, Cancer Res., 46:5302.

Yang, T., Craise, M., Mei, M., Tobias, C., 1985, Neoplastic transformation by heavy charged particles, Radiat.Res., 104S:177.

Hieber, L., Ponsel, G., Trieschieber, K., Fenn, S., and Kellerer, A.M., 1989, Neoplastic transformation of mouse C3H10T1/2 and Syrian hamster embryo cells by heavy ions, Adv. Space Res., 9:(10)141.

Reznikoff, C.A., Bertram, J.S., Brankow, D.W., and Heidelberger, C., 1973, Quantitative and qualitative studies of chemical transformation of cloned C3H mouse embryo cells sensitive to postconfluence inhibition of cell division, Cancer Res., 33:3239.

Shuin, T., Billings, P.C., Lillehaug, J.R., Patierno, S.R., Roy-Burman, P., and Landolph, J.R., 1986, Enhanced expression of c-myc and decreased expression of c-fos protooncogenes in chemically and radiation-transformed C3H/10T1/2 Cl 8 mouse embryo cell lines, Cancer Res., 46:5302.

Yang, T.C., Craise, L.M., Mei, M., and Tobias, C.A., 1985, Neoplastic cell transformation by heavy charged particles, Radiat. Res., 104:S177.

ON THE MECHANISM AND CONSEQUENCES OF RADIATION-INDUCED GENE AMPLIFICATION IN MAMMALIAN CELLS

Christine Lücke-Huhle

Nuclear Research Center
Institute of Genetics and Toxicology
P.O. Box 3640,
D-7500 Karlsruhe 1, Germany

In addition to lesions measured at the site of energy deposition, there are radiation-induced cellular responses occuring very distantly from the site of primary DNA damage - such as gene activations and gene amplification. This report shows, that various types of radiation (low LET γ-rays, high LET α-particles or Auger electrons as well as UV-light) induce gene amplification in mammalian cells and that this process generates mutations, chromosomal aberrations and translocations which are supposed to be involved in all steps of the multistep process of carcinogenesis.

INTRODUCTION

Normally all genes of a mammalian genome are replicated once during S-phase of the mitotic cycle, supplying daughter cells with 2 copies of each gene (one from the father and one from the mother) after cell division. However, quantitative changes in gene copy number are found rather frequently in biology. They range from polytenization of whole chromosomes (caused by the process of endoreduplication) to selective gene amplification. The amplified gene exists then in more than 2 copies per cell, it can be 10, 100 and even 1000 copies. The extra copies either remain integrated in the genome or are cut out and distributed between daughter cells by chance. The presence of multiple copies of one particular gene leads to an increase of the synthesis of the protein the gene codes for.

Upto about 1968 it was believed, that gene amplification plays an important role only in developmental biology. For instance, the ribosomal genes in amphibian oocytes are amplified, so that after fertilisation of the egg, protein synthesis can start immediately with high speed (Ritossa, 1968; Brown and David, 1968).

In the mid 70s, however, gene amplification was recognized in somatic cells and in cell culture. It was found to be related to the acquisition of resistance to various drugs employed in cancer therapy, specially to the cytostaticum methotrexate (MTX). MTX binds and therewith inhibits the enzyme dihydrofolate reductase (DHFR) which catalizes the reduction of dihydrofolate to tetrahydrofolate, an important step in the nucleotide metabolism. Patients treated with MTX become resistant to the drug and Biedler and Spengler (1976) were the first who found that the tumor cells of these MTX-resistant patients exhibited 2 extraordinary types of chromosomal aberrations, the so called double minutes (DM's: these are pairs of small acentric chromosomes which show a circular structure by electron microscopy) and homogeneously staining regions (HSR's: long chromosome segments with little evidence of G-banding).

Biological Effects and Physics of Solar and Galactic Cosmic Radiation, Part A, Edited by C.E. Swenberg *et al.*, Plenum Press, New York, 1993

Schimke and his group at Stanford reported in 1978 that resistance to MTX in cultured mouse cells is associated with the amplification of the target gene DHFR. Such gene amplifications could either be stable (HSR's were found) or unstable (DM's were found, which are easily lost in the absence of selective pressure).

In 1979 Wahl and Stark (Wahl et al., 1979), showed that resistance to an inhibitor of aspartyl transcarbamylase was the result of amplification of the gene coding for this enzyme. Now we know, that gene amplification is a common mechanism for acquisition of drug resistance in bacteria (Anderson and Roth, 1977), plants (Shah et al., 1986), insects (DeCicco and Spradling, 1984) and mammalian cells (reviews: Schimke, 1984a; Stark et al., 1989). Table 1 shows examples of drug resistance in somatic mammalian cells by amplification of the target gene.

Table 1. Examples of Drug Resistance in Somatic Mammalian Cells by Amplification of the Target Gene

DRUG	AMPLIFIED GENE
- methotrexate (MTX)	dihydrofolate reductase (DHFR)
- Cadmium	metallothionein I gene
- β-aspartylhydroxamate (β-AHA)	asparagine synthetase
- N-phosphonacetyl-L-aspartate (PALA)	carbamyl phosphate synthetase aspartate transcarbamylase dihydroorotase (CAD)
- hypoxanthin, amenopterin, thymidine (HAT)	hypoxanthine-guanine-phosphoribosyltransferase (HGPRT)

Additional interest in the phenomenon of gene amplification came from the fact that the typical chromosomal aberrations found after gene amplification (HSR's and DM's) have been found in a variety of human tumors and tumor cell lines. About 20-30% of all human tumors possess amplified oncogenes. Oncogenes, first discovered in retroviruses, are responsible for cell transformation caused by these viruses. Today we know that oncogenes exist as proto-oncogenes in all mammalian cells and that they have important regulatory functions. The questions which arise with respect to oncogene amplification are: How are these sequences amplified and what is their role in the malignant process?

Amplified genes are not only found in many tumours of uncertain origin, but also after treatment of cells or organisms with DNA damaging agents, most of which are carcinogenic. A compilation of such agents is given in Table 2.

We are interested in gene amplification induced by radiation and its relevance in radiation-induced carcinogenesis. Our interest has been focussed on the following questions. (1) Is gene amplification inducible by radiation? (2) Are there differences in efficiency between low and high LET radiation? (3) What are the genomic consequences of amplified DNA sequences? (4) What are the initial events in the signal transfer pathway following DNA damage and leading to gene amplification? (5) Can diploid human cells amplify genes as easily as rodent cell lines?

SIMIAN VIRUS 40 SEQUENCES AS INDICATOR GENE FOR RADIATION INDUCED DNA AMPLIFICATION

For addressing the first four questions we use a model system: Simian virus 40 (SV40)-transformed Chinese hamster embryo cells (Co631; courtesy, S. Lavi, Tel Aviv). These cells contain 5 viral copies integrated in the hamster genome. Hamster cells are semipermissive for SV40, that means they do not produce intact virus particles, but the viral DNA is replicated under cell cycle control only once during S-phase and therefore can be used

Table 2. Agents Which Induce Amplification of Specific DNA Sequences

1. CHEMICAL CARCINOGENS (Lavi and Etkin, 1981)
   MNNG (N-methyl-N'-nitro-N-nitrosoguanidin)
   DMBA (7,12-dimethylbenz(α)anthracene)
   methylnitrosourea
   ethylmethanesulfonate
   aflatoxin B1
   benzo(α)pyrene

2. RADIATION (Lücke-Huhle et al., 1986, 1988)
   ^{60}Co-γ-irradiation
   ^{241}Am-α-particles
   ^{125}I-UdR incorporation
   UV-light

3. VIRAL CARCINOGENS (Schlehofer et al., 1983)
   Herpes Simplex Virus

4. CHEMOTHERAPEUTIC AGENTS (Pool et al., 1989)
   BAD = 4-(4-(bis(2-clorethyl)amino)-phenyl)-1-hydroxybutane-1,1-biphosphonic acid)
   BCMP = 3-(bis-(2-chlorethyl)amino)-4-methylphenyl-hydroxymethane-biphosphonic acid)
   4-OH-peroxy-cyclophosphamide
   phosphoramide mustard
   methotrexate

5. METABOLIC INHIBITORS AND OTHERS
   methotrexate
   aphidicolin
   cycloheximide
   hydroxyurea
   5-bromodeoxyuridine
   heat
   hypoxia

as an endogenous indicator gene. Other laboratories have used similar model systems based on polyoma-DNA in rat cells (Lambert et al., 1986) or adeno-associated virus (AAV) in various mammalian cell lines (Yalkinoglu et al., 1988). Amplification of viral DNA sequences can be followed at very early times after treatment in the majority of cells, and therefore these model systems are well suited for mechanistic studies.

Upon exposure of SV40-transformed Chinese hamster embryo cells to various types of external (^{60}Co-γ-rays, 241-Am-α-particles or UV-light) or internal radiation (caused by the decay of ^{125}I incorporated into DNA inform of I-UdR) SV40-DNA amplification occurs without leading to the production of intact virus.

The methodology employed in these experiments has been described in detail elsewhere (Lücke-Huhle et al., 1986; Ehrfeld et al., 1986). In short, cells are irradiated as monolayers and analyzed for changes in gene copy number by dispersed cell blotting or DNA blotting followed by highly sensitive DNA hybridization methods.

Radiation-induced SV40-DNA amplification is dose-dependent and increases with time after irradiation. Maximum SV40 amplification up to 25 fold is achieved at day 3 after exposure to α–particles or UV-light and at day 6 after γ-irradiation (Fig. 1A; Table 3).

SV40 amplification is a transient process: The number of amplified copies starts to decrease by day 4 to 7 after radiation. This time course is independent of the dose and might be explained by the fact that amplified SV40 sequences, not offering any growth advantage to the host genome, get lost during cell divisions.

In case of ^{125}Iodine SV40 amplification increased until the latest time point investigated (day 6) due to the radiation still emitted by the decay of the incorporated ^{125}Iodine. These

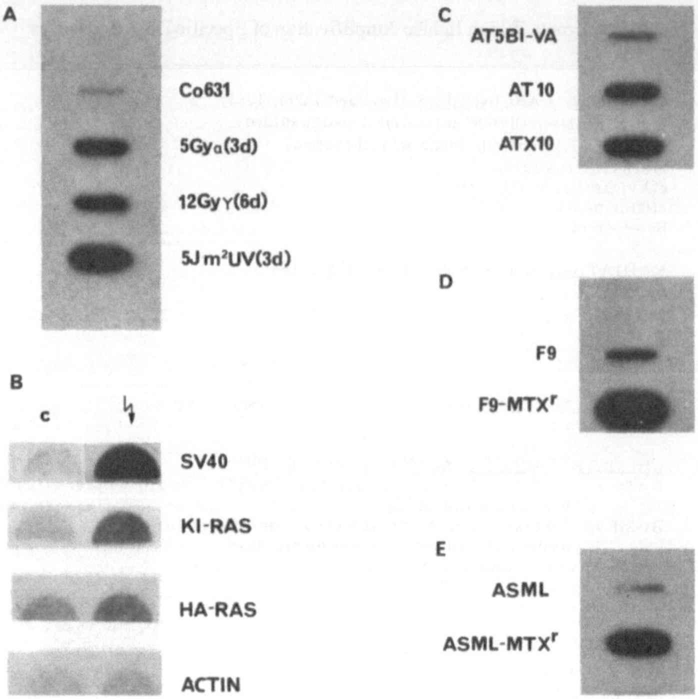

Figure 1. Autoradiograms of DNA Slot Blots (A,C,D,E: 1 μg DNA each) or of Dispersed Cell Blots (B: 2.5×10^5 cells each) Hybridized with Specific ^{32}P-labeled Gene Probes Show Gene Amplification after Exposure to Various Treatments. A: Co631 Hamster DNA Shows a 15 to 25 Fold Increase in SV40 Gene Copy Number at Day 3 after 5 Gy of α-particles or 5 J/m^2 UV Light and at Day 6 after 12 Gy of ^{60}Co-γ-irradiation. B: In Addition to a Rise in SV40 Gene Copy Number (13 fold), the Oncogenes Ki-ras and Ha-ras are Amplified 2 Fold, While α-actin as a Control Gene Shows no Amplification at Day 6 after Incorporation of ^{125}Iodine (Curved Arrow). C,D+E: After MTX Selection, MTX-resistant Cell Lines Show an Increase in DHFR Gene Copy Number. In Comparison to the Original AT5BI-VA Cells (1C), MTX-resistant Human AT Cells Exhibit a 10 Fold or 20 Fold DHFR Gene Amplification, Respectively, Depending Whether They Were Pretreated with γ-irradiation (ATX10) or not (AT10). Mouse Teratocarcinoma F9 Cells (1D) Increase Their DHFR Gene 40 Fold (F9-MTX^r) and BSp73 ASML Cells (1E) 13 Fold (ASML-MTX^r) after MTX Selection for 4 to 6 Months.

densely ionizing Auger electrons induce in addition to a 13 fold SV40 amplification, a 2 fold amplification of the cellular oncogenes Ki-ras and Ha-ras (Fig. 1B), the genes which have also been found amplified in radiation-induced animal tumors (Guerrero et al., 1984; Barbacid, 1986).

As mentioned before, SV40 amplification is dose-dependent. Increasing doses of radiation enhance SV40 amplification but concomitantly reduce cell survival. Since it is, of course, the surviving fraction of the treated cells which is of interest with respect to mutagenic and carcinogenic risk, it had to be excluded that gene amplification is simply the response of dying cells. In situ hybridization of surviving colonies and a comparison of their hybridization signal with that of untreated colonies of equal cell number (Lücke-Huhle et al., 1990b) demonstrates that all surviving clones had amplified their SV40 sequences as efficiently as the abortive colonies which had lost their capacity for sustained cell proliferation (Fig. 2).

Survivors and non-survivors amplify about equally well; that means that killing lesions are not identical with amplification-inducing lesions. This is in agreement with the finding that different types of radiation causing completely different spectra of DNA damage with different killing efficiencies, induce gene amplification equally well. A comparison of the amplification efficiency is, however, impeded by the different kinetics found for alpha and gamma rays,

Table 3. Maximum SV40 Amplification in Chinese Hamster Embryo Cells (Co631) after Exposure to Various Types of Radiation.

AGENT	DOSE	CELL KILL	SV40 AMPLIFICATION
^{60}Co-γ	2.5 Gy	20%	5 x
	8.0 Gy	90%	13 x
	12.0 Gy	98%	23 x
UV	2 J/m^2	50%	3 x
	5 J/m^2	95%	25 x
	15 J/m^2	>99%	4 x
^{241}Am-α	0.7 Gy	50%	2 x
	2.5 Gy	94%	5 x
	5.0 Gy	>99%	15 x
^{125}I-UdR	1.6x10^{-2} pCi/cell	20%	2 x
	4.8x10^{-2} pCi/cell	75%	3 x
	8.0x10^{-2} pCi/cell	93%	13 x

respectively. Calculations at day 3 after radiation would result in an RBE of 6 for alpha particles (Lücke-Huhle et al., 1986). However, if the maxima are considered, irrespective of time, no difference between alpha and gamma rays is observed: A dose of 2.5 Gy of alpha or gamma rays, respectively, yield a five fold SV40-amplification although the killing efficiency of alpha particles is higher by a factor of five (Table 3).

CONSEQUENCES OF AMPLIFIED DNA

The fate of the amplified copies has been followed by gel electrophoresis and Southern hybridization technique (Lücke-Huhle et al., 1990b). While in the DNA of untreated control cells all viral copies are integrated in the hamster genome, and therefore run within the high molecular weight fraction, part of the amplified SV40 copies induced by radiation gets excised (Fig. 3A).

In order to follow up the fate of amplified SV40 DNA in surviving cells, 36 individual colonies were grown up after exposure to 10 Gy of ^{60}Co-γ-irradiation. After 20-30 cell generations (about 20 days) all clones had lost most of the amplified SV40 copies (2 examples are shown in Fig. 3B), but 12 out of the 36 clones showed new integration sites of the SV40 sequences indicated by the appearance of new restriction fragments containing SV40 sequences (Fig. 3C).

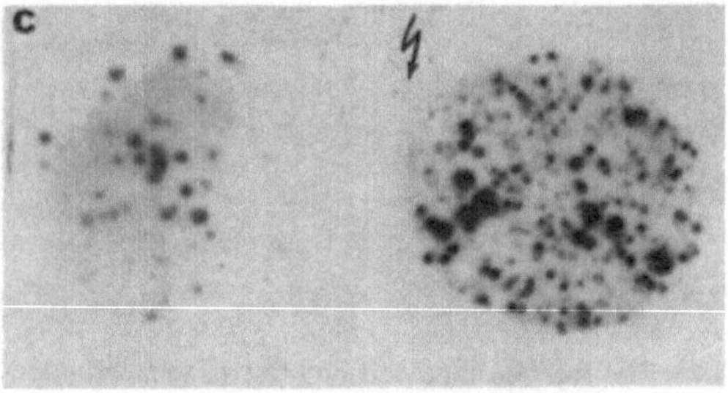

Figure 2. Autoradiograms of Clonal Blots from Unirradiated Cell Clones (C) or Colonies that Survived 10 Gy of ^{60}Co-γ-irradiation (Curved Arrow) Hybridized to ^{32}P-labeled SV40 DNA, Show SV40 Amplification for Surviving Colonies (Big Dark Spots = Stronger Signal as Compared to Spots of Control Colonies) and Abortive Colonies (Small Dark Spots, Normally Not Vsible after Crystal Violet Staining).

This finding suggests that radiation can generate mutations by inducing overreplication of chromosome segments that are then substrates of enzymatic rearrangements. DNA rearrangements by chromosomal transposition represents a particular intriguing mechanism, since it could bring oncogenes under cis control of a highly active chromosome region.

INFLUENCE OF GENOMIC INSTABILITY, CELL DEDIFFERENTIATION AND METASTATIC POTENTIAL OF A CELL ON DHFR GENE AMPLIFICATION

To answer the question whether gene amplification is one of the initial events in the multistep process of carcinogenesis we studied this phenomenon on human diploid skin fibroblasts. Our efforts to induce DHFR gene amplification by stepwise increasing the concentration of MTX in the culture medium, were without success for several years (Lücke-Huhle and Herrlich, 1986). However, in contrast to the genomically stable normal human cells a cell line originating from a patient suffering from ataxia telangiectasia (AT), amplified the DHFR genes more than 10 fold (Fig. 1C: AT10). After 6 months of MTX selection AT-cells could grow in concentrations of 100 μM MTX, a concentration higher by 4 orders of magnitude than the intial cell population would survive (Lücke-Huhle et al., 1987).

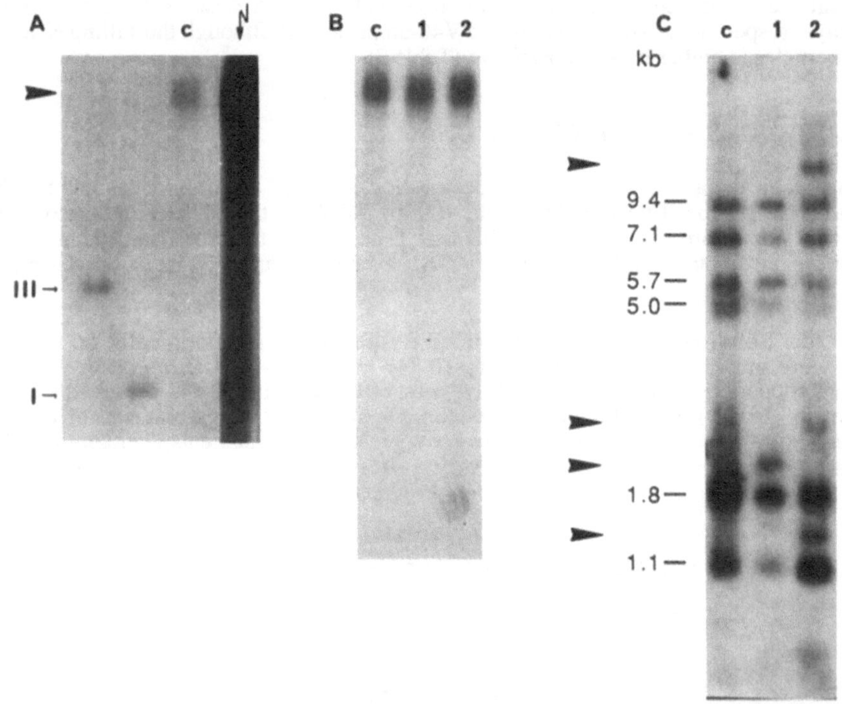

Figure 3. Autoradiograms Showing Southern Blots (10 μg DNA each) of Undigested (A+B) or HindIII Digested (C) DNA. DNA was isolated from untreated hamster cells (c) or From Cells 6 Days (A, Curved Arrow) and 20 Days (B1+2; C1+2) after Exposure to 10 Gy of ^{60}Co-γ-irradiation, Respectively. Arrow in A Marks DNA of High Molecular Weight. SV40 DNA Form I (supercoiled) and Form III (Eco RI cut) Were Used as Marker. The Band at the Position of SV40 DNA Form I and the Smear in the Irradiated Sample (A) Represents Cut out SV40 DNA of Unit Size and of Various Length, Respectively. In C, the Arrows Point to New Restriction Fragments Induced by γ-irradiation in the DNA of Survivors. Hind III Fragments of Control Cell DNA Containing SV40 Sequences are Indicated by Their Kb-value.

The amplification efficiency could be increased by a factor of two if the cells were exposed to γ-irradiation before the MTX selection procedure (Fig. 1C: ATX10).

At the chromosome level, untreated AT-cells already show increased numbers of spontaneous chromosomal aberrations as well as an enhanced response to induction of aberration by ionizing radiation. All these features are indicative of the genomic instability of cells from patients with ataxia telangiectasia (a repair deficiency syndrome with high cancer incidence: Lehman, 1982), a fact which made these cells especially interesting for the investigation of gene amplification.

Sister chromatid exchange (SCE) frequency (an indicator of DNA damage and its intrachromosomal recombination) rises from 0.21 to 0.44 SCE's/chromosome in the MTX-resistant cell lines as compared to the initial AT-cells without amplified DHFR copies. An augmentation of the genomic instability by acquisition of amplified gene copies is also revealed by an increase in chromosome rearrangements (from 4 to 18%, measured as exchange figures and ring chromosomes in the MTX-resistant lines), supporting the idea that the presence of amplified genes increases the genomic instability of these cells further (Speit and Lücke-Huhle, 1987).

The influence of dedifferentiation on the ability of a cell to amplify a specific DNA sequence has been studied in F9 mouse teratocarcinoma stem cells (Lücke-Huhle and Herrlich, 1991). These cells differentiate easily into an early embryonic cell type known as parietal endoderm in response to retinoic acid (given at a dose of 3.3×10^{-7}M at least for 6 days). Differentiation is documented by morphology changes, cell cycle prolongation and reduced expression of the cellular oncogene c-myc. Initially (prior to MTX contact), non-differentiated F9 teratocarcinoma cells and differentiated F9-RA endodermal cells are equally sensitive to MTX. Exposure of exponentially growing cultures to progressively increasing concentrations of MTX yields highly resistant cells within 4 months, but only from the non-differentiated F9 cultures. Resistance of the F9-MTX^r cells is associated with the acquisition of additional DHFR gene copies: a maximum of 85 copies per cell was found in the cells resistant to 1000 μM MTX (Fig. 1D).

Exposure to 5 Gy of ^{60}Co-γ-irradiation prior to MTX-selection enhances amplification in the undifferentiated F9 cells, while no amplification could be detected in the differentiated F9-RA cells under all conditions tested.

Undifferentiated F9 cells do not only amplify easily, their MTX-resistant cell clones also quickly lose their extra DHFR gene copies in the absence of MTX selection (95% within 3 months). Interestingly, this loss can be prevented by retinoic acid treatment. Thus it appears that differentiation of cells by retinoic acid prevents both: induced gene amplification and non-induced mobilization of DNA.

Clinical investigations have shown a correlation between the number of metastasis-positive lymph nodes in certain human tumor patients and the existence of amplified oncogenes in tumor cells. This makes amplification a reliable prognostic parameter (Slamon et al., 1989). To find out whether tumor cells with a high probability for metastatic progression also have a higher potential to amplify genes under selective pressure in cell culture, we examined a pair of tumor cell lines (BSp73 AS and BSp73 ASML) derived from a spontaneous rat pancreatic adenocarcinoma (BSp73: Matzku et al., 1983). BSp73 AS cells remain localized upon subcutaneous injection, while BSp73 ASML cells (originating from a lung metastasis) are highly metastatic if injected into isogenic rats. Exposure to stepwise increasing concentrations of MTX result in MTX-resistant ASML cells. Their ability to grow in 100 μM MTX is mainly due to a 13-fold amplification of their DHFR gene (Fig. 1E). The nonmetastasizing cells, however, within the same short period of time (4 months), do not show DHFR gene amplification and are unable to grow in medium containing high MTX-concentrations (Lücke-Huhle, unpublished). These results may simply reflect an increase in genomic instability with tumor progression. However, the consequences may specifically contribute to the metastatic behavior, in that amplified DNA will partly be reintegrated and thus will cause mutations in various genes from which the metastatic properties could be selected in vivo.

THE MECHANISM OF GENE AMPLIFICATION

Two general mechanisms can be envisaged: a) overreplication with recombination or b) unequal crossing over of sister chromatids. Overreplication involves multiple reinitiations of DNA synthesis within a single S-phase, generating onion skin like arrangements of DNA strands. Such structures can be resolved either in creating tandem arrays of the copies integrated in the genome as HSR's or as extrachromosomal circles evolving to form DM's.

For SV40-DNA amplification the onion skin model can be envisaged since amplification occurs inspite of complete division arrest following exposure to 5 Gy of a-particles (Lücke-Huhle, 1982). Unequal crossing over of sister chromatids would occur progressively during each mitosis, so cell division is needed. This alternative mechanism is more likely to play a role in the amplification of huge genes when a large number of origins of replication is involved.

Most of the mechanistic studies in gene amplification have been done on drug resistance. Here the relationship between amplification of a specific gene and overcoming growth constraint of the selecting agent by an increased production of the gene product, has been understood (Schimke, 1984b; Stark & Wahl, 1984). The physiological consequences of amplified cellular oncogenes, however, are not yet clear. They are found in a variety of human tumors (Schwab & Amler, 1990) and are commonly believed to be the basis for unlimitted growth in that they allow the host cell to escape growth control, to become mobile and invasive, or escape immune surveillance.

INITIATING EVENTS OF SV40-AMPLIFICATION

The fact that amplification has been found at doses of alpha particles as low as 0.7 Gy (a dose too low to hit the SV40 sequences directly in a large number of cells) suggested already that the site of primary DNA damage and the site of amplification are different entities and that these entities communicate by a trans-acting mechanism. The existence of a trans-acting cellular factor was confirmed by cell fusion experiments between irradiated and untreated cells carrying the indicator gene (Nomura & Oishi, 1984; Lambert et al., 1986; Lücke-Huhle & Herrlich, 1987; Ronai & Weinstein, 1988; Kleinberger et al., 1988). Amplification of the indicator gene in the heterokaryon indicates the transfer of a signal through the cytoplasm by a cellular factor which has been induced or activated by DNA damage.

Since overreplication requires and starts at an intact SV40 origin of replication it was supposed that the inducing factor acts at this site. By DNase I foot printing technique the binding of a protein in cell extracts from UV treated cells was detected at position 5207-5225, referred to as the early domain of the minimal origin of replication of SV40 (Lücke-Huhle et al., 1990a). By using a synthetic oligonucleotide (Fig. 4) comprising this domain for gel retardation experiments, the specific binding of a cellular protein has been detected (Lücke-Huhle et al., 1989). Binding activity is enhanced in extracts of irradiated cells (exposed to either γ-radiation, α-radiation or UV). A dose of 5 J/m^2 of UV enhances binding activity at least 10 fold within 2 min (Mai, Thesis 1991) and activity persisted for several hours. Point mutations of the binding motif decrease the binding efficiency in vitro considerably.

The binding of one or more cellular factor(s) to the early domain is absolutely necessary as the initial step in the process of DNA amplification. In an in vivo assay an excess of a double-stranded oligonucleotide comprising the early domain from position 5207 to 5225, added directly after irradiation, totally blocks radiation-induced SV40 amplification (Fig. 4) by competing for the factor. Oligonucleotides of small size, in our case 25 base pairs, are taken up by the cells from the culture medium if added in high concentration (30 μM). Thus we suppose that the competition experiment in vivo and the binding in vitro examined by gel bandshifts concern the same protein (Lücke-Huhle et al., 1989).

Characterization of the early domain binding complex by UV crosslinking and SDS-polyacrylamid gel electrophoresis shows three proteins of 67, 64 and 46 kD, but only

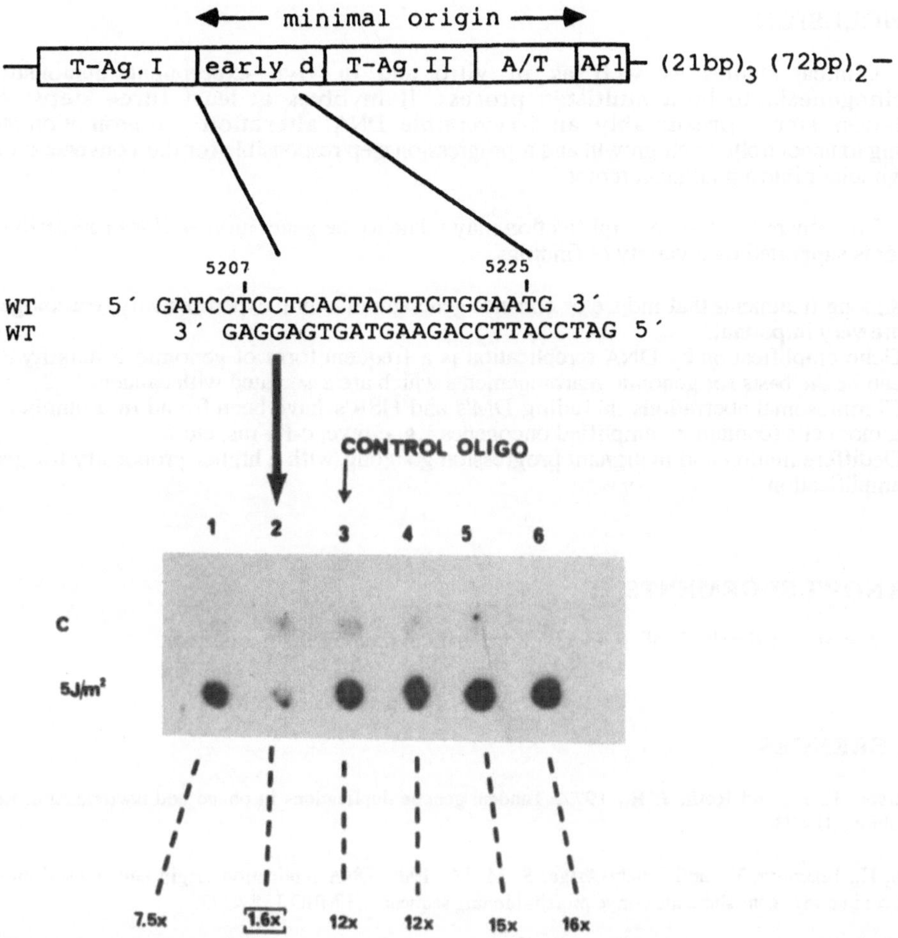

Figure 4. In Vivo Competition for a Cellular Binding Protein, Recognizing the Early Domain of the Minimal Origin of SV40, Inhibits Amplification. A Double Stranded Synthetic Oligonucleotide of 25 Base Pairs Comprising the Binding Site for the Cellular Protein Was Added to Either Untreated Cells (Lane C2) or to Cells Having Been Exposed to UV (5 J/m^2: Lane 2). All other Samples (lanes 1,4,5,6) Were Incubated with Normal Medium or Medium Containing an Unspecific Oligonucleotide of Equal Length (Lane 3). 3 Days after Irradiation Equal Numbers of Cells were Trapped on Nitrocellulose Paper and Hybridized to a 32-P-labelled SV40 Probe.

the binding activity of the first two proteins is inducible by DNA damaging agents (Mai et al., 1990).

Reports on increased SV40 replication in Myc expressing human lymphoma cells (Classon et al., 1987) as well as the result by Ariga et al. (1989) describing an origin of replication of c-myc which shows homology to the SV40 early domain and to which Myc protein binds itself, prompted us to study the involvement of c-Myc protein in gene amplification.

Using antibodies against Myc, a specific interference with the formation of the gel retardation complex was found suggesting that the formation of the DNA-protein complex, presumably essential for SV40 amplification, is or at least requires Myc. This is in agreement with our findings that cell lines like the dedifferentiated F9 mouse teratocarcinoma cells and the metastasis-positive BSp73 ASML rat cells with a high potential for induced gene amplification show overexpression of c-Myc.

CONCLUSION

Clinical studies as well as in vitro and in vivo experiments demonstrate carcinogenesis to be a multistep process. It involves at least three steps: An initiation step - presumably an irreversible DNA alteration - a promotion step leading to uncontrolled cell growth and a progression step responsible for the conversion of a benign tumor into a malignant tumor.

The concept that gene amplification may relate to the generation and/or progression of cancer is supported by a variety of findings.

1. Among treatments that induce or enhance gene amplification experimentally, carcinogens are very important.
2. Gene amplification by DNA rereplication is a frequent form of genomic instability and can be the basis for genomic rearrangements which are associated with cancer.
3. Chromosomal aberrations including DM's and HSR's have been found in a number of tumor cells (containing amplified oncogenes e.g. c-myc, c-Ki-ras, etc.).
4. Dedifferentiation and malignant progression go along with a higher propensity for gene amplification.

ACKNOWLEDGEMENTS

The author thanks Prof. P. Herrlich for critical reading the manuscript.

REFERENCES

Anderson, R. P., and Roth, J. R., 1977, Tandem genetic duplications in phage and bacteria,Ann. Rev. Microbiol., 31:473.

Ariga, H., Imamura, Y., and Iguchi-Ariga, S. M. M., 1989, DNA replication origin and transcriptional enhancer in c-myc gene share the c-myc protein-binding sequences, EMBO J., 8:4273.

Barbacid, M., 1986, Oncogenes and human cancer: cause or consequence?, Carcinogenesis, 7:1037.

Biedler, J. L., and Spengler, B. A., 1976, Methaphase chromosome anomaly: Association with drug resistance and cell-specific products, Science, 191:185.

Brown, D. D., and David, I. B., 1968, Specific gene amplification in oocytes, Science, 160:272.

Classon, M., Henriksson, M., Klein, G., and Hammaskjold, M.-L., 1987, Elevated c-myc expression facilitates the replication of SV40 DNA in human lymphoma cells, Nature, 330:272.

DeCicco, D. V., and Spradling, A. D., 1984, Localization of a cis-acting element responsible for the developmentally regulated amplification of drosophila chorion genes, Cell, 38:45.

Ehrfeld, A., Planas Bohne, F., and Lücke-Huhle, C., 1986, Amplification of oncogenes and integrated SV40 sequences in mammalian cells by the decay of incorporated Iodine-125, Radiation Res., 108:43.

Guerrero, I., Villasante, A., Carces, V., and Pellicer, A., 1984, Activation of a c-Ki-ras oncogene by somatic mutation in mouse lymphomas induced by gamma-radiation, Science, 225:1159.

Kleinberger, T., Flint, Y. B., Blank, M., Etkin, S., and Lavi, S., 1988, Carcinogen-induced trans activation of gene expression, Mol. Cell. Biol., 8:1366.

Lambert, M. E., Pellegrini, S., Gattoni-Celli, S., and Weinstein, I. B., 1986, Carcinogen induced asynchronous replication of polyoma DNA is mediated by a trans-acting factor, Carcinogenesis, 7:1011.

Lavi, S., and Etkin, S., 1981, Carcinogen-mediated induction of SV40 DNA synthesis in SV450 transformed Chinese hamster embryo cells, Carcinogenesis, 2:417.

Lehman, A. R., 1982, The cellular and molecular responses of ataxia telangiectasia cells to DNA damage, in: "Ataxia-Telangiectasia", B. A. Bridges, and D. G. Harnden, ed., John Wiley, New York.

Lücke-Huhle, C., 1982, Alpha-irradiation induced G2 delay - a period of cell recovery, Radiat. Res., 89:298.

Lücke-Huhle, C., and Herrlich, P., 1986, Gene amplification in mammalian cells after exposure to ionizing radiation and UV, in: "Radiation carcinogenesis and DNA alterations", F. J. Burns, A. C. Upton, and G. Silini, ed., Plenum Press, Amsterdam.

Lücke-Huhle, C., Pech, M., and Herrlich, P., 1986, Selective gene amplification in mammalian cells after exposure to ^{60}Co γ rays, ^{241}Am α particles, or UV light, Radiation Res., 106:345.

Lücke-Huhle, C., and Herrlich, P., 1987, Alpha-radiation-induced amplification of integrated SV40 sequences is mediated by a trans-acting mechanism, Int. J. Cancer, 39:94.

Lücke-Huhle, C., Hinrichs, S., and Speit, G., 1987, DHFR gene amplification in cultured skin fibroblasts of ataxia telangiectasia patients after methotrexate selection, Carcinogenesis, 8:1801.

Lücke-Huhle, C., Ehrfeld, A., and Rau, W., 1988, Gene amplification in Chinese hamster embryo cells by the decay of incorporated Iodine-125, in: "DNA Damage by Auger Emitters", K. F. Braverstock, D. E. Charlton, ed., Taylor & Francis, London-New York-Philadelphia.

Lücke-Huhle, C., Mai, S., and Herrlich, P., 1989, UV induced early-domain binding factor as the limiting component of Simian Virus 40 DNA amplification in rodent cells, Mol. Cell. Biol., 9:4812.

Lücke-Huhle, C., Gloss, B., and Herrlich, P., 1990a, Radiation-induced gene amplification in rodent and human cells, Acta Biol. Hung., 41:159.

Lücke-Huhle, C., Pech, M., and Herrlich, P., 1990b, SV40 DNA amplification and reintegration in surviving hamster cells after ^{60}Co γ-irradiation, Int. J. Radiat. Biol., 58:577.

Lücke-Huhle, C., and Herrlich, P., 1991, Retinoic-acid-induced differentiation prevents gene amplification in teratocarcinoma stem cells, Int. J. Cancer, 47:461.

Mai, S., Lücke-Huhle, C., Kaina, B., Rahmsdorf, H. J., Stein, B., Ponta, H., and Herrlich, P., 1990, Ionizing radiation induced formation of a replication origin binding complex involving the product of the cellular oncogene c-myc, in: "Ionizing Radiation Damage to DNA: Molecular Aspects", UCLA Symposia on Molecular and Cellular Biology, New Series, S. Wallace, and R. Painter, ed., Wiley-Liss, New York.

Mai, S., 1991, Mechanismen der Mutagen-induzierten zellulären Streßreaktionen: von DNA-Schädigung zu c-Myc-abhängiger Genamplifikation, Dissertationsschrift, Universität Karlsruhe,

Matzku, S., Komitowski, D., Mildenberger, M., and Zöller, M., 1983, Characterization of Bsp 73, a spontaneous rat tumor and its in vivo selected variants showing different metastasizing capacities, Inv. Met., 3:109.

Nomura, S., and Oishi, M., 1984, UV-irradiation induces an activity which stimulates Simian Virus 40 rescue upon cell fusion, Mol. Cell. Biol., 4:1159.

Pool, B. L., Yalkinoglu, A. ö., Klein, P., and Schlehofer, J. R., 1989, DNA amplification in genetic toxicology, Mutation Res., 213:61.

Ritossa, F. M., 1968, Unstable redundancy of genes for ribosomal RNA, Proc. Natl. Acad. Sci. U.S.A., 60:509.

Ronai, Z. A., and Weinstein, I. B., 1988, Identification of a UV-induced trans-acting protein that stimulates polyomavirus DNA replication, J. Virol., 62:1057.

Schimke, R. T., Alt, F. W., Kellems, R. F., Kaufmann, R. J., and Bertino, J. R., 1978, Amplification of dihydrofolate reductase genes in methotrexate-resistant cultured mouse cells and drug resistance in cultured murine cells, Cold Spr. Harb. Symp. Quant. Biol., 42:649.

Schimke, R. T., 1984a, Gene amplification in cultured animal cells, Cell, 37:705.

Schimke, R., T., 1984b, Gene amplification, drug resistance, and cancer, Cancer Res., 44:1735.

Schlehofer, J. R., Gissmann, L., Matz, B., and Zur Hausen, H., 1983, Herpes simplex virus-induced amplification of SV40 sequences in transformed Chinese hamster embryo cells, Int. J. Cancer, 32:99.

Schwab, M., and Amler, L. C., 1990, Amplification of cellular oncogenes. A predictor of clinical outcome in human cancer, Genes, Chromosomes & Cancer, 1:181.

Shah, D. M., Horsch, R. B., Klee, H. J., Kishore, G. M., Winter, J. A., Turner, N. E., Hironaka, C. M., Sanders, P. R., Gasser, S., Arykent, S., Siegel, N. R., Rogers, S. G., and Fowley, R. T., 1986, Engineering herbicide tolerance in transgenic plants, Science, 233:478.

Slamon, D. J., Godolphin, W., Jones, L. A., Holt, J. A., Wong, S. G., Keith, D. E., Levin, W. J., Stuart, S. G., Udove, J., Ullrich, A., and Press, M. F., 1989, Studies of the HER-2/neu proto-oncogene in human breast and ovarian cancer, Science, 244:707.

Speit, G., and Lücke-Huhle, C., 1987, Chromosomal changes associated with methotrexate-induced gene amplification in ataxia telangiectasia cells, Ann. Univ. Sarav. Med. Suppl., 7:302.

Stark, G. R., and Wahl, G. M., 1984, Gene amplification, Ann. Rev. Biochem., 53:447.

Stark, G. R., Debatisse, M., Giulotto, E., and Wahl, G. M., 1989, Recent progress in understanding mechanisms of mammalian DNA amplification, Cell, 57:901.

Wahl, G. M., Padgett, R. M., and Stark, G. R., 1979, Gene amplification causes over production of the first three enzymes of UMP synthesis in N(phosphoacetyl 1-aspartate)-resistant hamster cells, J. Biol. Chem., 254:8679.

Yalkinoglu, A. O., Heilbronn, R., Bürkle, A., Schlehofer, J. R., and zur Hausen, H., 1988, DNA amplification of adeno-associated virus as a response to cellular genotoxic stress, Cancer Res., 48:3123.

ALTERATION IN LIPID PEROXIDATION IN PLANT CELLS AFTER ACCELERATED ION IRRADIATION

A. Vasilenko, S. Zhadko and P.G. Sidorenko

Institute of Botany
Ukrainian Academy of Sciences
Repin Str. 2
254601 Kiev 4, Ukraine

SUMMARY

Superoxide dismutase activity (SOD), lipid peroxidation and the levels of adenylate metabolism were studied in cells of Nicotiana tabacum and root apex extracts of Pisum sativum irradiated with accelerated ions U-240, Institute for Nuclear Research, UAS, Kiev. Two groups of ions (protons and helium ions with energies 70 Mev and 100MeV,respectively) were used. The aim was to determine the possible relationship between the capacity of antioxidants, which detoxify potentially damaging forms of activated oxygen and the rate of lipid peroxidation by measuring the concentration of malondyaldehyde (MDA). Alteration in the levels of total antioxidants, lipid peroxidation product as well as the concentration of ATP and ADP in case proton and helium ion irradiation, occurred during the course of the radiational endogenous respnce.

INTRODUCTION

Plants being oxygen regenerates and a source of nutrient supply for the crews are a regular component of created closed ecological life-support subsystem (Kordyum et al., 1989; Midorikawa et al., 1989), and also represent convenient models in study of the radiation response to particle irradiation.

During irradiation, normal aerobic metabolism of the plant tissue can be seriously perturbed and dioxygen can be essentially activated by way of univalent reduction, yielding superoxide (O_2^-), hydrogen peroxide (H_2O_2) and the hydroxyl radical (*OH)(Spichalla, 1990). The biologically important forms of activated oxygen are highly reactive and can cause deleterious oxidation of the membrane phospholipids and nucleic acids. The damaging effects of activated oxygen are known to be modulated, if the radicals are scavenged by reacting with non protein traps and antioxidant enzymes like, superoxide dismutase (Blakely et al., 1989; Baraboy and Tchebotarev, 1986). SOD acts as a general protective agent to cells in plant tissue against the presence of damage by the superoxide anion and prevents this damages. SOD provides the ground state level of anion superoxide in plant and upsets a chain of the branch free-radical reactions in membranes catalyzing the dismutation of O_2^- to H_2O and 0_2. The plant cell membrane structure and properties (content of lipid peroxidation products, phospholipids and membrane viscosity) possesses also a certain sensitivity to the effects of space flight, like weightlessness (Polulakh et al., 1989).

Due to the significant deleterious influences of ionizing radiation on cell mitochondrial membranes (Polivoda and Konev, 1986; Seylanov et al., 1987) the adenylate status can also alter the normal metabolism of the plant tissues (Gerasimenko and Dvoretsky, 1989) and modulate the cell radiation responce.

In the study reported here we attempt to extend the results of biochemical investigations of the lipid peroxidation as well as the levels of SOD-enzyme activity and cellular energy relations by exposing two plant species of different developmental status to particle irradiation.

METHODOLOGY

Nicotiana tabacum cells cultured in vitro and seedling roots of Pisum sativum L. were used in these experiments. Experiments were carried out with exponentially growing cell populations of Nicotiana tabacum, as well as the root apices of Pisum sativum after four days germination. The cultured cells were maintained in the Murachige and Scoog (1962) medium.

Preparations similar to those in the accelerated ion experiments were irradiated in air with gamma-radiation from a ^{60}Co "Explorer" source as a control. Particle beams of the accelerated protons and helium ions with residual energies 0.95 keV/mkm and 9.34 keV/mkm were generated by the 95-in. isochronous cyclotron U-240, Institute for Nuclear Research, UAS, Kiev. Nicotiana tabacum cells in callus cultures (approximately 10^6 cells/sample) and Pisum sativum seedlings (approximately 250 apices/sample) were irradiated in air, at relative humidities and ambient temperature. The replacement of the irradiated plant samples were carried out by a manipulator and a remote control system. It was used in rapid freezing of the samples in a liquid nitrogen. Each frozen sample is maximally 3 mm thick, so that the whole sample will be frozen in less then 2 s. Thus within this time after irradiation the plant samples were deep-frozen. Within this time interval no changes were detectable in the content of metabolites such as adenine nucleotides which are know to respond rapidly to external perturbations (Goller et al., 1982).

For analysis of ATP and ADP levels after irradiation we applied the bioluminescence method using the luciferase and the piruatkinase enzyme systems. For this purposes we used the method of extraction and the assay of adenine nucleotide content as described by Malik and Thimann (1980) and Kimmich et al. (1975). For transferring of ADP into ATP we mixed 2 mL of the extract and 0.2 mL of solution containing 50 mM phosphoenolpiruat, 100 mM tris-HCl (pH=7.4), 35 mM KCl, 6 mM $MgCl_2$ and 150 units/mM piruat kinase "'Serva".

SOD was assayed photochemically (Beauchamp and Fridovich, 1971; Rabinovich and Sclan, 1980) using the modification of Chanway and Runeckles (1984). The 3-mL reaction mixture contained 50 mM potassium phosphate buffer, pH=7.8, 13 mM methionine, 63 mkM p-nitro blue tetrasolium, 2 mkM riboflavin, 0.1 mkM EDTA, and 0-200 mkM crude extract. Reactions were carried out under illumination from a single 20-W "cool-white" fluorescent lamp. The initial rate of reaction, as measured by the difference in increase in absorbance at 560 nm in presence and absence of enzyme extract was proportional to the amount of enzyme. Assays were carried out using three different volumes of enzyme extract. All assays contained an internal standard of humane erithrocyte SOD "Rostepidcomplex". In order to determine the specification this assay for SOD, analysis was carried out in the presence of KCN (1 mM), which is a specific SOD inhibitor, and with 0.2 mL of a chloroform-ethanol solution (3:5 v/v), which does not inhibit SOD (Rabinovich and Sclan, 1980).

The total antioxidant activity (AOA) was determined by the difference in decrease in absorbance at 517 nm of a stable radical dinitrophenol picril hydrasil in the absence and the presence of AOA extract, which was previously heated for 5 min at 95°C.

Malonic dialdehyde content (as a lipid peroxidation product) in tissue extracts was estimated after precipitation in 5% tiobarbitur acid, were heated for 10 min at 90°C and their absorbance was measured at 535 nm (Hodgson and Raison, 1991).

RESULTS AND DISCUSSION

In this paper, we report on ATP/ADP ratio levels associated with the cytosol of two species of plants. The most likely mediators of the interaction between respiration and

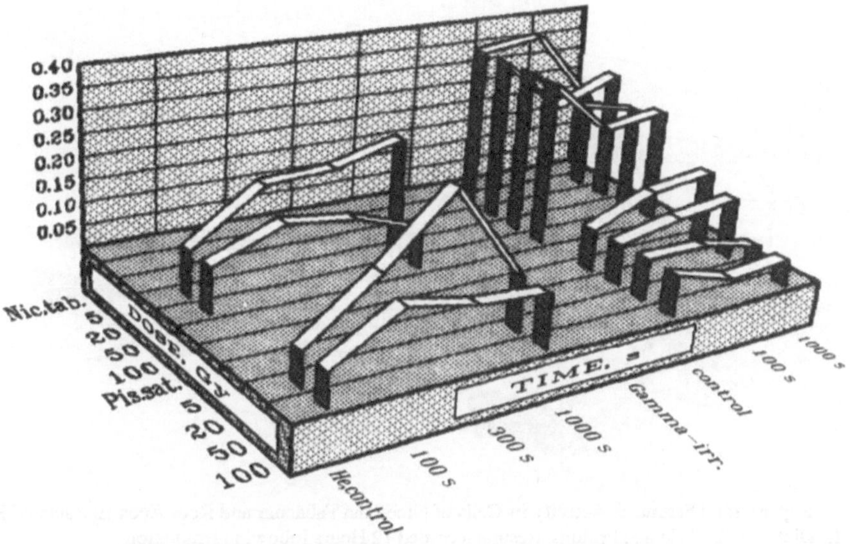

Figure 1. The ATP and ADP Content and the Adenylate Ratio Nicotiana Tabacum Cultured Cells and Pisum Sativum Root Apices for Gamma- and Helium Ion Irradiation, During the Course of Early Stages of Biochemical Response to the Radiation.

cytosolic effects are the adenylates. It indicates that the cytosolic energy state is the important regulatory parameter for organellar production and export of energy. The detail of the subsequent ATP and ADP assay are as described by Kimmich et al. (1975). This assay is based on the bioluminescence luciferin- luciferase reaction which requires ATP with single photons are counted.

Our primary interest was focused on possible alterations of adenylate ratio after helium irradiation at different dose rates. During 10^2 s of treatment, significant changes in the adenylate ratios are observed (Goller et al., 1982). About 300 s after irradiation (Gerasimenko and Dvoretsky, 1989) or clinostating (Vasilenko, 1986), changes in the ATP level can be detected.

Figure 1 shows the results of the ATP/ADP ratio measurements for the two plant species irradiated with ^{60}Co gamma-rays and accelerated helium ions (the mean of n= 3 to 5 separate experiments). The cell response for the two species of plants to sparce ionizing radiation are shown in the right part of Figure 1. Here the ATP/ADP ratios are plotted as a function of gamma-ray dose. The adenylate ratios are about 1.5 times higher for the Pisum sativum cells irradiated with gamma-rays in dose 5 and 20 Gy then for unirradiated samples. The ATP/ADP ratios are similar for the two species of plant tissues at about 10^3 s after irradiation for dose 50 and 100 Gy gamma-rays. Left part of the Fig. 1 indicates all adenylate ratio at time points following accelerated helium irradiation of the plant species. Each group reflect the ATP/ADP ratio versus the dose of the irradiation and the time after the treatment.

Quantitation of oxygen free radicals mediated damage in plant tissue is difficult. However, it is comparatively easy to quantitative endogenous antioxidants, which detoxify potentially damaging forms of activated oxygen and upset the lipid peroxidation promotion. High constitutive levels or high induced levels of total antioxidants and SOD-enzyme in a plant cells may provide resistance to particle irradiation and decrease late radiational effects (Blakely et al., 1989).

SOD-enzyme level in the extract from the plant species was estimated in a mixture containing riboflavin and EDTA, which produces the O_2^-. The SOD levels in the extracts of the plant species versus dose of sparce ionizing radiation are shown in the right part of

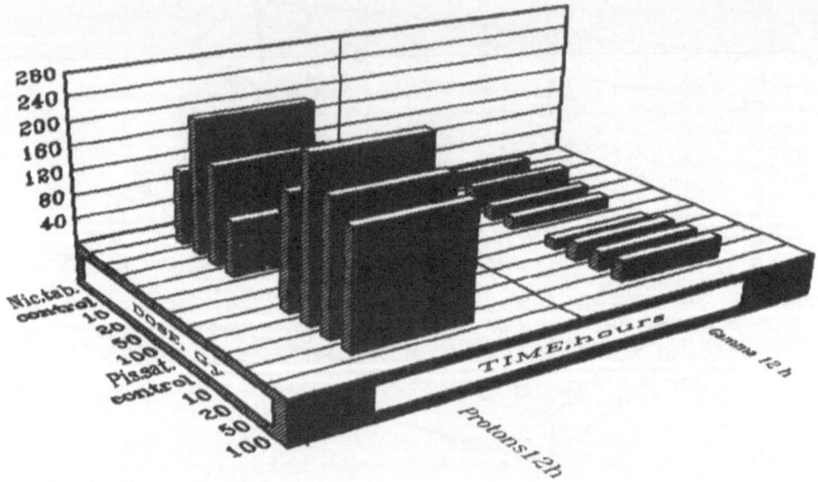

Figure 2 . Superoxide Dismutase Activity in Cells of Nicotiana Tabacum and Root Apex Extracts of Pisum Sativum Irradiated with ^{60}Co and Protons. Results recorded 12 Hours following Irradiation.

Figure 2. Left part of the Figure 2 indicates the SOD-enzyme levels we obtained 700 min after irradiation with accelerated helium ions. It revealed that the levels of SOD-enzyme decreased with increasing of the dose of irradiation for the extracts of Pisum sativum seedling roots and for Nicotiana tabacum cultured cells after ^{60}Co irradiation.

The results of the vulnerability of the membrane lipids of plant cells in vivo and in vitro (seedling roots of Pisum sativum and cultured cells of Nicotiana tabacum) to peroxidation processes during photon and particle irradiation and the level of total antioxidants are shown in the Fig.3. Results indicate that ionizing radiation enhances the chains of peroxidation reactions.

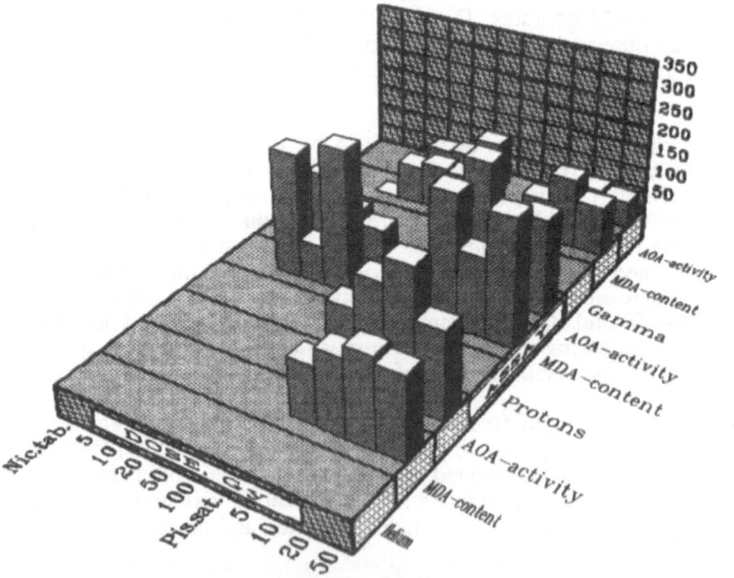

Figure 3. Lipid Peroxidation as TBA Reactivity and Total Antioxidant Content in Cells of Nicotiana Tabacum and Root Apex Extracts of Pisum Sativum Irradiated with ^{60}Co Gamma-rays, Protons and Helium Ions Occurred in 12 Hours following Irradiation.

REFERENCES

Baraboy, V.A. and Tchebotarev, E.E., 1986. The problem of peroxidation in biology, Radiobiology, 27, 591-597.

Beauchamp, C.O. and Fridovich, I., 1971. Superoxide dismutase: improved assay and an assay applicable of tylamide gels, Anal. Biochem., 44: 276-287.

Blakely, E., Chang, P., Lommel, L., Bjornstad, K., Dixon,M., Tobias, C., Kummar, K.,and Blakely, W.F., 1989. Cell-cycle radiation response: role of intracellular factors, Adv. Space Res., 9: 177-186.

Chanway, C.P., and Runeckles, V.C., 1984. The role of super oxide dismutase in the susceptibility of bean leaves to ozone injury,Canadian J. Botany, 62: 236-240.

Gerasimenko, I.V., Dvoretsky, A.I.,1989. Alteration of the late system energetic charge of the plant cell at ray pathology. In: Proceedings of the Ist All union radiobiological congress. -Pushchino, 5: 1023-1024.

Goller, M., Hampp, R., Ziegler, H., 1982. Regulation of the cytosolic adenylate ratio as determined by rapid fractionation of mesophyll protoplasts of oat. Effect of electron transfer inhibitors and upcouplers, Planta, 156: 255-264.

Hodgson, R.A.J., Raison, J.K., 1991. Lipid peroxidation and superoxide dismutase activity in relation to photoinhibition induced by chilling in moderate light, Planta, 185: 215-219.

Kimmich, G.A., Landless, A., Brand, G.S., 1975. Assay of picomole amounts of ATP, ADP andAMP using the luciferase enzyme system, Analit. Biochem., 68: 187-206.

Kordyum, E.L., Sidorenko, P.G., Klimchuk, D.A.,Martin, G.M. Zhadko, S.I., Vasilenko, A.I.,and Sytnik, K.M., 1989. Prospects of studies in space phytobiology, in: Preprint of the 40th Congress of the IAF, Malaga, Spain, AF/ IAA-89: 578.

McCord, J.M., Fridovich, I., 1969. Superoxide dismutase. An enzymatic function for erythrocuprein (hemocuprein), J. Biol. Chem., 44: 276-287.

Midorikava, Y., Fujii, T., Terai, M., Omasa, K.,Nitta, K., 1989. A food/nutrient supply planfor lunar base CELLS, in: Preprint of the40th Congress of the IAF, Malaga, Spain IAF/IAA - 89: 579.

Puppo, A., Herrada, G., Rigand, J., 1991. Lipid per-oxidation in peribacterial membranes from fronch bean nodules, Plant Physiol., 96: 826-830.

Polivoda, B.I., Konev, V.V., 1986. Correlation between membrane and genetic effects of LPO, Radiobiology, 26: 803-805.

Polulakh, Yu.A., Zhadko, S.I., Klimchuk, D.A.,Baraboy,V.A., Alpatov, A.N., and Sytnik,K.M., 1989. Plant cell plasma membrane structure and properties under clinostatting, Adv.Space Res., 9: 71-74.

Rabinowitch, H.D., Sclan, D., Superoxidedismutase,a possible protective agent against sunscaldin tomatoes (Lucersicon escentum Mill.),Planta, 148: 162-167.

Seylanov, A.S., Konev, V.V., Popov, G.A., 1987. Effects of radiomodifiers on lipid peroxidation andstructural-functional state of mitochondria, Radiobiology, 27: 854-869.

Swenberg, C.E., 1988. DNA and radioprotection, in:Terrestrial space radiation and its biological effects, ed. by McCormack, P., Swenberg, C.E., and Bücker, H., Plenum, New York, 456-67.

Vasilenko, A.I., 1986. ATP content of the cultured cells of Haplopappus gracilis (Nutt) A. Grey underclinostating, Ukrainian J. Botany, 43: 84-85.

TUMOUR INDUCTION IN ANIMALS AND THE RADIATION RISK FOR MAN

J.J. Broerse[1,2], J. Davelaar,[2] and C. Zurcher[3]

[1] Institute of Applied Radiobiology and Immunology TNO
2280 HV Rijswijk
[2] Department of Clinical Oncology, University Hospital
2300 RC Leiden
[3] Institute of Ageing and Vascular Research TNO
2300 AK Leiden, The Netherlands

INTRODUCTION

It is now generally recognized that carcinogenesis is the most important detrimental effect in man after low dose irradiation. The carcinogenic action of ionizing radiation is clearly demonstrated after exposure to relatively high doses (in excess of 0.5 Gy) but for the lower dose levels reliable direct observations are not available.

Animal experiments can be performed under standardized and controlled conditions and can provide insights on the following aspects: 1) the shapes of the dose-response relationships for tumour induction; 2) determination of the relative biological effectiveness (RBE) as a function of radiation quality; and 3) the influence of fractionation or protraction of the dose on tumour initiation and development. A common procedure in the analysis of carcinogenesis studies is to express the incidence of tumours in animals as a fraction of the total number of animals in the various cohorts. Such an approach has the disadvantage that the time dependence of the tumour incidence is disregarded and that the group of animals at risk could decrease with time. Actuarial methods, such as the Kaplan and Meier (1958) product limit estimate, allow corrections for loss of animals from the experiment due to intercurrent death or other reasons. Subsequently, the carcinogenic effects can be described by applying non-parametric models e.g. the proportional hazards model (Kellerer and Chmelevsky, 1982) or analytical models e.g. the Weibull distribution (Broerse et al., 1986).

The risks of total-body irradiation with large doses of X-rays and fission neutrons were investigated by keeping a group of long-term surviving monkeys from an experiment on acute effects under continuous observation. The incidence and classification of the radiation-induced tumours are reported with reference to that in the non-irradiated control group.

The mammary gland is one of the tissues with a relatively high susceptibility for radiation carcinogenesis in the western countries. For studies in experimental animals this end-point further offers the advantage of early detection by palpation. Results on mammary carcinogenesis under different conditions are discussed in relation to tumour induction data obtained in other tissues.

In its most recent recommendations, the ICRP (1991) has estimated the probability of a fatal cancer by relying mainly on studies of the Japanse survivors of the atomic bomb explosions. The lifetime cancer risk was derived by considering the accumulated data up to 1985, the new dosimetry and projection to lifetime for exposures to high doses and high dose rates. The extrapolation to the area of low doses and low dose rates depends to a great extent on the shape of the dose-effect relationships. One of the possible approaches is the use of a linear relationship, associated with a reduction factor for low doses and low dose rates. The newest probability coefficients for stochastic effects are summarized in conjunction with the tissue weighting factors. The increased risks coefficients for tumour induction have resulted in lower dose limits for occcupational exposure.

CARCINOGENESIS IN MONKEYS AFTER HIGH-DOSE IRRADIATION

Studies on acute and late effects in non-human primates are relevant for man, since the radiation effects in both species show similarities. The response of rhesus monkeys after exposure to relatively high doses of fission neutrons and X-rays and the protective effect of autologous bone marrow transplantation have been investigated (Broerse et al., 1978). At the time of the irradiation, the approximate age of the animals (see Fig. 1) was three years. Special provisions were made to achieve an irradiation with a dose distribution which is as homogeneous as possible. For the non-grafted monkeys, the LD50 values for X-ray and neutron-irradiated monkeys are 5 and 2.6 Gy, respectively, resulting in an RBE of

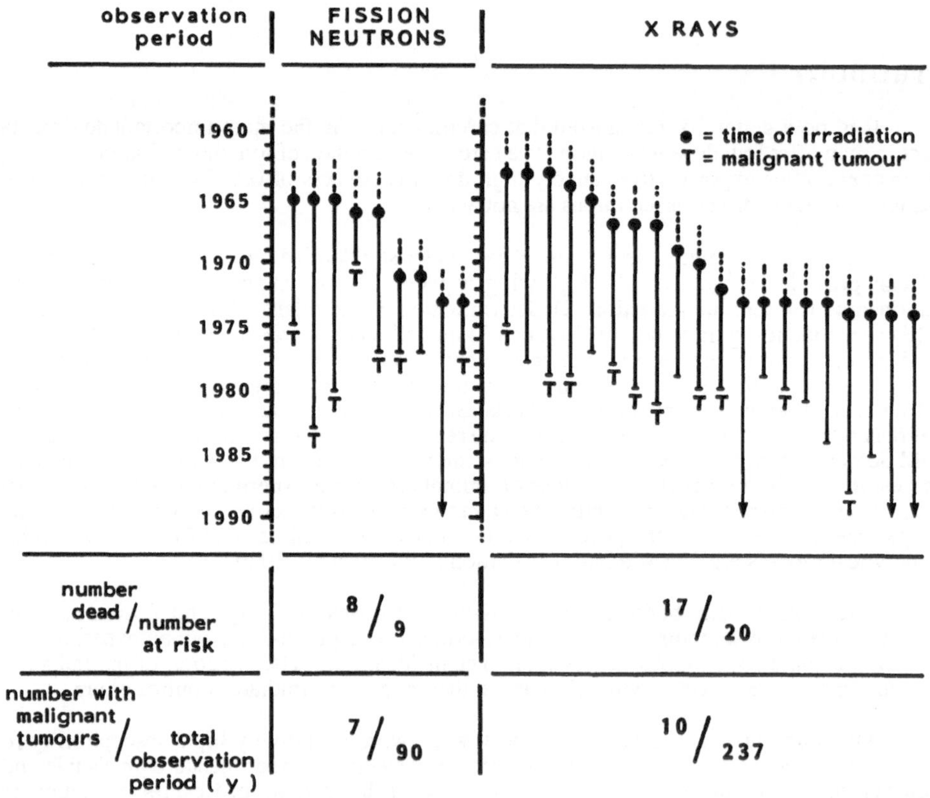

Figure 1. Tumour Incidence and Post-Irradiation Observation Periods for Long-Term Surviving Rhesus Monkeys after Whole-Body Irradiation and Autologous Bone Marrow Transplantation. The Dashed Portion of the Lines Indicate the Approximate Age of Monkeys Before Entering the Colony. Lines Ending in Cross Bars Signify Death and Arrow Heads Indicate that the Monkeys are Still Alive.

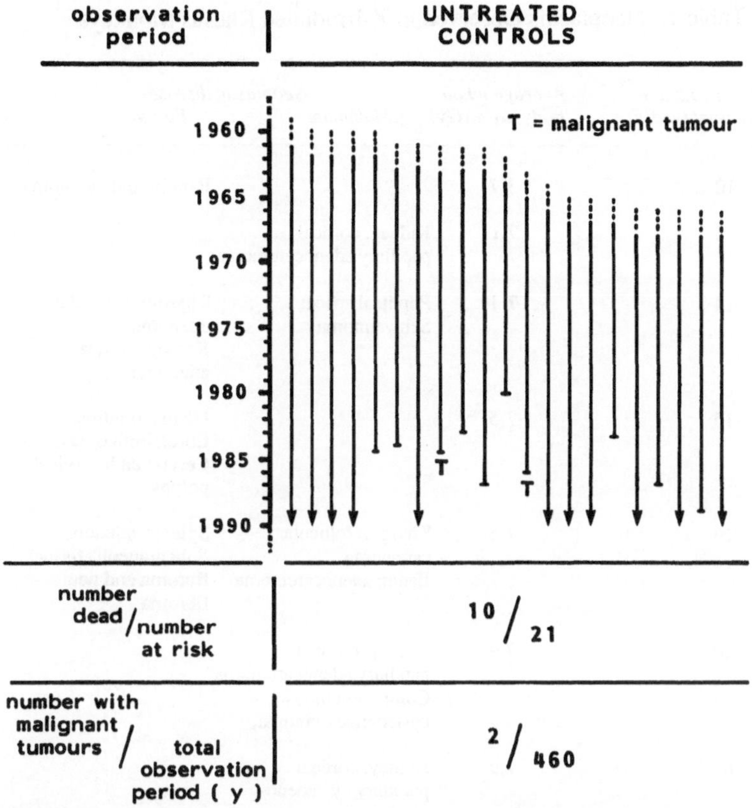

Figure 2. Tumour Incidence and Observation Periods in Untreated Rhesus Monkeys Which Served as an Age-Matched Control Group for the Animals Irradiated with X-Rays or Fast Neutrons.

approximately 2 for the occurrence of the haemopoietic syndrome. Protection by autologous bone marrow transplantation was demonstrated for doses up to 8 Gy of X-rays and 4.4 Gy of fission neutrons. For higher doses the bone marrow grafting was no longer effective and the majority of animals died within seven days with severe damage to the small and large intestine. Application of haemopoietic growth factors and optimal clinical supportive care are presently considered as a useful alternative for the autologous bone marrow transplantation (Wielenga, 1990).

The stochastic risks of total body irradiation are investigated by keeping the long-term surviving monkeys from the above described experiment under continuous observation. Rhesus monkeys of a comparable age distribution were maintained under identical conditions of housing and nutrition to serve as a control group. The animals were inspected frequently. In some cases the animals were killed when moribund. As soon as possible after the death of an animal, a complete necropsy was performed. The average absorbed doses in the animals were 6.7 Gy and 3.4 Gy, respectively, for the monkeys irradiated with X-rays and fission neutrons. The data on longevity and tumour incidence in the irradiated monkeys and in the control group are presented in Figures 1 and 2, respectively.

The minimum latency periods for neoplastic diseases were 7.5 years for X-irradiated animals and 4 years after neutron irradiation. In the X-irradiated group 9 out of 20 monkeys, and in the neutron group 7 out of 9 animals died with malignant tumours. A survey of the benign and malignant tumours after various treatment modalities is presented in Tables 1 and 2. The incidence of intracranial tumours is in accordance with the results of Yochmowitz et al. (1985), who performed a lifetime study in monkeys after single total body exposure to mono-

Table 1. Neoplasms Observed in X-Irradiated Rhesus Monkeys.

Sex	*Postirradiation interval (years)*	*Average whole body dose (Gy)*	*Neoplastic disease* *Malignant*	*Benign*
m	10	3.7		Parathyroid: adenoma
m	13	7.1	Kidney: cortical papillary adenocarcinoma	
f	14	7.1	Popliteal space: Schwannoma	Thyroid: follicular adenoma, Kidney: cortical adenoma.
f	15	7.5		Uterus: multiple fibroleiomyomas, Cervix: endocervical polyps
m	16	7.5	Thyroid: follicular carcinoma Ileum: adenocarcinoma	Spleen: splenoma Subcutaneous tissues: fibroma and neuro-fibroma
m	10	7.9	Kidney: cortical papillary adenocarcinoma Colon: mucinous cystadenocarcinoma,	
m	8	7.9	Kidney: cortical papillary cystadeno-carcinoma	
f	7	7.9	Kidney: cortical papillary cystadeno-carcinoma	
m	14	7.9	Multiple myeloma Kidney: cortical carcinoma	
m	12	8.0	Elbow: glomus tumour, metastatic to lung, kidney, meniges. Maxilla: osteosarcoma	Kidney: cortical adenoma

energetic protons with doses between 1 and 8 Gy. The induction of malignant glomus tumours is of interest, since this tumour type is rarely observed in man. In the group of 21 untreated control monkeys, two animals died at the age of 23 years with uterine cervical cancer and gastric cancer, respectively. Although the total number of animals included in our study is rather limited, it is noteworthy that the mortality with cancer (two out of ten) in the group of unirradiated monkeys, resembles the situation in the aging human population. As indicated in the figures, the results can be analyzed in terms of the number of animals developing tumours per group as a function of the total observation period for the entire group and the average absorbed dose. Thus in the neutron-irradiated group, seven monkeys developed malignant tumours in a total observation period of 90 monkeys-years. For the X-irradiated group this number is equal to 10 per 237 monkey-years and for the control group, 2 per 460 monkey-years. In this way the following risk factors for tumour induction: $(10/237 - 2/460) \cdot 6.7^{-1} = 57 \times 10^{-4}$ year^{-1} Gy^{-1} for X-rays and $(7/90 - 2/460) \cdot 3.4^{-1} = 216 \times 10^{-4}$ year^{-1} Gy^{-1} for fission neutrons and an approximate RBE of 4 can be derived.

Table 2. Neoplasms Observed in Rhesus Monkeys after Irradiation with Fission Neutrons.

Sex	*Post irrad. interval (year)*	*Average whole body dose (Gy)*	*Neoplastic disease* *Malignant*	*Benign*
m	10	2.3	Sacral area: glomus tumour metastatic to lung	Lip: fibroma Pancreas: islet cell adenoma.
m	15	2.6	Scrotum: glomus tumour Kidney: cortical carcinoma	Subcutis: cavernous hemangioma, neuro-fibroma, Schwannoma Adrenal: pheochromo-cytoma Pancreas: multiple insulinomas
m	18	3.5	Humerus: synovial sarcoma Kidney: cortical carcinoma metastatic to lung, liver lymph node Liver: carcinoma Thyroid: follicular carcinoma	Gingiva: fibrous epulis
m	12	3.5	Kidney: cortical papillary carcinoma Skull: (os frontale): osteosarcoma	Spleen: splenoma
m	4	3.8	Humerus: teleangiectatic osteosarcoma Mandible: glomus tumour, metastatic to liver	
m	6	4.1	Brain: astrocytoma	Skin: ossifying fibroma
f	4	4.4	Brain: glioblastoma	

It should be realized that these risk factors are calculated on the assumption of a linear dependency of the dose-effect relationship; that they are derived from tumour incidence data obtained at relatively high doses and that they pertain to risk factors per monkey-year. Furthermore, this approach does not take into account the time-dependence of the tumour appearance.

These drawbacks can partly be overcome by assessing the cumulative tumour rate from data for survival without evidence of tumour in course of time (see Fig. 3). The cumulative hazard H(t) is related to the survival function S(t) as follows:

$$H(t) = -\ln S(t) = [(t-\gamma)/\alpha(D)]\beta$$

where α is the time-scale parameter, β the shape parameter and γ the time offset parameter, which indicates the time prior to which tumours have not been observed.

The cumulative hazards at 15 years past irradiation amount to 0.84 and 0.31 for the neutron and X-ray experiments,respectively leading to an RBE of $(0.84 \times 3.4^{-1})/(0.31 \times 6.7^{-1}) =$ 5. This RBE value is similar to the one derived from the total observation period for the groups under consideration. Due to difference in the time dependence for tumour induction in the X-ray and neutron irradiated monkeys, the RBE will, however, vary from 15 at 10 years to a lower value of 2 at the maximum follow-up time of 20 years.

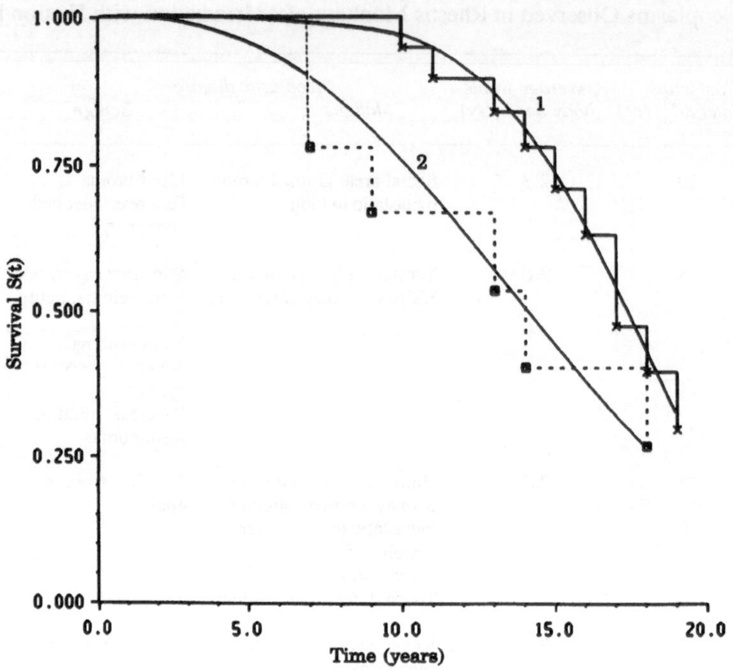

Figure 3. The Survival without Evidence of Tumour as a Function of Time for the Two Groups of Rhesus Monkeys, Irradiated with Average Doses of 6.7 Gy X-rays (Curve 1) and 3.4 Gy Fission Neutrons (Curve 2).

RADIATION CARCINOGENESIS IN RODENTS

Although the appearance of a malignant lesion will always require histopathological confirmation, tumours of the mammary gland in the rat offer the advantage of early detection. In view of expected variations in susceptibility it is useful to study the carcinogenic effect in different strains. Even in the same rat strain derived from different breeding stocks, variations in susceptibility can be anticipated. Outbred Sprague Dawley rats from an American source used at Brookhaven (Shellabarger et al., 1980) showed a larger and more rapid mammary neoplastic response to radiation than did partially inbred Sprague Dawley rats from a Dutch source (Van Zwieten, 1984). Mammary tumorigenesis has been studied at TNO in three rat strains, notably Sprague Dawley, WAG/Rij and BN/BiRij under different experimental conditions.

Although the results of some experiments are still under investigation, a number of findings can already be summarized. Considerable differences in susceptibility have been observed between the three rat strains, however, only in the WAG/Rij rats an appreciable number of carcinomas was induced by radiation (Van Zwieten, 1984). Estrogens are powerful promotors of both spontaneous and radiation induced mammary tumours in some rat strains, but not in others. After hormone treatment the tumours appear earlier than in the parallel groups without the hormone, and the proportion of rats with malignant tumours increases considerably after hormone treatment (Broerse et al., 1987).

Analysis of the data by hazard functions has resulted in linear dose-response curves for fibroadenomas in Sprague Dawley rats and for fibroadenomas and carcinomas in WAG/Rij rats after irradiation with 0.5 MeV neutrons and X-rays. It should be recognized, however, that the dose-response curves for the X-irradiations do not comprise experimental points for dose levels below 0.2 Gy. On the basis of the linear dose response curves RBE-values between 7 and 15 were derived for tumour induction by 0.5 MeV neutrons (Broerse et al., 1985). It is a universal finding that the highest RBE-values are observed for neutrons in this energy region. Identical sets of tumour induction data have been analyzed with a parametric Weibull function (Broerse et al., 1985) optimized via the maximum likelihood or chi-square method, as well as with the non-parametric proportional hazards model (Davelaar et al., 1991). The results using the two methods were similar, indicating the appropriateness of both models.

In order to investigate the risk of repeated exposure to very small doses of radiation as encountered in screening programs by mammography, the induction of mammary tumours has been studied in the WAG/Rij strain after fractionated irradiations with relatively low doses of gamma rays. The effects of fractionated irradiation with Cs-137 gamma rays for 120 fractions of 2.5 mGy and 10 mGy (interval 12 h) are compared with those of an acute exposure with doses of 0.3 Gy and 1.2 Gy. In general, the animals are introduced to the exposure regimen in their early maturity at an age of 8 weeks. In view of a possible dependence of susceptibility on age and developmental stage of the mammary glands, two additional groups were irradiated with single doses at the age, reached at the end of the fractionated irradiation, notably 17 weeks. The experiments were performed with normal animals and animals in which estradiol-17β (E_2) pellets were implanted subcutaneously at the age of 6 weeks.

The time dependence for survival of rats without evidence of either benign or malignant tumours has been analyzed. As an example, the occurrence of mammary carcinomas after irradiation with 1.2 Gy gamma-radiation is shown in Figure 4. The results of the non-parametric analysis with a proportional hazards model are shown in Figure 5 (Davelaar et al., 1991). The values for the hazards for carcinomas as well as for the fibroadenomas after irradiation with 0.3 Gy total dose are too small to be of statistical significance. For the 1.2 Gy exposure, the results are comparable for the carcinomas and the fibroadenomas, in that the single dose irradiation at 17 weeks of age is less tumourigenic than the single dose at 8 weeks of age. For both tumour types, fractionated irradiation results in significantly fewer tumours than after the single exposures. A mathematical analysis of the relative hazard reveals quadratic dose-response curves without a significant linear component for the induction of carcinomas. The susceptibility for tumour induction is considerably reduced when the irradiation is performed at an older age. On the basis of this experiment a DDREF (dose and dose rate effectiveness factor) of 4.5 ± 2 can be derived.

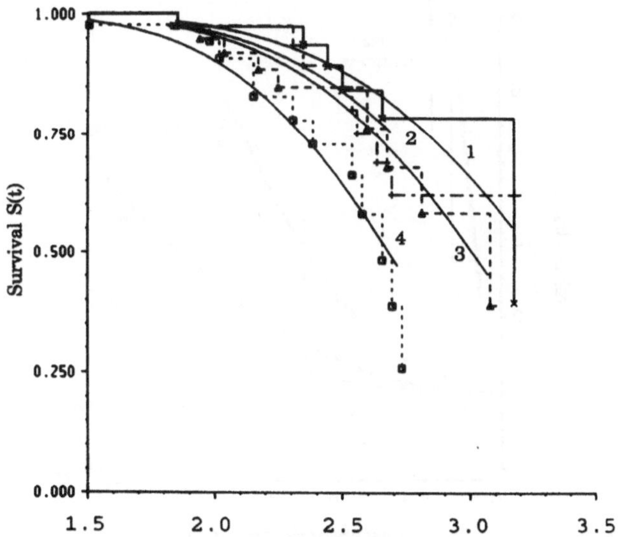

Figure 4. Probability of Surviving without Evidence of Mammary Carcinomas in WAG/Rij Rats after Single Dose Irradiations at 8 Weeks (Curve 4) and 17 Weeks of Age (Curve 3) and after Fractionated Irradiation (Curve 2) with 1.2 Gy Gamma Rays in Comparison with Controls (Curve 1).

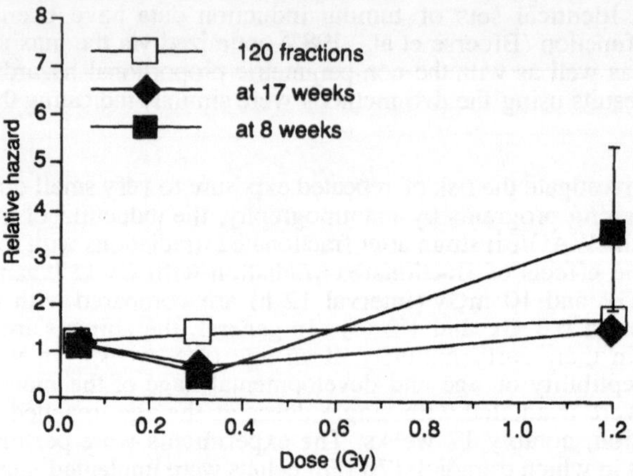

Figure 5. Relative Hazard for Mammary Carcinomas in Three Cohorts WAG/Rij Rats (Single Dose at 8 and 17 Weeks and 120 Fractions, No Hormones are Administered) as a Function of Dose Analyzed with the Proportional Hazard Method.

The RBE values reported earlier (Broerse et al., 1985) for induction of mammary tumours at TNO were based upon a linear dose-response curve for neutrons and X-rays. On the basis of actuarial survival curves the relative hazard as a function of the total absorbed dose was calculated as $\eta D) = \{\alpha(0)/\alpha(D)\}\beta$. Appreciably higher RBE values would be obtained, when the results of the gamma exposure were used as a base line (see Fig. 6). It should be recognized, however, that such an increase in RBE will be caused by the lower efficiency of low-LET radiation rather than by an increase in efficiency of the neutron irradiation at low doses.

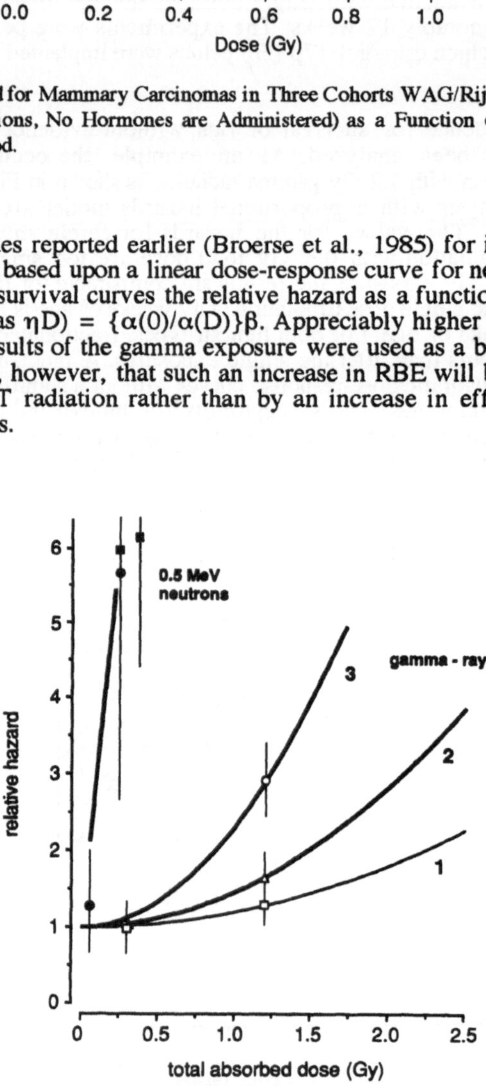

Figure 6. Relative Hazard as a Function of the Total Absorbed Dose for the Induction of Mammary Carcinomas in WAG/Rij Rats after Irradiation with 0.5 MeV Neutrons and Cs-137 Gamma-rays (Curve 1: Fractionated Irradiation, Curve 2: Single Dose Irradiation at 17 Weeks of Age and Curve 3: Single Dose Irradiation at 8 Weeks of Age).

EFFECTS OF FRACTIONATION OR PROTRACTION OF THE DOSE ADMINISTRATION

When the radiation dose is administered in a number of fractions or at a reduced dose rate, the biological response is different from that obtained after single acute doses. For cell survival, the modifying processes include repair of sublethal damage, repopulation and redistribution over the cell cycle. For tumour induction the nature of modifying mechanisms is not yet well understood, and it is rather difficult to interpret all experimental results obtained after fractionated or protracted exposures.

In general, for low-LET radiation the effectiveness is reduced with respect to the acute exposure when fractionation is applied. However, for more densely ionizing radiations, a number of experiments have been reported where a greater biological effect was observed after a protracted exposure than after the same total dose delivered in an acute exposure. This inverse dose rate effect was first demonstrated by Hill et al. (1982) for the induction of oncogenic transformation in vitro by fission spectrum neutrons. In subsequent experiments with neutrons (Balcer-Kubiczek et al., 1988, Saran et al., 1991) and α-particles (Hieber et al., 1989) the existence of such an effect was contradicted. In an evaluation of the experimental data, Brenner and Hall (1990) derived some common features. The largest enhancement was observed for fission neutrons at dose rates below 5 mGy/min, while little or no enhancement was observed at dose rates above this level. The inverse dose-rate effect seemed to be confined to radiations of intermediate LET and appeared most prominent at doses around 0.2 Gy. Rossi and Kellerer (1986) postulated that the dose rate dependence of oncogenic transformation by neutrons might be due to variation of response during the cell cycle.

Studies on in vivo radiation carcinogenesis have also shown that for high-LET radiation the effects of fractionation or protraction are different for various tumour types (see Table 3). For the induction of ovarian tumours in mice, Ullrich (1984) observed that fission

Table 3. The Effect of Fractionation or Reduction of Dose Rate on Tumour Induction or Longevity in Experimental Animals after High-LET Irradiation.

	*Change in effectiveness**
Mammary tumour	
Ullrich (1984)	+
Vogel and Dickson (1982)	+
Broerse et al. (1985)	=
Ovarian tumours	
Ullrich (1984)	-
Pulmonary tumours	
Ullrich (1984)	=
Lundgren et al. (1987)	+
Little et al. (1987)	=
Myeloid leukaemia	
Huiskamp (1991)	=
Osteosarcomas	
Müller et al. (1990)	+
Lifeshortening in mice	
Thomson et al. (1985)	+
Maisin et al. (1988)	=

* Enhanced (+), reduced (-) or equal (=) with respect to single -or high- dose rate irradiation.

neutron irradiation was less effective when delivered at low dose rates in comparison with high dose rates. However, the mammary carcinogenic effect of neutrons was enhanced at low dose rates, a finding similar to that of Vogel and Dickson (1982). Studies on the induction of mammary carcinoma in WAG/Rij rats (Broerse et al., 1985) after single and fractionated irradiations with X-rays and 0.5 MeV neutrons indicate that for equal total absorbed dose the tumours appeared at approximately the same age. It should be noted that the experimental studies on mammary carcinogenesis are generally based on whole-body irradiations of the animals. The induction of mammary cancer can easily be modified by hormonal factors, and it might well be that specific endocrinological effects caused by the irradiation, influence mammary tumour induction to a lesser extent when fractionated or protracted exposures are studied.

The induction of lung cancer after single or protracted irradiation with α-particles was investigated independently in two species. Lundgren et al. (1987) studied the effect of inhalation of Pu-239 oxide in mice after single or repeated exposure. For similar cumulative doses from α-particles , an approximately four times greater incidence of pulmonary tumours was observed than for the single-inhalation exposure. The effect of dose rate on the induction of lung cancer in Syrian hamsters was studied by Little et al. (1985) after intratracheal instillation of Po-210. Protraction of the γ-irradiation over 120 days was slightly more carcinogenic at lower total lung doses but slightly less carcinogenic at higher doses, in comparison to an exposure limited to a 10-day period.

RISK FOR RADIATION-INDUCED CANCER IN MAN

Information on the carcinogenic risks in man due to individual exposure stems from the atomic bomb explosions in Japan, and from medical applications such as the spondylitic patients and of women treated for cervix cancer. These three groups comprise an appreciable number of persons (14.000 for the spondylitic patients; up to 76.000 for the Japanese survivors). The inhabitants of Hiroshima and Nagasaki were exposed to an instantaneous whole body dose. The medical studies concern fractionated irradiations. The total observation period for the first group is longer than the follow-up time for the medical studies. The risk factors derived from the Japanese studies are higher than those resulting from medical applications. However, in its most recent recommendations the ICRP (1991) has estimated the probability of a fatal cancer by relying mainly on the studies of the Japanse survivors of the atomic bombs and their assessment by expert groups such as UNSCEAR and BEIR. These committees have estimated the lifetime cancer risk by considering the accumulated data up to 1985, the new dosimetry system (DS 86) and projection to lifetime by a multiplicative or

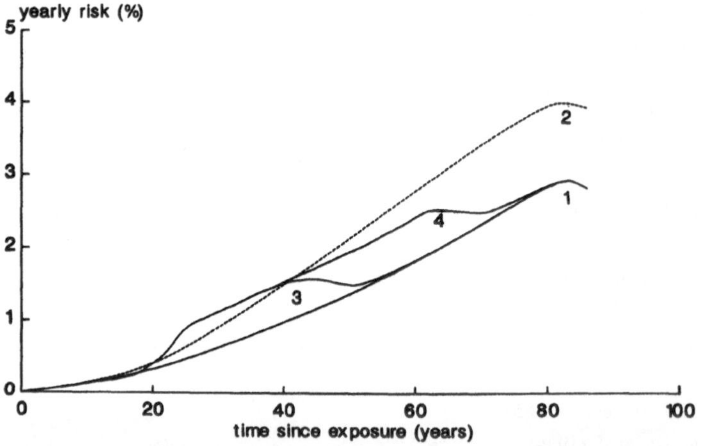

Figure 7. The Yearly Risk for Tumour Induction According to the Multiplicative Risk Model (Curve 2) and the Additive Risk Model with Short (Curve 3) and Long (Curve 4) Risk Period as a Function of Time after Exposure in Comparison with the Natural Incidence (Curve 1).

Table 4. Estimated Excess Cancer Risks in a UK Population of All Ages and Both Sexes Associated with Exposure to Low-LET Radiation (Stather et al., 1988).

Cancer type	Fatal cancer risks, 10^{-2} Gy^{-1} based on lifetime projection of relative risks		based on excess deaths to date in Japan		ICRP 1977 adults only
	HDR	LDR	HDR	LDR	
leukaemia*	0.28	0.28	0.84	0.28	0.2
breast**	1.1	0.55	0.42	0.21	0.25
lung*	3.5	1.2	1.15	0.38	0.2
thyroid**	0.075	0.025	0.055	0.02	0.05
bone**	0.15	0.05	0.15	0.05	0.05
liver**	0.45	0.15	0.23	0.08	0.1
LLI/colon*	1.1	0.37	0.38	0.13	0.1
stomach*	0.73	0.24	0.14	0.05	0.1
other organs*	5.0	1.63	0.565	0.20	0.2
total*	12.9	4.5	3.93	1.4	1.25

HDR high dose rate (>100 mGy d^{-1}).
LDR low dose rate (<100 mGy d^{-1})
* based on Japanese A-bomb survivors.
** based on other studies.

modified multiplicative model for high dose rate exposure. The yearly risk for tumour induction according to the multiplicative and the additive risk model are schematically depicted in Figure 7. As indicated by Stather et al. (1988) the risk factors show appreciable variations depending on the type of analysis. In Table 4 the effect of extrapolating the observed radiation risk over the remainder of the life-span is compared with that derived solely from excess deaths observed to date in the Japanese population. The ICRP has recognized that the use of the multiplicative model may overestimate the probability of cancer incidence at older ages because the multiplying factor may not persist over the whole span of life. In its recent recommendations (ICRP,1991) the ICRP stated a fatal probability coefficient of 8 x 10^{-2} Sv^{-1} for the sum of all malignancies in a population of both sexes and of working age at high dose and high dose rates of low-LET radiation. The corresponding value for the whole population including children is estimated as 10 x 10^{-2} Sv^{-1}.

The epidemiological studies in man are based on data for induced malignancies after irradiation with doses in excess of 0.5 Gy. The risk factors obtained at these relatively high dose values must be extrapolated to the low dose region relevant for radiation protection. For this purpose, the ICRP introduced the dose and dose-rate effectiveness factor (DDREF). On the basis of animal experiments for individual tumour types and for life shortening in animals, NCRP (1980) and UNSCEAR (1988) suggested the use of a DDREF between 2 and 10, the implications being that the effect varies for different types of tumours. Information from human epidemiology on dose-response relationships and dose rate effects is limited and subject to many uncertainties. The ICRP (1991) has decided to use a factor of 2, recognizing that the choice of this value is somewhat arbitrary and may be conservative. The use of a DDREF of 2 leads to nominal risk coefficients for workers of 4 x 10^{-2} Sv^{-1} and for the whole population of 5 x 10^{-2} Sv^{-1} at low dose and low dose rates. The estimation of the ICRP (1991) for the total detriment due to occurrence of stochastic effects is summarized in Table 5.

Table 5. Nominal Probability Coefficients for Stochastic Effects According to ICRP (1991).

	Detriment (10^{-2} Sv^{-1})			
Exposed population	*fatal cancer*	*non fatal cancer*	*severe hereditary effects*	*total*
adult workers	4.0	0.8	0.8	5.6
whole population	5.0	1.0	1.3	7.4

On the basis of the risk factors for cancer in individual organs, the ICRP introduced already in their 1977 recommendations the concept of the weighting factor. This weighting factor, w_T, represents the proportion of the stochastic risk resulting from a specific tissue or organ with respect to the total risk when the body is irradiated uniformly. The renewed estimates for probability of fatal cancers in different organs and the total risks for the Japanese population have resulted in a revision for the tissue weighting factors (see Table 6). For the thyroid, bone, skin and liver elevated but not significant risks are reported from the Japanese data. Risk factors for these tissues have been based on other studies (ICRP, 1991). The fatal probability coefficients for colon and stomach suggested in 1991 are considerably larger than the 1977 recomendations, whereas the new risk factor for the breast is somewhat lower than the earlier estimate. These unexpected deviations have to be attributed to the use of the multiplicative risk model applied only on the Japanese survivors. It may be questioned whether an increase in the weighting factor from 0.06 to 0.12 for colon and stomach and a decrease in the weighting factor from 0.15 to 0.05 for the breast should be applied to the

Table 6. Probability Coefficients for Stochastic Effects in the Whole Population per 10^4 Persons per Sv and Tissue Weighting Factors (ICRP, 1991).

Tissue or organ	*Fatal cancer*	*Severe hereditary disorders*	*Tissue weighting factor w_T*	
			ICRP 1991	*old value*
gonads		100	0.20	0.25
bone marrow	50		0.12	0.12
colon	85		0.12	
lung	85		0.12	0.12
stomach	110		0.12	
bladder	30		0.05	
breast	20		0.05	0.15
liver	15		0.05	
oesophagus	30		0.05	
thyroid	8		0.05	0.03
skin	2		0.01	
bone surface	5		0.01	0.03
remainder	50		0.05	0.30
total	500	100	1.00	1.00

human population on a world-wide scale, in view of the striking differences in the natural incidence of these tumours between various parts of the world. Concerning the choice of a suitable dose and dose rate effectiveness factor and rational tissue weighting factors, experiments with animals can provide the missing link information.

On the basis of the new risk factors the ICRP (1991) has recommended new dose limits for different categories of the population (see Table 7). For occupational exposure (radiation workers) the maximum effective dose equivalent should not exceed 20 mSv per year. In addition to the limit of the effective dose (stochastic limit), the yearly limit for the lens of the eye is 150 mSv, for skin, hands and feet 500 mSv (deterministic limit). For the non-radiation workers the deterministic limits are 1/10 of the limits for the radiation workers with the exception of the extremities. The limits for the extremities are equal for radiation and non-radiation workers. For any member of the public the ICRP (1991) has recommended a limit of 1 mSv per year for the effective dose.

Table 7. Recommended Dose Limits According to ICRP (1991)

Application	*Dose limit* *Occupational*	*Public*
effective dose	20 mSv per year averaged over defined periods of 5 years	1 mSv in a year
annual dose equivalent in		
the lens of the eye	150 mSv	15 mSv
the skin	500 mSv	50 mSv
the hands and feet	500 mSv	--

CONCLUSIONS

The risk of inducing neoplastic late effects after total-body irradiation with relatively high doses has been demonstrated in subhuman primates. As a consequence of the studies in primates, it can be recognized that there is a strong need to screen regularly for secondary tumours, which may arise in patients who previously received high-dose total-body irradiation.

Concerning the nature of dose-response relationships for tumour induction a general consensus has been reached about a linear dependence on the dose for high-LET radiation. For low-LET radiation, however, approximately pure quadratic dose-response curves have been reported for a number of endpoints. Apparently, repair processes at the (sub)cellular level and systemic factors can modify the primary process of cancer initiation.

In general for low-LET radiation, the sparing effect of fractionated or protracted irradiation is demonstrated. However, for high-LET radiations the animal studies do not provide an unambiguous answer. It can not be excluded that systemic factor of immunological or endocrinological nature can influence the tumorigenic response in a different way after single or fractionated whole-body irradiations.

The new risk estimates of the ICRP (1991) for radiation-induced cancer in man are derived from studies of the Japanese survivors considering the revised dosimetry for exposure due to the atomic bomb explosions, a longer observation period and the projection to lifetime by a multiplicative model. The revision is mainly based on an increase in the number of solid malignancies from 135 observed in 1975 to a number of 260 in 1985 on a total population of 90.000 exposed persons. The new risk coefficient for fatal cancers of 5 x 10-2 Sv-1 at low dose and low dose rates (ICCRP, 1991) is probably an overestimation (Broerse and Dennis, 1990). The new dose limits for different categories of the population can therefore be classified as conservatively safe recommendations.

ACKNOWLEDGEMENT

The studies on late effects in rhesus monkeys have been supported by CEC grant no BI6-D-075-NL and the analysis of animal carcinogenesis data by CEC grant BI6-D-219-NL.

REFERENCES

Balcer-Kubiczek, E.K., Harrison, G.H., Zeman, G.H., Mattson, P.J. and Kunska, A. Lack of inverse dose-rate effect on fission neutron induced transformation of C3H10T1/2 cells. Int. J. Radiat. Biol. 54, 531-536, 1988.

Brenner, D.J. and Hall, E.J. The inverse dose rate effect for oncogenic transformation by neutrons and charged particles. A plausible interpretation consistent with published data. Int. J. Radiat. Biol. 58, 745-758, 1990.

Broerse, J.J., Van Bekkum,D.W., Hollander, C.F. and Davids, J.A.G. Mortality of monkeys after exposure to fission neutrons and the effect of autologous bone marrow transplantation. Int. J. Radiat. Biol. 34, 253-264, 1978.

Broerse, J.J., Hennen, L.A. and Van Zwieten, M.J. Radiation carcinogenesis in experimental animals and its implications for radiation protection. Int. J. Radiat. Biol. 48, 167-187, 1985.

Broerse, J.J. Hennen, L.A. and Solleveld, H.A.. Actuarial analysis of the hazard for mammary carcinogenesis in different rat strains after X- and neutron-irradiation. Leuk. Res. 10, 749-754, 1986.

Broerse, J.J., Hennen, L.A., Klapwijk, W.M. and Sonneveld, H.A. Mammary carcinogenesis in different rat strains after irradiation and hormone administration. Int. J. Radiat. Biol. 51, 1091-1100, 1987.

Broerse, J.J. and Dennis, J.A. Dosimetric aspects of exposure of the population to ionizing radiation. Int. J. Radiat. Biol. 57, 633-645, 1990.

J. Davelaar, J. Weeda and Broerse, J.J.. Analysis of animal carcinogenesis data by various mathematical methods. Radiat. Environ. Bioph. 30, 249-252, 1991.

Hieber, L., Ponsel, G., Roos, H., Feen, S., Fromke, E. and Kellerer, A.M. Absence of a dose-rate effect in the transformation of C3H10T1/2 cells by α-particles. Int. J. Radiat. Biol. 52, 859-869, 1989.

Hill, C.K., Buonoguro, F.J., Myers, C.P., Han, A. and Elkind, M.M. Fission-spectrum neutrons at reduced dose rate enhance neoplastic transformation. Nature, 298, 67-68, 1982.

Huiskamp, R. Acute myeloid leukemia induction in CBA/H mice by irradiation with fission neutrons as a function of exposure rate. Radiat. Environ. Biophys. 30, 213-215, 1991.

ICRP, 1990. Recommendations of the International Commission on Radiological Protection, ICRP Publication 60, Annals of the ICRP, 20, no. 1-3, 1991.

Kaplan, E.L. and Meier, P. Non-parametric estimation for incomplete observations. J. Am. Statist. Ass. 53, 457-481, 1958.

Kellerer, A.M. and Chmelevsky, D. Analysis of tumor rates and incidences - a survey of concepts and methods. In: Neutron Carcinogenesis. EUR 8084. pp. 209-231. Eds. J.J. Broerse and G.B. Gerber. Commission of the European Communities 1982.

Little, J.B., Kennedy, A.R. and McGandy, R.B.. Effects of dose rate on the induction of experimental lung cancer in hamsters by α-radiation. Radiat. Res. 103, 293-299, 1985.

Lundgren, D.W., Gillett, N.A., Hahn, F.F., Griffith, W.C. and McClellan, R.O. Effects of protraction of the a-dose to the lungs of mice by repeated inhalation exposure to aerosols of 239PuO2. Radiat. Res. 111, 201 - 224, 1987.

Maisin, J.R., Wambersie, A., Gerber, G.B., Mattelin, G., Lambiet-Collier, N., De Coster, B. and Gueulette, J. Life-shortening and disease incidence in C57B1 mice after single and fractionated gamma and high-energy neutron exposure. Radiat. Res. 113, 300-317, 1988.

Müller, W.A., Luz, A., Murray, A.B., and Linzner, U. Induction of lymphoma and osteosarcoma in mice by single and protracted low α-doses. Health Physics, 59, 305-310, 1990.

Rossi, H.H. and Kellerer, A.M. The dose rate dependence of oncogenic transformation by neutrons may be due to variation of response during the cell cycle. Int. J. Radiat. Biol. 50, 353-361, 1986.

Saran, A., Pazzaglia, S., Coppola, M., Rebessi, S., Di Majo, V., Garavini, M. and Covelli, V. Absence of a dose-fractionation effect on neoplastic transformation induced by fission-spectrum neutrons in C3H10T1/2 cells. Radiat. Res. 126, 343-348, 1991.

Shellabarger, C.J., Chmelevsky, D. and Kellerer, A.M. Induction of mammary neoplasms in the Sprague - Dawley rat by 430-keV neutrons and X-rays. J. Natl. Cancer Inst., 64, 821-833, 1980.

Stather, J.W., Muirhead, C.R., Edwards, A.A., Harrison, J.D., Lloyd, D.C. and Wood, N.R. Health models developed from the 1988 UNSCEAR Report, N.R.P.B. Report R226 (National Radiological Protection Board), 1988.

Thomson, J.F., Williamson, F.S. and Grahn, D. Life shortening in mice exposed to fission neutrons and gamma-rays. Radiat. Res. 104, 420-428, 1985.

Ullrich, R.L. Tumor induction in Balb/c mice after fractionated or protracted exposures to fission-spectrum neutrons. Radiat. Res. 97, 587-597, 1984.

UNSCEAR, Sources, Effects and Risk of Ionizing Radiation. Report to the General Assembly, with annexes (New York, United Nations), 1988.

Vogel, H.H. and Dickson, H.W. Mammary neoplasia in Sprague-Dawley rats following acute and protracted irradiation. In: Neutron Carcinogenesis. EUR 8084, pp. 135-154. Eds. J.J. Broerse and G.B. Gerber. Commission of the European Communities, 1982.

J.J. Wielenga. Hemopoietic stem cells in rhesus monkeys - surface antigens, radiosensitivity and responses to GM-CSF. Thesis, Rotterdam, 1990.

Yochmowitz, M.G., Wood, D.W. and Salmon, Y.L. Seventeen-year mortality experience of proton radiation in Macaca mulatta. Radiat. Res. 102, 14-34, 1985.

Van Zwieten, M.J.. The rat as animal model in breast cancer research. Martinus Nijhoff Publishers. Dordrecht/Boston, 1984.

PROGRESS IN THE EXTRAPOLATION OF RADIATION CATARACTOGENESIS DATA ACROSS LONGER-LIVED MAMMALIAN SPECIES

A. B. Cox*, Y. L. Salmon*, A. C. Lee**, J. T. Lett**, G. R. Williams**, J. J. Broerse***, G. Wagemaker***, and Dennis D. Leavitt****

*Armstrong Laboratory, Directed Energy Division
Brooks AFB, TX 78235-5301 USA
** Department of Radiological Health Sciences
Colorado State University, Fort Collins, CO 80523 USA
***Institute of Applied Radiobiology TNO
2280 HV, Rijswijk, The Netherlands
****Department of Radiology
University of Utah Medical Center
Salt Lake City, UT 84132 USA

INTRODUCTION

Induction of cataracts from exposure of astronauts to 'ambient' galactic and episodic solar particulate radiations is considered to be one of the primary risks of extended missions beyond the protection of the terrestrial magnetosphere. Exposures to high fluxes of solar protons during the projected mission to Mars, for example, could result in vision-impairing lenticular opacities late in the life span after the career missions of the astronauts are over (Lett *et al.*, 1991).

Quantitative assessments of cataractogenic risks for astronauts from both densely and sparsely ionizing radiations requires extrapolation across species of results from mammalian models that simulate humans. Such models do not include short-lived rodents when late degenerative cataractogenesis is concerned. Data obtained from such longer-lived species as the New Zealand white (NZW) rabbit (*Oryctolagus cuniculus*, median life span in captivity = 5-7 years), the beagle dog (*Canis familiaris*, median life span in captivity = 13-14 years) and the rhesus monkey (*Macaca mulatta*, median life span in captivity = ~24 years), which have been normalized for differences in subjective scoring indices, will be used to simulate cataractogenic profiles for humans.

Proton-induced cataracts in the Fischer-344 rat (*Rattus norvegicus*, median life span = ~2 years) also have been studied by our group. From the standpoint of radiation cataractogenesis, the fundamental objective of the rodent project was to examine the validity of our scoring system with a short-lived animal model that exhibits a very high incidence of senile cataracts.

The most recent data on radiation cataractogenesis in New Zealand white rabbits have been published by Cox *et al.* (1992), and are not shown here. Our primary purpose now is to present data from additional animal models to illustrate the potential for cross-species extrapolation from present and future data bases.

Biological Effects and Physics of Solar and Galactic Cosmic Radiation, Part A, Edited by C.E. Swenberg *et al.*, Plenum Press, New York, 1993

MATERIALS AND METHODS

Irradiation Conditions

Fischer-344 rats were exposed to 55-MeV protons (spread-out Bragg peak) at the Harvard Cyclotron Laboratory. The animals were shielded such that only the eyes, brains and proximate tissues were given single acute doses of 0, 2.0, 4.0, 8.5 or 18.0 Gy. Full details of the experimental protocol will appear elsewhere.

Beagle dogs were irradiated (whole-body) at various ages with ^{60}Co γ-photons for a life-span study completed several years ago, e.g. see Lett *et al.* (1988). The animals discussed here received 100 R (0.83 Gy) at 70 days post-partum.

As part of a joint study initiated in 1964 by the National Aeronautics and Space Administration (NASA) and the United States Air Force (USAF), rhesus monkeys were exposed to single whole-body doses of protons ranging in energy from 10-2300 MeV. Developments of radiation sequelae in surviving subjects are being followed throughout their life spans, see Dalrymple *et al.* (1991), Lett *et al.* (1988), and Wood (1991) for reviews of the project, and Leavitt (1991) and Hardy (1991) for reviews of irradiation protocols and dosimetry.

Rhesus monkeys irradiated (whole body) with X-rays at the Radiobiological Institute, TNO, Rijswijk, The Netherlands, and "rescued" with autologous bone marrow transplants, continue to be observed throughout their life spans (Broerse *et al.* 1991). Additional animals received total body exposures and were rescued via treatments with such agents as granulocyte-macrophage colony-stimulating factor (GM-CSF) or interleukin-3 (IL-3). Among the radiation effects monitored as the animals aged were cataracts (Sonneveld *et al.* 1979), but it was not until 1991 that cataracts were graded with the system described by Keng *et al.* (1982).

Cataract Evaluations

Subjects' pupils were dilated with mydriatics and cataracts were graded in all the control and irradiated animals using the scoring system described by Keng *et al.* (1982). See also Cox *et al.* (1992) and Lett *et al.* (1991) for further details of examinations of each species studied.

New Proton Dosimetry Calculations

The subjects in the joint NASA-USAF long-term radiation effects study of protons in rhesus monkeys recently have been designated as the "Delayed Effects Colony" (or DEC). As late effects, especially brain tumors (e.g., see Wood, 1991), developed in the animals exposed to 55-MeV protons, a re-evaluation of the dosimetry for proton energies ranging from 10-110 MeV was initiated (e.g., see Leavitt, 1991) in order to characterize more accurately the doses to irradiated brains. Concomitantly, new depth-dose calculations were made for primate eyes exposed to proton energies of 10, 32, 55 and 110 MeV (Leavitt, 1990).

Details of the dosimetry calculations for the primate brain may be found in the article by Leavitt (1991). In addition to the data treatments discussed in that paper, reference points were entered in 1-mm increments along a line bisecting the eye and used to evaluate the depth doses through the eye (Leavitt, 1990).

RESULTS

Proton-induced Cataracts in Fischer-344 Rats

Rat lenses were graded starting at times approximately 10 weeks following exposure

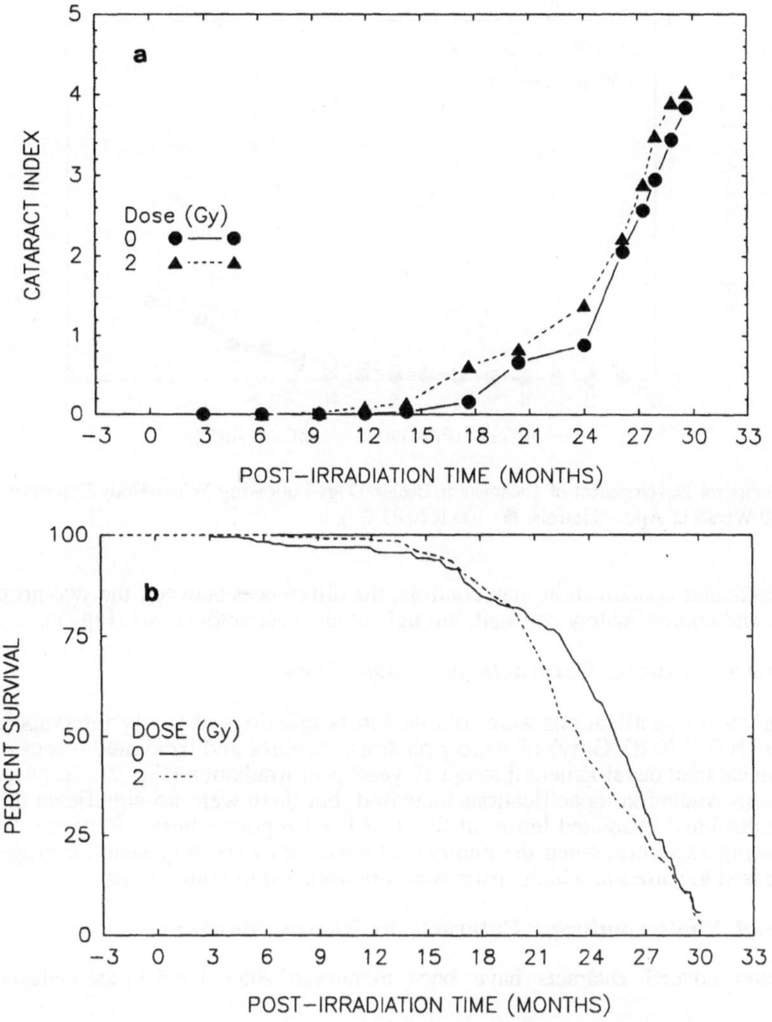

Figure 1. (a). Temporal Development of Cataracts in Male Fischer-344 Rats Following Exposure of the Head (Brain, Eyes and Proximate Tissues only) to Protons at 10 Weeks of Age; (b). Survival Curves for the Two Rat Populations; Doses are Indicated in the Figure.

to protons, but, for current purposes, only the data from animals given 2 Gray of protons (Fig. 1) will be examined in detail. Animals exposed to 18 Gy and to 8.5 Gy of protons began to develop cataracts soon after irradiation, but control lenses did not begin to exhibit significant levels of opacification until 80+ weeks post-irradiation (Fig. 1). Complete lenticular opacification occurred in the 18-Gy and 8.5-Gy subjects before lenticular opacifications began to be manifested in control subjects. Rats which received 4 Gy showed some increased degree of lenticular opacification as early as 28 weeks post-irradiation, but they began to show serious levels of opacification only around 60 weeks post-irradiation. Subsequent cataract development in those animals proceeded at a rate comparable to that of controls.

The lenses of rats exposed to 2 Gy of protons remained essentially clear until 50 weeks post-irradiation (Fig. 1). Although, on average, the 2-Gy animals exhibited a higher

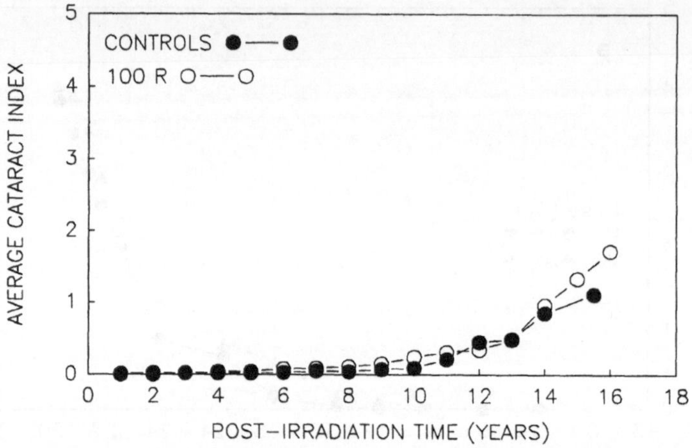

Figure 2. Temporal Development of Cataracts in Beagle Dogs Following Whole-Body Exposure to ^{60}Co γ Photons at 10 Weeks of Age. Controls, ●; 100 R (0.83 Gy), o.

degree of lenticular opacification than controls, the differences between the two groups were very small, and approximately constant, throughout late cataractogenesis (Fig. 1).

^{60}Co γ-photon-induced Cataracts in Beagle Dogs

Lenticular opacifications were followed in beagle dogs at yearly intervals following exposure to 100 R (0.83 Gray) of ^{60}Co γ photons. Control and irradiated lenses exhibited insignificant cataract development through 10 years post-irradiation (Fig. 2). Between 10 and 14 years post-irradiation, opacifications increased, but there were no significant differences between control and irradiated lenses at the dose level reported here. Between 14 and 16 years following exposure, when the numbers of survivors were very small, average cataract levels increased to values at which vision was compromised in both groups.

Proton- and Photon-induced Cataracts in Rhesus Monkeys

Proton-induced cataracts have been monitored since 1985 in the Delayed Effects

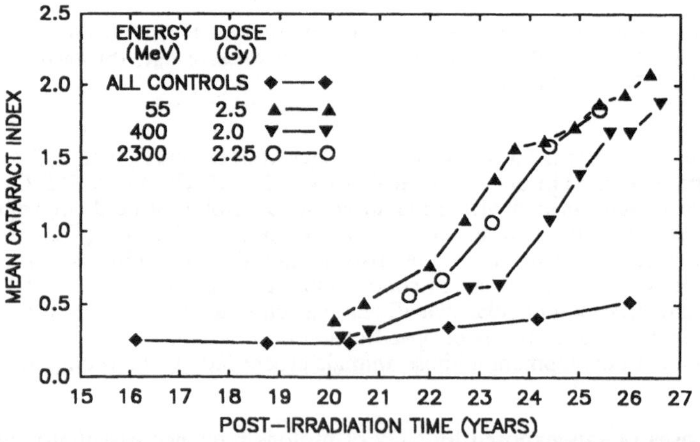

Figure 3. Temporal Development of Cataracts in Rhesus Monkeys Following Whole-Body Exposure to 55-, 400- or 2300-MeV Protons at ~2 Years of Age. Symbols for Each Dose Group Studied, Including Controls, Appear in the Figure. Note that the Ordinate Scale is Expanded (Range 0-2.5) Compared to the Other Cataract Figures in This Article (Range 0-5).

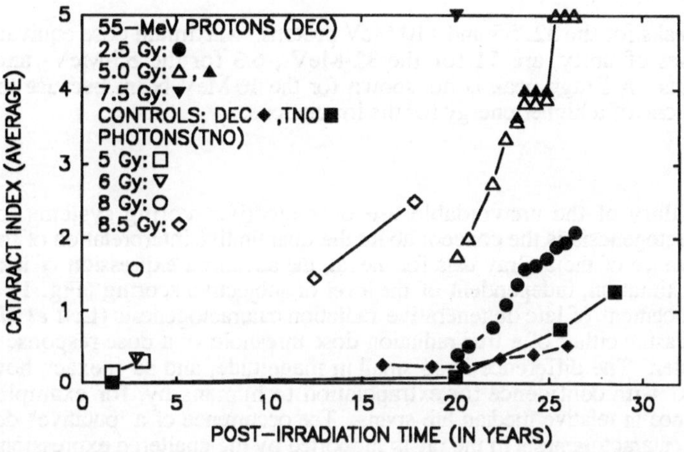

Figure 4. **Temporal Development of Cataracts in Rhesus Monkeys Following Whole-Body Exposure to Protons (Energy=55 MeV) or X Rays (300 kV or 6 MV) at ~2 Years of Age. The Proton-Irradiated Animals are Being Followed in the USA (DEC), and the Photon-Irradiated Subjects Are Being Followed in The Netherlands (TNO). Symbols for Each Dose Group Studied Appear in the Figure.**

Colony of rhesus monkeys at the Armstrong Laboratory (Cox *et al.*, 1992). Cataract data from three energy groups (55, 400 and 2300 MeV) which received doses to their lenses ranging from 2.0-2.5 Gray are shown in Fig. 3 together with data from unirradiated control animals. In Fig. 4, selected data from monkeys which received exposures to 55-MeV protons in the USA are plotted with data from the macaques exposed to 300 kV or 6 MV X-rays in The Netherlands.

Proton Dose Distributions Across the Monkey Lens

Four incident proton energies were used to calculate the dose distributions to the rhesus monkey eyes while the animals were rotating about their principal axes in the proton field. The static field depth doses for four proton energies (10, 32, 55 and 110 MeV), as determined from published and unpublished reports (Williams *et al.* 1966; Burton *et al.* 1970), are shown in Fig. 5. This figure illustrates the differences in the ranges of proton beams (from 1 cm for 32 MeV to nearly 10.5 cm for 110 MeV), and the increased doses due

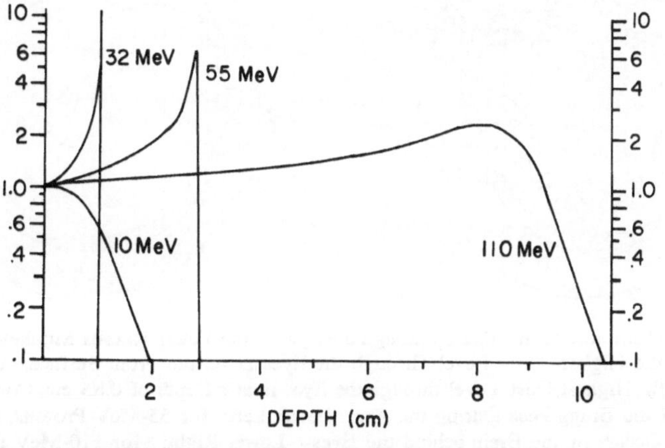

Figure 5. **Static Field Proton Depth Doses for 10-MeV, 32-MeV, 55-MeV and 110-MeV Protons. Each Curve is Normalized to a Unit Dose on the Phantom Surface.**

to the Bragg peaks for the 32, 55 and 110 MeV protons. Maximum dose equivalents, relative to surface doses of unity, are 11 for the 32-MeV-, 6.3 for the 55-MeV- and 2.4 for the 110-MeV-beams. A Bragg peak is not shown for the 10 MeV beam because this beam was degraded from one of a higher energy for the irradiations.

Discussion

A corollary of the unavoidable use of subjective scoring systems for studies of radiation cataractogenesis is the concern about the quantitative interpretation of the results. In the specific instance of the 2-Gray data for the rat, the advanced expression of late cataracts is, to a first approximation, independent of the level of subjective scoring (Fig. 1). Hence, that temporal advancement of late degenerative radiation cataractogenesis (Lett *et al.*, 1991) is a quantitative measure either of a true radiation dose threshold or a dose response curve with a shallow shoulder. The difference is so small in magnitude, and so inexact, however, that it cannot be used with confidence for extrapolation to humans by, for example, use of the 30-fold difference in relative median life spans. The occurrence of a "putative" dose threshold for late proton cataractogenesis in the rat is supported by the unaltered expression of lenticular opacification throughout the life span of ^{60}Co γ-irradiated beagle dogs shown in Fig. 2.

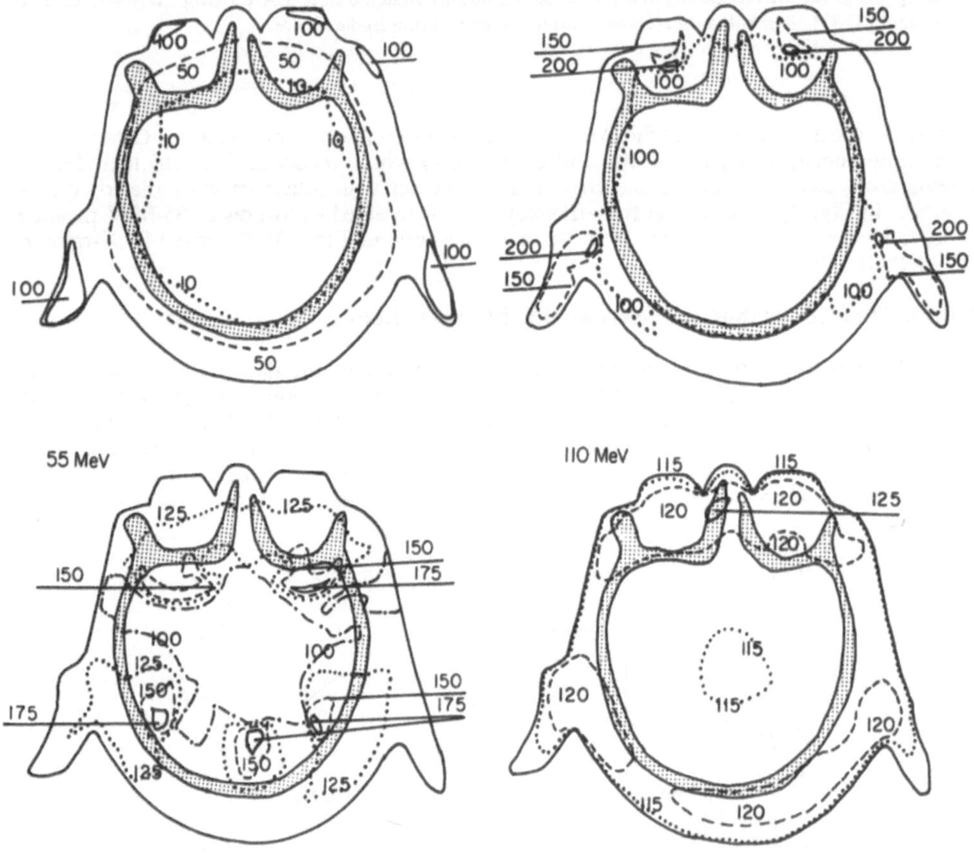

Figure 6. Isodose Distributions in a Plane Through the Eyes of the Young Rhesus Monkey. Upper Left: for 10-MeV Protons, the Highest Dose Level Through the Eyes is on the Front Surface. Upper Right: for 32-MeV Protons, the Highest Dose Level through the Eyes is at a Depth of 0.85 cm, Corresponding to the Depth of Focus of the Bragg Peak During the Arc. Lower Left: for 55-MeV Protons, the Greatest Dose within this Plane Occurs in the Brain behind the Eyes. Lower Right: for 110-MeV Protons, the Dose Distribution is Relatively Uniform Across the Entire Plane.

Although examinations of the proton-irradiated monkeys are still in progress, the acceleration of late cataractogenesis caused by 2 Gray of protons in that primate model seems to be ≥ 5 years (Figs. 3 and 4). From prior, but preliminary, analyses of our data and those of others (Lett *et al.*, 1991), proton RBE's (RBE = relative biological effectiveness) at the doses examined are ~1.5. That evaluation, when coupled with analysis of radiation cataractogenesis caused in the NZW rabbit by heavy ions (Lett *et al.*, 1991; Shinn *et al.*, 1991) permit reiteration of our previous conclusions (Lett *et al.*, 1991; Lett *et al.*, 1992): 1) The risk of overall radiation cataractogenesis in humans from exposure to HZE particles (HZE = relativistic atomic nuclei with an atomic number greater than 2) and for "ambient" protons during the projected Mars mission is negligible, and the radiation guidelines in NCRP Report Number 98 (1989) are adequate in that regard; 2) Significant, but as yet unquantified risks, will occur if astronauts on the Mars mission receive high proton exposures during periods of intense solar activity; 3) Short-lived rodent models do not seem to be completely satisfactory for simulating late radiation cataractogenesis in humans.

It is the belief of the authors that through collaborative approaches a quantitative assessment of cataractogenic risk for the Mars mission and other extended missions in space will be possible by the end of the decade. In this regard, the extension of the dose analyses (Figs. 5 and 6) to include LET distributions (LET = linear energy transfer) will permit better evaluations of the variations in animal anatomy (e.g. thickness of eyelids), see Fig. 6), on the cataractogenic effects of protons with short ranges (e.g. 10 and 32 MeV).

ACKNOWLEDGMENTS

The contributions of the following individuals are gratefully acknowledged: Dr. J. W. Fanton, Dr. D. H. Wood, Mr. K. A. Hardy, Mr. A. B. Smith, Mr. M. Hall, Mr. T. B. Casey and Mr. H. Wiersema. This research was supported by NASA Grant NAG 9-10, USAF Contract F-33615-85-C-4514 and grants from the US Food and Drug Administration to Colorado State University, NASA Contract #T-13215 to the Armstrong Laboratory, USAF Contract F-33615-87-D-0627/0009 to Texas A & M Research Foundation, Inc., and by CEC grant # B16-D-075-NL to the Institute of Applied Radiobiology and Immunology.

The animals involved in these studies were procured, maintained and used in accordance with the Animal Welfare Act and the Guide for the Care and Use of Laboratory Animals (Institute of Laboratory Animal Resources - National Research Council).

Opinions or assertions presented are the private views of the authors and are not to be construed as official or as reflecting the views of the Department of the Air Force or the Department of Defense.

REFERENCES

Broerse, J. J., van Bekkum D. W., Zoetlief, J, and Zurcher, C., 1991, Relative biological effectiveness for neutron carcinogenesis in monkeys and rats, Radiat. Res. 128: S128-S135.

Burton, B. S., Parker, C. V., Boles, L. A., Alexander, E. F. and Nelson, J. B., 1970, Simulation and measurement of solar-flare proton exposures, USAF School of Aerospace Medicine Technical Report number SAM-TR-70-38.

Cox, A. B., Lee, A. C., and Lett, J. T., 1988, Delayed effects of proton irradiation in the lens and integument: a primate model, *in:* Terrestrial Space Radiation and its Biological Effects, P. D. McCormack, C. E. Swenberg and H. Bücker, eds., pp. 415-422, Plenum Press, New York.

Cox, A. B., Lee, A. C., Williams, G. R. and Lett, J. T., 1992, Late cataractogenesis in primates and lagomorphs after exposure to particulate radiations, Adv. Space Res. 12: (2)379-(2)384.

Dalrymple, G. V., Lindsay, I. R., Mitchell, J. C., and Hardy, K. A., 1991, A review of the USAF/NASA proton bioeffects project: rationale and acute effects, Radiat. Res. 126: 117-119.

Hardy, K. A., 1991, Dosimetry methods used in the studies of the effects of protons on primates: a review, Radiat. Res. 126:120-126.

Keng, P. C., Lee, A. C., Cox, A. B., Bergtold, D. S., and Lett, J. T., 1982, Effects of heavy ions on rabbit tissues: cataractogenesis, Int. J. Radiat. Biol. 41: 127-137.

Leavitt, Dennis D., 1990, Proton arc doses to the primate head, USAF School of Aerospace Medicine Technical Report #TP-90-24, December 1990.

Leavitt, Dennis D., 1991, Analysis of primate head irradiation with 55-MeV protons, Radiat. Res. 126: 127-131.

Lett, J. T., Cox, A. B., and Lee, A. C., 1988, Delayed effects of proton irradiation in the lens and integument: a primate model, *in:* Terrestrial Space Radiation and its Biological Effects, P. D. McCormack, C. E. Swenberg and H. Bücker, eds., pp. 393-413, Plenum Press, New York.

Lett J. T., Lee, A. C., and Cox, A. B., 1991, Late cataractogenesis in rhesus monkeys irradiated with protons and radiogenic cataract in other species, Radiat. Res. 126: 147-156.

Lett, J. T., Lee, A. C. and Cox, A. B., 1992, Risks of radiation cataracts from interplanetry space missions, Acta Astronautica, Proceedings of 9th IAA Man-in Space Symposium, (in press).

National Council on Radiation Protection and Measurements, 1989, Guidance on Radiation Received in Space Activities, NCRP Report No. 98.

Shinn, J. L., Wilson, J. W., Cox, A. B. and Lett, J.T., 1991, A study of lens opacification for a Mars mission, Proc. 21st Int. Conf. Environm. Systems, SAE Tech. Paper Series #911354.

Sonneveld, P., Peperkamp, E., and van Bekkum, D. W., 1979, Incidence of cataracts in rhesus monkeys treated with whole-body irradiation, Radiology 117: 227-229.

Williams, G. H., Hall, J. D. and Morgan, I. L., 1966, Whole-body irradiation of primates with protons of energies to 400 MeV, Radiat. Res. 28: 372-389.

Wood, D. H., 1991, Long-term mortality and cancer risk in irradiated rhesus monkeys, Radiat. Res. 126: 132-140.

EFFECTS OF LINEAR ENERGY TRANSFER ON THE FORMATION AND FATE OF RADIATION DAMAGE TO THE PHOTORECEPTOR CELL COMPLEMENT OF THE RABBIT RETINA: IMPLICATIONS FOR THE PROJECTED MANNED MISSION TO MARS

J.T. Lett and G.R. Williams

Department of Radiological Health Sciences
Colorado State University
Fort Collins, CO 80523

Science is built up with facts, as a house is with stones. But a collection of facts is no more a science than a heap of stones is a house.

Jules Henri Poincare'

INTRODUCTION

According to the deliberations of prestigious review committees over the past two decades, astronauts and space workers who will participate in extended missions beyond the protection of the geomagnetic field should be concerned about three general types of risks from exposure to external radiations (NCRP 1989). Two of those risks, induction of cancers and cataracts, are characterized sufficiently well in qualitative scientific terms that meaningful evaluations of the dose relationships ought to proceed without unexpected complication. About the third risk, damage to the central nervous system (CNS), little is known even of the general scope of the problem let alone the medical consequences of damage to specific tissues, and therein lies the real difficulty of risk assessment for the CNS. Therefore, that aspect of the situation will be examined in some detail here and particularly from the position of the student or young investigator who may wish to address a problem that could require a scientific lifetime for solution.

In the interests of brevity, a review (Lett *et al.*, 1988) in the publication resulting from the previous NATO meeting on the biological effects of space radiation will be used as the main reference for, and an extended introduction to, this article. Summaries of the review material are given here together with more anecdotal information so that the young investigator can make a pragmatic appraisal of scientific life "in the trenches". Furthermore, since the characteristics of the space radiation environment and its dosimetry are discussed in detail elsewhere in the current volume, it is assumed that the reader (student) is well versed in those matters, or can easily become so. For analogous reasons, the primary sources given for information on, and references to, other pertinent subjects will be articles in this volume wherever possible. Unless stated otherwise, all general reference here to the space environment will be restricted to the circumstances of the Mars mission. Effects of solar activity on the 'ambient' spectrum of galactic cosmic rays will be ignored.

From the standpoint of radiation risk to the CNS, the term "extended manned (interplanetary) mission", will be synonymous, for the next quarter of a century or so, with

Biological Effects and Physics of Solar and Galactic Cosmic Radiation, Part A, Edited by C.E. Swenberg *et al.*, Plenum Press, New York, 1993

the projected manned mission to Mars. Given the current status of the global economy and plausible expectations for improvements in cancer therapy and so forth before the year 2019, consideration here of extended missions later than the Mars mission would be impractical. Only the implications for the Mars mission are examined henceforth.

The proposed orbit(s) for Space Station Freedom do(es) not evoke serious concerns about medical radiation risks (Cucinotta *et al.*, 1991), and shelters can be constructed on the moon to protect humans from space radiations during habitation of lunar bases and exploration of the lunar surface. Travellers to Mars, however, will receive only limited shielding against relativistic heavy ions inside the spacecraft because of engineering and fiscal constraints (Curtis, 1993), and prior to extravehicular activity (EVA) they cannot anticipate reliable prediction (warning) of the magnitudes of relativistic proton showers resulting from periods of intense solar activity (Heckman ,1993; Shea and Smart, 1993).

Of the three major types of putative radiation risks to humans from exposure to the heavy ion components of galactic cosmic rays during the Mars mission, only impairment of the CNS is likely to cause deleterious dysfunction during the mission. Damage caused to the complements of post-mitotic, terminally differentiated, neuronal cells in organized tissues of the CNS by the densely ionizing tracks of heavy charged particles can affect electrical function in ways analogous to its effects on the circuitry of computers (McNulty *et al.*, 1993). Subsequently, the biological components may die and so cause permanent disruptions in the circuitry; but unlike computer chips (circuit boards), neuronal cells cannot be replaced. Note that the adjective neuronal will be used in a generic sense here to describe the types of cells under discussion, even though some of them are not actually neurons.

RADIATIONS AND DOSE RATES IN INTERPLANETARY SPACE AND GROUND-BASED EXPERIMENTS

Environmental ambient fluxes of heavy ions in space are so low that the doses deposited by them consist of localized, isolated, events at the cellular level that are widely separated in tissues by space and time (Curtis, 1993). Much the same is true also for the ambient fluxes of sparsely ionizing radiations (X-, γ-rays, protons) despite the much larger numbers of more randomly distributed events. Even though sparsely ionizing radiations do cause densely ionizing events (clusters of ionizations) (Lett, 1990), the cells along the tracks of heavy charged particles through tissues can receive localized doses that, depending on the actual ion, can easily exceed 1 Gy (Lett, 1990). As the linear energy transfer from an ion increases above circa 200 keV μm^{-1}, with increasing Z and/or decreasing speed, the probability is enhanced for the formation of localized lesions (mini- or microlesions) involving rows of cells along the particle track in organized tissues (Lett *et al.*,1988; NCRP, 1989). Those are the events most likely to cause deleterious instantaneous dysfunctions of organized electrical circuitry and subsequent disruptions from permanent damage to, and hence losses of, the post-mitotic cells which constitute that circuitry in neuronal tissues. In this way, exposure to a low flux of heavy ions can represent a hazard for the CNS that a much larger flux of sparsely ionizing radiations does not.

Exposures of the CNS to large showers of relativistic protons during periods of heightened solar activity are special cases, but on the basis of observations of the retina (rabbit, monkey) and cerebellum (dog), and the accumulated experience from radiation therapy of the CNS with sparsely ionizing radiations, damage to systems of proliferating cells may well be far more deleterious in those instances (Lett *et al.* 1988; NCRP, 1989). Hence, the emphasis here is on damage from heavy ions.

When a large dose of sparsely ionizing radiation is delivered to tissues at a low dose rate, or in several fractions that are widely separated temporally, the radiobiological effect usually is much smaller than when the same dose is given acutely. Generally, the maximum tissue sparing achieved by fractionation or protraction of dose rate is 2-3 fold; but the factor tends to unity as the dose decreases toward zero. On the other hand, dose fractionation (protraction) for densely ionizing radiations may actually enhance the biological effect. During the Mars mission, however, the macroscopic doses of galactic heavy ions are so low that, for the reasons given above (Curtis, 1993), the biological effects will be similar to those from a single acute dose and the enhancement will be ≤ 1.1, (D. Brenner, personal communication). One important experimental consequence of this situation is that ground-based experiments designed to adumbrate the terrestrial response of a tissue can utilize single acute doses of heavy ions, and so preclude a very large drain on the resources of heavy-ion accelerators. Another benefit relates to an experimental question concerned with space radiation health; namely, which animal models should be used to simulate human radiation responses to heavy ions? Consider the extreme case: if shorter-lived rodents were used in ground-based experiments that involved the actual fluences and fluxes of heavy ions anticipated on the Mars mission, most of the animals would die from natural aging before the experiment could be completed.

One interesting aspect of this whole situation is that modern theories of cellular radiation lethality predict that the tissue sparing will also be circa 1 for the ambient fluxes and doses of *sparsely* ionizing radiations on the Mars mission, a prediction in keeping with many experimental results (NCRP, 1989). Large showers of relativistic protons again represent a special case, because they often arrive in more than one burst (Heckman, 1993; Shea and Smart, 1992) and the tissue sparing can be approximated better with the response to the first few dose fractions of regimens used routinely in radiation therapy.

RECENT ADVANCES IN CELL SURVIVAL THEORY AND DNA RADIATION CHEMISTRY

No matter how applied they may be, scientific experiments must strive to address underlying mechanisms through hypotheses, the subsequent formulation of rigorous theories and the eludication of fundamental natural laws. By so doing they accommodate also the admonition of Poncairé, see *Introduction*. Experiments conducted in this laboratory over the past twenty years on the responses of the CNS to ionizing radiations had several basic objectives apart from their obvious practical intent. Ostensibly, the experiments were designed to examine the fate of cells that have neither to replicate their DNA nor to divide after irradiation. At a more fundamental level, the series of experiments with the CNS are part of a broader experimental approach designed to facilitate the formulation of a general theory of cellular radiosensitivity and ultimately a theory of tissue radiosensitivity. These objectives are discussed in detail elsewhere (Lett 1990, 1992; Lett and Peters, 1992), and only more limited objectives are examined further here. It is sufficient to note now that classical theories of cellular radiosensitivity generally have failed because they have ignored cellular functions for processing DNA damage and their controls (Figure 1).

Proliferating cells die from the very necessity of engagement before division in the processing of DNA damage caused by the deposition of radiant energy in deoxyribonucleoproteins because the processes are not completely efficacious (Lett, 1990; 1992). Brief consideration of the current understanding of chemical and biochemical mechanisms involved in cellular radiation sensitivity (Figure 1) is given next.

Until about five years ago, it was widely accepted that cellular lethality caused by exposure to ionizing radiations resulted mainly from free radicals (mostly °OH) formed in nuclear water. Some fraction of those OH radicals formed within 5-10 nm of nucleochromatin were thought to be able to migrate to the cellular DNA and cause lethal damage in the giant macromolecules. The process was called indirect action and support for its operation came mostly from interpretations of the effects of so-called radical scavengers on the survival curves of cells exposed to sparsely ionizing radiations (Lett ,1990; 1992).

The opposing viewpoint was that cellular lethality resulted mostly from energy deposited directly in DNA, its hydration cell and perhaps adjacent hydrated proteins, i.e., deoxyribonucleoproteins. Cellular lethality thus was thought to be caused primarily by direct action. Migrating free radicals, e.g., •OH, played little role in the process. This concept found its genesis in the interpretations of measurements of double strand breaks (DSB) caused in DNA when fibers were hydrated with different levels of water (Lett and Alexander 1961) and evidence of the transfer of energy from protein to DNA in irradiated deoxyribonucleoprotein (Alexander *et al.*, 1961). Support for a unique radiation chemistry in the DNA hydration shell has become so widespread in recent years that there is now strong

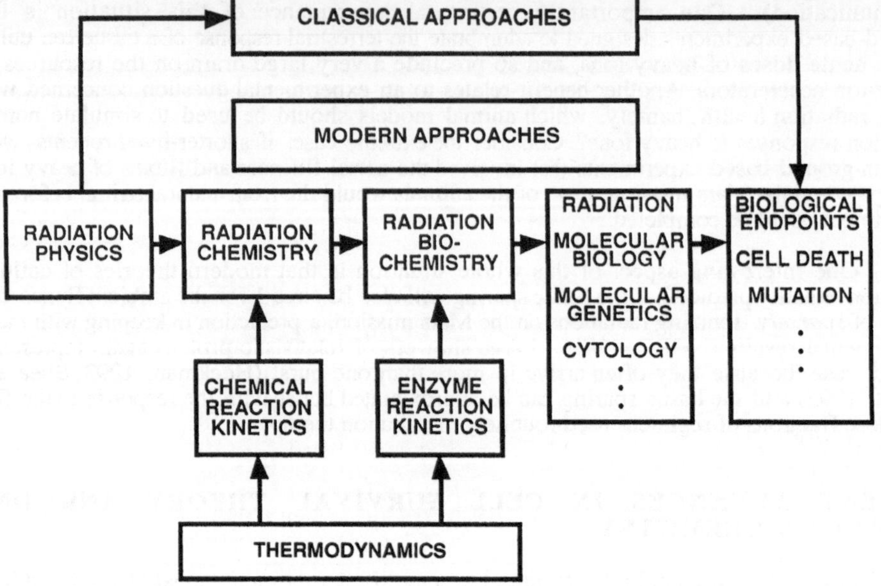

Figure 1. Scientific Disciplines Involved in Classical and Modern Theoretical Approaches for Explaining the Lethal Effects of Ionizing Radiations on Cells. Roles for Thermodynamics are Given Only for Areas of Interest here. As was Demonstrated Conclusively in the Last Decade, Authors who Made Incorrect Use of Such Classical Approaches as General Target Theory to Explain Cellular Radiotoxicity Did so Despite Specific Caveats to the Contrary Given by the Original Progenitors of the Theory. By so Doing, Those Authors, in Essence Ignored Cellular Processing functions for (DNA) Damage and the Genetics of Their Controls. Taken with Permission from Lett (1992).

sentiment for the abandonment of the terms direct and indirect action, especially since, for certain kinds of DNA damage, the same chemical changes are produced if damaged DNA reacts with (hydration) water or OH radicals react with undamaged DNA (Fielden and O'Neill 1991; Lett, 1992).

From evidence produced by a variety of chemical techniques, the structure of water in the hydration shell of DNA is known to be quite different from that of surrounding water in the cell nucleus; hydration water facilitates the formation of DNA damage by charge (energy) transfer reactions (Fielden and O'Neill 1991; Lett, 1990; 1992). On the basis of several kinds of chemical criteria, energy absorbed by the (anhydrous) DNA twin helix and energy absorbed by the hydration shell (equal mass of water) are about equally effective in causing damage, including DSB, in DNA (Lett, 1990; 1992). Energy absorbed by hydration water is transferred to the DNA probably by mechanisms that involve $H_2O^{\bullet +}$ (radical ions). In the water surrounding hydrated DNA, $H_2O^{\bullet +}$ reacts with H_2O to form $^{\bullet}OH$. Confirmation of the results and ideas first presented more than 30 years ago (Alexander and Lett, 1967; Lett, 1990; Lett and Alexander, 1961) has extended to the notions that DSB are induced by sparsely ionizing radiations in hydrated DNA (chromatin) by clusters of ionizations, from secondary electrons, etc., and that DSB are formed by them in cells as discrete events of low probability which need not involve $^{\bullet}OH$ attack.

LINEAR ENERGY TRANSFER, QUALITY FACTOR AND RELATIVE BIOLOGICAL EFFECTIVENESS

Characteristic changes in cell survival curves (Blakely *et al.*, 1984), the efficacies of DSB induction and the induction frequencies of breaks in prematurely condensed chromosomes (PCC) are produced by heavy charged particles as the ionization densities along

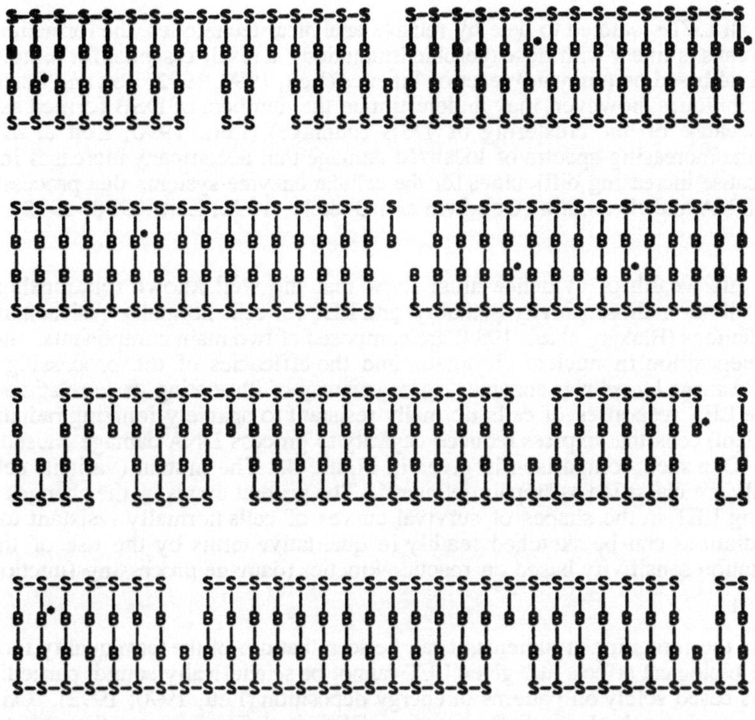

Figure 2. Clustering of Typical Double-Strand-Break Events in DNA Irradiated with ^{56}Fe Ions (600 MeV u^{-1}; LET = 175 keV μm^{-1}) as Simulated by Computer Using Monte Carlo Techniques (A. Chatterjee, Personal Communication). Base Damage at Sites where Release Does not Occur is Indicated by B*. For Further Details, See Chatterjee and Holley (1991) and Holley and Chatterjee (1991).

their tracks increase (Lett, 1990; 1992). The energy deposited usually is expressed as the linear energy transfer per unit length of track (LET_∞) in keV μm^{-1}. Increases in the above effects occur until the LET_∞ reaches circa one to few hundred keV μm^{-1} (depending on the atomic number, Z, of the ion) and, thereafter, the efficiencies decline until they fall below the values found for sparsely ionizing radiations (Lett 1990, 1992). To a first approximation, the induction frequencies for *all* other known kinds of DNA (chromatin) damage *decrease* with LET_∞ , a result taken as powerful evidence for the involvement of DSB and/or chromatin breaks in cellular radiosensitivity. For a variety of reasons, however, LET_∞ does not provide an adequate description of the pattern of energy deposition around the particle track (Lett, 1990), so in the absence currently of an effective alternative, LET is used here to acknowledge that inadequacy and maintain continuity with usage in the literature.

As the LET of the incident radiation increases, localized damage to DNA (deoxyribonucleoprotein) becomes more extensive. Terms such as DSB can only be generic because the damage associated with the physical break in the DNA duplex will occur in numerous forms (Lett, 1990; 1992). Variations in the damage to DNA and chromatin with LET are being simulated by computer with Monte Carlo procedures by Chatterjee and his colleagues (Chatterjee and Holley, 1991; Holley and Chatterjee, 1991; Lett, 1992) at the Lawrence Berkeley Laboratory (LBL). An example is given in Figure 2 of the findings of those authors for the ^{56}Fe ion, which probably is the galactic heavy ion of most biological interest for the Mars mission. Qualitatively, the effect of increasing LET is roughly analogous to ionization clusters occurring closer and closer together until they become continuous. A maximum frequency of DSB induction is reached, which is unique for ions of a given Z (Lett, 1990; 1992), but with further increase in LET the DSB induction frequency declines (Lett, 1990; 1992) because of the increased probability of ion recombination and/or radical recombination in the volume of energy deposition along the particle track and the consequent "wastage" of energy.

For all LET's studied to date by reliable analytical techniques, the formation of DSB and PCC breaks is linear with dose (aerobic irradiation), a result essential for simple theories of cell survival based on (enzyme) reaction kinetics (Lett, 1990, 1992; Lett and Peters, 1992). All those techniques, however, may underestimate the numbers of DSB formed as the LET increases because of the clustering of DSB (damage) (Lett, 1990; Lett *et al.*, 1987). Currently, the increasing spectra of localized damage that accompany increases in LET are thought to cause increasing difficulties for the cellular enzyme systems that process radiation damage in DNA and chromatin (Goodwin and Blakely, 1992; Lett, 1990, 1992), e.g., see Figure 3.

All this wealth of evidence must show that the well known relationships among patterns of cellular radiosensitivity (lethality) and LET for cells normally resistant to sparsely ionizing radiations (Blakely *et al.*, 1984) are composed of two main components: the patterns of energy deposition in nuclear chromatin and the efficacies of the processing of DNA (chromatin) damage by cellular enzymes. A cogent way of illustrating these relationships is to compare the LET responses of cells normally resistant to sparsely ionizing radiations with mutant (variant) cells that express reduced capacity to process DNA damage caused by those radiations. One such comparison is given in Figure 4. The mutant (variant) cells in that figure usually are classified as "repair deficient". The general trends of the changes produced by increasing LET in the shapes of survival curves of cells normally resistant to sparsely ionizing radiations can be sketched readily in qualitative terms by the use of theories of cellular radiation sensitivity based on reaction kinetics (damage processing functions) (Lett, 1992).

From the foregoing arguments, it can be seen that use of the term quality factor, Q, to describe the biological effects at a given LET cannot be scientifically sound, particularly if its magnitude is based solely on patterns of energy deposition (Lett, 1990; 1992). On the other hand, the term relative biological effectiveness, RBE, is defined rigorously, although RBE cannot have constant value even for a given ion at a given energy since the response of any given cell will depend on a variety of factors that includes its ability to process (DNA) radiation damage. Many 'normal' cells or tissues have roughly similar RBE's (Q's) simply

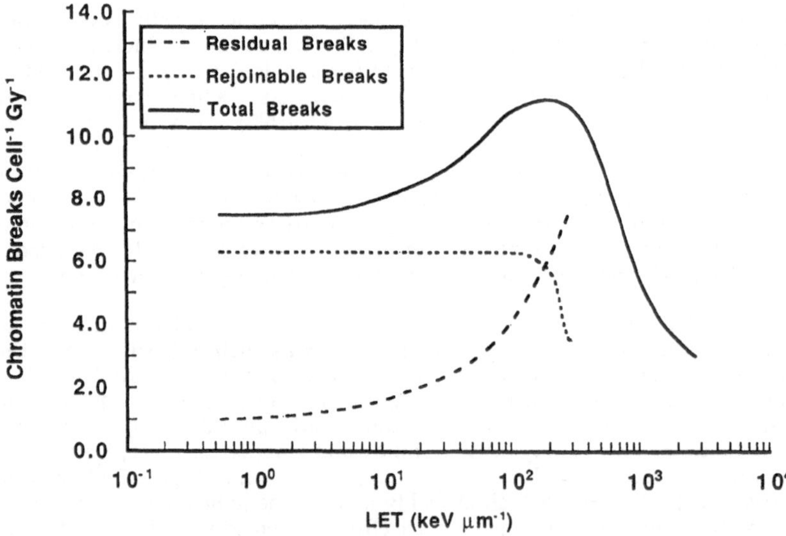

Figure 3. Effect of Linear Energy Transfer (LET) of the Formation and Processing of Chromatin Breaks in Proliferative Mammalian Cells Irradiated in the G_1 phase of the Cell Cycle and Maintained there Until Processing Ceased. Redrawn with Permission from Goodwin and Blakely (1992) and Published in Lett (1992) from which it is Reproduced, with Permission, here.

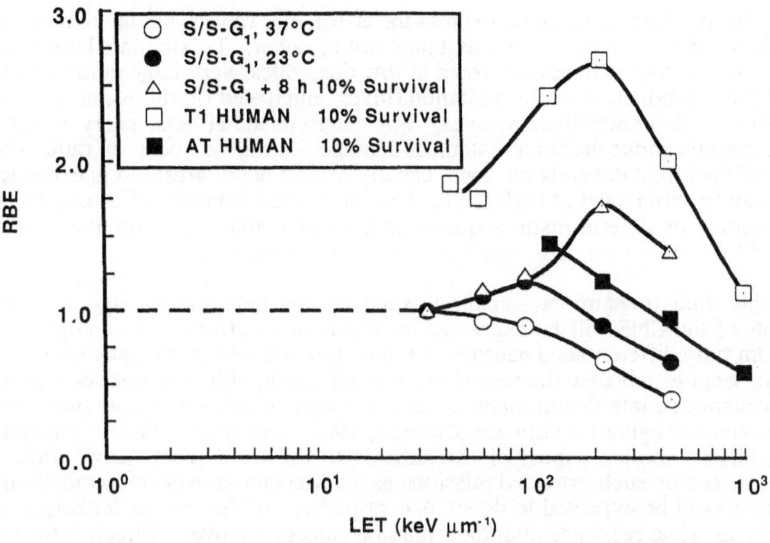

Figure 4. Effect of linear Energy Transfer (LET) on the Relative Biological Effectiveness (RBE) for Cellular Lethality. The L5178Y S/S Murine Leukemic Lymphoblast Is the Most Radiosensitive, Proliferative Cell Known and Shows an Unusual Cell-Cycle Response to X-Photons. That Response is Essentially Flat Through the First Half of the Cycle and a Peak of Radioresistance Occurs in Late S (- early G_2) at G_1 + 8 hours. In G_1, Cellular Radiosensitivity to X Photons is Temperature Sensitive and an Immediate Post-Irradiation Treatment at 23° C (12 hours) Causes Survival to Increase because of Enhanced Recovery (Processing). Damage Processing is Reduced Progressively as the LET is Increased, and by 475 keV μm^{-1} (^{93}Nb ions) It Is Effectively Eliminated, Hypothermic Recovery Does not Occur and the Radiation Response Through the Cell Cycle Is Essentially Constant. Ataxia Telangiectasia (AT) Is a Human Disease that Is Accompanied by, as yet Unidentified, Defects in Damage Processing. For Current Purposes, AT Cells May Be Considered to Have Proficiencies for (DNA)Damage Processing Between Those of Normal Human Cells and the Murine S/S Variant. Taken with Permission from Lett (1992).

because, all other things being equal, they have roughly the same abilities to process DNA damage caused by ionizing radiations.

TISSUE AND ANIMAL MODELS

Many years of radiotherapeutic practice, the continuing lifespan study of the survivors from the atomic blasts at Hiroshima and Nagasaki, other less comprehensive sources of human data, and a variety of animal studies, have provided a wealth of information on the radiation responses of tissues to sparsely ionizing radiations (NCRP, 1989). Much of the information, however, was not obtained for low doses of radiations given at low dose rates, and extrapolations of extant data to those conditions is still the subject of controversy (NCRP, 1989). By contrast, the extant data for heavy ions are relatively few and most of them come from experiments with animal models given acute doses of radiation.

Uncertainties associated with the simulation of effects produced by low doses of radiation delivered at low dose rates stem from two sources: 1. The effects will be small in number and/or in kind; 2. Most effects are likely to occur late in life and so the responses of animal models must be examined throughout their residual lifespans.

Radiation responses are generally classified into two categories: stochastic and deterministic (previously called nonstochastic) (Lett *et al.*, 1988; NCRP, 1989). Stochastic effects, by definition, can be examined by statistical methods; some dose-dependent ***number*** (*fraction*) of the irradiated population express the effect, e.g. cancer, but the remainder do not. Dose thresholds do not occur, or are assumed not to occur. To simulate human stochastic responses to low radiation doses delivered at low dose rates, very large numbers of animals are needed but reproduction of the radiation fluxes anticipated on the Mars mission would seem pointless with a short-lived species. For deterministic effects, *every* member of the population exhibits about the same extent of damage for a given dose of radiation and the *magnitude* of the effect depends on dose, usually after a dose threshold is exceeded. That threshold may be eliminated at high LET. Use of limited numbers of a long-lived animal model in studies of deterministic sequelae is both plausible and effective. Appropriate utilization of low, protracted, radiation fluxes is sound if impractical.

At this time, it seems reasonable to suppose that radiation damage to neuronal cell complements of the CNS will be expressed as deterministic effects--for example, tumors do not arise from fully differentiated neurons. Furthermore, many deterministic effects produced in mammals seem to advance the sequela of normal aging, although not necessarily by the same mechanisms, so late deterministic responses from radiation exposure have been called radiation accelerated aging (Rubin and Casarett, 1968; Lett *et al.*, 1988). That analogy, if correct, highlights another aspect of the hazards to the CNS from heavy ions. Should astronauts engaged on such extended missions as the Mars mission develop radiation-induced tumors, they would be expected to do so, except perhaps in the case of leukemia and some other tumors, as a late response after their mission careers are over. Effective treatments for cancer may be anticipated during the next half century, but the likelihood of the discovery of a cure for aging during that period of time is more remote.

Chromosomal aberrations long have been thought to be an overt expression of DNA damage caused by ionizing radiations and the principal cause of 'mitotic' death for proliferating cells. Liver is mostly a nondividing tissue in which defective (or dying) cells are replaced by regeneration (Rubin and Casarett, 1968). When the regeneration of new cells was induced by various agents or by partial hepatectomy of aging livers, or livers showing no overt signs of damage long after irradiation, cells with chromosomal aberrations were observed e.g., (Crowley and Curtis, 1963). The inferences from those findings were that DNA damage accumulated in erstwhile nondividing cells with age and that DNA damage could persist or accumulate in them long after irradiation. Those seminal experiments provided the impetus for our investigations.

Tissue systems in which damaged or destroyed cells can be replaced by regeneration (stem cells?) would not have been suitable for our objectives. Estimations of residual DNA damage in irradiated cells would have been undermined by the replenishment with healthy (undamaged) cells and subsequent losses of defective cells could not have been examined by reductions in size of the postnatal cell complements. Attention was focused, therefore, on the complements of neuronal cells in the CNS for which differentiation is complete at birth, or shortly thereafter, and in which cells are never replaced. In this case, of course, the persistence or accumulation of DNA damage, etc., could not be confirmed by the expression of chromosomal aberrations because a way had not been found (nor subsequently has been found) to stimulate fully differentiated neuronal cells into, and through, the cell cycle and so replicate their DNA before undergoing division.

WORKING HYPOTHESIS

Most proliferating systems of cells must exhibit lethality after any dose of ionizing radiation because the initial slopes of their survival curves are finite not zero (Alper, 1979). Incomplete or inadequate processing of DNA damage is most likely to account for those responses, and an innate reason for the deficiency in the processing rather than simple lack of time before DNA replication or division should be revealed in some way by terminally differentiated, post-mitotic cells. Thus there were four parts to the working hypothesis for our experimental program on radiation damage to the CNS and they have not altered over the past two decades. Implicit to it also was the question: are the efficacies of DNA processing mechanisms a function of age in nondividing tissues?

1. Residual DNA damage could be detectable weeks or months after irradiation or be

manifested much later at times generally associated with the appearance of overt symptoms of natural aging.

At the time our experiments were begun, examination of the former possibility was well beyond the resolution of extant techniques for measuring DNA damage (Lett, 1990), a situation that still persists today for DSB but not single-strand breaks (SSB) or total- strand breaks (TSB) in DNA (Lett, 1990).

2. If DNA damage persisted after irradiation, neuronal cells could be killed by it and their losses from a fixed complement of terminally differentiated, post-mitotic, neuronal cells could be determined at reasonable temporal intervals.

3. If DNA damage accumulated late in the lifespan, then some level of damage could kill cells, their losses should increase in the latter part of the lifespan and the result could be an enhancement of the responses (depletion) of normally aging cells.

4. Increases in LET should reduce the extent of DNA processing by neuronal cells and hence influence cellular radiosensitivity in one or more of the ways described above.

SUMMARY OF THE FIRST TWO DECADES OF THE RESEARCH PROGRAM

At the outset, our program essentially was the whole research field (Lett, 1974) and progression from experiment to experiment often was based on serendipity. Nonetheless, the difficulties were surmounted, albeit sporadically, through the generous assistance of scientists at this university and officials at the National Institute and Neurological Diseases and Stroke, the National Institute on Aging and the National Aeronautics and Space Administration.

Donation of aging beagle dogs culled from an experimental colony permitted examination of the fully differentiated, post-mitotic, neurons in the internal granular layer of the cerebellum of a relatively long-lived, mammalian species (median lifespan, 13-14 years) (Wheeler and Lett, 1974). The results were promising but limited by the resolution of even the most sensitive technique available at the time for measuring DNA damage in nondividing cells, which obviously cannot be labelled with radioactive precursors. Furthermore, tissue biopsies of a standard size (and hence putative numbers of cells) could not be obtained surgically from the cerebellum, DNA damage was introduced by the preparative procedures, and electrical function was not amenable to facile examination.

Replacement of cerebellar neuronal model with that of the sensory cells of the retinas of New Zealand white (NZW) rabbits solved many of the expeimental problems. The retina also was of overt practical interest because of the flicker-flash phenomenon, yet the median lifespan of the NZW rabbits (5-7 years) was still adequate for the basic objectives. Gross electrical responses could be examined readily as the *a* and *b* waves of the electroretinogram (ERG), and the DNA from *whole* photoreceptor cell complements (rods only) of >90% purity could be obtained by simple surgical procedures followed by rapid isolation of cell nuclei. Isolated retinas could be stored in liquid nitrogen for many years without affecting the integrity of photoreceptor DNA. Concomitantly, the technique of reorienting gradient zonal ultracentrifugation was being improved to the point where DNA damage induced by 0.05-0.10 Gy of sparsely ionizing radiations could be detected routinely (Lett, 1990).

Many of the experiments envisaged were possible only because of the development of an high energy machine, the BEVALAC at the LBL, able to generate beams of relativistic heavy ions at fluxes suitable for radiation therapy and with ranges of many centimeters in mammalian tissues (Blakely *et al.*, 1984). Even then progression in the animal research program was dictated, in large measure, by further developments at the BEVALAC that permitted the irradiation of mammalian tissues with the heavier ions thought to be especially important for space radiation biology; so the sequence of ions used thus far has been ^{20}Ne, ^{40}Ar and ^{56}Fe. For the investigations of L5178Y S/S cells, see Figure 4, which were started much later, all the ions (^{20}Ne, ^{28}Si, ^{40}Ar, ^{56}Fe, ^{93}Nb) were available at the time the experiments were designed.

In the rabbit experiments, the incident energies of the ions (^{20}Ne, ^{40}Ar, ^{56}Fe: 365, 530, 460 MeV μ^{-1}, respectively) were such that both retinas in the animal were exposed to nearly identical parts of the Bragg plateau region of energy deposition despite the curvatures of the retinas in situ, and thus narrow, well defined, ranges in LET were employed (35 ± 3, 90 ± 5, 220 ± 31 keV μm^{-1}, respectively). The heavy ion beam was aligned and collimated so that only the eyes and proximate tissues were exposed. Animals so irradiated with moderate doses of ^{20}Ne, ^{40}Ar and ^{56}Fe ions lived essentially normal lifespans and developed relatively low levels of lenticular opacifications and other radiation sequelae (Lett *et al.*, 1988; 1991).

Baseline measurements of initial DNA damage, for either the cerebellar neurons or the retinal photoreceptor cells, had to involve irradiation in vitro because processing of DNA damage occurred in situ during the interval post-irradiation needed for tissue isolation (Lett *et al.*, 1986; 1987). One major advantage of the retinal model was that whole excised retinas could be set at appropriate intervals in a "submarine" perpendicular to the heavy ion beam throughout, and beyond, the whole of the Bragg curve (range), and irradiated at 0°C (Lett *et al.*, 1987).

Normally Aging Animals

Colonies of NZW rabbits were followed from the ages of 6-10 weeks throughout the median lifespan and up to, for the oldest individual, *circa* 9 years (Lett *et al.*, 1988). Within the current level of resolution of reorienting gradient zonal ultracentrifugation, retinal DNA showed no change in size during the first half of the median lifespan. During the second half of the median lifespan, however, the size of the photoreceptor DNA began to decrease, and continued to do so in animals that lived beyond that age (Lett *et al.*, 1988). Losses of cells from the retinas began only after the appearance of DNA damage.

Animals Exposed to ^{60}Co γ-photons and Relativistic ^{20}Ne and ^{40}Ar Ions

Sequelae resulting from the exposure of optic and proximate tissues of young NZW rabbits to ^{60}Co γ-photons and relativistic ^{20}Ne and ^{40}Ar ions at Bragg plateau energies have been documented in more than 30 communications. Although the results from the retinal studies have been presented to date mostly in preliminary fashion, e.g., Lett *et al.*, (1986, 1987, 1988), they will be reported in detail as the current experiments with ^{56}Fe ions approach completion. Unscheduled reductions in the funding for the rabbit experiments over past decade were absorbed by shelving the retinal studies because the other principal practical objective, namely evaluations of late radiation cataractogenesis in long-lived mammalian species (Cox *et al.*, 1993; Lett *et al.*, 1991), utilized non-invasive analyses, and so did not require sacrifices of previously irradiated animals. The long hiatus in the retinal studies ended with the recent implementation by NASA of the Space Radiation Health Program; however, fiscal uncertainties still cloud the futures of high energy machines capable of generating relativistic heavy ions. Nevertheless, the results available now should be sufficiently intriguing to inspire young investigators to brave the uncertain future and broaden investigations of damage to the CNS by using molecular biological and neurobiological techniques.

Irradiation of young animals with ^{60}Co γ-rays (acute doses) induced damage in photoreceptor DNA that seemingly was restored by cellular processing functions within a few days without detectable loss of photoreceptor cells. Analogous responses were observed even after large acute doses of ^{20}Ne and ^{40}Ar ions (Figures 5 and 6) but the time needed for the maximum extent of processing lengthened with dose and LET. Note that if the irradiated DNA returns to control size (within the resolution of the analytical technique) all the DSB and SSB must have been eliminated in some way. *Accurate* rejoining of all the DNA breaks or restitution of other kinds of DNA damage in mammalian cells, i.e., repair proper, cannot be determined yet by *any* known analytical technique (Lett, 1990).

Normal aging of the NZW rabbit is accompanied by increasing reduction in the size of photoreceptor DNA followed by increasingly larger losses from the photoreceptor cell complement during the second half of the median lifespan. Decreases in DNA size with age, as the percentage of control size (young animals), can be described by a parabolic function.

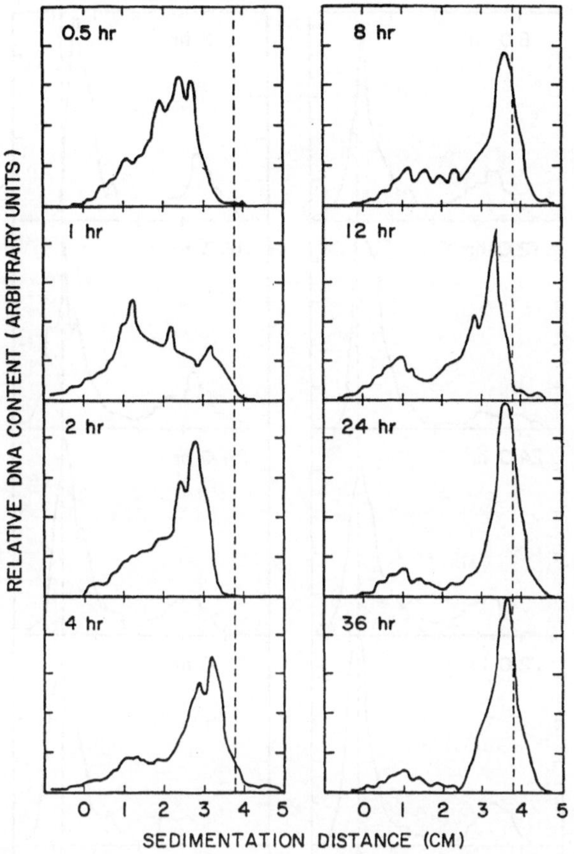

Figure 5. Restoration in Vivo of the Chromatin DNA Structures in the Photoreceptor Cell Complement of the Leporine Retina Following a Large Dose (25.3 Gy) of ^{40}Ar Ions (LET: 90 keV μm^{-1}. The Sedimentation Profiles (by Reorienting Gradient Zonal Ultracentrifugation Through Alkaline Sucrose Gradients) of the DNA from the Whole Photoreceptor Cell Complement of the Retina (Individual Animals), Which Register Total Strand Breaks in the DNA, Return Nearly to Control Size (Broken Line) in 36 Hours. Sedimentation Distance is a Function of DNA Size. Taken from Bergtold (1983).

After exposure to ^{60}Co γ-rays, the onset of the reduction in DNA size and subsequent losses of photoreceptor cells occurred also, but they were were advanced temporally in a dose-dependent manner.

Irradiation with ^{20}Ne and ^{40}Ar ions elicited comparable responses from the photoreceptor cells; advances of the 'late' changes were dependent on dose and LET. Given those effects, the question now was whether low doses of heavier ions, and especially ^{56}Fe, advanced the late degeneration of seemingly restored photoreceptor DNA, and the subsequent cell losses, into time frames comparable to those anticipated for extended manned missions in space.

Very large doses of ^{60}Co γ-rays, ^{20}Ne and ^{40}Ar ions (~50 Gy) eliminate the ERG of the retina (Lett *et al.*, 1988), a response essentially independent of LET (0.3 -95 keV μm^{-1}). Such threshold doses are roughly comparable to those which 'knock out' the optical nerve. This drastic effect on the retina rapidly resulted in blindness: little DNA processing occurred after doses above the threshold dose; extensive DNA processing occurred at doses below it. No effort was made to explore the possibility of a late radiation response for the ERG following moderate doses because of the interference that late radiation cataracts would cause to the standardized light flashes used to stimulate the retina.

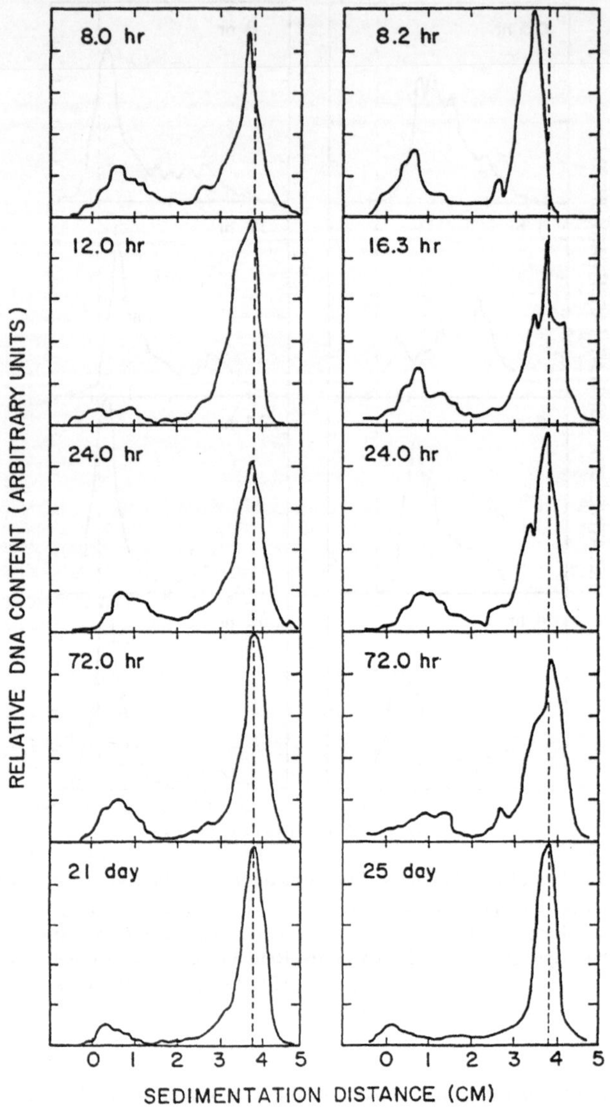

Figure 6. Sedimentation Profiles of DNA from Complete Photoreceptor Cell Complements of Leporine Retinas (see caption to Figure 5) Return to, and are Indistinguishable from, Control Values after Circa 3 weeks Following Irradiation of Optical and Proximate Tissues of Rabbits to Moderate Doses of Heavy Ions. Left Column: 4.7 Gy of ^{40}Ar Ions (LET: 90 keV μm^{-1}). Right Column: 10 Gy of ^{20}Ne Ions (LET: 35 keV μm^{-1}). Taken from Bergtold (1983).

Animals Exposed to Relativistic ^{56}Fe Ions: Initial Results

Shortly after the experiments with ^{56}Fe ions were begun, fiscal reductions imposed such curtailments of them that much of the temporal precision was lost and most of the earliest DNA analyses had to be made *circa* two years after irradiation (Table 1). The DNA sedimentation profiles from photoreceptor cells of irradiated animals were not significantly different from those of control animals of the same age at any time studied throughout their subsequent lifespans. Yet, significant losses of cells did occur even by two years (Tables 1

Table 1. Mean Numbers of Photoreceptor Cells ($x10^{-7}$) in the Retinas of Rabbits Locally Irradiated with ^{56}Fe-Ions at circa Ten Weeks of Age.

Dose	Animal Age (Months)			
(Gy)	Young[1]	Middle-Aged[1]	Old[1a]	Statistical Significance[2]
	≤ 24	25-35	53-80	
0	6.00 ± 0.05 (53)	6.00[3]	5.50 ± 0.08 (9)	p < 0.05, young controls: old controls
0.05	5.93 ± 0.04 (29)	--	--	N/S versus young controls
0.10	6.08 ± 0.05 (13)	--	--	N/S versus young controls
0.25	5.93 ± 0.07 (12)	--	--	N/S versus young controls
0.50	[a]5.74 ± 0.08 (5)	--	5.46 ± 0.25 (6)	[a]N/S versus young controls; [b]N/S versus old controls
1.00	--	--	4.90± 0.22 (9)	p < 0.05 versus old controls
2.00	[a]5.60 ± 0.12 (4)	--	[b]4.57 ± 0.12 (8)	[a]p < 0.05 versus young controls; [b]p < 0.05 versus old controls
3.00	5.64 ± 0.05 (17)	--	--	p <0.05 versus young controls
4.00	--	[a]5.35 ± 0.09 (12)	[b]4.92 ± 0.11 (3)	[a]p <0.05 versus young controls; [b]p < 0.05 versus old controls
5.00	[a]5.11 ± 0.23 (5)	[b]5.20 ± 0.10 (10)	--	[a]p <0.05 versus young controls; [b]p <0.05 versus young controls

Numbers in brackets are the numbers of retinas examined. Data are expressed as mean ± SEM.
[1]By linear regression analysis: Cell number ($x10^{-7}$) = 5.97 - 0.152 D (Gy), p <0.01; [1a]Cell number ($x10^{-7}$) = 5.55 - 0.51 D (Gy),p < 0.05. Note: Surviving animals that had received 4.0 Gy were sacrificed at circa 4 years post-irradiation because of high levels of cataracts; results from those animals were not included in the regression analysis, which thus represents effects of doses ≤ 2 Gy expressed circa 6 years post-irradiation.
[2]One way analysis of variance by Student-Newman-Keuls Multiple Comparison Test.
[3]Data not available. Previous earlier lifespan experiments showed no detectable cell losses by this age.

and 2). Currently, various possibe interpretations these responses are being explored through re-evaluations of many data accrued over almost two decades from the experiments with ^{60}Co γ-rays, ^{20}Ne ions and ^{40}Ar ions. This provisional approach utilizes linear regression analysis, which is a rough approximation at best, but the results are illustrative. Best fits for losses of cells at circa two years postirradiation as a function of dose and LET are shown in Table 2 together with those at later times for ^{56}Fe ions. Correlations (r^2) are weak for ^{20}Ne ions and ^{60}Co γ-rays (data reported elsewhere) but strengthen as the LET increases. The re-evaluations of the photoreceptor cell losses from the earlier rabbit experiments were prompted also by the recent findings of analogous, but very rapid, losses from the retina of the rat after irradiation with ^{56}Fe ions but not ^{40}Ar ions (Krebs *et al.*, 1990).

In an experiment begun in 1992 year, which involved exposure of rabbits aged 3-3.5 years in order to evaluate the effect of age at the time of irradiation on biological response (Lett *et al.*, 1988), animals given 3 Gy of ^{56}Fe, c.f., Table 1, were examined at 0.02 - 14 days after irradiation. Losses of photoreceptor cells were not detected, although processing functions produced the anticipated changes in the DNA sedimentation profiles.

Table 2. Losses of Photoreceptor Cells from Retinas of Rabbits Locally Irradiated with Heavy Ions at circa Ten Weeks of Age.

(a) Losses at circa Two Years Post-Irradiation			
Radiation	LET keV μm^{-1}	Percent Loss Gy^{-1}	# Retinas Examined
Controls	-	0	80
^{60}Co γ-photons	0.3	0^*	75
^{20}Ne ions	35	1.0	$60^{\dagger}$
^{40}Ar ions	90	1.7	155
^{56}Fe ions	220	2.5	138

(b) Losses at Different Times after Exposure to ^{56}Fe Ions		
Time Post-Irradiation (Yrs)	Percent Loss Gy^{-1}	# Retinas Examined
Circa 2	2.5	138
Circa 2-4	2.7	25
Circa 6	9.2	29

*Doses $\leq$ 10 Gy; †Analysis still in progress.
Data for heavy ions circa 2 years post-irradiation were obtained by linear regression analyses:

^{20}Ne: Percent cells remaining = 100 - 1.0 D(Gy); r^2 = 0.79
^{40}Ar: Percent cells remaining = 100 - 1.7 D(Gy); r^2 = 0.87
^{56}Fe: Percent cells remaining = 100 - 2.5 D(Gy); r^2 = 0.94

DISCUSSION

Damage induced in the DNA of fully differentiated, post-mitotic, photoreceptor cells in the rabbit retina by ionizing radiations with LET's in the range 0.3 - 220 keV μm^{-1} is comparable to that found in proliferating cells (Lett, 1990; Lett*et al.*, 1987), and probably is processed in analogous ways by DNA enzyme systems. All that is known to date is consistent with recent advances in the radiation chemistry of DNA (deoxyribonucleoprotein) and theories of cellular radiation sensitivity based on (enzyme) reaction kinetics (Lett, 1990; 1992). Also the results are consistent with the notion that nondividing cells survive much longer than dividing cells after irradiation because they do not have to replicate DNA containing residual, or improperly processed, DNA damage before they attempt to divide. The reasons for the late accumulation of DNA damage and subsequent cell losses are not known.

Although the transition probably is continuous (Tables 1 and 2), the effects of ^{60}Co γ-photons, ^{20}Ne ions and ^{40}Ar ions on the photoreceptor cells on the rabbit retina seem to be qualitatively different to those following exposure to ^{56}Fe ions. Significant numbers of cells are lost early in the lifespan yet the remainder show no detectable signs of DNA damage. The late loss of cells also occurred, even though prior accumulation of DNA damage could not be confirmed with the limited number of retinas available at long times post-irradiation. These results would be internally consistent if, at an LET of 220 keV μm^{-1}, some cells died relatively quickly, and certainly in less than two years, as a symptom of the ionization density along the tracks of ^{56}Fe ions vis-a-vis those of ^{40}Ar ions and ^{20}Ne ions. Such a response could signal the onset of the formation of mini- or microlesions, which is postulated to occur at LET's around 200 keV μm^{-1} (Lett *et al*, 1998).

Damage to, and losses of, electrical function in the CNS represent serious potential radiation hazards for the crew of the Mars mission, especially if those effects are caused *during the mission* by the particle tracks of galactic heavy ions and the mini-lesions induced

RADIATION CATARACTOGENESIS IN LONG-LIVED ANIMAL MODELS
Implications for the Projected Mars Mission
• Anticipated Exposures to Galactic Heavy Ions will not produce Significant Lenticular Opacification in Astronauts during the Mars Mission.
• Anticipated Exposures to Galactic Heavy Ions are unlikely to produce Significant Lenticular Opacification throughout the Lifespans of Astronauts engaged in the Mars Mission.
• Exposures to Ambient Protons will not involve Significant Risk of Radiation Cataract throughout the Lifespans of Astronauts engaged in the Mars Mission.
• Exposures to High Proton Fluxes from Large Solar Events that are not lethal will pose Significant Risk of Late Degenerative Radiation Cataracts for Astronauts engaged in the Mars Mission.

Figure 7. Recommendations Concerning Risks of Radiation Cataractogenesis for Astronauts Participating in the Projected Manned Mission to Mars.

Radiation Risks for the Projected Mars Mission
Recommended Order of Research Priorities for Ground-Based Experiments
• Late Damage to the CNS
• Early Damage to the CNS
• Late Degenerative Radiation Cataractogenesis
• Carcinogenesis

Figure 8. Overall Recommendations Concerning Research Priorities for Ground-based Experiments for Evaluation of the Radiation Risks for Astronauts from Exposure to Galactic Heavy Ions during the Projected Manned Mission to Mars. Note: 1) Evaluations Should Be Made with Animal Models from Long(er)-lived Species. 2) Use of Dose Fractionation for Heavy Charged Particles is Unnecessary. This will Reduce Substantially the Costs of Beam Time at Heavy-Ion Facilities.

by them. In view of this, and the conclusions from companion studies of another deterministic effect, viz., late radiation cataractogenesis in long-lived mammalian species (Cox *et al.*, 1993), we recommend that the priorities for research on radiation risks for the Mars mission be as indicated by Figures 7 and 8.

Obviously, the research on the CNS must be raised to levels of sophistication needed for evaluations of the effects of radiations on the functions of neuronal cells. The way is likely to be long and hard.

ACKNOWLEDGEMENTS

This article was prepared under the auspices of the National Aeronautics and Space Administration through grant #NAG 9-10. The research could not have been conducted without support for the operation of the BEVALAC from the Department of Energy.

The animals involved in these studies were procured, maintained and used in accordance with the Animal Welfare Act and the Guide for the Care and Use of Laboratory Animals (Institute of Laboratory Animal Resources - National Research Council).

REFERENCES

Alexander, P. and Lett, J.T., 1967, Effects of ionizing radiations on biological macromolecules, Comprehensive Biochem., 27: 267-356.

Alexander, P., Lett, J.T., and Ormerod, M.G., 1961, Energy transfer in irradiated deoxyribonucleic acid and nucleoprotein. Biochem. Biophys., 51: 207-209.

Alper, T., 1979, Cellular Radiobiology, Cambridge University Press, Cambridge.

Blakely, E.A., Ngo, F.Q.H., Curtis, S.B. and Tobias, C.A., 1984, Heavy-ion radiobiology: cellular studies, Adv. Radiat. Biol., 11: 295-389.

Bergtold, David S., 1983, PhD Dissertation, Effects of heavy ions on NZW rabbit photoreceptor cell DNA. Colorado State University.

Chatterjee, A. and Holley, W.R., 1991, Energy deposition mechanisms and biochemical aspects of DNA strand breaks by ionizing radiation, Int. J. Quant. Chem., 39: 709-727.

Cox, A.B., Salmon, Y.L., Lee, A.C., Lett, J.T., Williams, G.R., Broerse, J.J., Wagemaker, G., Leavitt, D.D., 1993, Cataractogenesis data across longer-lived mammalian species, in: Biological Effects and Physics of Solar and Galactic Cosmic Radiation, C.E. Swenberg, G. Horneck and E.G. Stassinopoulos, eds, pp. 177-184, Plenum Press, New York.

Cucinotta, F.A., Atwell, W., Weyland, M., Hardy, A.C., Wilson, J.W., Townsend, L.W., Shinn, J.L., Katz, R., 1991, Radiation risk predictions for Space Station Freedom Orbits. NASA Technical Paper 3098. National Aeronautics and Space Administration, Washington, D.C.

Crowley, C. and Curtis, H.J., 1963, The development of somatic mutations in mice with age, Proc. Nat. Acad. Sci. USA, 49: 626-628.

Curtis, S.B., 1993, Relating space radiation environments to risk estimates, in: Biological Effects and Physics of Solar and Galactic Cosmic Radiation, C.E. Swenberg, G. Horneck and E.G. Stassinopoulos, eds, pp. 817-830, Plenum Press, New York.

Fielden, E.M. and O'Neill, P., 1991, The Early effects of radiation on DNA. NATO ASI Series H: Cell Biology, Vol. 54 Springer-Verlag, Berlin.

Goodwin, E.H. and Blakely, E.A., 1992, Heavy ion-induced chromosomal damage and repair, Adv. Space. Res. 12 (2): 81-89.

Heckman, G., 1993, Relating space radiation environments to risk estimates. in: Biological Effects and Physics of Solar and Galactic Cosmic Radiation, C.E. Swenberg, G. Horneck and E.G. Stassinopoulos, eds, pp. 89-100, Plenum Press, New York.

Holley, W.R. and Chatterjee, A., 1991, The application of chemical models to cellular DNA damage. NATO ASI Series H: Cell Biology, E.M. Fielden, P. O'Neill eds., Vol. 54 Springer-Verlang, Berlin, pp. 196-209.

Krebs, W., Krebs, I., and Worgul, B.V., 1990, Effect of accelerated iron ions on the retina. Radiat. Res., 123: 213-219.

Lett, J.T., 1974, Neuron isolation from the CNS as determined by DNA damage. DHEW #(NIH)74-296, Survey Report on Aging Nervous System. Edited by G. Maletta, pp. 47-57.

Lett, J.T., 1990, Damage to DNA and chromatin structure from ionizing radiations, and the radiation sensitivities of mammalian cells, Prog. Nucleic Acid Res. and Mol. Biol., 39: 305-352.

Lett, J.T., 1992, Damage to cellular DNA from particulate radiations, the efficacy of its processing and the radiosensitivity of mammalian cells: emphasis on DNA double strand breaks and chromatin breaks, Radiat. Environ. Biophy., 31: 257-277.

Lett, J.T. and Alexander, P., 1961, Crosslinking and degradation of deoxyribonucleic acid gels with varying water contents when irradiated with electrons, Radiat. Res., 15: 362-373.

Lett, J.T., Cox, A.B., and Bergtold, D.S., 1986, Cellular and tissue reponses to heavy ions: basic considerations. Radiat. Environ. Biophys., 25: 1-12.

Lett, J.T., Cox, A.B., and Lee, A.C., 1988, Selected examples in degenerative late effects caused by particulate radiations in normal tissues, in: Terrestrial Space Radiation and its Biological Effects, P.D. McCormack, C.E. Swenberg and H. Bücker, eds., pp. 393-413, Plenum Press, New York.

Lett, J.T., Keng, P.C., Bergtold, D.S. and Howard, J., 1987, Effects of heavy ions on rabbit tissues: induction of DNA strand breaks in retinal photoreceptor cells by high doses of radiation. Radiat. Environ. Biophys. 26, 23-36.

Lett, J.T., Lee, A.C. and Cox, A.B., 1991, Late cataractogenesis in rhesus monkeys irradiated with protons and radiogenic cataract in other species, Radiat. Res., 126: 147-156.

Lett, J.T. and Peters E.L., 1992, Deoxyribonucleoprotein structure and radiation injury: cellular radiosensitivity is determined by LET∞ - dependent DNA damage in hydrated deoxyribonucleoproteins and the extent of its repair. Adv. Space Res., 12(2): 51-58.

McNulty, P.J., Roth, D.R., Beauvais, W.J., Abdel-Kader, W.G. and Stassinopoulos, E.G., 1993, Relating space radiation environments to risk estimates. in: Biological Effects and Physics of Solar and Galactic Cosmic Radiation, C.E. Swenberg, G. Horneck and E.G. Stassinopoulos, eds, pp. 165-180, Plenum Press, New York.

National Council on Radiation Protection and Measurements, 1989, Guidance on Radiation Received in Space Activities, NCRP Report No. 98.

Rubin, P. and Casarett, G.W., 1968, Clinical Radiation Pathology, Saunders, Philadelphia.

Shea, M.A. and Smart, D.F., 1993, Relating space radiation environments to risk estimates. in: Biological Effects and Physics of Solar and Galactic Cosmic Radiation, C.E. Swenberg, G. Horneck and E.G. Stassinopoulos, eds, pp. 37-72, Plenum Press, New York.

Wheeler, K.T. and Lett, J.T., 1974, On the possibility that DNA repair is related to age in non-dividing cells, Proc. Nat. Acad. Sci. USA, 71: 1862-1865.

EFFECTS OF NEON IONS, NEUTRONS AND X RAYS ON SMALL INTESTINE

K.E. Carr*, J.S. McCullough*, R.St.C. Gilmore*, B. Abbas**, S.P. Hume+, A.C. Nelson++, T.L. Hayes#, and E.J. Ainsworth##

*School of Biomedical Science/Anatomy, The Queen's University of Belfast, Medical Biology Centre, 97 Lisburn Road, Belfast BT9 7BL N. Ireland
**School of Life Sciences, Kingston Polytechnic, Penrhyn Road Kingston-upon-Thames, Surrey, England
+MRC Cyclotron Unit, Hammersmith Hospital, Ducane Road, London
++NeoPath, Bellevue, Washington, USA
#Lawrence Berkeley Laboratory, Berkeley, California, USA
##Armed Forces Radiobiology Research Institute, Bethesda, Maryland USA

SUMMARY

This article reports on the response of the mouse small intestine to irradiation by 670 MeV /amu neon ions and 225 kVp X rays, using doses expected to give the same effect in terms of microcolony survival. Pieces of small intestine were examined by scanning and transmission electron microscopy. Using resin histology the numbers of several intestinal parameters were assessed and entered into a data display to summarise the overall gut damage.

The results underline the fact that more forms of morphological damage are detected in integrated biological systems when multiple cell types are investigated and multiple techniques used. Irradiation by neon ions produces structural changes, some of which are similar in quality to those seen after neutron treatment. However, the pattern of response is not identical to that produced by neutrons, indicating that each modality produces its own response profile.

INTRODUCTION

Radiation interacts with biological tissues through medical diagnosis and treatment (Galland and Spencer, 1990, Inskip et al, 1990, Parker et al, 1984, Vaeth and Meyer, 1989) and through environmental exposure of various types (Christaldi et al, 1991, Frome et al, 1990, Gomez et al, 1987, Lang and Raunemma, 1991), including space flight (Conklin and Hagan, 1987, Fry and Lett, 1988, Letaw et al, 1987, Lett et al, 1991, Townsend et al, 1991). The radiation environment to which space crews are exposed includes photons, and energetic charged particles, such as electrons and protons, together with helium and heavier ions. These particles come from trapped particle radiation, galactic cosmic rays or solar particle events (NCRP report, 1989). In addition, the crew will also be at risk from secondary radiation produced after interaction of these particles with the space craft or in their bodies: such radiations will include neutrons and photons as well as heavier particles.

The biological effects of any radiation can be described in several ways, ranging from theoretical modelling to lethality measurements. Between these two extremes, radiation-induced biological effects fall into several levels of response, including changes at molecular,

Biological Effects and Physics of Solar and Galactic Cosmic Radiation, Part A, Edited by C.E. Swenberg *et al.*, Plenum Press, New York, 1993

cellular, tissue, organ or whole body level. Since organs comprise several types of tissues and cells, it is plain that space radiobiology requires some of its research to be aimed at understanding the integrative aspects of radiation damage on whole organs, in addition to studies of the changes seen in those populations of cells which might become involved in mutagenesis or carcinogenesis. This need to understand the effect of radiation in integrated biological systems is in line with current work, where more than one parameter is often studied after the application of various experimental procedures including radiation (Carr et al, 1991, Argenzio et al, 1990, Goodlad and Wright, 1990, Mayhew et al, 1990, Ray et al, 1989, Saxena et al, 1991, Vigneulle et al, 1990, Wright et al, 1989). Specific multi-parameter topics of interest have included stromal-epithelial communication (Anderson et al, 1990), the pathophysiology of the radiation response of multiple cell populations (Rubin, 1989) and potential interrelationships between different responses (Lehnert et al, 1991).

The subject of the response of a multicellular organ has already been addressed by previous studies of the morphological changes in various tissues or cell populations, in clinical reports or papers on research on topics relevant to tumour therapy (Busch, 1990), where changes in the non-proliferative compartments are seen. It has also been addressed in some reports of the effects of heavy ions, where there have been reports of structural changes in the lens (Worgul et al, 1989, Carr et al, 1991, Riley et al, 1991) or in various aspects of the small intestine (Carr et al, 1987, 1990, Fatemi et al, 1985). The main subject of this paper is the effect of neon ions, but the observed changes will be compared with the alterations seen after neutron or X-irradiation, in order to highlight any parts of the response which may be characteristic of high linear energy transfer (LET) radiations. Compared with other organs, the small intestine has the great research advantage of containing easily identified areas where the four main tissue types of the body can be found with epithelium lining the luminal aspect, connective tissue underlying it and on the outer aspect of the wall, muscle forming a contractile layer in the wall, and nerve plexus profiles easily seen in relation to the other tissues. It is therefore possible to draw general conclusions about the response of these tissues, although there are indications that extrapolations to other organs must take into account the fact that the response of any cell type to irradiation may be affected by the physicochemical environment in the organ within which it is situated.

The data reported in the current paper can be regarded as testing two proposals -

1. that different cell types respond differently to irradiation as revealed by a study of the resin histology and electron microscopy of the cellular and subcellular responses of neon ion and X-irradiated small intestine

2. that a data display can be constructed to lay out the changes in the different tissues and cells and summarise the multiple tissue response.

MATERIALS AND METHODS

Animals

Small intestine was obtained from LAF1 female mice killed by cervical dislocation or overdosage of Nembutal 3 days after irradiation. Sections of small intestine, approximately 2 to 2.5 cm in length, were removed and fixed in 2.5% glutaraldehyde containing sodium cacodylate buffer (pH 7.4) and sucrose.

Tissues were washed subsequently in sodium cacodylate buffer at 4°C. For microscopic observation, all four mice were examined for each group. For quantitative work, a minimum of three mice were sampled per group.

Irradiation

The mice were injected with 1.25 mg sodium nembutal solution in normal saline 20-30 min before irradiation. They were then irradiated with either 225 kVp X rays (mean dose 16.6 Gy, two at 16.0 and two at 17.2 Gy, dose rate 140-160 cGy per min) or neon ions (20Ne): these ions were delivered at the plateau portion of the Bragg curve, energy 670 MeV

/amu, mean dose 11.0 Gy (two mice at 10.0 Gy two at 12.4 Gy), dose rate 75-150 cGy per min. The X ray and neon doses were expected to produce equal decrease in crypts the number of surviving based on an assumed relative biological effectiveness (RBE) of about 1.5 (1.4-1.6).

Topographical histology, resin histology and electron microscopy

Preparation for SEM involved specimens being critical point dried from CO_2, sputter coated with gold and examined in a JEOL T300 or 840A scanning electron microscope.

For resin histology, the tissue was usually post-fixed in buffered 1% osmium tetroxide. The specimens were then dehydrated in a graded series of ethanol and embedded in either Spurr's or Epon 812 substitute resin. Specimens were prepared in one of two ways, firstly as full circumferences of the gut tube and secondly in such a way that the villi could be cut in cross sections. The sections were stained with toluidine blue, methylene blue and basic fuchsin (Aparicio and Marsden, 1969) or silver.

For transmission electron microscopy, areas of interest were chosen after examination of the semi-thin sections. Ultra-thin sections (70-90 nm) were cut on a Reichert-Jung Ultracut-E, after the blocks had been trimmed. These sections were stained with uranyl acetate and lead citrate and then examined in a JEOL 100 CXII transmission electron microscope operated at an accelerating voltage of 80 kV. Images were recorded on Kodak electron microscope cut film (No. 4489).

Response of different parameters

The procedure for data collection has already been reported (Carr et al, 1991). Six sections per animal and three to four animals per experimental point were sampled and the numbers of villi, crypts, mitotic figures, enterocytes, goblet, endocrine, Paneth and villous stromal cells, submucosal arterioles, inner and outer muscle nuclei, Auerbach's and Meissner's plexus profiles were counted. Either a Hewlett Packard 86 B image analysis system or an Optilab image analysis software package (Graftek) was used to measure areas of epithelium, muscle, nerve and connective tissue within sections. The data for each parameter were compared independently by subjecting the results to a one way ANOVA and computing multiple comparisons between pairs of means using the Scheffe-F method (Sokal and Rohlf, 1981). In order to compare the changes in percentage response between different parameters, the raw data were expressed as a percentage of the relevant control figure. Using the Mann-Whitney U test (Sokal and Rohlf, 1981), significantly different responses between pairs of parameters were assessed.

Data Display for comparison of effects of neon ions and X rays

In order to summarise the effects for the whole organ for the two experiments, a data display was prepared (Carr et al, 1991). This contains a control-referenced deviation ratio for each parameter: those which differed from their own control are marked with an asterisk and the direction of deviation from control is indicated by an arrow. The larger of the two numbers from control and irradiated groups was used as the denominator to ensure that deviation ratios were always lower than 1. As well as forming part of the data display, the counts were also entered into a mathematical equation, to produce a summary figure, the Morphological Index:

$$M^{\circ}I = {}^{x}\Sigma_{n=1} \{ TWF_n \cdot TD_n \cdot [{}^{y}\Pi_{m=1}{}^{m} CD] \cdot [{}^{m}UD]\}$$

where Morphological Index ($M^{\circ}I$) = the total figure of structural change for any specimen or group, based on entering data for tissue weighting factors, tissue deviations and cell deviations into a mathematical model.

Tissue Weighting Factor (TWF) = the proportion of total section area made up of the relevant tissue (epithelium, connective, muscle, nerve).

Tissue Deviation (TD) = the amount by which the experimental tissue deviates from control. It is the ratio of experimental to control specimens with regard to tissue areas.

Cell Deviation (CD) = the amount by which the number of cells or structural features of a certain type in the experimental situation deviates from that of control. It is the ratio of experimental to control counts.

Ultrastructural Deviation (UD) = ratio of treated to control empirical scores for ultrastructural features. The values m and n are the arbitrary running totals in the multiplicative and summation figures respectively, with each having their final value limited to y and x respectively, x being the number of tissue compartments, in this case four, and y being the total number of cell types or structural units in any one tissue.

The Morphological Index is calculated from the data display by multiplying all the relevant figures within each tissue row, since it is assumed that all these figures are interdependent: this produced a subtotal Index for each tissue. A total Index is prepared by adding these four Tissue Indices, since it is assumed that each can be regarded as an independent compartment. Indices are prepared for individual control animals to show variation within the control group. Initially Indices are prepared using pooled data for all points. Indices are also prepared for individual animals, so that each group can be represented by a mean figure and standard error bars. A comparison of different points on this display indicates whether differences in total Index between points are likely to be significant.

Comparison with effects of X rays and neutrons

The data were compared with the corresponding results for small intestinal specimens collected three days after treatment with 10 Gy X rays or 5 Gy neutrons, the doses being chosen to produce identical crypt survival in the two groups. Nine female (10-12 weeks old) CFLP mice (Hacking & Churchill Ltd) were used in the experiment. They were randomly divided into three groups of three mice i.e. untreated control, X-irradiated and neutron irradiated. For irradiation, unanaesthetized, unrestrained mice were placed in a Perspex box. One group received a whole-body dose of 5 Gy cyclotron-produced neutrons (mean energy 7.5 MeV, with a gamma-ray contamination of 4%) at a dose rate of 0.45 Gy min-1 and a beam current of 100 μA. The second irradiated group received a whole body dose of 10 Gy X rays (from a 250 kVp Marconi X ray set with an HVL of 1.2 mm copper) at a dose rate of 2.35 Gy min-1. Again, an RBE of about 2 was assumed based on crypt-microcolony survival 3-4 d after irradiation (Hornsey, 1970).

Portions of duodenum were removed from irradiated animals 6 h and 1, 3 d after treatment. The tissues were fixed in 5% glutaraldehyde, washed in 0.1 M sodium cacodylate buffer, and dehydrated through a graded ethanol series to propylene oxide, before embedding in Epon and preparation as above for the collection of data on full circumference counts and areas.

RESULTS

Resin histology and electron microscopy

(i) Topographical histology

Examination with scanning electron microscopy showed that the key feature of the topographical histology of the surface was the presence, after neon ions but not X-rays, of collared crypts, with more prominent openings than are seen in control specimens (Figure 1).

(ii) Light microscopy of resin histology sections

Examination of the sections showed that both radiations produced intestinal specimens that were grossly distorted compared with the control material (Figures 2, 3). Crypts with distorted shoulders were seen. The presence of greater damage at the epithelial/stromal boundary after neon ion treatment was revealed by semi-quantitative analysis of the changes seen at this interface in silver stained sections. The damage was greater in the villous and intervillous shoulders after neon ion treatment, although no difference was seen between the

Figure 1. Scanning Electron Micrograph of a Region of Neon Irradiated Small Intestine Showing a Collared Crypt(c) amongst Villi (v). Bar = 100 μm.

two groups of interfaces in the cryptal region. The responses of the villous capillaries were also different, with 16.6 Gy X-irradiation producing vasodilation, while neon ion treatment led to vasoconstriction.

(iii) Ultrastructure

Transmission electron microscopy revealed damage in several parts of the villi (Figure

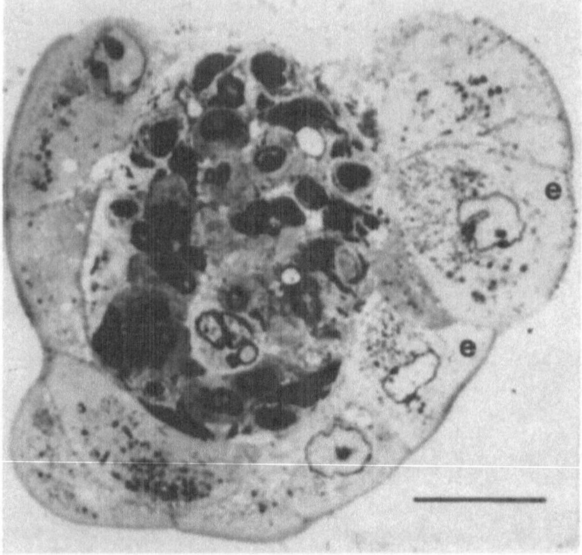

Figure 2. Light Micrograph of a Villus in Cross Section from Neon Irradiated Mouse Small Intestine. Note the Enlarged, Lightly Staining Epithelial Cells (e) and the fact that Stromal Blood Capillaries are so Constricted as to Make them Difficult to Identify. Bar = 25 μm.

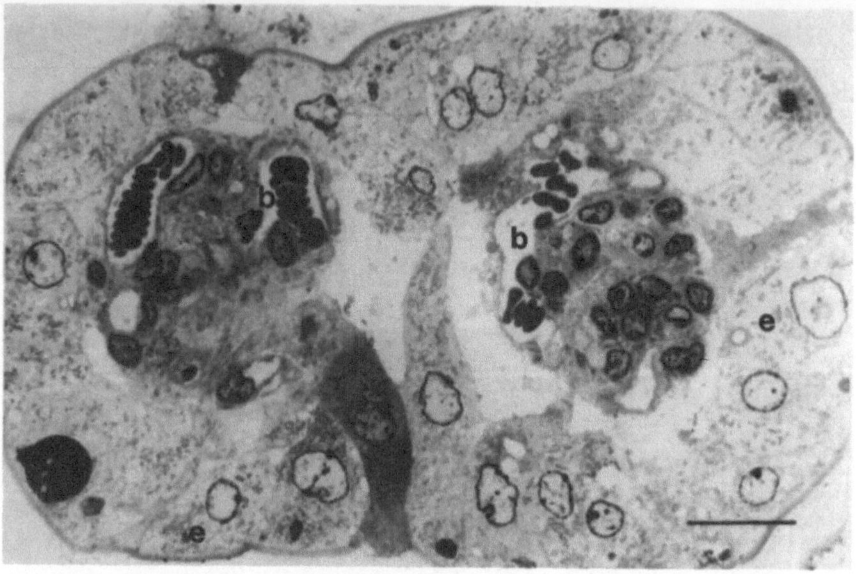

Figure 3. Light Micrograph of a Villus in Cross Section from 16.6 Gy X-irradiated Mouse Small Intestine. Epithelial Cells (e) are Similar to those Following Neon Irradiation but Stromal Blood Capillaries (b) Containing Numerous Erythrocytes are Clearly Visible because of their Dilation. Bar = 25 μm.

4). Epithelial changes included the presence of irregularly-shaped nuclei, cytoplasmic vacuolisation and displacement of organelles: some of these changes appeared more severe after neon irradiation. The greater damage seen after this treatment at the epithelial/stromal boundary in silver stained resin sections was reflected in the presence of a more irregular

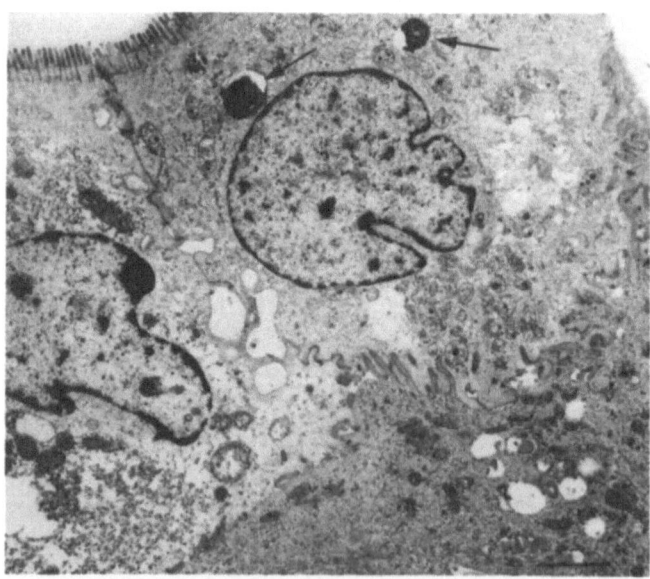

Figure 4. Transmission Electron Micrograph of Epithelial Cells from a Neon Irradiated Villus. The Cells are Cuboidal in Shape and have Large Irregular Euchromatic Nuclei and Electron Dense Inclusions (Arrows). Most of the Inter-epithelial Contacts are Intact although Some Gaps are Present between the Cells. Bar = 2 μm.

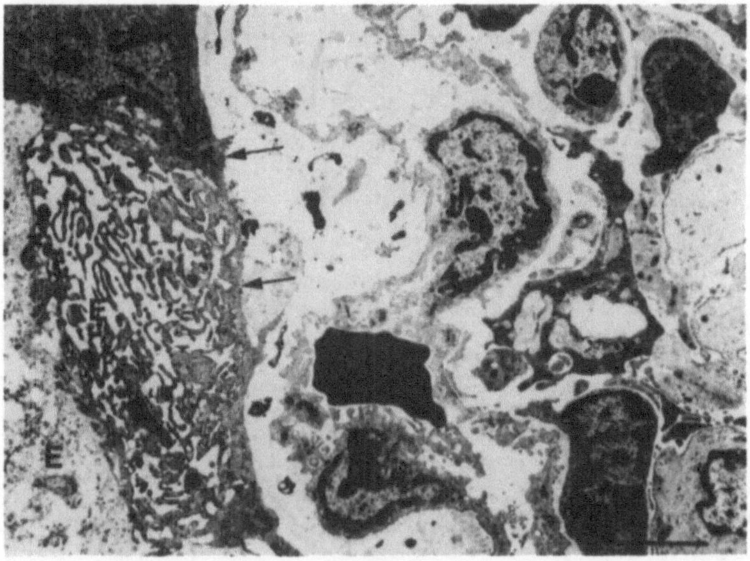

Figure 5. Transmission Electron Micrograph of the Epithelial/Stromal Boundary from a Neon Irradiated Villus. Note that the Epithelium (E) has Begun to Separate from the Stroma along Their Boundary (Arrows). Bar = 3 μm.

stromal boundary at the ultrastructural level (Figure 5). Subepithelial fibroblasts were also more abnormal after neon ion treatment. Thickening and irregularities in the endothelial cells were also more marked after neon ion treatment.

The pattern of ultrastructural change was submitted to a subjective scoring system (Carr et al, 1992), the results being included along with the other quantitative data below.

Different response of intestinal parameters

Quantification of the response of different parameters (Figures 6 and 7) shows that both types of radiation produce non-uniform changes in the various parameters, but not to exactly the same extent. X-irradiation produced a clear difference in the responses in the villous and cryptal compartments (Figure 7). The neon ion responses, although somewhat similar, are less clear-cut in the difference they show between cryptal and villous changes (Figure 6), with a greater spread in the values reported for the number of crypts per circumference and the endocrine value overlapping with some of the villous as well as with the cryptal ranges.

Both treatments produce changes in non-epithelial compartments, with villous stromal cells affected by both types of irradiation, while the submucosal (Meissner's) plexus profiles are only affected by the neon ion schedule.

Data display for all parameters

The data display summarising the different effects produced by X rays and neon ions (Figure 8) confirms the presence of marked changes in the epithelial compartment, with alterations also in villous stromal cells for both treatments and in nerve after neon ion treatment only. The insertion of the ultrastructural information does not depress the Epithelial Index much further, since there was already so much damage seen by light microscopy. With the connective tissue compartment, however, the insertion of the ultrastructural data produces scores which illustrate the presence of more substantial damage after the neon ion irradiation, depressing further the relevant Tissue Index.

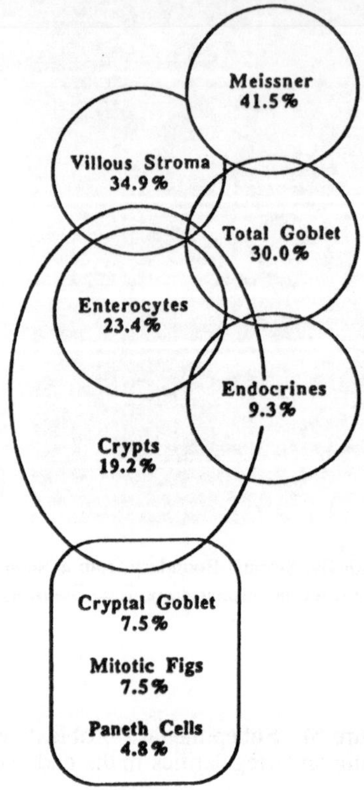

Figure 6. Diagram of the Comparative Percentage Response of those Parameters which Show a Significant Difference from Control Following Neon Irradiation. Non-overlap between Parameters Implies a Significantly Different Response after Testing with the Mann Whitney U procedure.

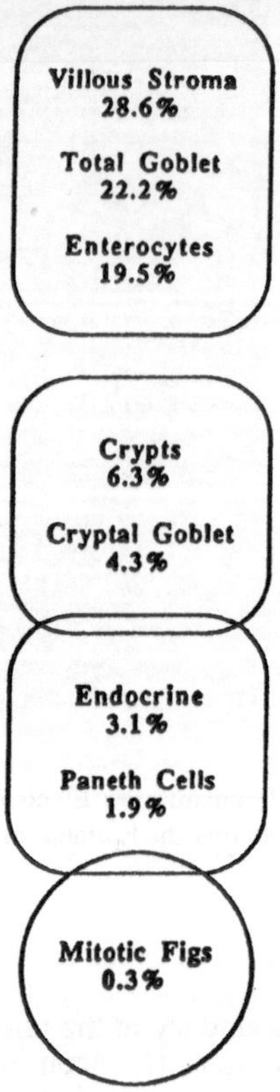

Figure 7. Diagram of the Comparative Percentage Response of those Parameters which Show a Significant Difference from Control Following 16.6 Gy X-irradiation. Non-overlap between Parameters Implies a Significantly Different Response after Testing with the Mann Whitney U procedure.

Any differences between the changes induced by neon ions and neutrons can be appreciated by comparing the data in Figure 8 with that of Figure 9 and Table 1, which summarises the changes produced by 5 Gy neutrons and 10 Gy X rays, 3 days after treatment. However, examination of the data for other time points (Carr et al, 1991) makes it clear that the neutron treatment produced substantially more damage to the neuromuscular compartment throughout the time studied, whereas such damage as is seen in nervous tissue after X–irradiation is only seen at one time point: study of the earlier changes shows that it is possible that this change may be secondary to vascular effects, which are not seen after neutron irradiation. Integrating this information with the neon ion experiment, it appears that higher LET radiations such as neon ions or neutrons may produce more changes in the neuromuscular parts of the intestinal wall.

DISCUSSION

The data allow discussion in turn of the two proposals as stated in the Introduction.

1. The suggestion that more damage would be revealed by the use of more sophisticated morphological techniques appears intrinsically obvious, since the higher

	TWF	TD	CD				2 Level Tissue Index	UD	3 Level Tissue Index
E	31.48	0.47	Enterocyte *↓ 0.23	Goblet *↓ 0.30	Endocrine *↓ 0.09	Paneth *↓ 0.05	0.005	0.40	0.002
CT	55.38	0.53	Villous Stromal Cells *↓ (0.35)^2		Submucosal Arterioles (0.94)^2		3.18	0.50	1.59
M	10.23	0.88	Outer Muscle Nuclei (0.49)^2		Inner Muscle Nuclei (0.98)^2		2.08	ND	2.08
N	2.91	1.00	Auerbach (0.79)^2		Meissner *↓ (0.41)^2		0.31	ND	0.31
Neon 11.2 Gy					2 Level M°I (analytical)		5.57	3 Level M°I (empirical)	3.98

	TWF	TD	CD				2 Level Tissue Index	UD	3 Level Tissue Index
E	37.45	0.78	Enterocyte *↓ 0.19	Goblet *↓ 0.22	Endocrine *↓ 0.03	Paneth 0.02	0.001	0.40	0.0004
CT	48.92	0.45	Villous Stromal Cells *↓ (0.29)^2		Submucosal Arterioles (0.79)^2		1.16	0.60	0.70
M	11.37	0.58	Outer Muscle Nuclei (0.32)^2		Inner Muscle Nuclei (0.70)^2		0.33	ND	0.33
N	2.27	1.00	Auerbach (0.94)^2		Meissner (0.95)^2		1.81	ND	1.81
X-Ray 16.6 Gy					2 Level M°I (analytical)		3.30	3 Level M°I (empirical)	2.84

Figure 8. Morphological Index Data Display for 11.2 Gy Neon and 16.6 Gy X-ray of Mouse Small Intestine. The data Include a Tissue Weighting Factor (TWF), which Reflects the Prominence of Some Tissues by Comparison with Others and Which is the Proportion of the Total Section Area Made up by Each Tissue (E, Epithelium; CT, Connective Tissue; M, Muscle; N, Nerve). Other Data Points Include Tissue, Cell and Ultrastructural Deviations (TD, CD, UD) which are All Ratios of Control and Irradiated Assessments of Each Parameter. A two Level Tissue Index is Calculated for Each Tissue by Multiplying TWF, TD and CD and the Three Level Index by Multiplying TWF, TD, CD and UD. Total Two Level or Three Level Morphological Indices are Obtained by Adding the Appropriate Tissue Indices. The Two Level Figure is an Analytical Index, Based on the Data Measured Using Objective Criteria. The Ultrastructural Deviation is, however, Based on a Subjective Score, Giving an Empirical Index. Asterisks Mark Parameters Whose Control and Irradiated Values are Significantly Different: Arrows Show the Direction of Deviation from Control.

resolution available with resin histology, SEM and TEM should allow better visualisation of the exact status of the tissues and cells after treatment. However, it does not follow that cells seen to be damaged on examination of wax embedded sections will necessarily show greater damage when examined using these more time-consuming techniques. However, previous work on the effects of neutrons has shown the presence of unexpected changes in villous shape (Carr et al, 1983), the presence of giant cells and collared crypts (Carr et al, 1981a) and the presence of ultrastructural damage in smooth muscle cells where Morphological Index counting had shown loss of cells (Carr et al, 1992). The X-irradiated material examined in the current experiments revealed findings that could have been expected from the previous literature, in that there were epithelial changes (Quastler, 1956, Lieb et al, 1977) and also

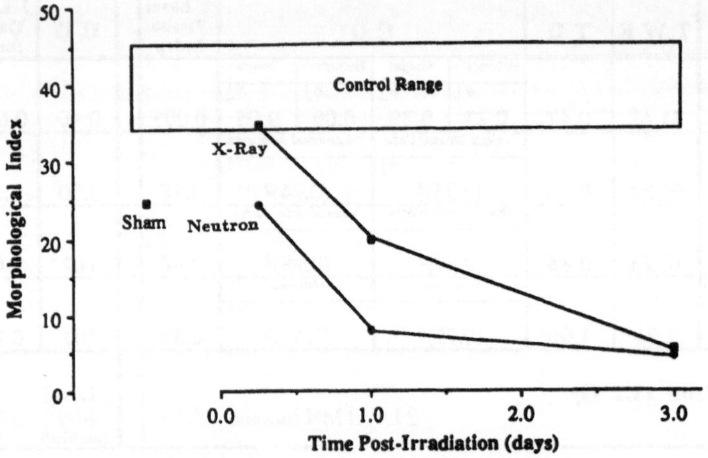

Figure 9. Graph of Morphological Indices for 5 Gy Neutron and 10 Gy X-irradiated Mouse Small Intestine, 6 h, 1 Day and 3 Days after Treatment. Note that for Both Treatments, the Morphological Indices Decline with Time.

alterations in the vascular components of the villi (Abbas et al, 1990a, 1990b).

The neon ion irradiated intestine contained collared crypts, also found after neutron irradiation (Carr et al, 1981a) and iron ion treatment (Carr et al, 1990). These structures imply the presence of very specifically localised damage concentrated in a small volume of the organ. It was previously suggested that if the definition of a 'tunnel lesion' was altered to include the concept of damage occurring at a specific point in the X, Y and Z axes, then for small intestine three days after treatment, such a lesion could have the appearance of a collared crypt.

Resin histology of the neon ion irradiated material showed that the basement membrane region was disturbed, a situation which leads to atypical contact between the epithelial and stromal compartments. It has been previously noted that this may lead to the presence of cell/cell contact between the two tissue types (Carr et al, 1981b, 1990, Fatemi et al, 1985) and epithelial/stromal stripping has also been reported in radiation-treated gut, with a

Table 1. Summary of Changes to Intestinal Parameters and their Likely Effects 3 Days after 10 Gy X ray or 5 Gy Neutron Irradiation.

Parameter	Effect
Enterocytes Goblet cells	Malabsorption and ulcer risk, impairment of gut function
Endocrine cells	Impairment of digestion control,proliferation, muscle tone etc.
Villous stromal cells	Villous shape, onward movement of absorbed molecules, defence
Auerbach's plexus	Changes in muscle tone of intestinal wall and blood vessels etc.
Mitotic figures	Indication of impairment of stem cell proliferative capacity
Crypts	Impaired repopulation of villous epithelium

greater loss of contact happening after neutron than after higher doses of X-irradiation (Carr et al, 1992). Ultrastructural examination of the neon irradiated material showed signs of damage in the epithelial and stromal compartments of the wall, as previously described by Fatemi et al, (1985) one year after treatment.

Given that the crypt counting assays could not differentiate between the two treatments at the dose and time points used, it is clear that the use of the various other microscopic techniques does produce extra information, which in some cases allows for differentiation between the two forms of irradiation.

Study of the different responses of the cell types to radiation reveals that patterns for the two treatments are consistent with previous data for other radiation schedules. There is confirmation in the neon ion pattern of response of the previous report (Carr et al, 1991) that the response of individual cryptal cell types may not necessarily be identical, seen here for goblet and endocrine cells, which respond to different extents. The neon ion response pattern likewise is consistent with the limited cryptal/villous separation of responses seen at the moderate doses used in a study of the effects of neutron treatment (Carr et al, 1991).

2. The data display for the comparison of the neon ion and X-irradiated specimens summarises well the information collected. There is already so much damage at the light microscope level that the insertion of the ultrastructural information does not add much, unlike the situation for neutron irradiation, where the insertion of the ultrastructural results for the 7 day time point brings the overall structural Index figure down from the control band to a level, consistent with the presence of residual damage (Carr et al, 1992).

When the data for the neon ion experiment are compared with those for neutrons, some comments are first required on the differences between the two experiments, which include variations in the strain of mouse, in the part of the small intestine sampled and in the selection of material for ultrastructural examination. The variations reflect the fact that the experiments were carried out at different times, when different aspects of the biology of the radiation response in the gut were highlighted. Despite the differences listed, there are still many common features. All of the data used are control-referenced within the relevant experiment and comparison across the two experiments shows that the control levels for the different parameters are identical for all parameters except mitotic activity and the two nerve plexuses.

Comparison of the effects of neon ions, neutrons and their corresponding X-irradiated controls three days after treatment shows that all groups have suffered changes in the enterocyte and goblet cell populations, which could lead to less efficient absorption and risk of gastrointestinal ulceration. In addition, all groups have depletion of the endocrine cell population (Wyatt et al, 1987), implicating humoral involvement (Vigneulle et al, 1989), which could have several serious side-effects, such as impairment in digestion, proliferation and muscle tone. A further common feature of interest is the decrease in the number of villous stromal cells, assumed to reflect a change in the wandering cell population, which is bone marrow derived and involved in the defence mechanisms of the gut: this response is likely to be linked to bone marrow effects (Mason et al, 1989).

The main difference between the neon ion and the corresponding X ray control is, however, the change in nervous tissue. When the neutron experiment is compared with its own X-irradiated control, 3 days after treatment, nerve damage is seen in both, albeit in another situation in the wall. However, the neutron irradiated specimens show changes in the neuromuscular compartment at more than one time point (Carr et al, 1991, 1992), perhaps following on from earlier vascular damage, not seen in neutron irradiated material. Whatever the reason for the change in nervous tissue after neon ion irradiation, it may be that the changes reported in the current papers in nerve profiles at these early time points could lead to the changes reported in the external laminae of nerves 1 year after irradiation (Fatemi et al, 1985).

The overall conclusion is that neon ion treatment produced changes in several structural components of the small intestinal wall, in a way which may be indicative of possible physiological disturbances during exposure and may also lead to likely late effects.

Comparison with effects of neutrons and preliminary work on the effects of other heavy ion species lead to the tentative conclusion that all radiation types damage epithelium, but that higher LET radiations are more likely to produce early changes in the neuromuscular tissues. However, the degree of damage depends on the type of these high LET radiations. It is therefore important that a study is made of the effects of each type of radiation and that all tissue and cell types in the wall are studied.

Future experiments are also planned to test empirically the correlation between the Morphological Index and the functional ability of the intestine (examples of this would be absorption and resistance to infection).

REFERENCES

Abbas, B., Boyle, F.C., Wilson, D.J., Nelson, A.C., and Carr, K.E., 1990a, Radiation induced changes in the blood capillaries of rat duodenal villi: a corrosion cast, light and transmission electron microscopical study, J. Submicrosc. Cytol. Pathol., 22 (1): 63.

Abbas, B., Hume, S.P., McCullough, J.S., Wilson, D.J., Stewart, P.C.,and Carr, K.E., 1990b, Early morphological changes in blood capillaries of mouse duodenal villi induced by X-irradiation, J. Submicrosc. Cytol. Pathol., 22 (4): 609.

Anderson, T.L., Gorstein, F., and Osteen, K.G., 1990, Stromal-epithelial cell communication, growth factors, and tissue regulation, Lab. Invest., 62: 519.

Aparicio, S.R., and Marsden, P., 1969, A rapid methylene blue - basic fuchsin stain for semi-thin sections of peripheral nerve and other tissues, J. Microsc., 89: 139.

Argenzio, R.A., Liaus, J.A., Levy, M.L., Menten, D.J., Lecce, J.G.,and Rowell, D.W., 1990, Villous atrophy, crypt hyperplasia, cellular infiltration and impaired glucose + NA, Gastroenterol., 98: 1129.

Busch, D.B., 1990, Pathology of the radiation-damaged bowel, in: "Radiation Enteritis," Galland, R.B., and Spencer, J., eds, Edward Arnold, London.

Carr, K.E., Hamlet, R., and Watt, C., 1981a, Scanning electron microscopy, autolysis and irradiation as technique for studying small intestinal morphology, J.Microsc.,123: 161.

Carr, K.E., Toner, P.G., McLay, A.L.C. and Hamlet, R., 1981b, The ultrastructure of some gastrointestinal lesions in experimental animals and man, Scand. J. Gastroenterol., 16, Suppl. 70: 107-128.

Carr, K.E., Hamlet, R., Nias, A.H.W., and Watt, C., 1983, Damage to the surface of the small intestinal villus: an objective scale of assessment of the effects of single and fractionated radiation doses, Br. J. Radiol., 56: 467.

Carr, K.E., Hamlet, R., Nias, A.H.W., and Watt, C., 1984, Morphological differences in the response of mouse small intestine to radiobiologically equivalent doses of X and neutron irradiation, Scanning Electron Microsc., I: 445.

Carr, K.E., Hayes, T.L., Indran, M., Bastacky, S.J., McAlinden, G., Ainsworth, E.J., and Ellis, S., 1987, Morphological criteria for comparing effects of X rays and neon ions on mouse small intestine, Scanning Microsc., Vol. 1(2): 799.

Carr, K.E., Hayes, T.L., Abbas, B., and Ainsworth, E.J., 1990, Collared crypts in irradiated small intestine, J. Submicrosc. Cytol. Pathol., 22 (2): 265.

Carr, K.E., McCullough, J.S., Nunn, S., Hume, S.P., and Nelson, A.C., 1991, Neutron and X ray effects on small intestine summarized by using a mathematical model or paradigm, Proc. Roy. Soc., Lond. B, 243: 187.

Carr, K.E., McCullough, J.S., Nelson, A.C., Hume, S.P., Nunn, S., and Kamel, H.H.M., 1992, Relationship between villous shape and mural structure in neutron irradiated small intestine, Scanning Electron Microsc. Int., in press.

Christaldi, M., Ieradi, L.A., Mascanzoni, D., and Mattei, T., 1991, Environmental impact of the Chernobyl accident: mutagenesis in bank voles from Sweden, Int. J. Radiat. Biol., 59(1): 31.

Conklin, J.J., and Hagan, M.P., 1987, Research issues for radiation protection for man during prolonged spaceflight, in: "Advances in Radiation Biology," 13th edition, Lett, J.T., Ehmann, U.K., and Cox, A.B., eds, Academic Press, London.

Fatemi, S.H., Antosh, M., Cullan, G.M., and Sharp, J.G., 1985, Late ultrastructural effects of heavy ions and gamma irradiation in the gastrointestinal tract of the mouse, Virchows Arch. (Cell Pathol.), 43: 325.

Frome, E.L., Cragle, D.C., and McLain, R.W., 1990, Poisson regression analysis of the mortality among a cohort of World War II Nuclear Industry Workers, Radiat. Res., 123: 138.

Fry, R.J.M., and Lett, J.T., 1988, Radiation hazards in space put in perspective, Nature, 335: 305.

Galland, R.B., and Spencer, J., 1990, "Radiation Enteritis," Edward Arnold, London,

Gomez, L.S., Yayanos, A.A., and Jackson, D.W., 1987, Subsealed disposal of high-level nuclear wastes, in: "Advances in Radiation Biology," 13th edition, Lett, J.T., Ehmann, U.K., and Cox, A.B., eds, Academic Press, London.

Goodlad, R.A., and Wright, N.A., 1990, Changes in intestinal cell proliferation, absorptive capacity and structure in young, adult and old rats, J. Anat., 173: 109.

Hornsey, S., 1970, The relative biological effectiveness of fast neutrons for intestinal damage, Radiology, 97: 649.

Inskip, P.D., Monson, R.R., Wagoner, J.K., Strovall, M., Davis, F.G., Kleinerman, R.A., and Boice, J.D., 1990, Cancer mortality following radium treatment for uterine bleeding, Radiat. Res., 123: 331.

Lang, S., and Raunemma, T., 1991, Behaviour of neutron-activated uranium dioxide dust particles in the gastrointestinal tract of the rat, Radiat. Res., 126: 273.

Lehnert, B.E., Dethloff, L.A., Finkelstein, J.N., and Van der Kogel, A.J., 1991, Temporal sequence of early alterations in rat lung following thoracic X-irradiation, Int. J. Radiat. Biol., 60: 657.

Letaw, J.R., Silberkerg, R., and Tsao, C.H., 1987, Radiation hazards on space missions, Nature, 330: 709.

Lett, J.T., Lee, A.C., and Cox, A.B., 1991, Late cataractogenesis in Rhesus monkeys irradiated with protons and radiogenic cataract in other species, Radiat. Res., 126: 147.

Lieb, R.J., McDonald, T.F., and McKenney, J.R., 1977, Fine structural effects of 1200 R. abdominal X-irradiation on rat intestinal epithelium, Radiat. Res., 70: 575.

Mason, K.A., Withers, H.R., McBride, W.H., Davis, C.A., and Smather, J.B., 1989, Comparison of the gastrointestinal syndrome after total-body or total-abdominal irradiation, Radiat. Res.,117: 480.

Mayhew, T.M., Dantzer, V., Elbrond, V.S., and Skadhauge, E., 1990, A sampling scheme intended for tandem measurements of sodium transport and microvillous surface area in the coprodaeal epithelium of hens on high- and low-salt diets, J. Anat.,173: 19.

NCRP Report, 1989, "Guidance on radiation received in space activities," National Council on Radiation Protection and Measurements, Bethesda.

Parker, R.P., Smith, P.H.S., and Taylor, D.M., 1984, "Basic Science of Nuclear Medicine," 2nd edition, Churchill Livingstone, Edinburgh.

Quastler, H., 1956, The nature of intestinal radiation death, Radiat. Res., 4: 303.

Ray, M., Dinda, P.K., and Beck, I.T., 1989, Mechanism of ethanol-induced jejunal microvasculature and morphologic changes in the dog, Gastroenterol. , 96: 345.

Riley, E.F., Lindgren, A.L., Andersen, A.L., Miller, R.C., and Ainsworth, E.J., 1991, Relative cataractogenic effects of X rays, fission-spectrum neutrons, and 56Fe particles: a comparison with mitotic effects, Radiat. Res., 125: 298.

Rubin, P., 1989, Law and order of sensitivity. Absolute versus relative, in: "Radiation Tolerance of Normal Tissues," Vaeth, J.M., and Meyer, J.L., eds, Karger, Basel.

Saxena, S.K., Thompson, J.S., Crouse, D.A., and Sharp, J.G., 1991, Epithelial cell proliferation and uptake of radiolabeled urogastron in the intestinal tissues following abdominal irradiation in the mouse, Radiat. Res., 128: 37.

Sokal, R.R., and Rohlf, F.J., 1981, "The Principles and Practice of Statistics in Biological Research," 2nd edition, W.H. Freeman and Co., San Francisco.

Townsend, L.W., Shinn, J., and Wilson, J.W., 1991, Interplanetary crew exposure estimates for the August 1972 and October 1989 solar particle events, Radiat. Res., 126: 108.

Vaeth, J.M., and Meyer, J.L., 1989, "Radiation Tolerance of Normal Tissues," Karger, Basel.

Vigneulle, R.M., Vriesendorp, H.M., Taylor, P., Burns, W., and Pelkey, T., 1989, Survival after total-body irradiation. I. Effects of partial small bowel shielding, Radiat. Res., 119: 313.

Vigneulle, R.M., Herrera, J., Gage, T., MacVittie, T.J., Taylor, P., Zeman, G., Nold, J.B., and Dubois, A., 1990, Non-uniform irradiation of the canine intestine. I. Effects., Radiat. Res.,121:46.

Worgul, B.V., Medvedovsky, C., Powers-Risius, P., and Alpen, E.,1989, Accelerated Heavy Ions and the Lens, IV. Biomicroscopic and cytopathological analyses of the lenses of mice irradiated with 600 MeV/amu 56Fe Ions, Radiat Res., 120: 280.

Wright, N.A., Carter, J., and Irwin, M., 1989, The measurement of villus cell population size in the mouse small intestine in normal and abnormal states: a comparison of absolute measurements with morphometric estimators in sectioned immersion-fixed material, Cell Tissue Kinet., 22: 405

Wyatt, M.G., Hume, S.P., Carr, K.E., and Marigold, J.G.L., 1987, A preliminary study of the role of gastrointestinal cells in the maintenance of villous structure following X-irradiation, Scanning Electron Microsc., I: 291.

MUCOPOLYSACCHARIDE VASCULAR COATING RELATIONSHIP TO ENVIRONMENTAL FACTORS

Delbert E. Philpott and Katharine Kato

NASA, Ames Research Center, MS 239-14
Moffett Field, CA 94035-1000

ABSTRACT

The lumen of blood vessels is coated with an acid mucopolysaccharide (AMPS) which is sensitive to environmental factors, i e., radiation, endotoxin, breathing 100% oxygen and aging. Loss of the AMPS results in hypercoagulability as shown by blockage of the femoral artery. Biochemical examination of trypsin treated blood vessels identified mannose, glactose, glucosamine and neuramic acid. The presence of these sugars in the perfusate almost doubled after 800 rads of X-rays. Attempts to use the stainable coating as a marker for radiation failed. The non-irradiated areas were equally effected. However, this proved the radiation was not the primary cause of the coating loss, but was due to a radiation by-product carried in the blood. The coating is quite sensitive to the effects of radiation. Twenty four hours after 10 rads of X-rays 23% of the coating was removed and 48 hours later 45% was missing, 200 rads removes 85-90%. Experiments for prediction of AMPS loss on long term space flights and its effects should be planned.

INTRODUCTION

For many years it was believed that there was no biochemical lining on the lumen of the endocapillary cells. Blood flowing through the capillaries was assumed to be in direct contact with the endothelial cell membranes. This notion was refuted by Luft in 1966 when he found an endocapillary layer lining the lumen of capillaries in mice fixed with glutaraldehyde and osmium tetroxide solutions containing Ruthenium Red. From the semi-selective binding properties and known histochemical reactions of the Ruthenium Red dye, Luft (1966, 1971) concluded that the lumen coating was probably an acid mucopolysaccharide (AMPS) made electron opaque by selectively-binding Ruthenium-osmium complexes. The presence of this AMPS coating on the endothelial cells was corroborated by Shirakama and Cohen in 1972. They used a Ruthenium Red-osmium fixation technique similar to that of Luft to preserve the coating on the rabbit myocardial capillaries.

Earlier observers noted changes in capillary circulation as a result of X-irradiation. Haymaker et al. (1968) exposed the brains of monkeys to 2000 rads at a dose rate of 50 rads/min. The brain tissue was examined with the light microscope between thirteen and seventeen weeks later. Haymaker observed aneurysms in the capillaries, and peculiarities in the endothelial and other vascular cells. He concluded that the vascular permeability had probably been altered, stating: "Certain vessels in the cerebral cortex were, as far as could be determined, of normal appearance, and yet plasmatic material could be found escaping from them, inundating and destroying tissue, we would assume that more changes occur in vessels than meet the eye, also that, although vessels in the irradiated central nervous system appear morphologically intact, they can be nonetheless functionally defective."

Chambers claimed in 1940 that he saw a cell coat with the dissecting microscope which he assumed served as a cell cement. His observations were largely ignored at the time. More than a decade later, Wislocki (1951) visualized a coating using Periodic Acid Schiff (PAS) staining of stratified squamous epithelial cells and lateral surfaces of intestinal epithelium, areas previously believed to be uncoated. Indirect evidence of an extraneous adhesive cell coat was provided by Moscona (1952) who showed that cells lost their adhesiveness after being trypsinized. Rambourg, et al. (1967) in their review reported over fifty cell types in the rat as having stainable cell surfaces as seen with PAS staining. They also found that all of the cell surfaces stained using Mowry's colloidal iron (1963), a stain specific for acidic carbohydrate stains devised for the electron microscope; e.g., periodic acid with aldehydes and periodic acid-chromic acid-silver methanamine. Golgi saccules and inner surfaces of cytoplasmic vesicles were stained as well as cell membranes. They also stained cell surfaces for electron microscopic study with phosphotungstic acid.

Cervos-Navarro (1964) studied the brain cortex from an irradiated rabbit (1500-2500 rads Co-60 source) 6 and 12 months after exposure using the electron microscope, and noted damage to the capillary endothelial cells and the capillary basement membrane. Kristensson et al. (1967) studied the human spinal cord exposed to an estimated 5100R of X-irradiation. They noted a thickening of the walls of capillaries, arterioles and arteries under the light microscope. Dilated capillaries and flattened endothelial cells were also seen. Strel'Tsova, et al., (1968) observed changes in the kidneys of dogs exposed to neutron irradiation. They concluded that the radiation injury of the vascular system in the kidney resulted in the observed pathology. Maisin (1970) observed the ultrastructural changes in the lungs of mice exposed to whole-chest X-irradiation (2000 rads), over a period of 15 months. In the early phase of radiation damage (few hours to one month), edema was found in the interstitium, and clusters of platelets were found opposed to the capillary endothelium. Endothelial cells of the small capillaries showed the most intensive early damage. Permeability of capillaries to horseradish peroxidase was increased and some capillary edema was noted. Eight to fifteen months after exposure (delayed response), Maisin reported that the basal membrane of the capillaries was often very thick and edematous. Maisin concluded that functional changes in the capillary permeability occur very soon after irradiation.

Philpott, et al., (1973) have also observed this mucopolysaccharide (MPS) coating on endothelial cells (Fig. 1,2) in many different tissues (heart, lung, kidney, brain etc.) taken from young rats (one month, 120 gm) fixed with Ruthenium Red following Luft's technique (1971). In addition, it was observed that young rats subjected to 2400 rads of whole- body X-irradiation and sacrificed 24 and 48 hours later were essentially devoid of the MPS coating when examined with Luft's technique (Fig. 3). Marcum and co-workers (1986) demonstrated anticoagulative properties on the surface of aortic endothelial cells. They used cloned bovine aortic endothelial cells to synthesize anticoagulantly active heparan sulfate proteoglycans. Their binding studies using 125I-labeled antithrombin demonstrated that these proteoglycans are located on the surface of the cloned bovine aortic endothelial cells.

It seems quite plausible that the early disappearance of the MPS coating observed by Philpott et al., (1973,1974b) could be intimately related to the change in capillary permeability noted by so many observers of radiation damage to the capillary systems in a variety of tissues. In addition, Matsumura, et al. (1966) observed a 20% decrease in viscosity in a buffered solution of hyaluronic acid (an AMPS found in bacteria and animals) exposed to 2000 rads of X- irradiation, thus indicating depolymerization.

Polysaccharides, in general, represent ubiquitous macromolecules which are located throughout the body and play a role in health and disease. One such compound is heparin, an anticoagulant which is located in the mast cells around blood vessels and very similar in structure to the coating material. It is not unreasonable to suppose that the acid MPS coating on the lumen also functions as an anticoagulant. It's disappearance might explain Maisin's (1970) observation of clusters of platelets oppossed to the capilary endothelium. Cartilage contains another polysaccharide, chondroitin sulfate, while heart valves contain both

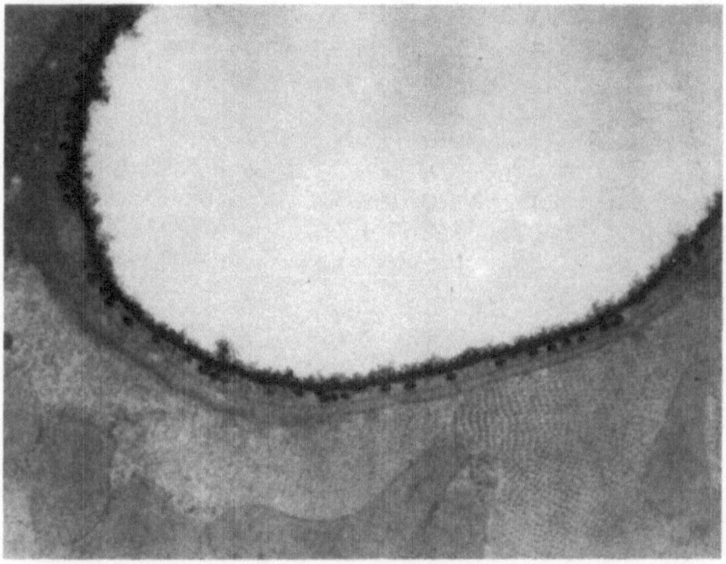

Figure 1. Normal Mouse Heart Capillary Stained with Ruthenium Red (RR). Note the Dark RR Material Lining the Lumen. 13,000X.

chondroitin sulfate A and B, plus hyaluronic acid. Cornea contains keratosulfate. Normal human plasma contains AMPS (Bassiouni 1955), but AMPS from the blood of patients with rheumatic fever or rheumatoid arthritis have anticoagulant activity not found in normals. Clinical use of MPS may have consequences. Bauer (1983) tested three MPS used to treat rheumatology. They found, Arteparon, to have anticoagulant properties and indicated attention should be paid to its use. Kirk and Dyerby (1957) isolated MPS containing mainly chondroitin sulfate A and B. The average yield from subjects under age 59 was 4.8 mg/g of wet tissue while a decrease of one half to 2.4 mg/g occurred in subjects 60-70 years old. Acidic MPS, which does not originate from the urinary tract or the prostate gland, has been

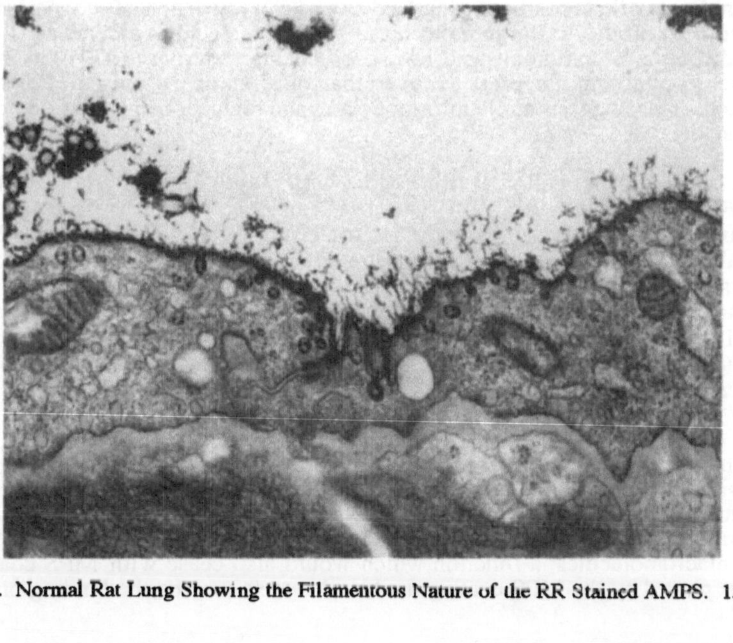

Figure 2. Normal Rat Lung Showing the Filamentous Nature of the RR Stained AMPS. 15,000X.

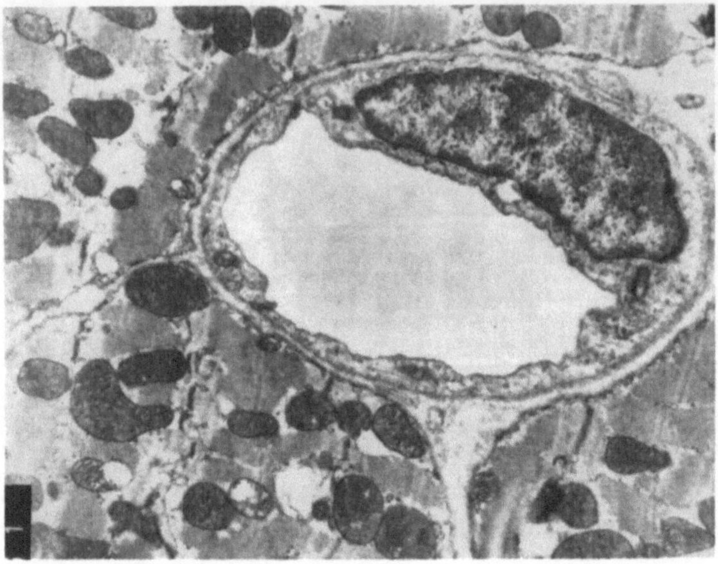

Figure 3. Rat Heart 48 Hours after 400 Rads of X-rays. RR Treated. Note Absence of Any AMPS Coating. 10,000X.

isolated from normal human urine. This AMPS is strongly metachromatic with toluidine blue and is hydrolyzed by testicular hyaluronidase. Electrophoresis of normal urine separated out three AMPS, and all three had some anticoagulant activity of the heparin type (Heremans, et al. 1959).

Hyaluronic acid, possibly the most studied MPS, shares many properties with the other MPS and forms part of the ground substance or matrix surrounding the cells. One of its suggested roles is to bind water in the interstitial spaces and to hold cells together in a jelly-like matrix (Meyer 1947) and another to resist compression and provide lubrication. The long polysaccharide molecule zig-zags back and forth, forming a loose but tangled "ball-of-string" network. It has been suggested as a role in restriction of movement of viruses and selective permeability of large molecules. Hyaluronic acid at a concentration of only 0.1% already forms a continuous meshwork of chains. Laurent (1977) has measured a two-fold increase in osmotic pressure over that of albumin in only a 1% solution of hyaluronic acid. This suggests a role in osmotic pressure regulation and is of obvious interest in capillary studies.

The cationic dyes, colloidal iron (Mowry 1963), thorium (Rambourg 1967), and Ruthenium Red (RR), (Luft 1966,1971) have been employed to visualize the cell coating and results had indicated the presence of acidic groups. Using these dyes plus Alcian Blue mixed with glutaraldehyde, Behnke (1968) demonstrated a MPS coating on red blood cells and platelets. His work with thorotrast indicated that this coat was on the order of 130 angstroms thick. He used trypsin digestion and showed that this treatment removed RR stained material, but colloidal iron and thorotrast showed binding to the cell surface. Luft's technique (1966) using RR has been the most widely used, yielding a dense coating easily visible in the electron microscope.

The properties of MPS, e.g., anticoagulant, osmotic pressure regulator, lubricant, mechanical barrier, etc., suggested several important roles for this structure along the lumen of blood vessels (Philpott 1973). It is hypothesized that reduction in the amount of coating, or its total removal, might very well cause osmotic pressure changes which in turn would alter capillary permeability. The network of intermeshed molecular strands could selectively restrict passage of macromolecules, a function which would also cease with MPS coat removal. Further roles played by the MPS coating the blood vessels may be the binding of water and anti-shock. In addition, this MPS coat could play a role in the laminar flow observed between blood cells and the capillary wall.

Studies have been carried out to test the effect of various stressors on the blood vessel lining. Aging may be one of them. Strokes are, in general, a disease of the elderly. To shed some possible light on this problem, young and old rats were examined for MPS coating on the vessels of the brain and heart (Philpott 1973). The MPS coating was thinner and spotty in the old animals. Endotoxin also removes the MPS coat An injection of 10 or 20 mg/Kg of E coli endotoxin completely removes the coating in 24 hours in rats (Philpott 1974a). The breathing of 100% oxygen (Philpott 1973) and exposure to X-rays (Philpott 1974b) also removes this coating.

Interestingly, Davies and Gamble (1977) found that abdominal radiation of 500 to 1000 rads increased the permeability of the intestinal vasculature 24 hours after exposure. Willoughby (1960) found the same result in his radiation studies. Therefore, if radiation decreases the MPS coating and at the same time increases permeability, a very good case for cause and effect exists. It seems that this blood vessel coating may play several key roles in the normal maintenance of function and permeability of the blood vessels.

In view of the fact that the blood vessel coating may play a role in altering coagulation, the reports of Skylab III and Salyut 4 (Kimsey 1977) are most interesting. There was a post-flight decrease of 20-30% in fibrinogen and a two-fold increase in fibrinogen split products in two of the Skylab III crewmen and results from Salyut-4, 2 and 7 days post-flight, showed an increase in the total coagulative activity of the blood as indicated by an increase in the heparin tolerance of the blood, by an increase (48-69%) in the fibrinolytic activity of the blood, and by an increase in recalcification. It was suggested that the increase in thrombogenic properties might be related to dehydration of the body and that the increase in fibrinolytic activity might be related to stress effects. Hypercoagulability could be a problem, especially if crew members are drawn from older age groups where strokes are more common.

The duration of space flights is increasing and there is serious talk about a mission to Mars. The MPS coating of cells and especially blood vessels appears to be susceptible to conditions of space flight. By visualizing and measuring the MPS coating on the blood vessels, its role in normal animals and in those exposed to simulated and/or spaceflight conditions can be elucidated. Immobilization, radiation, altered G levels and endotoxins appear to affect the coating. Combinations of these stresses should also be studied to help quantitate the tissue response. The coating also appears to play a role in relation to capillary permeability and alterations in coagulation time.

Hypokinesia studies have investigated changes in general MPS metabolism and production, but have not covered the effects on the MPS coating of the blood vessels. There is a lack of information concerning this coating in relation to spaceflight problems.

METHODS

Six C-57 black mice were used for each radiation response study and six more were used as controls. Tissue staining of the MPS utilized Luft's method (1966). Six 120 gm rats were used to test coagulation times in each category of 1) controls, 2) endotoxin injection and 3) 400 and 800 rads of X-irradiation. Coagulation times were determined 10 and 24 hours after each treatment.

Disappearance of the MPS after X-irradiation was plotted after exposing heart tissue to 10, 40, 50, 100, 200, 400, 500, 1000, 1500, 2000 and 4000 rads of whole body irradiation. One hundred capillaries were chosen at random for each tissue and graded for the presence of the lumenal MPS coating after 24 and 48 hours.

Animals were irradiated in head only and chest only p=09 areas by X-rays and Neon particles to see if the removal of the MPS would be restricted to only these areas. Irradiated and non-irradiated areas were then selected for examination to test the theory that RR could be used as a selective marker for radiation.

Tissues were fixed by perfusion using 7% glutaraldehyde. The tissue was then sliced at 200 microns on a Vibratome and immersed in RR osmium solution. The following day selected tissues were rinsed briefly with buffer, and processed by routine embedding procedures. The energy dispensive X-ray analyzer attachment on the electron microscope was used for Ruthenium identification in the sections.

Luminal coating material was removed, collected and identified by the following procedure. Six young control rats, 180-210 gms, were perfused, first with 5-10 ml of Ringers and then with Ringer's containing trypsin. The trypsin flush removed the coating and the perfusate containing the MPS coating was collected. The MPS was removed by trypsin dissolved in 0.001 M HCl, then it was added to 0.01 cadodylate buffer to give 0.1 mg/ml of trypsin. Supernatant was collected and spun down. Perfusate was preserved in 1/3 ethyl alcohol, 2/3 trypsin solution. The method is repeated without trypsin for control. Following the trypsin flush, the animal was perfused with glutaraldehyde and following embedding and sectioning, the vessels were checked for coating. Gas liquid chromatography (GLC) was used to identify the components.

Using six rats for each dose, in-vivo coagulation time studies were carried out by placing five half-hitch loops of thread over each isolated femoral vein exposed on the rat hind leg (leaving the same space between each loop). One pull on the thread simultaneously tied off four sections of blood vessel, Fig. 4. By cutting open the vessel sections at timed intervals, in-vivo coagulation time was established.

Two month, eight month and two year old rats were treated with 10 and 20 mg/Kg of E. Coli endotoxin I.P. Tissues were examined after 10 and 24 hours. Heart and skeletal muscle, lung, liver, kidney, spleen, brain, retina, trachea, intestine, salivary gland, adrenal gland, gingiva and carotid artery were prepared for examination of the lumenal coating. Rats were also kept in 100% oxygen and 1 atmosphere for 72 hours.

RESULTS AND DISCUSSION

The eight month old rats proved the most resistant to the endotoxin treatment. This probably reflects the period when they are physiologically the strongest. Ten and twenty four hours of treatment resulted in increased loss of the MPS and increased damage to the capillaries and basement membrane. In general, the MPS loss preceded the cellular damage. The endotoxin affects the MPS luminal coating in a progressive manner in all of the tissues examined. While the lumen of the younger control rats had a continuous coating of the MPS, the two year old controls showed irregular aggregation of the MPS and general loss of thickness. Consequently, the two year old controls were not considered the best choice for general studies.

Reference to Table 1, shows the average reduction of in-vivo coagulation times after both endotoxin and irradiation. Since these treatments remove MPS, this indicates the MPS is at least partly responsible for anti-coagulative properties within blood vessels. GLC analysis was done (by Dr. M. Mathews, U. Chicago) for sugars identified manose, glactose,

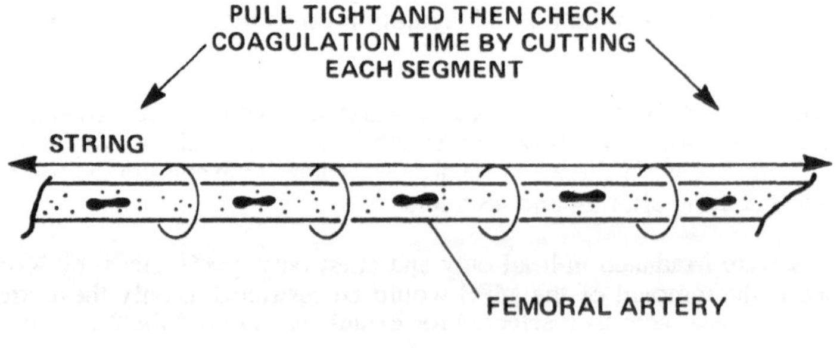

Figure 4. Schematic of in-situ Method Used to Determine Coagulation Times after X-irradiation of the Rat.

Table 1. Graph Showing the Loss of the AMPS after 24 and 48 Hours. The 2000 Rad Exposure is Not Shown as the AMPS is Essentially Removed.

Average Coagulation Time

Controls		E. Coli Endotoxin	X-irradiation		
27 secs	10 mg/Kg;	10 hrs=20 secs	400rads;	10 hrs	17 secs
	10 mg/Kg;	24 hrs=15 secs	400rads;	24 hrs	15 secs
	20 mg/Kg;	10 hrs=20 secs	800rads;	10 hrs	18 secs
	20 mg/Kg;	24 hrs=20 secs	800rads;	24 hrs	16 secs

Table 2. Results of Gas Liquid Chromatography Analysis for Carbohydrate Components (mg) in Normal and Tadiation Exposed Rats.

	CONTROL mg	RADIATED 800 rads mg
MANNOSE	1.90	3.0
GLACTOSE	1.80	4.4
GLUCOSAMINE	2.52	3.6
NEURAMINIC ACID	3.08	6.5

glucosamine and N-acetylneurominc acid. Reference to Table 2, shows the amount found in controls as compared to animals exposed to 800 rads of X- rays four hours after irradiation.

The MPS interposes itself between the endothelium of the blood vessel walls and the blood elements. As such, any transcapillary exchange must take place through this layer. There are several possibilities for its function. The layer consists of long chain molecules which reduces friction when an object passes over a surface. A more important feature may be the resemblance of the MPS structure to heparin. The calcium binding of heparin confers anti-coagulant properties and the MPS also has calcium binding properties due to their sulfate and carboxyl groups. Ofosu, et. al. (1987) increased the sulfation of heparin sulfate and dermatan sulfate and found increasing the sulfation increased the anticoagulant activity. Thus the presence of the MPS would confer an important property to the blood vessels. Also, any MPS sloughed off from the lumen would be carried in the blood stream, its presence possibility decreasing the rate of coagulation. Disappearance of the MPS could have serious consequences. Since this coating also decreases with age, it may be associated with the increase of strokes in older people.

The susceptibility of the MPS to radiation has implications for the amount which is safe to absorb at any one time. More studies should be done on the recovery rate of the MPS coating after the various environmental insults.

The MPS coating appears quite sensitive to radiation. Reference to the graph (Fig. 5) shows that 85 to 90% of the coating is removed after 200 rads. The other tissues that had been prepared were briefly examined for their response and similar results were evident. This indicates the MPS is quite susceptible to depolymerization after irradiation.

Attempts to use the disappearance of the coating as a selective marker for the location of specific areas which had been irradiated failed. Both X-rays and Neon particle irradiation were used. Regardless of the area irradiated, all of the blood vessels in all the tissues examined had lost an equal amount of coating. Since the blood is circulating during the exposures and part of it is continuously passing through the area being irradiated, all of the blood is exposed. It is known that by-products, especially peroxides, are formed during irradiation and this may be the answer. Any by-products from the irradiation would have continuous access to the coating and could depolymerize the MPS. This result is strong

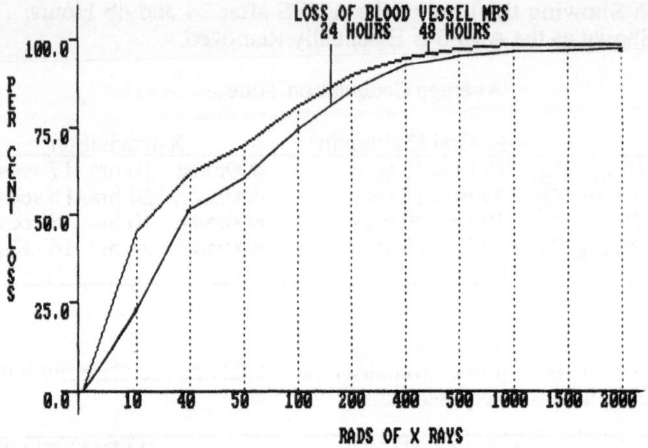

Figure 5. Graph showing the loss of the AMPS after 24 and 48 hours. The 4000 rad exposure is not shown as the AMPS is essentially removed.

evidence that deploymerization is an indirect effect of the radiation and may give clues to the real mechanism of its sensitivity.

Past investigations have visualized the coating, but techniques have not been quantitative or easily reproducible. The fluorescein method of Chernov (1965) measures capillary permeability, but it often only penetrates 3 or 4 cells deep. Ferritin labeling (Bruns and Palade 1968) does a better job at permeability and perhaps can be adapted to quantitating the lining layer and the change in permeability.

Environmental factors have been shown to effect MPS metabolism and production. A study by Mitsenko (1976) showed changes in MPS metabolism in individuals working in a hot environment. An accumulation of hexoses, seromucoids,ciruloplasmin in the blood and a decrease in fibrinolytic activity were seen. These changes affect components of the MPS lining, and the change in fibrinolytic activity may also show a correlation.

Fluid electrolyte levels and metabolism have been shown to be altered in space and during bed rest. Potapov (1977) suggested that investigation of MPS metabolism involved in binding and transport of fluids and inorganic ions should reveal new factors involved in impairing fluid and electrolyte metabolism during hypokinesia. Whether or not the MPS lining of blood vessels is involved should be investigated. His studies utilizing hypokinesia as a stressor reveals that phosphoric ethers of hexoses, which are immediate precursors of structural components of MPS, are used essentially to meet energy requirements of the organism and are therefore not available for the synthesis of hexosamines and hexuronic acid present in MPS.

SUMMARY

MPS layer exists which covers the luminal wall of blood vessels. This lining is thought to play a role in capillary permeability, coagulation, and lubrication of the vessel wall. It is also affected by radiation, infection, drugs, increased oxygen and age. All of these or combinations thereof can occur in space in addition to weightlessness. Reports from Salyut 4 disclosed an increase in total coagulative activity of the blood; reports from Skylab III also indicated coagulative changes. The mechanisms responsible for these changes remain unknown. Changes in MPS coating of blood vessels may play a significant role. It was suggested that the observed changes may be due to dehydration, but the role of MPS's was not known then. It appears that MPS may play quite a role in the reaction of the body to space flight.

A prediction model based on MPS coating changes could be developed to estimate the relevance for human long-term spaceflight exposure. Functional and structural relationships in the blood vessel wall and adjacent cellular structures should be delineated. In particular, clotting, fibrinogen leakage, alteration of pinocytotic vesicle size and number, and permeabililty need to be elucidated. Space flight experiments need to be planned in order to accurately determine the combined effects present in space.

Our early studies indicated that age reduces the coating thickness. An older group of astronauts may have a higher risk of thromboembolism in conjunction with thinning of the MPS layer if age is combined with radiation and/or infection, increasing the body's burden to the point of vascular malfunction. The susceptibility of the MPS to radiation has implications for the amount which is safe to absorb at any one time. Work should be done on the recovery rate of the MPS coating after the various environmental insults.

REFERENCES

Bassiouni, M. 1955, Studies on the acid mucopolysaccharides of the white cells in rheumatic and other diseases showing its similarity to the acic mucopolysaccharides of amyloid, Ann. Rheumatic Diseases, 14: 288-292.

Behnke, O. 1968, Electron microscopical observations on the surface coating of human blood platelets, J. Ultrastructure Res., 24: 51.

Bruns, R.R. and Palade, G.E. 1968, Studies on blood capillaries. II Transport of ferritin molecules across the wall of muscle capillaries. J. Cell Biol. 37: 277.

Bauer, F., Schulz, P., Reber, G. and Bouvier, C.A. 1983, Anticoagulant properties of three mucopolysaccharides used in rheumatology, Thromb. Haemost. 50(3): 652-655.

Cervos-Navaro, J. 1964, Spatveranderrungen des Zentralnervensystems nach Scha delbestrahlung, Proc. 3rd European Regional Conf. Electron Microscopy, 71-74.

Chambers, R. 1940, The Relation of the extraneous coats to the organization and permeability of cell membranes. Cold Spring Harbor Symp. Quart. Biol., 8: 144-153.

Chernov, G.A., Sheremet, Z.I. and Lenskaya, R.V. 1965, Effect of irradiation on vascular permeability and on blood mucopolysaccharide and serotonin levels. Fed. Proc. Trans. Supp. 24(6): 974-976.

Davies, R.W. and Gamble, J. 1977, Changes in the rate of transduction of vascular fluid in the isolated rat mesentary following irradiation. The Phys. Soc. 266(1): 71P-72P.

Haymaker, W. 1968, Delayed radiation response in the brains of monkeys exposed to X- and Gamma Rays, J. Neuropath. Exp. Neurol. 27: 50-78.

Heremans, J.F., Vaerman, J.P. and Heremans, M.Th. 1959, Acid mucopolysaccharides of normal urine. Nature, 183: 1606.

Kimsey, S.L. 1977, Hematology and immunology studies. Chapter 28, Biomedical results from Skylab. R.S. Johnston and L.F. Dietlein, ed., NASA SP-377.

Kirk, J.E. and Dyerby, M. 1957, Mucopolysaccharides of human arterial tissue. I. Isolation of Mucopolysaccharide material. J. Gerontol.,12: 20-23.

Kristensson, K. et al. 1967, Delayed radiation lesions of the human spinal cord. Acta Neuropatholigica, 9: 34-44.

Laurent, T.C., et al. 1977, Diffusion of macromolecules through compartments containing polysaccharides. Bibl. Anat. 15(1): 489-482.

Luft, J. 1966, Fine structure of capillary and endocapillary layer as revealed by Ruthenium Red. Fed. Proc, 25: 1773-1783.

Luft, J. 1971, Ruthenium Red and Violet, I: Chemistry, Purification, methods of use for electron microscopy and mechanism of action. Anat. Rec. 171: 347-368.

Maisin, J. 1970, The ultrastructure of the lung of mice exposed to a supra-lethal dose of ionizing radiation on the thorax. Rad. Res., 44: 545-564.

Marcum, J.A., Atha, D.H., Fritze, L.M., Nawroth, P., Stern, D. and Rosenberg, R.D. (1986). Cloned bovine aortic endothelial cells synthesize anticoagulantly active heparin sulfate proteoglycan, J. Biol. Chem. 261(16): 7507-7517.

Matsumura, et al. 1966, Depolymerization of hyaluronic acid by autooxidants and radiations. Rad. Res. 28: 735-752.

Meyer, K. 1947, Bilological significance of hyaluronic acid and hyaluronidase. Phys. Rev. 27: 335-359.

Mitsenko, M.D. 1976, Characteristics of changes in humoral indicators of mucopolysaccharide metabolism and in blood coagulation of workers of hot shops and practical importance of detection of these changes. Ter. Arkh. 48(6): 91-6.

Moscona, A. 1952, Cell suspensions from organ rudiments of chick embryos. Exp. Cell Res., 3: 535-539.

Mowry, R. 1963, The special value of methods that color both acidic and vicinal hydroxyl groups in the histochemical study of mucins. With revised directions for the colloidal iron stain, the use of alcian blue G8X and the combination with the PAS reaction. Ann. N.Y. Acad. Sci. 106: 402-423.

Ofosu, F.A., Modi, G.J., Blajohman, M.A., Buchanan, M.R. and Johnson, E.A, 1987, Increased sulphation improves the anticoagulant activities of heparan sulphate and dermatan sulphate, Biochem. J. 248(3): 889-896.

Philpott, D., Takahashi, A. and Turnbill, C. 1973, The disappearance of the acid mucopolysaccharide coating in blood vessels exposed to oxygen, radiation and in old age and possible implications, Proc. Elec. Micro. Soc. 31: 398- 399.

Philpott, D. and Takahashi, A. 1974a, Endotoxin alteration of the mucopolysaccharide blood vessel coating in rats. Proc. Elec. Micro. Soc. of Am., 32: 128-129.

Philpott, D. and Takahashi, A. 1974b, The disappearance of the blood vessel lining after X-irradiation. International Congress on Elec. Micro., Vol II, 550-551.

Potapov, P.P. 1977, Mucopolysaccharides and collagen of tissues in hypokinetic rats. Kosmicheskaya Biologiya I Aviakosmicheskaya Meditsina. No 3: 44:48.

Rambourg, A. and Leblond, C.P. 1967, Electron microscopic observations on the carbohydrate-rich cell coat present at the surface of cells in the rat, J. Cell Biol., 32: 27-53.

Shirakama, T. and Cohen, A. 1972, The role of mucopolysaccharides in vesicle architecture and endothelial transport. J. Cell Biol., 52: 198-206.

Strel'Tsova, V. et al. 1968, Radiation pathology of the kidneys. Arkh. Pathol., 30(2): 25-33.

Willoughby, D.A. 1960, Alteration of intestinal vasculature permeability after irradiation. Br. J. Radiol. 33: 515-519.

Wislocki, G.B., Fawcett, D. and Dempsey, E.W. 1951, Staining of stratified squamous epithelium of mucous membranes and skin of man and monkey by the periodic acid schiff method. Anat. Rec. 110: 359-375.

AN EVALUATION OF THE RELATIVE BEHAVIORAL TOXICITY OF HEAVY PARTICLES

Bernard M. Rabin[1,2], Walter A. Hunt[1], James A. Joseph[1], Thomas K. Dalton[1] and Sathasiva B. Kandasamy[1]

[1] Behavioral Sciences Department
Armed Forces Radiobiology Research Institute
Bethesda, MD 20889 U.S.A.

[2] Department of Psychology
University of Maryland Baltimore County
Baltimore, MD 21228 U.S.A.

INTRODUCTION

As astronauts leave the magnetic field of the earth, they will be exposed to radiation qualities and doses that differ from those experienced in low-earth orbit. Research using low LET radiation (gamma rays or high-energy electrons) indicates that large doses of these types of radiations (> 20-40 Gy) are needed to produce significant decrements in performance (Bogo et al., 1989; Hunt, 1983; Mickley et al., 1988). It is unlikely that astronauts will be exposed to such high doses (Mullen et al., 1988; McCormack, 1988). However, the extent to which exposure to heavy particles outside the magnetosphere can affect behavior has not been extensively studied. In contrast, recent research suggests that exposure to low doses of HZE particles can produce effects on behavior. Exposing rats to ^{56}Fe particles (600 MeV/amu) at doses as low as 10 cGy, produces significant alterations in dopaminergic function in the striatum and in the motor behavior that depends upon the proper functioning of these neurons (Joseph et al., 1988; Hunt et al., 1988, 1989, 1990). Similarly, exposure to 50 cGy of ^{56}Fe particles produces destruction of hippocampal neurons and an associated decrease in the speed of maze running (Philpott & Miguel, 1986).

The research reviewed here was designed to evaluate the effects of exposure to HZE particles on several behavioral endpoints compared to that of lower LET types of radiation (e.g., gamma rays or high-energy electrons). These experiments evaluated the behavioral toxicity of HZE particles using the CTA paradigm in rats and emesis in ferrets as the behavioral endpoints.

RADIATION-INDUCED TASTE AVERSION LEARNING IN RATS

A CTA is produced when a novel taste solution is paired with a toxic unconditioned stimulus, such that the organism will avoid ingestion of a normally preferred solution at a subsequent presentation. Because the CTA is extremely sensitive to low doses of a variety of toxic and self-administered stimuli, it is a standard procedure for evaluating behavioral toxicity (Riley & Tuck, 1985).

To produce a CTA rats were first placed on a 23.5-hr water deprivation schedule for 5 days (Rabin et al., 1989, 1991). On the conditioning day (Day 6), the rats were presented with a 10% sucrose solution for 30 min and their intake recorded. Immediately following the drinking period, rats were irradiated. On the test day (Day 7), the rats were again given the 10% sucrose solution and intake recorded.

Exposure to heavy particles was done using the BEVALAC at Lawrence Berkeley Laboratory. All exposures were in the plateau region of the Bragg curve. Exposure to protons was done using the linear accelerator at Brookhaven National Laboratory. Exposure to gamma rays (^{60}Co), high-energy electrons, bremsstrahlung and fission spectrum neutrons was done utilizing the sources at AFRRI. Dose rates were typically in the range of 10-100 cGy/min. Fission spectrum neutrons were produced using the TRIGA reactor with a neutron:gamma ratio of 20:1.

Fig. 1 shows that there were no differences in the acquisition of a CTA between gamma rays (^{60}Co), high-energy electrons (18.5 MeV), protons (155 MeV) or bremsstrahlung (4 MVp, average energy). Similarly (Fig. 2), except at the 50 cGy dose, there were no significant differences in the behavioral toxicity of the heavy particles ^{40}Ar (670 MeV/amu), ^{20}Ne (520 MeV/amu) and ^{4}He (165 MeV/amu). In contrast, the behavioral toxicity of ^{56}Fe particles (600 MeV/amu) was significantly greater than that of any other type of radiation tested. The behavioral toxicity of fission spectrum neutrons was intermediate between that of ^{56}Fe particles and other heavy particles.

The behavioral toxicity of the different types of radiations can be placed into three groups based upon their ED50 (the dose which produced a 50% reduction in test day sucrose intake) and which show no overlap in their 95% confidence limits (Table 1). In the first group are ^{60}Co photons, electrons, protons and bremsstrahlung, and the heavy particles ^{4}He, ^{20}Ne and ^{40}Ar. Although there is some variability in the ED50s, which ranged from 88-121 cGy, the 95% confidence intervals show considerable overlap. The LETs of these particles range from ~0.2 keV/μm for high-energy electrons to ~85 keV/μm for ^{40}Ar. Within this range, there was no apparent relationship between LET and behavioral toxicity measured using the CTA paradigm.

At the next level are fission spectrum neutrons with an ED_{50} of 40 cGy. The 95% confidence levels (37/61 cGy) do not overlap those of the other types of radiation. This indicates that the behavioral toxicity of fission spectrum neutrons is significantly greater than that of the types of radiations included within the first group. Previous research (cf., Ainsworth, 1986; Leith et al., 1983) has shown that fission neutrons have a very high RBE

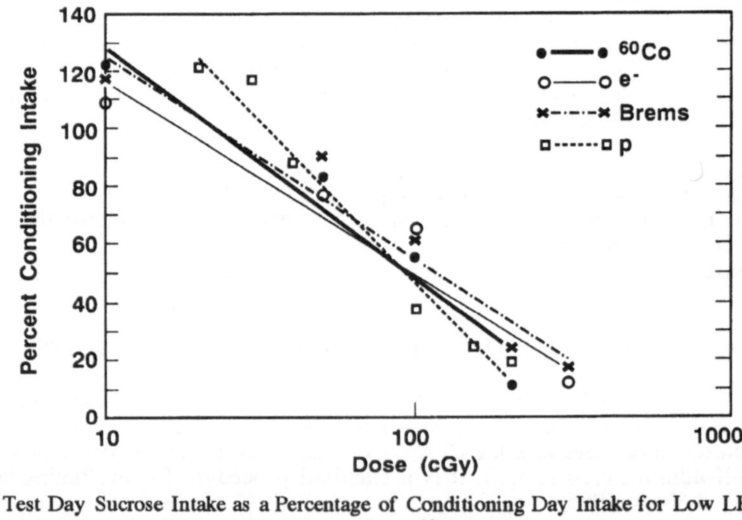

Figure 1. Test Day Sucrose Intake as a Percentage of Conditioning Day Intake for Low LET Radiation. Redrawn from Rabin et al., 1989, 1991. Abbreviations: ^{60}Co, Cobalt; e-, Electrons; P, Protons; B, Bremsstrahlung.

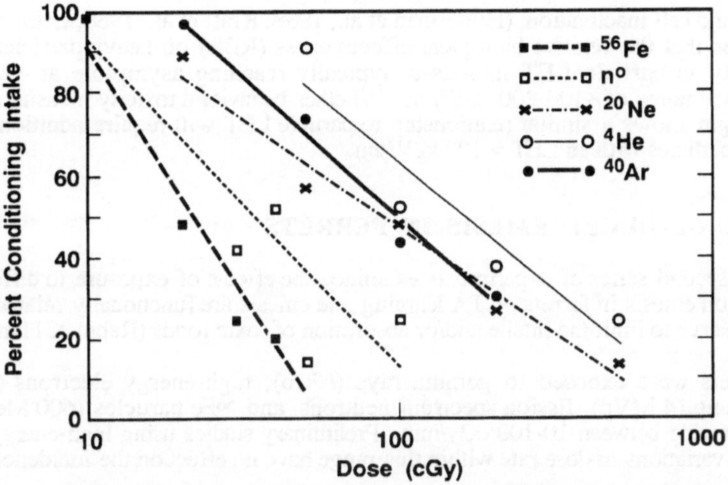

Figure 2. Test Day Sucrose Intake as a Percentage of Conditioning Day Intake Following Exposure to Heavy Particles and Fission Spectrum Neutrons. Redrawn from Rabin et al., 1989, 1991. Abbreviations: ^{56}Fe, Iron; n, Neutrons; ^{20}Ne, Neon; ^{4}He, Helium; ^{40}Ar, Argon.

for many biological endpoints, including V12 cell killing, carcinogenesis and life shortening in mice. It is particularly interesting to note that the present results show that neutrons are also highly effective for a totally unrelated behavioral endpoint.

The most behaviorally toxic radiation was ^{56}Fe. The ED_{50} for CTA acquisition following exposure to ^{56}Fe particles was significantly lower (21 cGy; 95% confidence intervals: 19/23 cGy) than that for fission spectrum neutrons or for the other types of radiation tested. The high behavioral toxicity of ^{56}Fe particles may be related to its LET (~190 keV/μm). Previous research using a variety of biological endpoints, including chromosomal

Table 1. Relative Behavioral Effectiveness (ED_{50}) of Different Radiations in Producing a CTA in Rats

Radiation	LET∞[a] (keV/μm)	ED_{50} (cGy)	95% Confidence Limits (cGy) Lower	Upper
Group 1[b]				
^{4}He (165 MeV/Amu)	2.1[c]	121	95	168
Bremsstrahlung[d]	0.3[e]	118	100	138
Proton (155 MeV)	0.3[e]	105	85	134
^{40}Ar (670 MeV/Amu)	85[c]	98	79	129
Electron (18.5 MeV)	0.2[e]	95	79	116
^{60}Co	0.3[e]	94	82	109
^{20}Ne (522 MeV/Amu)	28[c]	88	65	117
Group 2				
Fission Neutrons	65[f]	47	37	61
Group 3				
^{56}Fe (600MeV/Amu)	190[c]	21	19	23

[a]Unrestricted, absorbed dose-average values of LET
[b]Groups defined by overlapping 95% confidence intervals.
[c]Bevalac Handbook
[d]Bremsstrahlung photons produced with 18 MeV electrons and tantalum foils. Average energy approximately 4 MeV.
[e]ICRU 16
[f]Neutron Component of mixed neutron/gamma field. Quantity reported is calculated from: $L\infty_{,D} = 8/9 y_D$

aberrations and cell inactivation, (Hutterman et al., 1988; Kraft et al., 1988; Leith et al.,1983) has indicated that the relative biological effectiveness (RBE) of heavy particles increases exponentially as particle LET increases, typically reaching asymptote as particle LET approaches the range of ~100-200 keV/μm. Whether behavioral toxicity measured using the CTA paradigm shows a similar relationship to particle LET will require additional research using other particles with an LET > 100 keV/μm.

RADIATION-INDUCED EMESIS IN FERRETS

The second series of experiments examined the effects of exposure to different types of radiation on emesis in ferrets. CTA learning and emesis are functionally related behaviors in that both serve to limit the intake and/or absorption of toxic foods (Rabin & Hunt, 1986).

Ferrets were exposed to gamma rays (^{60}Co), high-energy electrons (18 MeV), bremsstrahlung (4 MVp), fission spectrum neutrons, and ^{56}Fe particles (600 MeV/amu) at dose rates ranging between 10-100 cGy/min. Preliminary studies using high-energy electrons indicate that variations in dose rate within this range have no effect on the incidence of emesis in ferrets.

The fission spectrum neutrons were produced by a TRIGA reactor which provided a neutron:gamma ratio of 6.5:1 measured at the midline of an acrylic phantom. Exposure to 20-60 cGy ^{56}Fe particles was done using the BEVALAC at LBL. The nominal extraction energy of the particles was 600 MeV/amu. However, because of the thickness of the animal and the holder, it is possible that the exit dose could have been as much as 25% greater than the entrance dose because of an increase in the LET of the ^{56}Fe beam at this depth.

The relative effectiveness of different types of radiation in producing emesis is summarized in Fig. 3 and Table 2. The radiations tested can be divided into two general groups. The least effective types of radiation are those which have the lowest LET: ^{60}Co gamma rays, high-energy electrons and bremsstrahlung. Significantly more effective in producing emesis are fission spectrum neutrons and high-energy ^{56}Fe particles.

While this pattern is similar to that observed using CTA acquisition as the behavioral endpoint, there are some critical differences when emesis is used as the endpoint. First, in contrast to the results using the CTA, exposure to ^{60}Co photons was significantly more effective in producing emesis than was exposure to high-energy electrons. Second, there were no apparent differences in the RBE of ^{56}Fe particles and fission spectrum neurons for producing emesis, but the dose-response curve for neutrons is not well defined.

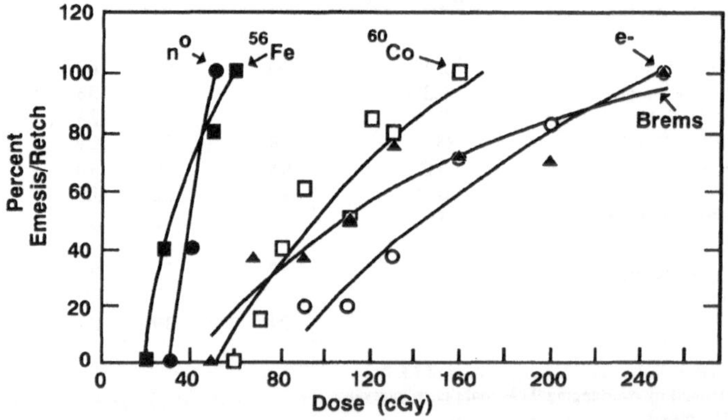

Figure 3. Dose-response Curves for Retching and/or Emesis in Ferrets Following Exposure to Different Types of Radiation. Abbreviations as in Figures 1 and 2.

Table 2. Relative Behavioral Effectiveness (ED50) of Different Radiations in Producing Emesis in Ferrets.

Radiation	LET∞[a] (keV/μm)	ED_{50} (cGy)	95% Confidence Limits (cGy) Lower	Upper
Electron (18 MeV)	0.2[b]	138	109	169
Bremsstrahlung[c]	0.3[b]	124	67	147
^{60}Co	0.3[b]	95	83	109
Fission Neutrons	65[d]	40	[e]	[e]
^{56}Fe (600 MeV/Amu)	190[f]	35	25	46

[a]Unrestricted, absorbed dose-average values of LET.
[b]ICRU 16
[c]Bremsstrahlung photons produced with 18 MeV electrons and tantalum foils. Average energy approximately 4 MeV.
[d]Neutron component of mixed neutron/gamma field. Quantity reported is calculated from: $L_{\infty,D} = 8/9y_D$.
[e]Confidence intervals could not be calculated because only a single dose produced a response other than 0 or 100%.
[f]Bevalac Handbook

Thus, as with CTA in rats, radiation LET is not completely predictive of behavioral toxicity measured using emesis in ferrets as the endpoint. Given the relative similarity in LET between ^{60}Co gamma rays (LET = ~0.3 keV/μm) and high-energy electrons (LET = ~0.2 keV/μm), the significant difference in RBE for emesis was not expected, particularly because there was no difference in the effectiveness of these radiations in producing CTA learning. Similarly, given the relatively large difference in LET between fission spectrum neutrons (LET = ~70 keV/μm) and ^{56}Fe particles (LET = ~190 keV/μm), the failure to observe significant differences in the effectiveness of the radiations in producing emesis was not anticipated.

DISCUSSION

Outside the magnetosphere astronauts will be increasingly exposed to different types and doses of radiation than those experienced in low-earth orbit. To predict and understand the potential effects of such exposures on performance requires research using a variety of behavioral endpoints in order to determine the range of behaviors that may be affected by exposure to HZE particles and the mechanisms that produce these effects.

The results of the experiments reviewed in this paper indicate that HZE particles are behaviorally toxic stimuli that can affect the behavior of the organism. However, the specific behavioral toxicity of most types of radiation tested in this series of experiments is relatively low. Therefore, with the exception of large solar particle events, it is unlikely that astronauts would be exposed to doses of radiation (Mullen et al., 1988; McCormack, 1988), including protons or bremsstrahlung, that would significantly affect their performance.

In contrast the high behavioral toxicity of low doses (~10-20 cGy) of high-energy iron particles suggests the possibility of significant effects on the performance capabilities of astronauts. Dose estimates for heavy particles for astronauts on missions outside the magnetic field of the earth are in the range of 20-49 cSv/year, depending upon the level of shielding and the solar cycle (Mullen et al. 1988; McCormack, 1988). Our data indicate that the possible toxicological consequences of exposure (e.g., nausea and emesis) may produce effects which could affect the performance of astronauts. The extent to which the present results can be extrapolated to the space radiation environment is limited by the fact that human sensitivity to heavy particles is unknown and that the doses were delivered within a short time period. While the fluence of these particles in space is low, the neurons upon which behavior ultimately depends are postmitotic, which means that repair of damaged neurons is not a likely occurrence. As such, it is possible that the cumulative effects of repeated hits from these particles over long duration missions may be sufficient to affect the ability of astronauts to successfully complete their assigned missions.

A better evaluation of the effects of exposure to HZE particles will require additional research using a variety of behavioral and neurochemical endpoints and research designed to determine the mechanisms by which exposure to HZE particles can affect behavior.

ACKNOWLEDGEMENTS

The authors wish to acknowledge the assistance of Drs. E. John Ainsworth, Patricia Durbin, Bernhard Ludewigt and the staff at the Lawrence Berkeley Laboratory, and of Drs. Tom Ward, Jack Rothmann Lewis Snead and the staff at the Brookhaven National Laboratory, without whose help these studies could not have been undertaken. We also wish to acknowledge the support of the Computer Science Center Facilities of the University of Maryland Baltimore County. This research was supported by the Armed Forces Radiobiology Research Institute, Defense Nuclear Agency, under work unit 00157. Views presented in this paper are those of the authors; no endorsement by the Defense Nuclear Agency has been given or should be inferred. This research was conducted according to the principles described in the Guide for the Care and Use of Laboratory Animals prepared by the Institute of Laboratory Animal Research, National Research Council.

REFERENCES

Ainsworth, E. J., 1986. Early and late mammalian responses to heavy charged particles. Adv. Space Res. 6, 153-165.

Bevatron/Bevalac User's Handbook: Biology and Medicine, 1985. Lawrence Berkeley Laboratory, Berkeley, CA.

Bogo, V., Zeman, G. H., and Dooley, M., 1989. Radiation quality and rat motor performance. Radiat. Res. 118, 341-352.

Hunt, W. A., Joseph, J. A., and Rabin, B. M., 1989. Behavioral and neurochemical abnormalities after exposure to low doses of high-energy iron particles. Adv. Space Res. 9, 333-336.

Hunt, W. A., Dalton, T. K., Joseph, J. A., and Rabin, B. M., 1990. Reduction of 3-methoxytyramine concentrations in the caudate nucleus after exposure to high-energy iron particles: Evidence for deficits in dopaminergic neurons. Radiat. Res. 121, 169-174.

Hunt, W. A., Rabin, B. M., Joseph, J. A., Dalton, T. K., Murray, W. E., Jr., and Stevens, S. A., 1988. "Effects of iron particles on behavior and brain function". P. D. McCormack, C. E. Swenberg, and H. Bücher, eds., pp. 537-551, Plenum Press, New York.

Hüttermann, J., Schaefer, A., and Kraft, G., 1989. The influence of radiation quality on the formation of DNA breaks. Adv. Space Res. 9, 35-44.

International Commission on Radiological Units and Measurements (ICRU), 1970. Report 16, Washington, DC.

Joseph, J. A., Hunt, W. A., Rabin, B. M., and Dalton, T. K., 1988. Correlative motor behavioral and striatal dopaminergic alterations induced by 56-Fe. "Terrestrial Space Radiation and its Biological Effects". P. D. McCormack, C. E. Swenberg, and H. Bücher, eds., pp. 553-571, Plenum Press, New York.

Kraft, G., Kraft-Weyrather, W., Ritter, S., Scholz, M., and Stanton, J., 1989. Cellular and subcellular effect of heavy ions: A comparison of the induction of strand breaks and chromosomal aberrations with the incidence of inactivation and mutation. Adv. Space Res. 9, 59-72.

Leith, J. T., Ainsworth, E. J., and Alpen, A. L., 1983. Heavy ion radiobiology: Normal tissue studies. "Advances in Radiobiology" J. T. Lett, ed., pp. 191-236, Academic Press, New York, 1983.

McCormack, P. D., 1988. Radiation hazards in low earth orbit, polar orbit, geosynchronous orbit, and deep space. "Terrestrial Space Radiation and its Biological Effects" P. D. McCormack, C. E. Swenberg, and H. Bücher, eds., pp. 71-98, Plenum Press, New York.

Mickley, G. A., Bogo, V. A., Landauer, M. R., and Mele, P. C., 1988. Current trends in behavioral radiobiology, "Terrestrial Space Radiation and its Biological Effects", P. D. McCormack, C. E. Swenberg, and H. Bücher, eds., pp. 517-536, Plenum Press, New York.

Mullen, E. G., Gussenhoven, M. S., and Hardy, D. A., 1988. The space radiation environment at 840 km. "Terrestrial Space Radiation and its Biological Effects", P. D. McCormack, C. E. Swenberg, and H. Bücher, eds., pp. 41-60, Plenum Press, New York.

Philpott, D. E., and Miquel, J., 1986. Longterm effects of low doses of 56Fe ions on the brain and retina of the mouse: Ultrastructural and behavioral studies. Adv. Space Res. 6, 233-242.

Rabin, B. M., Hunt, W. A., Joseph, J. A., Dalton, T. K., and Kandasamy, S. B., 1991. Relationship between LET and behavioral toxicity in rats following exposure to protons and heavy particles. Radiation Research, 128, 216-221.

Rabin, B. M., and Hunt, W. A., 1986. Mechanisms of radiation-induced conditioned taste aversion learning. Neurosci. Biobehav. Rev. 10, 55-64.

Rabin, B. M., Hunt, W. A., and Joseph, J. A., 1989. An assessment of the behavioral toxicity of high-energy iron particles compared to other qualities of radiation. Radiat. Res. 119, 113-122.

Riley, A. L., and Tuck, D. L., 1985. Conditioned taste aversions: A behavioral index of toxicity. Ann. N. Y. Acad. Sci. 443, 272-292.

A MODEL OF CELL DAMAGE IN SPACE FLIGHT

Robert Katz

University of Nebraska
Lincoln, Nebraska 68588-0150

F. A. Cucinotta, J. W. Wilson, and J. L. Shinn

NASA Langley Research Center
Hampton, Virginia 23665-5225

Duc M. Ngo

Department of Physics
Old Dominion University
Norfolk, Virginia 23508

ABSTRACT

Cell damage by high LET radiations has been described by a phenomenological model (track theory) for 20 years and more. Molecules of biological significance (dry enzymes and viruses) act as 1 hit detectors. Recent additions to the class of 1-hit detectors are E. Coli B, and the creation of both single and double strand breaks in SV-40 virus in EO buffer, where indirect effects predominate. The response of cells (survival, transformation, chromosome aberration) to these radiations is typically described by a 4-parameter model whose numerical values are determined by fitting the equations of the theory to experimental data at high dose (typically above 1 Gy) with bombardments with γ rays and HZE particle beams, of the widest possible dynamic range. Once these parameters are determined the model predicts cellular response in any radiation environment whose particle-energy spectrum is known. Perhaps the central importance of the present model is the ability to estimate the response of a complex environment with many components from a limited set of laboratory data. For example, we have calculated cell survival after neutron irradiation, with mixtures of neutrons and γ rays; cell survival and transformation after irradiation with HZE ions of different energies. The model does not yet include cellular repair. Although some hopeful approaches to repair dependence are now being developed. It does not include cancer induction, for the available data neither give the number of cells at risk or the number of cancers induced, and are thus not suited to our formulation.

Most recently NASA-Langley models of HZE beams, including projectile and target fragmentation, have been joined with the biological model. This combination has been tested against ground based radiobiological data for cell survival after irradiation with protons and HZE beams with good success. Where our earlier model failed downstream of the Bragg peak (for both protons and heavy ions) for want of a proper description of fragmentation the NASA-Langley model succeeds.

Based on this experimental validation of our procedures, we have initiated calculations of cellular damage in space flight from solar protons and galactic cosmic rays. Here we incorporate NASA models of cosmic rays, beam penetration, projectile and target fragmentation with track theory. The essential radiobiological theme is that knowledge of parameters extracted at high doses makes it possible for us to calculate the response of cells at the lowest possible doses of HZE particles when only intra track (ion-kill) effects are involved for which repair is known to be minimal. Our procedures here too have ground based experimental validation in recent work of Bettega et al. where measurements made of RBE with protons and alphas of the survival of C3H10T½ cells, at doses down to 0.01 Gy are consistent with our predictions based on survival measurements made at high doses with γ rays and HZE ions.

INTRODUCTION

Detectors of radiation differ according to whether single particle response is normally observed, as with nuclear emulsions, solid state nuclear track detectors, and scintillation counters, or whether the response is to beams of particles or photons in a gross macroscopic irradiation, as in radiobiology or in the alteration of bulk material properties by radiation. In the former case it is more natural to think in terms of track structure, while in the latter case one frequently refers to macroscopic dose (Katz, 1978). Response is then correlated to the physical description of these stimula. It is common to relate response to energy deposition (dose). Problems arise because response depends not only on total energy deposition but on the microscopic structure of that deposition and also on its time development. One analysis of these details is called microdosimetry, a subject that has stimulated many investigations. An alternate procedure favored here relates the observed effect to track structure for individual particles, which then may be related to macroscopic dose for gross irradiations. These perspectives are principally reported in the several Symposia on Microdosimetry sponsored by the Commission of European Communities.

The galactic cosmic ray (GCR) environment is the most complicated mixture of radiation components known. It is doubtful that the GCR will ever be adequately simulated in the laboratory for biological experiments. The primary role of track structure models will be to extrapolate laboratory response data to the GCR environment for the estimate of risk to biological tissues in space exposure. This is, we believe, to be a more practical approach to the issue of additivity of response of disparate components than the usual quality factor approach based on relative biological effectiveness (RBE) which has been used with limited success in terrestrial radiation protection. We will now discuss the current approaches to the question of additivity being pursued by various groups.

ENERGY DEPOSITION IN SMALL VOLUMES: MICRODOSIMETRY

One way to analyze the stimulus to biological systems is to examine the details of energy deposition in small volumes, sized to represent what are thought to be critical targets within the cell. Experimentally small gaseous proportional counters are used whose diameter, scaled to the density of tissue, is from micrometers to nanometers in unit density material. The critical targets are then considered to be either the nucleus of a mammalian cell or a chromosome, or a small region of DNA. The fluctuations of the energy deposited within the small target region is assumed related to biological response. A Monte Carlo simulation of a radiation field can yield a similar decomposition.

Even when one has a complete microdosimetric description of the radiation environment, the problem remains as to how that may be interpreted to predict the response of a detector. As yet we have no means of calibrating response in terms of the statistical distribution of energy depositions in small volumes. Nor do we know what volume is appropriate. It is on this level that microdosimetry has not been able to make extensive quantitative predictions, nor is it able to yield calculations of cross section. But the study has yielded many interesting insights into the structure of a radiation field (Goodhead, 1988). The small counter has found application in monitoring neutron beams used in radiotherapy and in other radiation fields including space and high altitude aircraft. Most instruments are used in practice to derive averages over quality factors discussed in section 5.

CROSS SECTION

A second approach has been to attempt to mimic the kind of logical structure used in experiments in physics; that is, to describe the relevant interactions through the concept of an interaction cross section. We imagine that a projectile passing down a channel 1 cm^2 in area interacts with a target located somewhere within that channel, and measure the fraction of successes after a large number of identical repeated trials. This probability is represented as though it is a geometrical target, as the cross section σ in cm^2 to the cross sectional area of the channel. We then speak of the action cross section even if the observed end point is achieved as a result of many internal changes stimulated by the initial interaction. Here we make no attempt to examine the internal processes mechanistically. The target is a black box. We know only the incident radiation and the observed end point. In radiobiology the concept of cross section is sometimes used in ways which depart from its original physical meaning. This can lead to misinterpretations of experimental data (Katz, 1990). Curtis et al. (1990) has recommended an additivity formalism based on a limited set of data for Harderian gland tumorgenesis using a cross section like formalism as an alternative to the use of RBE.

G VALUE

When a projectile impinges on a thin slice of matter containing N targets/cm^3, the number of observable events per cm of path length is $n = \sigma N$. If the energy deposited per cm of path length is L (LET = Linear Energy Transfer, or stopping power), the number of observed events per unit of energy deposited, the G value is $G = n/L = \sigma N/L$. The cross section is a function of the medium, the end point, and the character of the projectile: if a photon its energy, if a naked charged particle its charge and speed, if a nucleus partially clothed with electrons its effective charge and speed. This formulation of the G value has been used in the analysis of heavy ions radiolysis (Katz and Huang, 1989). In dealing with liquids where the meaning of N may be obscure we have calculated the G value for heavy ion bombardment from calculated values of the RBE and known G values for γ irradiation, as in the Fricke dosimeter (Katz, Sinclair and Waligorski, 1986). In other cases we have tried to relate N to a fitted target size. Nuclear collisions are here neglected except as a source of charged fragments.

If the G value is normalized to molecular weight, expressed as events/rad/Dalton rather than events/100 eV, we find it proportional to the RBE for dry 1-hit detectors for which the "target molecular weight" is equal to the true molecular weight (Katz, 1990).

RELATIVE EFFECTIVENESS; QUALITY FACTOR

When intense neutron environments became available, the existing body of biological response data was mainly for X-ray and γ-ray sources. The first efforts at protection attempted to scale the known X-ray and γ-ray risks according to equivalent neutron dose giving rise to the concepts of radiation quality and RBE. For the case of space radiations, we should like to relate the response of our detectors to energetic heavy ions, that is to high LET radiations, to their response to photons and electrons, that is to low LET radiations. In radiobiology the ratio of the dose of gamma rays to that of another radiation which produced the same observed end point is called the relative biological effectiveness, the RBE. In radiotherapy this quantity is frequently taken to be a property of the two radiation fields, but it depends on the dose level, the dose rate and the end point as well. An extension of this idea used in radiation protection is called the quality factor estimated as the upper limit of RBE values for a selected set of biological endpoints judged relevant to human risk and taken solely as a function of LET. An important unresolved question is whether an upper limit or maximum RBE is achieved at the low exposures of interest for radiation protection. The quality factor is used to convert a measured dose in Gray into an effective dose reported (not measured) in Sieverts. We note that the Sievert is not a directly measurable quantity and thus violates the underlying philosophy through which physical units are defined. The redeeming quality of the unit is that risk estimates based on Sieverts should be conservative providing the quality factor is adequately defined. The conservative nature of the method may also be an unacceptable burden in many operations, especially in space.

RADIATION QUALITY

An irradiation with photons leads to secondary electrons randomly dispersed through a medium. The initial energy spectrum of these electrons and their path lengths depends on the initial energy spectrum of the photons. An irradiation with a beam of heavy ions yields a random distribution of heavy ion paths, with the secondary electrons (delta rays) clustered around each ion's path (correlation effects), and having a different energy distribution. Hence the δ rays are not truly randomly distributed. They are clustered about the paths of heavy ions. This leads to a basic difference in the manner in which their effects are approached statistically (spatial correlations). In the present model we speak of "gamma-kill" to describe the effects of a random distribution of spatially uncorrelated electrons, and of "ion-kill" to describe the effects of the spatial correlations within single particle tracks. At high fluences of low LET ions, where only a fraction of the intersected targets is inactivated, we approximate the effect of the sparsely distributed and overlapping δ rays from several ions as due to randomly distributed electrons, and so we speak of gamma-kill as responsible for part of the effect from beams of some heavy ions.

With photons minutes may elapse for the traversal of secondary electrons from different photons through a target (uncorrelated temporal events). With heavy ions a single ion and its delta rays pass through a target in an extremely short time for the projectile moves at nearly the speed of light through a target whose diameter is of order 1 micron. These differences lead to a variation in response to radiations of different admixtures of photons and heavy ions at the same dose. When the temporal correlation time is on the order of the cell repair time then response also varies with dose rate or fractionation schedule. The dependence on dose rate or fraction schedule is due in part to radiation quality. The quality factor taken as the low dose rate limit RBE's in an attempt to normalize the biological effects of radiations of different quality. This assumes that it is logically correct to represent the response of a detector as a product of two separate factors, one of which is the dose while the other is the quality.

RADIAL DOSE DISTRIBUTION

For purposes of track structure calculations we need the radial distribution of dose, from delta rays and the primary interactions, about the path of an energetic charged particle (Waligorski, Hamm and Katz, 1986). We presently use an analytic representation of the results of a Monte Carlo calculation made for liquid water for this purpose. More recently we have extended this model to include some solids used as radiation detectors (Katz et al., 1990). Additional information about both theoretical and experimental determinations of the radial dose distribution may be found in a recent work by Katz and Varma (1990). We use this information in connection with the response of the detector to γ rays to find the radial distribution of effect around a particle's path. Since we interpret the response to be the probability for activating a target, we can make a map of the radial distribution of activated targets. If we are interested in the opacity of a track in nuclear emulsion we can calculate the attenuation of a beam of light in a microscope photometer, as in the study of cosmic ray tracks (Katz and Kobetich, 1969). Alternatively we can integrate the probability radially to yield the cross section (Waligorski, Loh and Katz, 1987) for the interaction of a single ion with the target (Katz, 1978).

HITTEDNESS

In radiobiology we are not yet able to measure the effect produced by the interaction of a single ion. Yet this is of central importance in estimating the effects of galactic cosmic rays in space flight. For this one needs to know the cross section. In track theory we wish to know whether a single particle, be it electron, proton, alpha particle, or whatever, is capable of inducing the tested end point with observable probability. We characterize these interactions through the concept of hittedness, borrowed from biological target theory (Dertinger and Jung, 1970). For a specific irradiation the appropriate hittedness is either the number of interactions between charged particles and target needed to induce the end point, or the number of incident particles which must bombard the target, whichever is smaller. If either a single particle or a single interaction leads to the event we will observe exponential response, as demanded by the cumulative Poisson distribution. If two electrons are required we expect to observe a response described by the 2-or-more hit cumulative Poisson distribution. But the inactivation may take place through the transit of a single α particle. In that case we expect to observe that the response to α particles is 1-or-more hit. We characterize the hittedness of a detector by its response to electrons or to gamma-rays. Experimentally if the response to the dose of gamma rays is exponential we speak of a 1-or-more hit detector.

THE 1-HIT DETECTOR

Most commonly radiation detectors can be described as 1-or-more hit detectors. We imagine the detector to be a collection of targets, sometimes explicit as in photographic emulsion, and sometimes implicit, as in a Fricke dosimeter, or in alanine. Each of these targets is capable of responding to the transit of a single electron of appropriate energy. The response is exponential; that is, it is linear at low dose and sublinear at high dose as the available targets tend to have been (in)activated. We speak of saturation or overkill at high dose. For 1-hit detectors the response to heavy ions is also exponential with dose or fluence.

To calculate the (in)activation cross section for a 1-hit detector we first find $P(D)$, the probability for target inactivation as a function of the dose D of γ rays. Next we fold this into the average radial dose distribution about an ions path, to find the probability for target inactivation $P(t)$ at radial distance t. We integrate $P(t)$ radially to find the cross section σ. When targets out to about 3 target diameters are all (in)activated we simply use the point distribution of dose in our calculations. If we must take into account effects closer than 3 target radii from the ion's path it is necessary to average the dose in the extended targets to accommodate the dose gradient, for the radial dose falls off essentially inversely as the square of the radial distance, out to a limit determined by the maximum δ ray penetration. This limiting distance, determined essentially by the speed of the ion, places an upper limit on the cross-section, observed experimentally in experiments with very heavy ions as "thin-down", so called because of the appearance of the tracks of heavy ions in electron sensitive emulsions, where the stopping end of a track looks like a sharpened pencil.

Typically for these detectors the cross section increases with an increase in $(Z^*/\beta)^2$ to a maximum (typically unrelated to target size) and then declines in thin-down.

The relative effectiveness defined biologically as RBE is equal to $\sigma E_0/L$, where E_0 is the 1/e dose or the dose for 37% survival. For 1-hit detectors the RBE never exceeds 1. The magnitude of the cross section is approximately determined by the radial distance at which the dose equals E_0. One may speak of the cross section as approximating the size of the damaged region, but it is inappropriate to speak of track size without specifying the end point. It is easy to estimate the cross section of a heavy ion with the grains of a nuclear emulsion from a microphotograph by estimating the radial distance at which about 63% of the grains are developed. There we note that for insensitive emulsions where the track resembles a string of beads, that the cross section is less than the grain size, while for a sensitive emulsion where the track resembles a hairy rope, that the cross section may be orders of magnitude greater than the grain size.

Our first venture into the 1-hit detector was made for dry enzymes and viruses (Butts and Katz, 1968). This was shortly followed by a model for the response of nuclear emulsions (Katz and Kobetich, 1969), scintillation counters (Katz and Kobetich, 1968), TLD's, and subsequently of alanine (Waligorski et al., 1989), and most recently for E. Coli B (Katz and Zachariah, 1991). There are indications that CR-39, used as an etchable track detector is also a 1-hit detector (Katz, 1984).

The global applicability of the model of the 1-hit detector to a wide variety of detectors whose mechanisms are vastly different from each other is at first thought rather astonishing. It arises simply from the fact that in each case the end point is stimulated by the passage of a single electron through the target volume.

SUPRALINEARITY; THE LINEAR QUADRATIC "MODEL"

If a system has both 1 hit and 2 hit targets having different radiosensitivities and populations, we must expect that response will be linear at low dose, quadratic at intermediate doses, and finally saturating at high dose. We call such a response supralinear and have proposed such a model to explain supralinearity in TLD-100. Note that the concept of a 2-hit target requires only that two incident electrons are required to stimulate the end point. The response may arise after processing as well as in the initial interactions, for the present model treats each detector as a black box. But if there are not two varieties of response we cannot understand supralinearity in this model. Nor can we understand how a detector whose response to γ rays is exponential can exhibit an RBE greater than 1 with heavy ions unless there are temporal effects hinging on the time difference between γ-ray and δ ray exposures. Such a time difference between hits appears explicitly in a kinetics model which may provide an approach to temporal effects.

In the same way we do not understand the basis of the linear quadratic formula widely used to fit radiobiological data, if there are not two types of targets within a cell. We note that the formula is simply the first two terms of a series expansion and is usually applied to data of very limited dynamic range. Further we note that there is no theory which can predict the values of either the linear or quadratic terms reliably, for radiations of different quality. Nevertheless these terms are liberally interpreted on the basis of such phrases as "could be" or "might be" though with equal validity one might insert "not".

Those who prefer to interpret data on the basis of hypothetical mechanisms (whose details are rarely accessible to experiment) object to our parametric formulations. Yet the model and its experimental parameters should not be dismissed lightly, for they may suggest mechanistic interpretations which supersede those presently popular (Goodhead, 1989). We need to be reminded that Newton's laws were stimulated by Kepler's phenomenology, and that quantum theory was stimulated by Planck's exercise in curve fitting.

THE CELL SURVIVAL MODEL

Biological cells require special consideration. For other detectors we assume, as in the case of nuclear emulsions, that there is a characteristic target size without internal structure. These are then characterized by the parameters E_o, the dose of γ rays at which there is an average of 1 hit per target, a_o, the target radius, and C, the hittedness. We sometimes introduce a dimensionless parameter κ proportional to $E_o a_o^2$. Biological cells have internal targets. We imagine the cells to resemble a bean bag in which the cell nucleus is the bag and the targets are the beans. We take it that the beans are 1-hit in character but that m of the beans must be (in)activated to generate the observed response. We also imagine that there are beans well distributed through the bag so that an energetic ion passing through the bag has the possibility for inactivating m beans. Such a model makes it possible to understand why flatted cells respond differently to α particles than rounded ones.

In this model the observed cross section is related to the size of the bean bag, the cell nucleus, while the variation of response with LET is related to the properties of the bean. To set up a model of cellular response we calculate the cross section for a hypothetical cluster of m overlapping beans (Katz, Sharma, and Homagyoonfar, 1972) and then assume that the cross section for the bean bag is proportional to that of the cluster. Since our model is based on the radial distribution of dose from δ rays it automatically predicts thindown. For mammalian cells the fitted value of the κ parameter suggests that the bean radius is about 1 micron, hinting that the target for cell killing may be a chromosome.

THE MATHEMATICAL MODEL

Detailed descriptions of the cellular track model have been given elsewhere. Here we present only the main concepts of the model and list the equations used in our calculations. Following our earlier studies of the appearance of particle tracks in nuclear emulsion (Katz and Kobetich, 1969), the model distinguishes between the "grain-count" regime where inactivations occur randomly along the particle's path, and the "track width" regime where the inactivations are distributed like a "hairy rope". The transition from the grain-count to track width regime trakes place in the neighborhood of $Z^{*2}/\kappa\beta^2$ of about 4; at lower values we are in the grain-count regime, at higher values in the track width regime. The quality κ is a parameter of the model which combines both the target size and the characteristic dose of gamma rays at which there is an average of one hit per target. As in nuclear emulsions we speak of a thin down regime where the cross section is limited by the kinematic constraint on δ ray energies, but has nothing to do with the Bragg peak in stopping power nor with the changing effective charge of a slowing down ion.

To accommodate for the capacity of cells to accumulate sublethal damage, two modes of inactivation are identified, namely "ion-kill" (or "intratrack") and "gamma-kill" (or "intertrack"). In these two inactivation modes it is the statistical character of the inactivation which is changing rather than the fundamental physical interaction. Effects are referred to dose rather than to the number of electrons passing through the nucleus. We do not find justification for considering the stopping end of an electron track as a source of ion kill nor for the radial separation of a heavy ion track into core producing ion kill and penumbra producing gamma kill.

The model leads to the use of Z^{*2}/β^2 as a plotting parameter superior to LET, now in wide use. At the stopping end of a track, at highest LET even this parameter fails for in the thin down regime the cross section depends on β, the relative speed of the ion. Here the cross section is sometimes plotted against the energy per unit mass, a related parameter. The model provides a basis for the meaning of low LET, based on the comparison of Z^{*2}/β^2 with κ. Similarly low dose means low compared to E_O. The model explains why plots of extrapolated cross section (from the tail of a survival curve) tend to be single valued functions of LET at low LET (because the response is dominated by gamma kill) and why they are multiple valued (with Z) at high LET (because the response is dominated by ion kill and thin down). It explains why plots of RBE vs LET for biological cells pass through a maximum (when about half the intersected cells are killed in ion kill). It predicts that the RBE for lighter ions will be greater than the RBE for heavy ions at the same LET and the same survival level. This is because of the structure of particle tracks. At the same LET the heavier ions move faster. Its δ rays are fewer but more energetic. Gamma-kill is more likely, reducing the RBE.

GAMMA KILL

Cells not inactivated in the ion-kill mode can be sublethally damaged by the δ-rays from the passing particle and then inactivated, in the gamma-kill mode, by cumulative addition of sublethal damage due to δ rays from other passing ions. Survival in the gamma-kill mode is taken to follow the m-target statistics of inactivation by secondary electrons from X-ray or gamma-ray photons.

MATHEMATICAL FORMALISM

In the *grain-count* regime the surviving fraction of a cellular population whose radiosensitivity parameters are m, E_0, σ_0 and κ, after track-segment irradiation with an ion dose D of a fluence of F particles of atomic number Z, effective charge value Z^*, relative speed β and stopping power L (LET_∞), is found from the expressions

$$N/N_0 = \Pi_i \times \Pi_\gamma \tag{1}$$

where the ion-kill mode survival probability is

$$\Pi_i = \exp(-\sigma F) \tag{2}$$

where the gamma-kill mode survival probability is

$$\Pi_\gamma = 1 - \left[1 - \exp\left(\frac{-D_\gamma}{E_0}\right)\right]^m \tag{3}$$

and the gamma-kill dose fraction is

$$D_\gamma = (1 - P)D \tag{4}$$

where

$$\frac{\sigma}{\sigma_0} = P = \left[1 - \exp\left(\frac{-Z^{*2}}{\kappa\beta^2}\right)\right]^m \tag{5}$$

In the *track-width* regime, where $P > 0.98$, we take

$$\Pi_\gamma = 1 \tag{6}$$

and find σ from the "track width" which increases linearly with Z^*/β while the inactivation cross section increases with Z^{*2}/β^2 up to the limit set by the maximum radial range of δ-rays. This is the "thin-down" region.

To find the cross section in the track width regime, including thin-down region, a separate calculation must be made. First we must find the "target cross section" S, for targets of radius a_0 found from κ and E_0 according to

$$\kappa = E_0 a_0^2 \times 5 \times 10^6 \mathrm{erg\ cm} \tag{7}$$

and having multi-target response to gamma rays characterized by E_0 and m found for the cell. This must be multiplied by the ratio of the plateau value of the cellular cross section σ_0 to the plateau value of the target cross section S_0 to yield the cellular action cross section in the track width regime (Katz et al. 1971). In this region we make the approximation that there is no gamma kill dose, though in the outer reaches of the track width some small fraction of the energy lost by the ion is deposited in the gamma-kill mode.

To calculate RBE at a given "kill" (transformation) or survival level we use the definition

$$RBE = \frac{D_x}{D} \tag{8}$$

where

$$D_x = -E_0 \left\{\ln\left[1 - (1 - N/N_0)^{1/m}\right]\right\} \tag{9}$$

is the X-ray dose after which this level obtains, and D is the corresponding ion dose.

All our calculations pertain to water so the ion dose is always

$$D = F\,L \tag{10}$$

Where "cross sections" and RBE's are calculated from the final slope of the survival curves, we refer to the cross section and the RBE as "extrapolated" and in the grain count regime we write

$$\sigma_{ext} = \sigma_0 P + (1 - P)L/E_0 \tag{11}$$

and

$$RBE_{ext} = (\sigma_0 E_0/L)P + (1 - P) \tag{12}$$

To calculate the effective charge value of an ion of atomic number Z moving with a relative velocity β we use the expression (Barkas, 1963)

$$Z^* = Z[1 - \exp(-125\beta Z^{-2/3})] \tag{13}$$

We calculate the stopping power and range in water of an ion of atomic number Z with the expressions

$$L(Z, E) = L(p, E)[Z^*/Z_p^*]^2 \tag{14}$$

where Z^* and Z_p^* are the effective charges of the ion and proton, respectively and $L(p, E)$ is the stopping power, in water, of a proton at the same energy/nucleon, E.

At low fluence, where ions are sufficiently far apart that inter-track effects are unlikely, we can neglect the contribution from gamma-kill. Under this circumstance the RBE is

$$RBE = E_0 \left(\frac{\sigma}{L}\right)^{1/m} D^{(1/m-1)} \tag{15}$$

This is applicable to low doses of neutrons as well as the effects of galactic cosmic rays (Katz and Cucinotta, 1991).

For mixed radiation fields, our model requires knowledge of the particle-energy spectrum of the radiation field. We then find the totality of effects due to ion kill, add together the gamma kill doses including the dose from gamma rays, find the ion kill survival and the gamma kill survival probabilities, and take their product to be the surviving fraction of irradiated cells. We have done this for neutrons admixed with gamma rays, for range modulated heavy ion beams, and most recently for cosmic rays.

CELL KILLING, CHROMOSOME ABERRATIONS, TRANSFORMATIONS

Our treatment of transformations (Waligorski, Sinclair and Katz, 1987) is based on data obtained with the BEVALAC accelerator by Yang et al. (1985). We use the same form of equations as for cell killing, and take cell killing and transformation to be independent processes that take place along the same particle track. Our parameters for Chinese hamster cells are based on the data of Skarsgard et al. (1967) while the parameters for tradescantia are based on the data of Underbrink et al. (1978). Here we have attempted to extract parameters from data obtained with x rays and neutron irradiations of two different energy spectra, ignoring possible gamma ray contamination. The parameters for T-1 cells of human origin are taken from Todd (1967). The separate sets of parameters for survival, chromosome aberration and for transformation, are shown in Table 1 (Katz and Huang, 1991). Where two sets of parameters are shown, the data do not permit a clear distinction between them.

TARGET FRAGMENTATION EFFECTS

High energy protons passing through tissue will occasionally suffer nuclear reactions that produce low energy, high LET ions from the tissue itself. The target fragments, in turn, will be a source of delta-ray production which should contribute to biological damage locally in the tissue matrix. The differential fluence (Wilson, 1977) describes the local source of target fragments

$$F_j = \frac{1}{L(Z_j, E)} \int_E^\infty \frac{d\Sigma_j(E')}{dE'} F_p(E_p) dE' \tag{16}$$

where j is the fragment label, L the stopping power or LET, Σ_j the macroscopic nuclear production cross section, and F_p the fluence of protons with energy E_p. An effective action cross section for the proton dressed by the target fragments (nuclear stars) is now written as

$$\sigma^* = \sigma_p(E_p) + \frac{1}{F_p(E_p)} \sum_j \int_0^\infty F_j(E_j)\sigma_j(E_j)dE_j \tag{17}$$

where σ_p and σ_j are given by the Katz formalism [Eq. (5)]. The gamma-kill dose for the proton plus target fragments is written

$$D_\gamma = D_{\gamma_p}(E_p) + \sum_j \int_0^\infty F_j(E_j)L(Z_j, E_j)\left[1 - P_j(E_j)\right] dE_j \tag{18}$$

The production energy spectra for the target fragments is expressed as (Wilson, et al., 1989)

$$\frac{d\Sigma_j}{dE} = \frac{\Sigma_j(E_p)\sqrt{E}}{\left(2\pi E_{0j}^3\right)^{1/2}} e^{-E/2E_{0j}} \tag{19}$$

where $3E_{0j}$ is the average energy of the fragment.

The fragmentation parameters used are discussed in Wilson et al. (1989). The light ion production cross sections are from the Bertini Monte Carlo results (Bertini, 1970) and the Silberberg-Tsao empirical model is used for the heavier fragments (Silberberg, et al., 1976). The average energy of the tissue fragments is related to the momentum width measured experimentally (Greiner et al., 1975), which Wilson et al. (1989) fits empirically. We note that the largest uncertainties exist for light ion production (A = 7 and 9) and for energies below 100 MeV. Elastic recoils and meson production, above several hundred MeV, will also

TABLE 1. PARAMETERS FOR SURVIVAL, ABERRATIONS, TRANSFORMATIONS

	m	κ	E_o Gy	σ_0 cm^2
CH2B$_2$ Chinese Hamster Cells				
Skarsgard et al. 1967				
survival	3	1100	1.82	4.3×10^{-7}
abnormal metaphases	3	900	1.82	3×10^{-7}
chromatid exchanges	2	1400	25	6.5×10^{-9}
C3H10T1/2 Mouse Cells				
Yang et al. 1985				
survival	3	750	1.7	5×10^{-7}
transformations	2	750	180	1.2×10^{-10}
transformations	or 3	475	50	7×10^{-11}
Tradescantia				
Underbrink et al. 1978				
survival	2	1000	2.1	3.5×10^{-7}
survival	or 1.5	1900	2.6	4.0×10^{-7}
Human T-1 Cells				
Todd, 1967				
survival (aerobic)	2.5	1000	1.7	6.7×10^{-7}
survival (hypoxic)	2.5	1300 (1450)	4.6 (5.2)	6.7×10^{-7}

contribute and should be added. The stopping power in tissue is from the work of Wilson (1983) based on the Ziegler analysis (Ziegler, 1980).

The solid line in Fig. 1 displays the fragment LET component from 1-GeV protons in tissue derived from eq. (16). The dotted and dashed lines show the contributions from proton fragment and alpha secondaries, respectively.

The proton action cross section for cell survival of Chinese hamster cells is shown in Fig. 2 versus the proton energy. The cellular response parameters are given in Table 1. Discussed below are comparisons to the data of Hall et al. (1978) for survival of Chinese hamster cells where the characteristic X-ray dose, E_o, is taken as 2.9 Gy, as found from their X-ray data and with the remaining parameters the same as given by Table 1. The dotted line in Fig. 2 shows the contributions from primary ionizations; the dashed line shows contributions from secondary ions. We note that the oxygen and nitrogen fragments contribute partially to the cross section in the track-width regime. The decreasing proton LET with increasing energy is seen to lead to complete domination by target fragments above about 50 MeV. The shape of the action cross section in Fig. 2 directly reflects the nuclear absorption cross section in tissue. We expect a further increase above several hundred MeV when meson production is included in the cross section. In Table 2, the individual contributions to the action cross section are shown for several proton energies. Secondary protons and alphas are dominant with a broad spectra of tissue fragments making non-negligible contributions. The primary proton makes up an insignificant fraction of the action cross section above 100 MeV, and the relatively slow change with energy of the nuclear production cross sections leads to a plateau in the action cross section at high proton energies.

Figure 3 shows the action cross section versus proton LET with the calculations of Fig. 2, extended down to 0.1 MeV, corresponding to high LET protons. The behavior of the cross section below 0.5 keV/μ shows the dominance of the tissue secondaries (nuclear stars). At about 0.2 keV/μ, the proton LET minimizes and then increases, which is the origin of the 'hook' in Fig. 3 at the lowest LET values. It would be interesting to test our results for the proton cross section by experiment. Results herein assume an equilibrium in the local secondary fluence spectra, and are sensitive to interface effects (Cucinotta et al., 1990) and the composition of the host media of the cell culture.

In Fig. 4 we show the proton gamma-kill dose divided by E_o versus proton energy. The primary ionization is the dotted line and the solid line includes the effects of fragments. Secondary ion production is seen to have a negligible effect on intertrack effects, except at the highest energies where a small contribution is seen.

Cellular parameters obtained for survival and neoplastic transformations of C3H10T1/2 cells obtained from the experiments of Yang et al. (1985) are given in Table 1. We note that the large uncertainties in the transformation data of Yang should lead to a similar uncertainty in the transformation parameters. Parameter sets were found from data for instanteaneous and delayed plating of the cells after the irradiation. Here only the delayed plating case is considered. General agreement with the measured RBE values was found using these parameter sets (Waligorski et al., 1987). The single-particle-inactivation cross section neglecting the target fragmentation of eq. (17) is shown in Figs. 5 and 6 for cell death and cell transformation, respectively, as a function of the energy of the passing ion. The target-fragmentation contribution [the second term in eq. (17)] for protons has been evaluated as shown in Figs. 7 and 8. For protons the effects of the target fragments [dashed line, second term in eq. (17)] dominate over the proton direct ionization (dotted line) at high energy. For high LET particles (low energy), the direct ionization dominates and target-fragmentation effects become negligible. A simple scaling by $A_j^{1/2}$ relates the proton target-fragment term to ions of mass A_j. The resulting effective-action cross sections for cell death and cell transformation are plotted in Figs. 9 and 10, respectively. We note that the low-energy ^{56}Fe component of the GCR spectra extends into the trackwidth regime where $\sigma > \sigma_0$ and is not represented in the present calculations.

TABLE 2. TARGET FRAGMENT CONTRIBUTIONS TO PROTON ACTION CROSS-SECTION FOR V79 CHINESE HAMSTER CELLS σ, 10^{-12} cm^2

Zf	Af	E_p (MeV)		
		10	100	1000
1	1	4.58	6.70	7.14
1	2	0.21	0.40	1.49
1	3	0.10	0.19	0.22
2	3	0.16	0.34	0.66
2	4	1.68	3.59	12.22
3	5	0.32	0.42	0.65
3	6	0.46	0.53	0.68
3	7	0.04	0.17	0.42
4	6	0.01	0.07	0.20
4	7	0.36	0.42	0.49
4	8	0.41	0.51	0.53
4	9	<0.01	0.04	0.09
5	8	<0.01	0.04	0.11
5	9	0.10	0.30	0.32
5	10	0.22	0.35	0.28
5	11	0.02	0.21	0.35
6	10	<0.01	0.03	0.08
6	11	0.04	0.41	0.32
6	12	0.50	1.00	0.63
6	13	0.14	0.32	0.24
6	14	<0.01	0.02	0.03
7	12	<0.01	0.01	0.04
7	13	0.03	0.09	0.07
6	14	<0.01	0.02	0.03
7	12	<0.01	0.01	0.04
7	13	0.03	0.09	0.07
7	14	1.11	0.82	0.37
7	15	0.02	0.24	0.41
8	14	<0.01	0.02	0.05
8	15	0.02	0.56	0.28
Primary		32.61	0.05	<0.01
Total		43.16	17.84	28.35

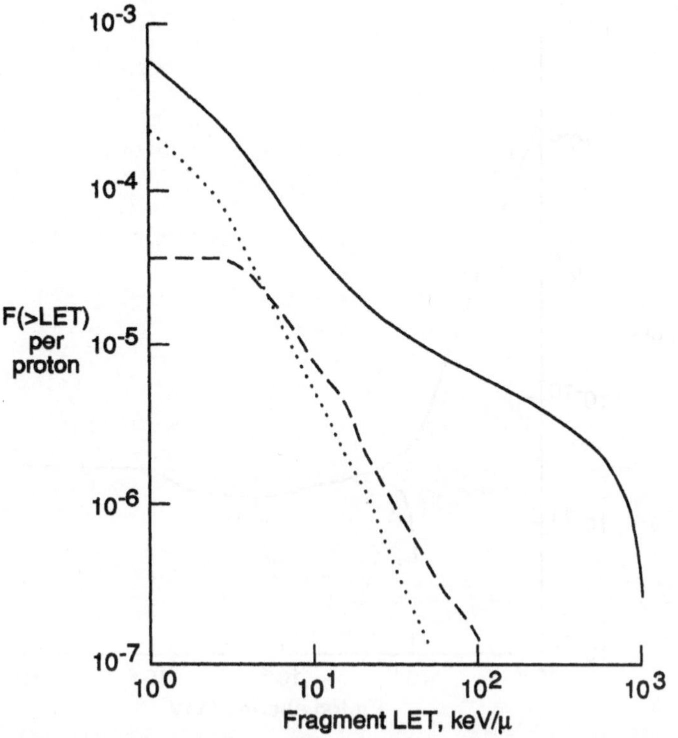

Fig. 1. Integral LET spectra for nuclear fragments produced by 1-GeV protons in water. Solid line, all fragments; dotted line, secondary protons; and dashed line, secondary alpha particles.

SURVIVAL CURVES AND THE PROTON RBE

Cell survival curves for 10-, 100-, and 1000-MeV protons are shown as a function of absorbed dose in Figs. 11–13. The solid line includes the target fragment terms and the dashed line neglects their contributions. Results show the importance of secondary production for increasing energy. We note, for example, that at 1000 MeV the increase in cell death due to the fragments does not lead to substantial changes in RBE at high doses as can be calculated from Fig. 13. It is in the initial portion of the survival curves where the ion-kill mode causes large differences in RBE when compared to gamma rays. The RBE versus dose is shown in Fig. 14 with all curves including the effects of target fragmentation. We note that the proton fluence is found as $F_p = 6.24\mathrm{D}/\mathrm{LET}$, with F_p in protons/μm^2, in Fig. 14. The rise in RBE at low dose or fluence, where single proton tracks dominate, is directly attributed to ion kill from both primary protons at 10 MeV and nuclear fragments at the higher energies. Not shown are RBE calculations neglecting the target fragments that are nearly identical to the 10 MeV results in Fig. 14, and are almost identical to unity for the 100- and 1000-MeV protons. The low dose behavior of the RBE can be seen from eq. (15), where for $m = 3$ (Table 1) as found from the data of Skarsgard (1967) an RBE dependence on $D^{-2/3}$ is found. This effect is supported experimentally as discussed below.

EFFECTIVENESS OF 160-MEV PLATEAU REGION PROTONS

Cell survival experiments have been performed at the Harvard Cyclotron for the purpose of determining the biological effectiveness of the protons. V79 Chinese hamster cells cultured *in vitro* were irradiated in the plateau region of the Bragg curve and in a spread-out Bragg peak by Hall et al. (1978). Here we compared the survival measurements and RBE determinations

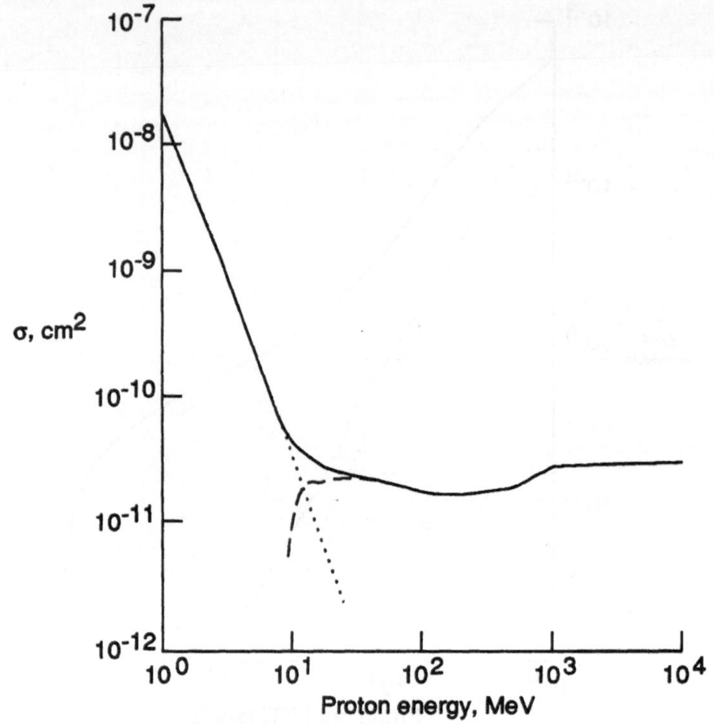

Fig. 2. Calculated values of the proton action cross section for survival of Chinese hamster cells versus the proton energy. Dotted line is primary proton contribution, dashed line is the tissue fragmentation contribution, and solid line is the total.

for attached cells in the plateau region of the 160-MeV proton Bragg curve using the track model and the high energy nucleon transport code BRYNTRN of Wilson et al. (1989).

The nucleon transport code BRYNTRN solves the coupled proton-neutron transport problem for high energies in the straight-ahead approximation with multiple-scattering and straggling effects ignored. Target fragments with $A > 1$ are transported using the production collision density as given in eq. (16). The Boltzman equations for proton and neutron transport are

$$\left[\frac{\partial}{\partial x} - \frac{\partial}{\partial E}L(Z_p, E) + \Sigma_p(E)\right]\Phi_p(x, E) = \sum_j \int_E^\infty f_{pj}(E, E')\Phi_j(x, E')dE' \tag{20}$$

and

$$\left[\frac{\partial}{\partial x} + \Sigma_n(E)\right]\Phi_n(x, E) = \sum_j \int_E^\infty f_{nj}(E, E')\Phi_j(x, E')dE' \tag{21}$$

where Φ_j is the differential flux of type j particles at x with energy E; $L(Z_j, E)$ is the proton stopping power; $\Sigma_p(E)$ and $\Sigma_n(E)$ are proton and neutron total cross sections, respectively; and $f_{ij}(E, E')$ represents the differential cross sections for elastic and inelastic processes. As described by Wilson and Lamkin (1975) the Boltzman eqs. (20) and (21) are solved using a characteristic transformation to reduce the problem to a set of coupled integral equations with boundary conditions at $x = 0$, which are then solved numerically. More details on the method of solution and the nuclear scattering data base are given by Wilson et al. (1989). The Bragg curve obtained from BRYNTRN for 160-MeV protons in water is shown in Fig. 15 as compared to the measurements of Verhey et al. (1979) (solid line). In Fig. 15, the squares represent the primary dose and the circles the total dose with secondary production included.

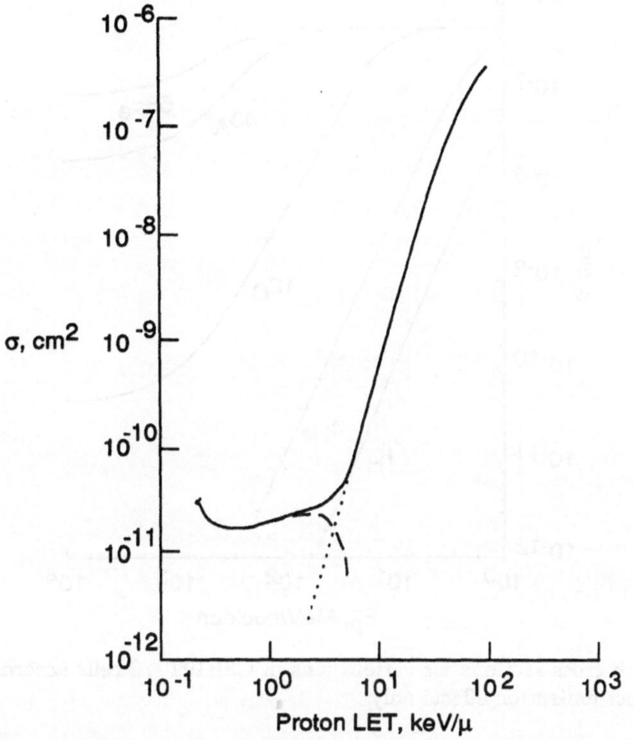

Fig. 3. Calculated values of the proton action cross section for survival of Chinese hamster cells versus the proton LET. Lines same as in Fig. 2.

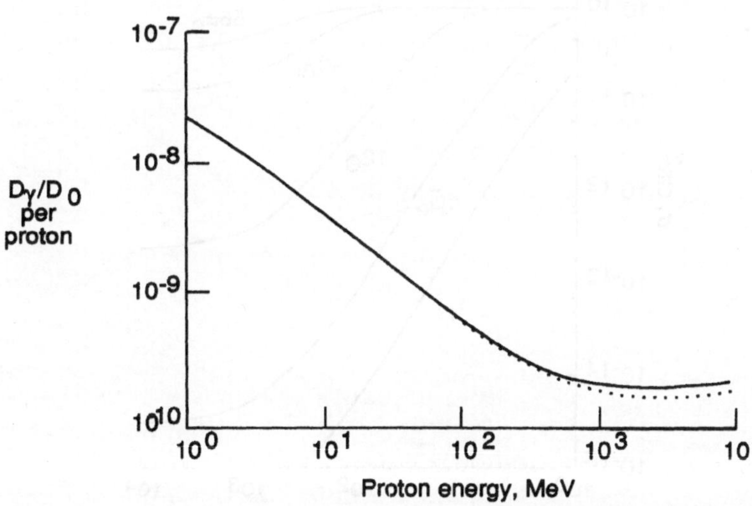

Fig. 4. Calculated values of the proton gamma-kill dose for survival of Chinese hamster cells versus the proton energy. The dotted line is the primary proton contribution and the solid includes contributions from fragmentation.

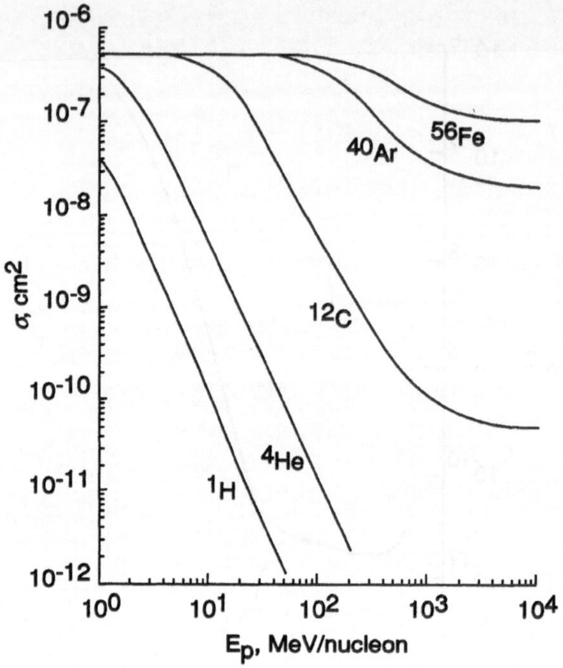

Fig. 5. Cell-death cross sections for various ions in C3H10T1/2 cells according to the Katz model for direct ionization effects only.

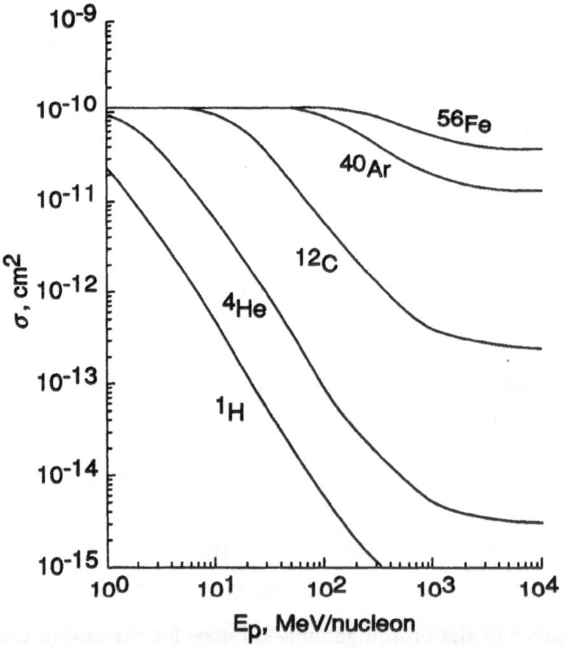

Fig. 6. Cell-transformation cross sections for various ions in C3H10T1/2 cells according to the Katz model.

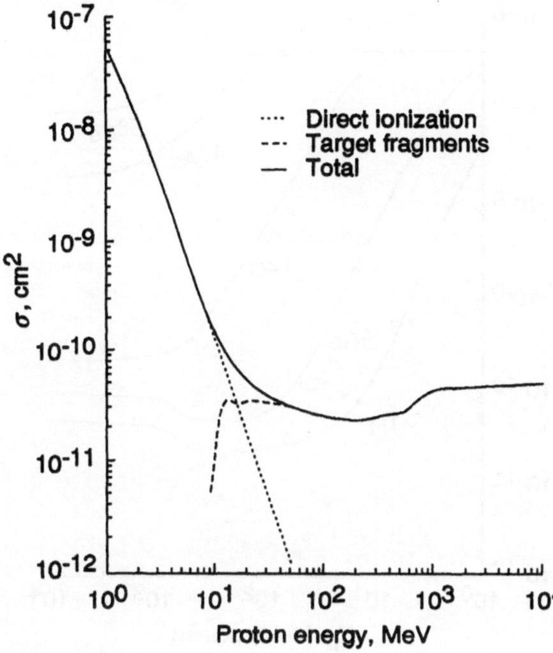

Fig. 7. Cell-death cross sections including effects of nuclear reactions for protons in C3H10T1/2 cells according to the Katz model.

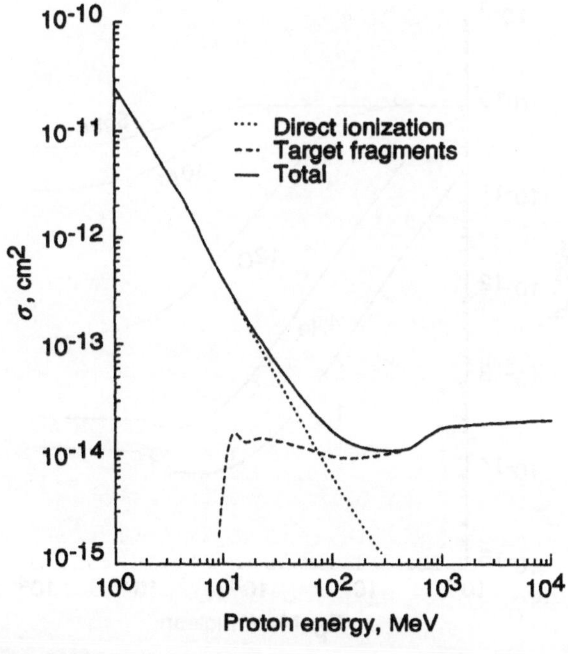

Fig. 8. Cell-transformation cross section including effects of nuclear reactions for protons in C3H10T1/2 cells according to the Katz model.

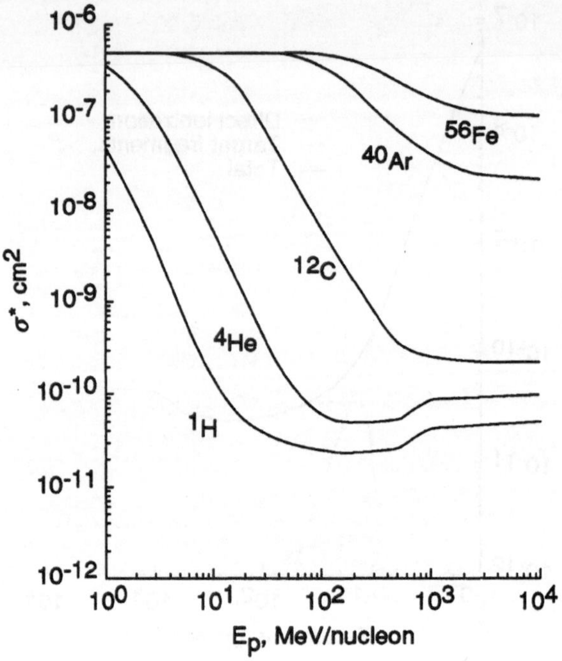

Fig. 9. Effective cell-death cross sections including effects of nuclear reactions for various ions in C3H10T1/2 cells.

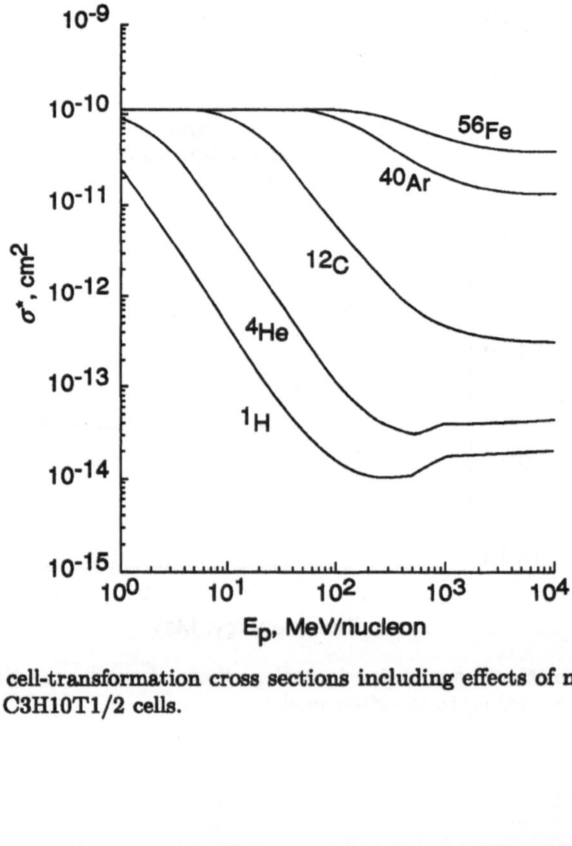

Fig. 10. Effective cell-transformation cross sections including effects of nuclear reactions for various ions in C3H10T1/2 cells.

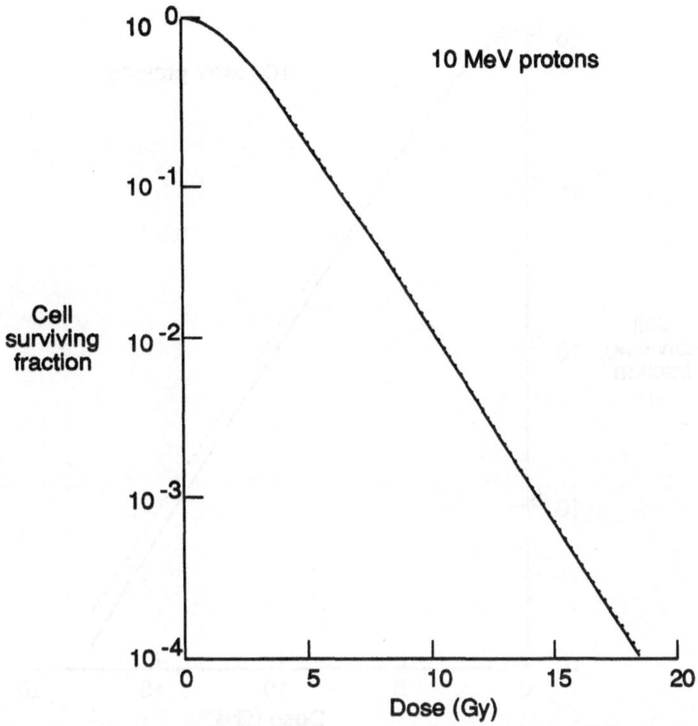

Fig. 11. Calculated cell surviving fraction of Chinese hamster cells for 10-MeV protons. The dashed line is the primary proton and the solid adds the effects of nuclear fragments.

Calculations are normalized to the peak of the experimental Bragg curve. Straggling and multiple-scattering effects, which are not included here, both contribute significantly at the peak of the Bragg curve. We consider the plateau region where the high energy assumptions are approximately true.

At energies of 160 MeV, nuclear recoils from elastic scattering provide a sizable correction to the secondary ion production represented by target fragmentation. Elastic nuclear scattering is represented by the Born term to the optical model renormalized to the total scattering cross section in the BRYNTRN code. This representation of elastic nuclear scattering is fairly accurate for integral quantities above 100 MeV, but breaks down at lower energies because of multiple scattering, nuclear medium corrections, and especially Coulomb effects. The correction to the proton action cross section from elastic scattering is shown in Table 3 for several energies and is included in the following comparisons.

Results for the surviving fraction of suspended V79 Chinese hamster cells irradiated in the plateau region of a 160-MeV proton beam are shown in Fig. 16. The dashed line is the fit to the γ-ray survival curve, the dotted line (barely distinguishable from the γ-response) is the contribution from primaries only, and the solid line calculations include the effects of nuclear reactions. The characteristic gamma-ray dose, E_O, is taken as 2.9 Gy to reproduce the experimental gamma-ray curve with the other response parameters given above. The dashed and dotted lines are nearly identical, indicating that high energy protons minus the effects of the nuclear force indeed act as gamma rays. Agreement with the data is fair indicating that the modeling of nuclear fragmentation made here is somewhat lacking. In Fig. 17 we compare our results for the proton RBE (solid line) against the values obtained using the analysis methods (vertical bars) of Kellerer and Brenot (1973) as discussed by Hall et al. (1978). The 'bare' proton RBE has a value of 1 (not shown), except at the lowest doses where a small

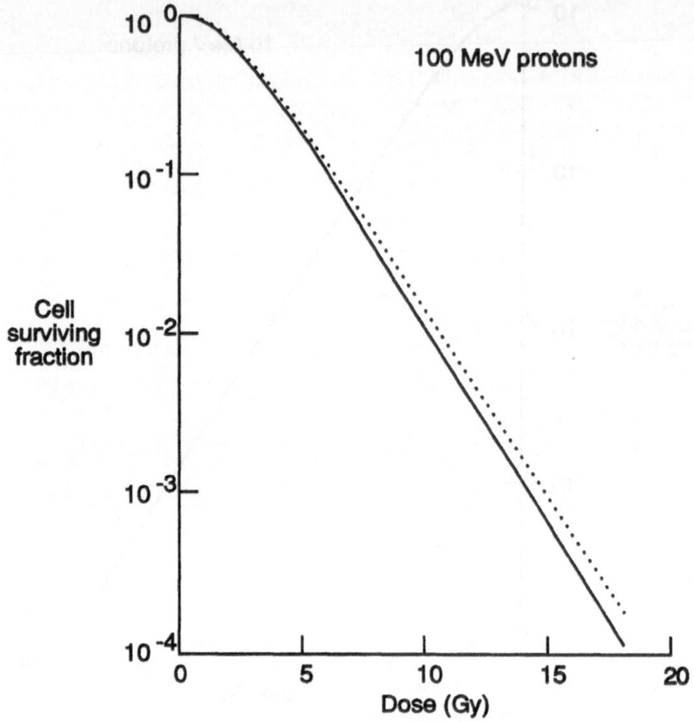

Fig. 12. Same as Fig. 11 for 100-MeV protons.

contribution from ion kill gives a slight increase. The calculations of the proton RBE presented in Fig. 17 show good agreement with the experimentally obtained values, with the increasing RBE at low doses shifted to lower values than experiment. A second analysis methodology which assumes only that the dose effect curve is convex but is otherwise shape independent was used to derive RBE values (Hall et al. 1978) and are shown in Fig. 18 in comparison to the present predictions. The RBE rise at low dose as $D^{-2/3}$ predicted from eq. (15) is clearly seen in the calculations (solid line) and the experimental analyses (dash line). Also shown in Fig. 18 by the dotted line are our calculations neglecting nuclear reactions which are almost exactly 1 for all doses.

CELL SURVIVAL IN HZE BEAMS

The HZE transport problem has been solved and related to the Bragg curve (Wilson 1978, 1983) for monoenergetic unidirectional ion beams. The Bragg curves we calculate also provide the values for fluence estimates for the exposure conditions of biological samples yet to be analyzed. Errors in the Bragg curve translates directly into errors in exposure levels for comparisons with experimental response data.

The relative positions along the Bragg curve where biological exposures were made are indicated in Fig. 19. The survival for aerobic and hypoxic T-1 cells of human origin have been calculated using the Katz parameters in Table 1 for several locations along the beam line within a water column for three different ion beams of C, Ne and Ar. The calculation includes both projectile as well as target fragments. The results for the C beam experiments (Blakeley et al. 1979) are shown in Fig. 20. The effects of overlapping δ-rays are clearly apparent except near the Bragg peak where the sigmoid appearance has all but disappeared. The sigmoid returns downstream from the Bragg peak where the overlapping δ-rays

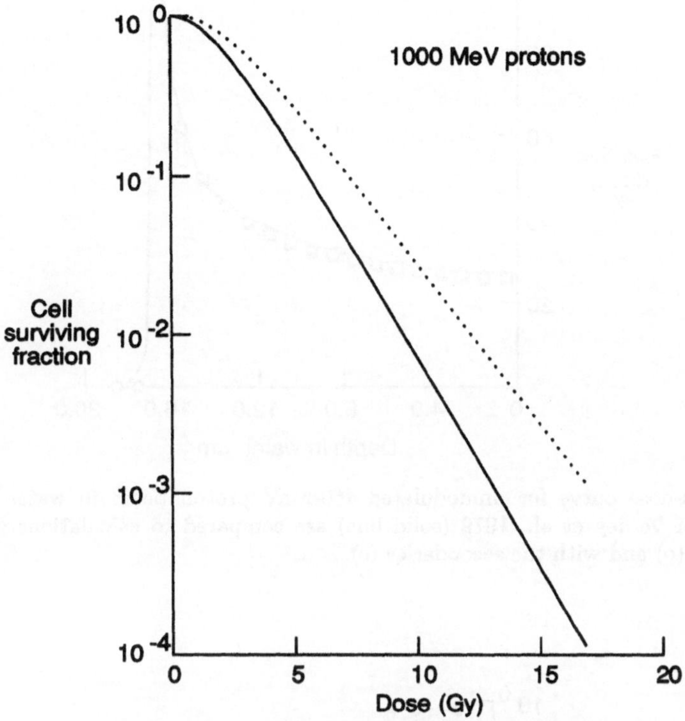

Fig. 13. Same as Fig. 12 for 1000-MeV protons.

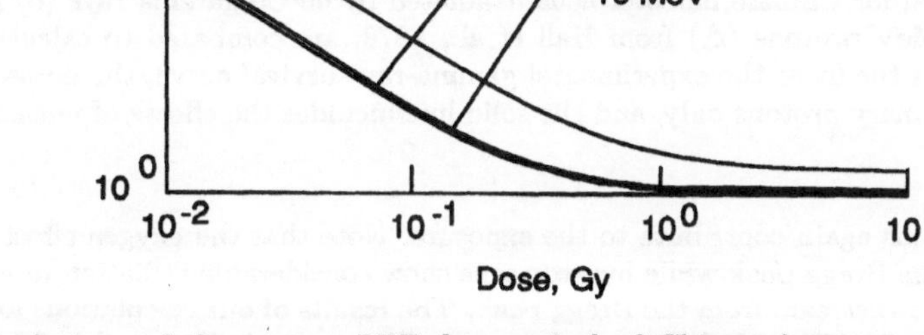

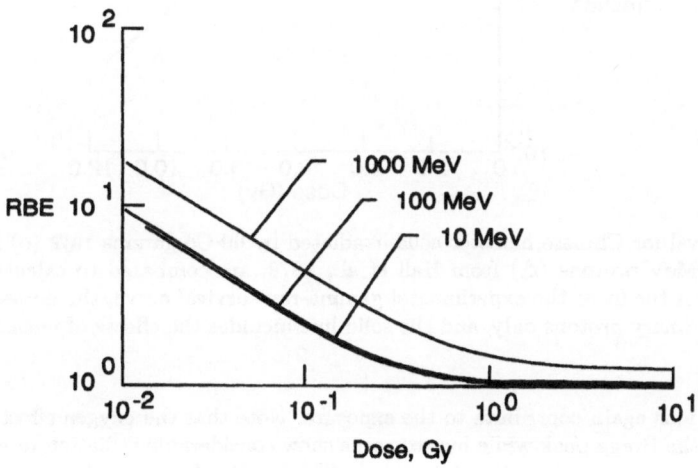

Fig. 14. Calculations of the proton RBE for survival of Chinese hamster cells versus the absorbed dose. In increasing order; 10-, 100-, 1000-MeV protons.

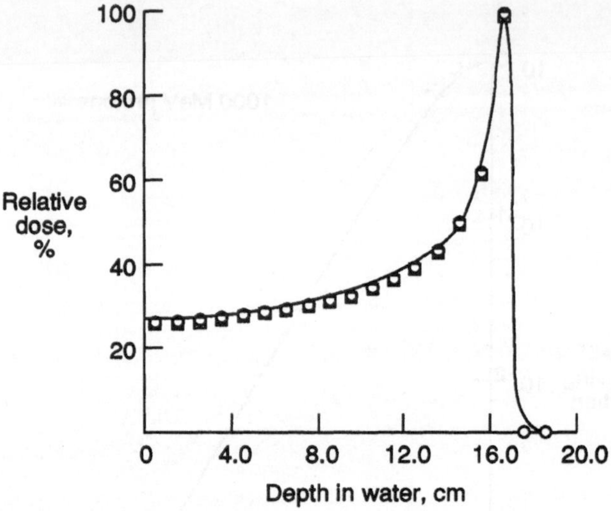

Fig. 15. Depth-dose curve for unmodulated 160-MeV proton beam in water. The measurements of Verhey et al., 1979 (solid line) are compared to calculations with nuclear secondaries (o) and with the secondaries (▫).

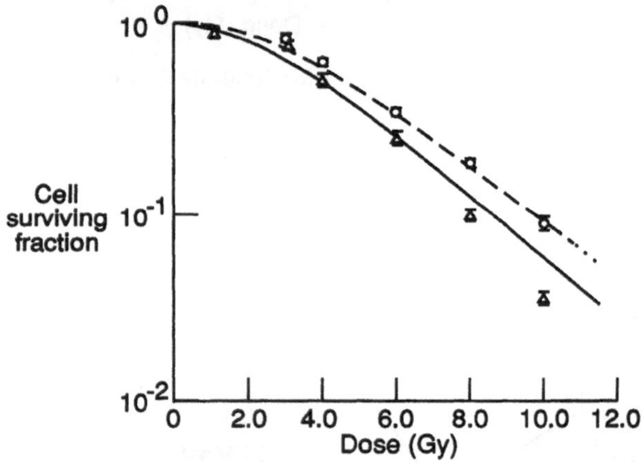

Fig. 16. Survival for Chinese hamster cells irradiated by 60-Co gamma rays (o) and plateau region 160-MeV protons (Δ) from Hall et al., 1978, are compared to calculations. The dashed line is the fit to the experimental gamma-ray survival curve, the dotted line is the result for primary protons only, and the solid line includes the effects of nuclear reactions.

from adjacent ions again contribute to the exposure. Note that the oxygen effect has all but vanished near the Bragg peak while hypoxic cells show considerable radiation resistance both upstream and downstream from the Bragg peak. The results of our calculations for Ne beams is shown in Fig. 21. The Ne beam results are qualitatively similar to those for C beams. The region over which the sigmoid appearance is suppressed is greatly expanded in the Bragg peak region. The oxygen enhancement is greatly diminished one full centimeter before and after the Bragg peak as can be seen in Figs. 21. This fact is of potential importance to radiation

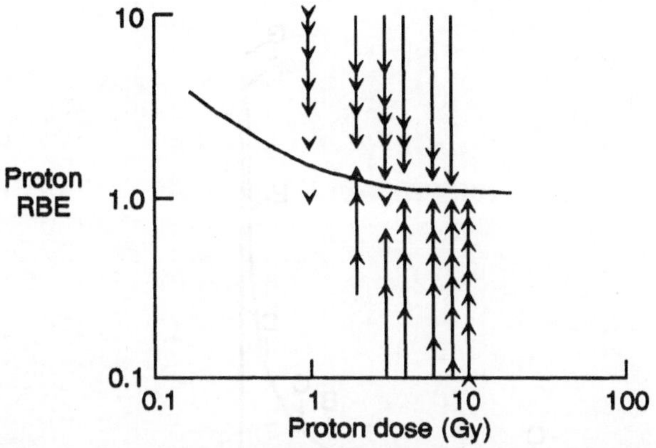

Fig. 17. Proton RBE versus proton dose for Chinese hamster cell survival in plateau region of 160-MeV proton Bragg curve. Experimental determination from Hall et al., 1978. The vertical bars denote RBE values that are excluded at 95% confidence level, while the detached arrows indicate values that are unlikely at a lower level of confidence. The most likely RBE values fall in the space between the vertical bars. The solid line is from the calculations where the effects of nuclear reactions are included.

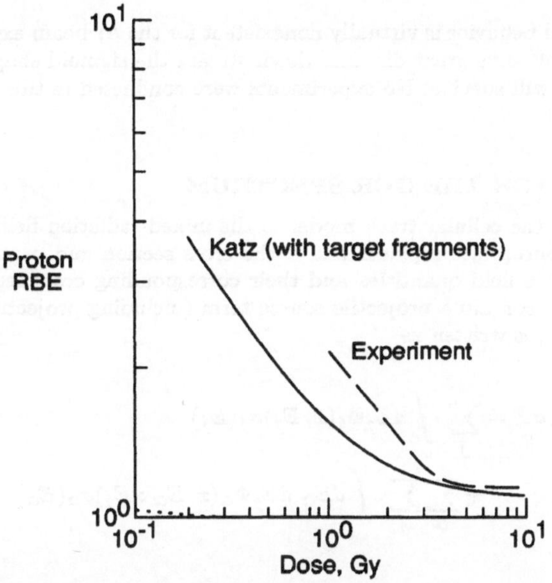

Fig. 18. Proton RBE versus proton dose for Chinese hamster cell survival in plateau region of 160-MeV proton Bragg curve. The dashed line is from experimental analysis at ref. 10 as discussed in text. The solid line is calculations which include nuclear reactions effects, and the dotted line without nuclear reactions.

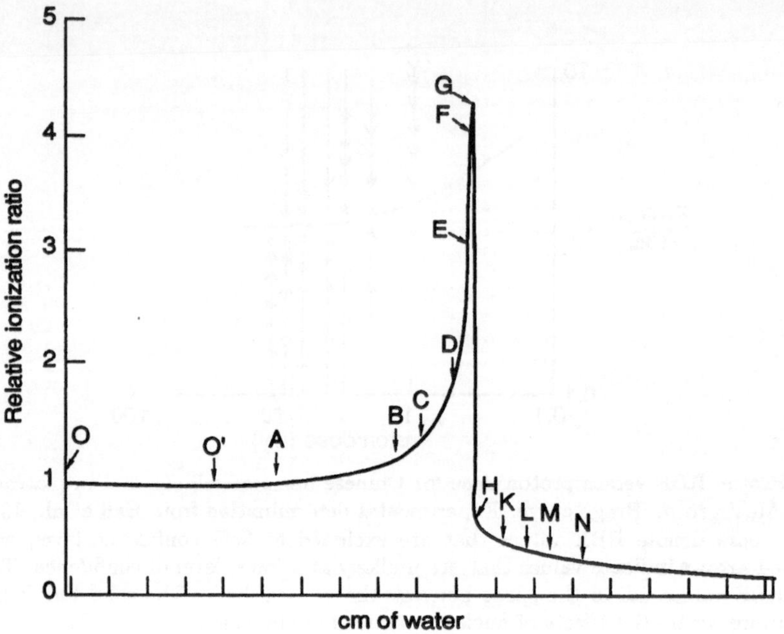

Fig. 19. Relative positions along beam line where biological experiments were performed by Blakeley et al. (1979).

therapy. The sigmoid behavior is virtually nonexistent for the Ar beam exposures as shown in Fig. 22. Obviously, at some great distance down stream the sigmoid shape will appear since only light fragments will survive. No experiments were conducted in this region.

CELL DAMAGE FOR THE GCR SPECTRUM

In order to apply the cellular track model to the mixed-radiation fields seen in space, we need to make the appropriate replacement of the cross section and particle fluence number (σF) with the particle field quantities and their corresponding cross sections. The ion-kill term, which will now contain a projectile source term (including projectile fragments) and a target fragment term, is written as

$$\sigma F = \sum_j \int dE_j \Phi_j(x, E_j)\sigma_j(E_j) + \sum_\alpha \sum_j \int dE_\alpha\, dE_j \Phi_\alpha(x, E_\alpha : E_j)\sigma_\alpha(E_\alpha) \tag{22}$$

where the second term is the contribution of nuclear fragments produced locally in the biological medium. This may be written in terms of an effective-action cross section σ^* for the passing ion, whose track is dressed by the local target fragments (nuclear stars), as

$$\sigma F = \sum_j \int dE_j \Phi_j(x, E_j)\; \sigma^*(E_j) \tag{23}$$

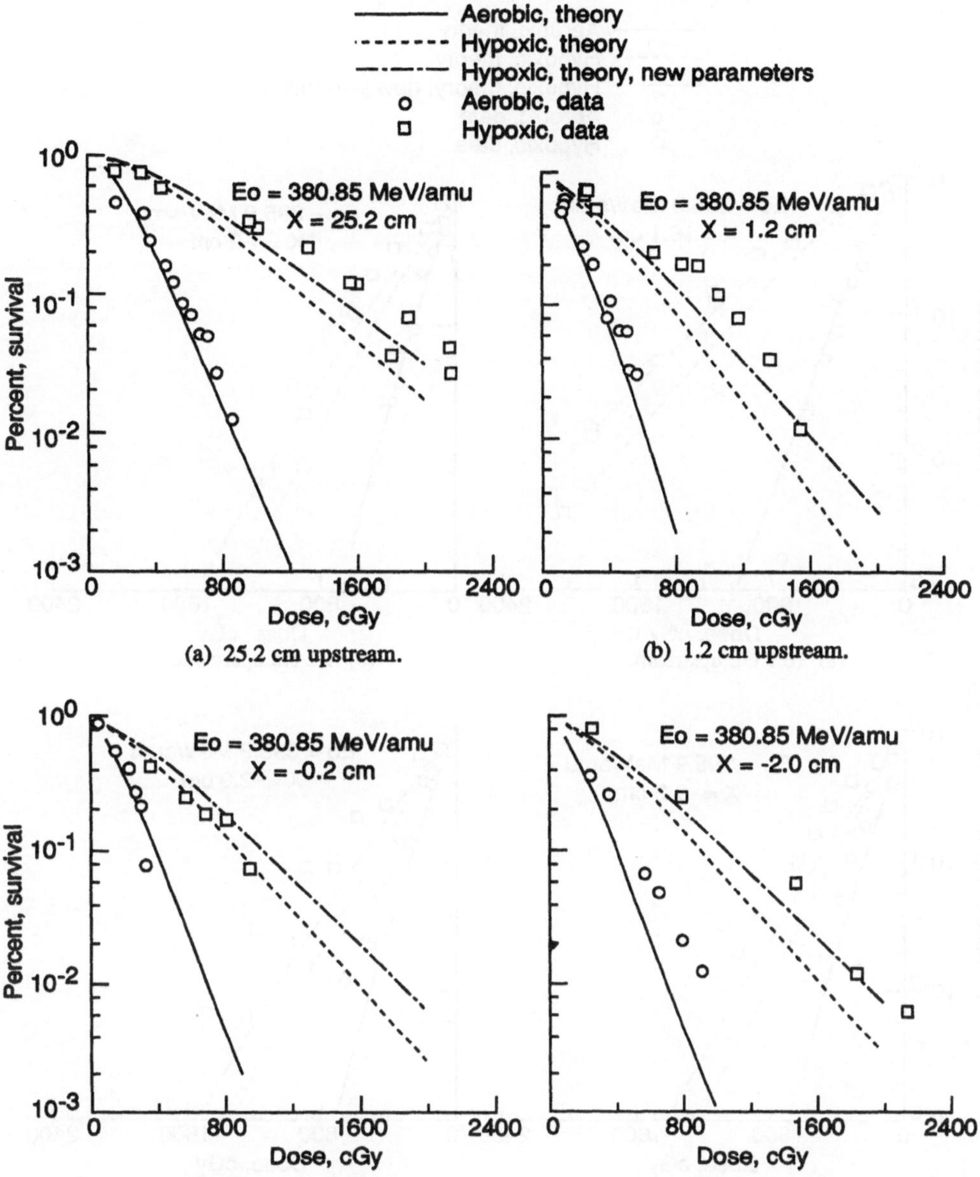

(a) 25.2 cm upstream.

(b) 1.2 cm upstream.

(c) 0.2 cm downstream.

(d) 2.0 cm downstream.

Fig. 20. Cell survival as a function of dosage at several locations relative to the Bragg peak in a ^{12}C beam.

The gamma-kill dose fraction becomes

$$D_\gamma = \sum_j \int dE_j \Phi_j(x, E_j)[1 - P_j(E_j)]S_j(E_j)$$
$$+ \sum_j \sum_\alpha \int dE_j \, dE_\alpha \Phi_\alpha(x, E_\alpha : E_j)$$
$$\times [1 - P_\alpha(E_\alpha)]S_\alpha(E_\alpha) \tag{24}$$

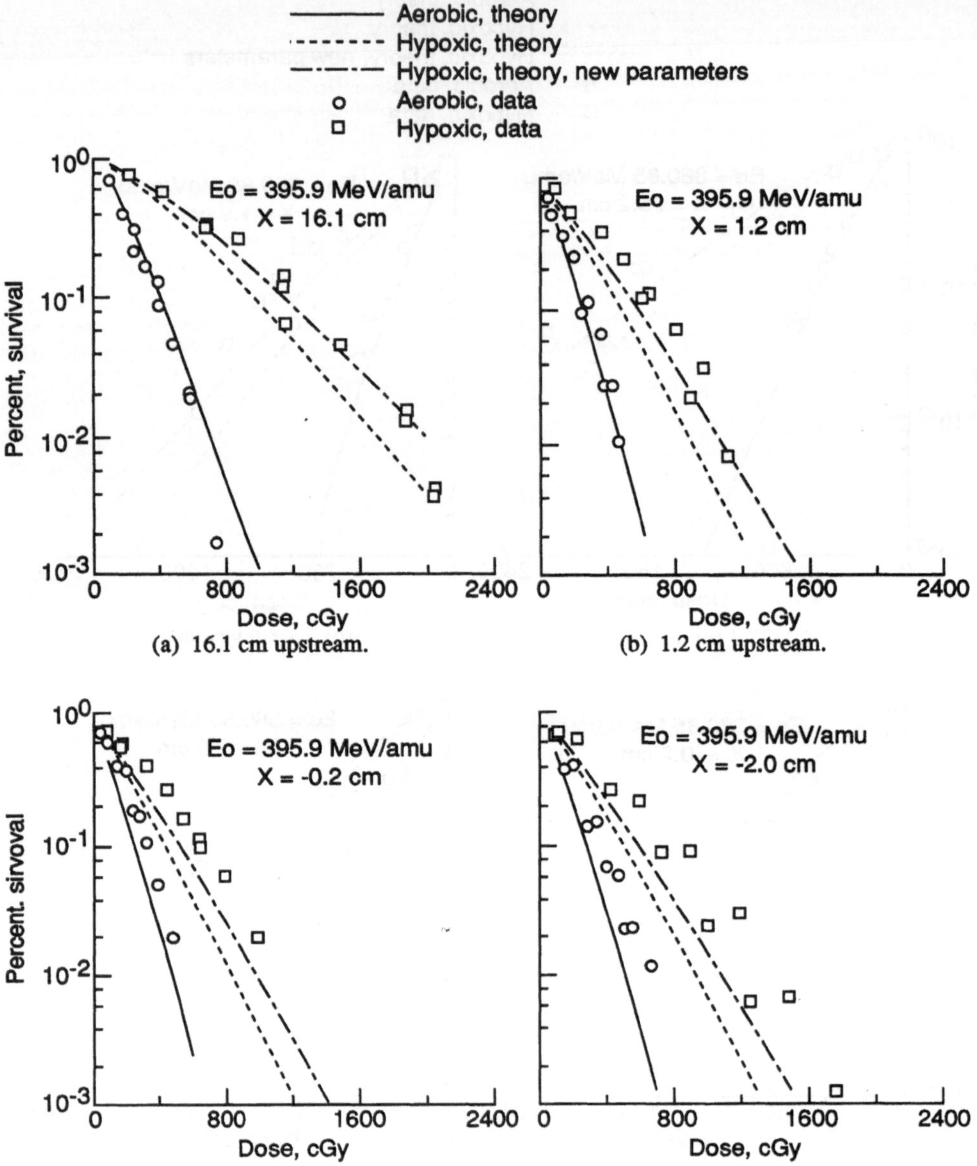

(a) 16.1 cm upstream.

(b) 1.2 cm upstream.

(c) 0.2 cm downstream.

(d) 2.0 cm downstream.

Fig. 21. Cell survival as a function of dosage at several locations relative to the Bragg peak in a ^{20}Ne beam.

Equations (22) and (24) are used in eqs. (2) and (3), respectively. The summations over all particle types in eqs. (22) and (24) represent the addition of probabilities from all ions in the radiation field that contribute to the end point under study.

The cellular track model was applied to predict the fraction of C3H10T1/2 cells killed or transformed for 1 year in deep space at solar minimum for typical spacecraft shielding. The GCR environment was taken from the Naval Research laboratory code (Adams et al. 1981). Aluminum shielding was considered with a local region of tissue for the cell cultures,

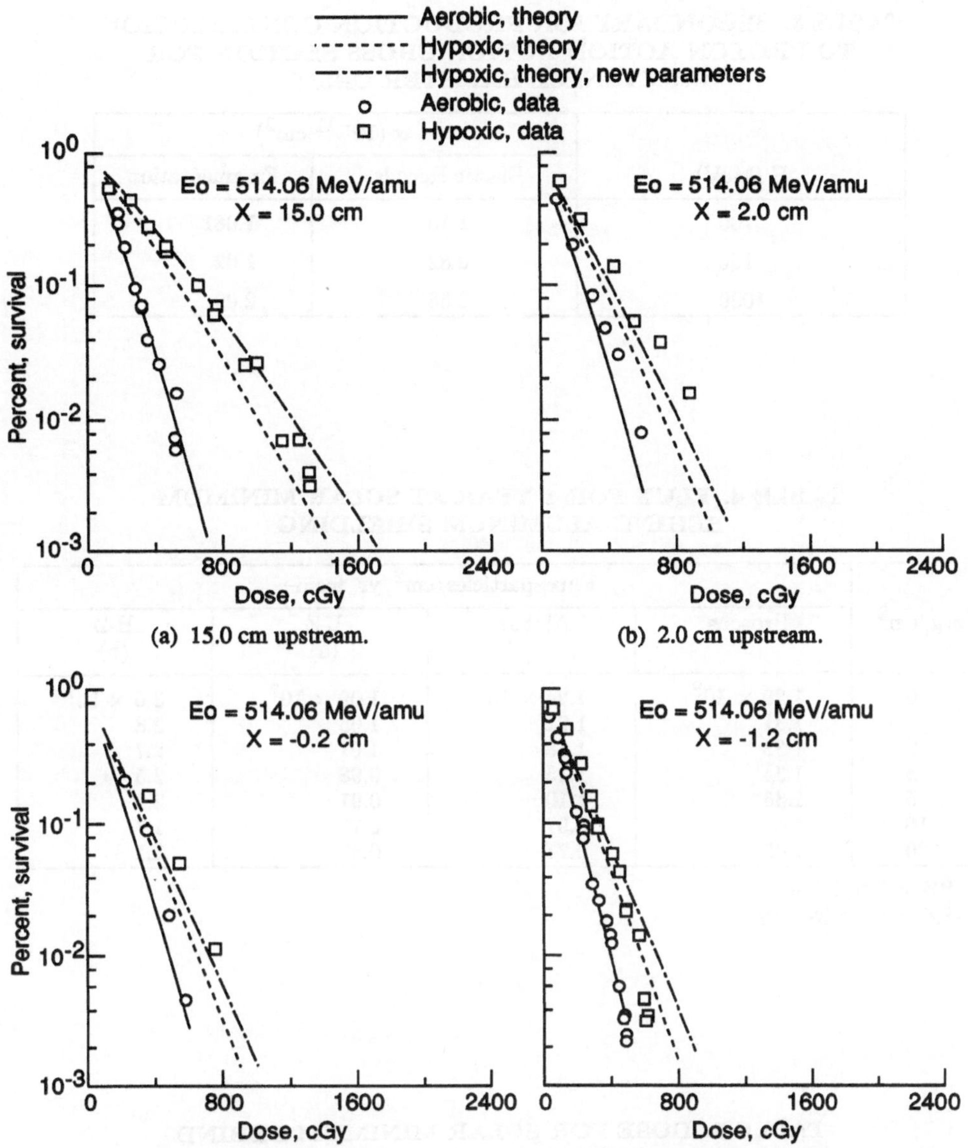

(a) 15.0 cm upstream.

(b) 2.0 cm upstream.

(c) 0.2 cm downstream.

(d) 1.2 cm downstream.

Fig. 22. Cell survival as a function of dosage at several locations relative to the Bragg peak in a ^{40}Ar beam.

Tables 4 and 5 contain individual particle fluences and absorbed doses, respectively, for the protons, alpha particles, $Z = 3$ to 9 ions (labeled H-Z) as determined by the Langley GCR code. Results for the fraction of C3H10T1/2 cells killed and transformed for 1 year at solar minimum are listed in Tables 6 and 7, respectively. The gamma-kill mode was found to be of negligible importance in the calculations, indicating that biological damage in deep space from GCR particles at the cellular level will indeed result from the action of single particles. The importance of the target terms in biological effects for low LET protons and alpha particles is quite apparent. The results also indicate that the HZE component of the GCR spectrum is

TABLE 3. SECONDARY ION PRODUCTION CONTRIBUTION TO PROTON ACTION ACTION CROSS SECTION FOR V79 CHINESE HAMSTER CELLS

E (MeV)	σ (10^{-11} cm^2)	
	Elastic Recoils	Fragmentation
100	1.13	1.081
150	0.82	1.02
1000	0.58	2.09

TABLE 4. FLUX FOR 1 YEAR AT SOLAR MINIMUM BEHIND ALUMINUM SHIELDING

x, g/cm^2	Flux, particles/cm^2/yr, from—			
	Protons	Alphas	L-Z (a)	H-Z (b)
0	1.29×10^8	1.24×10^7	1.09×10^7	3.0×10^7
1	1.31	1.21	1.05	2.8
2	1.33	1.18	1.01	2.7
3	1.34	1.15	0.98	2.5
5	1.36	1.10	0.91	2.2
10	1.40	0.97	0.77	1.7
20	1.43	0.77	0.57	1.1

[a] $Z = 3$ to 9 ions.
[b] $Z = 10$ to 28 ions.

TABLE 5. DOSE FOR SOLAR MINIMUM BEHIND ALUMINUM SHIELDING

x, g/cm^2	Dose, *c*Gy/yr, from				
	Protons	Alphas	L-Z (a)	H-Z (b)	Total
0	6.2	3.0	2.8	5.0	17.1
1	6.3	2.7	2.5	3.6	15.1
2	6.8	2.6	2.4	3.3	15.1
3	7.1	2.6	2.3	3.1	15.0
5	7.6	2.4	2.1	2.7	14.8
10	8.5	2.1	1.7	2.0	14.3
20	9.5	1.7	1.1	1.1	13.4

[a] $Z = 3$ to 9 ions.
[b] $Z = 10$ to 28 ions.

TABLE 6. FRACTION OF C3H10T1/2 CELLS KILLED IN DEEP SPACE FOR 1 YEAR AT SOLAR MINIMUM BEHIND ALUMINUM SHIELDING

x, g/cm^2	Fraction of cells killed of—					
	Protons	Alphas	L-Z (a)	H-Z (b)	Total	RBE
	Including Target Fragments					
0	1.35×10^{-2}	$.46 \times 10^{-2}$	$.57 \times 10^{-2}$	2.08×10^{-2}	4.46×10^{-2}	7.1
1	.76	.15	.43	1.84	3.18	7.0
2	.80	.14	.41	1.69	3.04	6.9
3	.83	.14	.38	1.55	2.90	6.8
5	.88	.14	.34	1.32	2.68	6.7
10	.95	.12	.25	0.91	2.22	6.5
20	1.02	.09	.15	0.49	1.74	6.2
	Without Target Fragments					
0	$.84 \times 10^{-2}$	$.37 \times 10^{-2}$	$.55 \times 10^{-2}$	2.08×10^{-2}	3.79×10^{-2}	6.7
1	.24	.06	.41	1.83	2.54	6.5
2	.28	.06	.39	1.68	2.41	6.3
3	.31	.06	.37	1.55	2.27	6.2
5	.35	.06	.33	1.31	2.04	6.1
10	.42	.05	.24	0.91	1.61	5.7
20	.49	.04	.14	0.48	1.15	5.3

[a] Z = 3 to 9 ions.
[b] Z = 10 to 28 ions.

TABLE 7. FRACTION OF C3H10T1/2 CELLS TRANSFORMED IN DEEP SPACE FOR 1 YEAR AT SOLAR MINIMUM BEHIND ALUMINUM SHIELDING

x, g/cm^2	Fraction of cells transformed—					
	Protons	Alphas	L-Z (a)	H-Z (b)	Total	RBE
	Including Target Fragments					
0	5.2×10^{-6}	2.0×10^{-6}	3.1×10^{-6}	7.5×10^{-6}	1.78×10^{-5}	6.4
1	3.5	1.0	2.7	6.7	1.39	6.4
2	3.7	1.0	2.6	6.2	1.35	6.3
3	3.9	0.9	2.4	5.7	1.29	6.3
5	4.2	0.9	2.2	4.9	1.22	6.2
10	4.7	0.8	1.7	3.5	1.06	6.0
20	5.2	0.6	1.1	2.0	.88	5.7
	Without Target Fragments					
0	3.2×10^{-6}	1.6×10^{-6}	3.1×10^{-6}	7.5×10^{-6}	1.53×10^{-5}	6.0
1	1.4	0.6	2.7	6.7	1.13	5.8
2	1.6	0.6	2.5	6.2	1.09	5.7
3	1.8	0.6	2.4	5.7	1.05	5.6
5	2.1	0.5	2.1	4.9	0.97	5.4
10	2.5	0.5	1.6	3.5	0.82	5.2
20	3.0	0.4	1.0	2.0	0.64	4.9

[a] Z = 3 to 9 ions.
[b] Z = 10 to 28 ions.

most damaging for small shielding depths. At large depths the HZE components break up and cause proton buildup with increasing shield depth. At large depths, the protons dominate the biological effects. In comparing individual charge components, we see that the H-Z particles have a reduced effectiveness for the transformation end point.

Also listed in Tables 6 and 7 are the values of RBE versus depth for the two end points. In Table 8 we present the present RBE values beside the average QF values taken from (Townsend et al. 1990) using the same transport code. The fact that RBE and QF are nearly equal at small depths is somewhat coincidental. We note that the quality factor is independent of the fluence level, which is not true for the Katz model. The Katz model indicates a substantial increase in risk, at higher shielding levels, than the ICRP 26 quality factors (ICRP 1977).

The RBE values show a simple scaling with exposure time for the GCR particles as can be seen from eqs. (8), (9), and (2) when ion kill dominates. Here we find for

$$\frac{N}{N_0} \cong 1 \tag{25}$$

with

$$\sigma F \ll 1 \tag{26}$$

that

$$\mathrm{RBE} = \frac{E_o}{\mathrm{LET}} \sigma^{1/m} F^{[-1+(1/m)]} \tag{27}$$

Then, scaling the RBE as a function of duration in deep space to the 1-year value for a duration period of τ (with $F = n\tau$) gives

$$\mathrm{RBE}(\tau) = (\tau/\tau_1)^{(-1+1/m)} \mathrm{RBE}(\tau_1) \tag{28}$$

As a result, a one-hit ($m = 1$) system RBE becomes fluence independent as expressed by

$$\mathrm{RBE}(\tau) = \mathrm{RBE}(\tau_1) \tag{29}$$

a two-hit ($m = 2$) system is expressed by

$$\mathrm{RBE}(\tau) = \frac{\mathrm{RBE}(\tau_1)}{(\tau/\tau_1)^{1/2}} \tag{30}$$

and a three-hit ($m = 3$) system is expressed by

$$\mathrm{RBE}(\tau) = \frac{\mathrm{RBE}(\tau_1)}{(\tau/\tau_1)^{2/3}} \tag{31}$$

Results of this scaling approximation agree quite well with calculations using the Katz model, as seen in Table 9 where values obtained using the approximations of eq. (27) are shown in parentheses as scaled from the 1-year RBE values taken from Table 8, and results of the calculations are shown without parentheses. The extremely large RBE values that would be obtained for small values of τ are due to the choice of energetic photons as the reference radiation.

SUMMARY AND CONCLUSIONS

Over the past 25 years Katz and coworkers developed a model of particle tracks which began with nuclear emulsions and subsequently was extended to other detectors and to the biological effects of high LET radiations. That model requires as input information knowledge of the particle-energy spectrum of the radiation environment as well as the dose of gamma rays. Calculations of the effects of beams of protons, of heavy ions, and of energetic neutrons have been hindered because of a lack of a model of such beams which included both

TABLE 8. COMPARISON OF ICRP 26 QUALITY FACTORS VERSUS RBE FOR CELL DEATH AND TRANSFORMATION

[One Year in Deep Space at Solar Minimum]

x, g/cm^2	QF	RBE for cell death	RBE for cell transformation
0	7.1	7.1	6.4
1	5.6	7.0	6.4
2	5.3	6.9	6.3
3	5.1	6.8	6.3
5	4.7	6.7	6.2
10	3.9	6.5	6.0
20	3.2	6.2	5.7

TABLE 9. RBE FOR CELL DEATH AND TRANSFORMATION OF C3H10T1/2 CELLS FOR GCR SPECTRUM AT SOLAR MINIMUM BEHIND ALUMINUM SHIELDING*

x, g/cm^2	RBE values for time periods of—		
	1 month	1 year	2 years
	Cell death		
0	33.2 (37.0)	7.1	4.8 (4.6)
1	33.2 (36.1)	7.0	4.7 (4.5)
3	32.4 (35.1)	6.8	4.5 (4.3)
	Cell Transformation		
0	22.3 (22.2)	6.4	4.6 (4.5)
1	22.0 (22.2)	6.4	4.5 (4.5)
3	21.6 (21.8)	6.3	4.4 (4.4)

*Values in parentheses scaled from 1 year value using eq. (18).

projectile and target fragmentation. A beam model created through the efforts of John Wilson (1977b, 1983) and his collaborators at NASA Langley Research Center remedied this neglect. Through it we have been able to validate both the track theory of biological effects and the beam model by comparison of our calculated radiobiological end points with ground based measurements for proton and heavy ion beams. Based on this validation we have initiated calculations of biological effects in space vehicles in selected orbits, incorporating knowledge of the distribution of solar and galactic cosmic rays to be encountered there (Cucinotta et al. 1990, 1991a, 1991b). We know of no other way to estimate the biological damage in space flight at the very low fluences of heavy ions to be encountered there.

The track structure model with the deterministic galactic cosmic ray (GCR) transport code predicts the fractions of cell death and neoplastic transformations for C3H10T1/2 cells in deep space behind typical spacecraft shielding. Results indicate that the level of damage from the GCR particles does not attenuate appreciably for large amounts of spacecraft shielding and that single particles acting in the ion-kill mode dominate the effects. The contribution from target fragments was seen to be important in assessing the biological effect of protons and alpha particles. The relative biological effectiveness (RBE) values obtained in this fluence dependent model were found to be more severe than the ICRP 26 quality factors. A simple scaling law with the duration time in space was found to account for the change in RBE with fluence for the uniform GCR background.

The results of our calculations of the RBE for both cell death and cell transformation are remarkably close, especially when considering the very large difference in radiosensitivity parameters for these end points and the huge difference in the fraction of affected cells. About 1000 times as many cells are killed as are transformed. Nevertheless, 90 percent of the cells survive the conditions calculated here, and of these about 1 or 2 in 100 000 are transformed. Yet, this is not an insignificant fraction when we consider the number of cells per cubic centimeter in tissue and speculate about the number of cells transformed by radiation that are likely to lead to cancer.

The cell population in tissue, about 10^9 per cubic centimeter, suggests that after 1 year of exposure to GCR at solar minimum there would be about 10^4 transformed cells per cubic centimeter in tissue if *in vitro* and *in vivo* transformation parameters are equal. Additionally, we do not know the minimum number of transformed cells that can be injected into a mouse to induce a cancer. Clearly, priority must be assigned to the investigation of these questions. If one or two transformed cells were to lead to cancer, as in leukemia, we could not tolerate an exposure in which the transformation fraction exceeded 10^{-9}.

ACKNOWLEDGMENT

Research at the University of Nebraska is supported by the U.S. Department of Energy. Support for one of us (DMN) is from NASA Grant NCCI-42.

REFERENCES

Adams, J. H., Jr.; Silberberg, R.; and Tsao, C. H.; 1981: *Cosmic Ray Effects on Microelectronics. Part I—The Near-Earth particle Environment.* NRL Memo. Rep. 4506-Pt. I, U.S. Navy, Aug. 1981. (Available from DTIC as AD A103 897).

Barkas, W. H.; 1963: Nuclear Research Emulsions, vol. 1, Academic Press, New York.

Blakely, E. A.; Tobias, C. A.; Yang, T. C. H.; Smith, K. C.; and Lyman, J. T.; 1979: Inactivation of human kidney cells by high energy monoenergetic heavy ion beams. *Radiat. Res.* 80:122–160.

Bertini, H. W.; 1970: *MECC7: a Monte Carlo Intranuclear Cascade and Evaporation Code.* Oak Ridge, Tennessee: Oak Ridge National Laboratory, Radiation Shielding Information Center.

Butts, J. J.; and Katz, R.; 1967: "Theory of RBE for Heavy Ion Bombardment of Dry Enzymes and Viruses.," *Radiat. Res.* 30:855–871.

Cucinotta, F. A.; Atwell, W.; Hardy, A. C.; Golightly, M. J.; Wilson, J. W.; Townsend, L. W.; Shinn, J.; Nealy, J. E.; and Katz, R.; 1990: "Predications of Cell Damage Rates for Lifesat Missions." NASA TM 102170.

Cucinotta, F. A.; Hajnal, F.; and Wilson, J. W.; 1990: Energy deposition at the bone-tissue interface from target fragments produced by high energy nucleons. *Health Phys.* 59:819–825.

Cucinotta, F. A.; Katz, R.; Wilson, J. W.; Townsend, L. W.; Nealy, J. E.; and Shinn, J. L.; 1991a: "Cellular Track Model of Biological Damage to Mammalian Cell Cultures from Galactic Cosmic Rays." NASA TP 3055.

Cucinotta, F. A.; Atwell, W.; Weyland, M.; Hardy, A. C.; Wilson, J. W.; Townsend, L. W.; Shinn, J. L.; and Katz, R.; 1991b: "Radiation Risk Predictions for Space Station Freedom Orbits." NASA TP 3098.

Cucinotta, F. A.; Katz, R.; Wilson, J. W.; Townsend, L. W.; Shinn, J. L.; and Hajnal, F.; 1991c: Biological Effectiveness of High Energy Protons: Target Fragmentation. *Radiat. Res.* 127.

Dertinger H.; and Jung, H.; 1970: *Molecular Radiation Biology*, Springer-Verlag, New York.

Curtis, S. B.; Townsend, L. W.; and Wilson, J. W.; 1990.

Goodhead, D. T.; 1988: "Spatial and Temporal Distribution of Energy," *Health Physics* 55:231–240.

Goodhead, D. T.; 1989: "Relationships of radiation track structure to biological effect: a re-interpretation of the parameters of the Katz model," *Nuclear Tracks and Radiation Measurements* 16, 177–184.

Greiner, D. E., et al.; 1975: Momentum distributions of isotopes produced by relativistic 12-C and 16-O projectiles. *Phys. Rev. Lett.* 35:152–155.

Hall, E. J.; Kellerer, A. M.; Rossi, H. H.; and Lam, Y. P.; 1978: The relative biological effectiveness of 160 MeV protons-II. *Int. J. Radiat. Onc. Bio. Phys.* 4, 1009–1013.

ICRP; 1977: *Recommendations of the International Commission on Radiological Protection.* ICRP Publ. 26, Pergamon Press, Jan. 17, 1977.

Janni, J. F.; 1982: "Proton Range-Energy Tables, 1 keV-10 GeV," *At. Data and Nucl. Data Tables* 27:147–529.

Katz, R.; 1978a: "Track Structure Theory in Radiobiology and in Radiation Detection," *Nucl. Track Detection* 2:1–28.

Katz, R.; 1978b: "High LET Constraints on Low LET Survival," *Phys. Med. Biol.* 23:909–916.

Katz, R.; 1984: "Formation of Etchable Tracks in Plastics," *Nuclear Tracks and Radiation Measurements* 8:1–4.

Katz, Robert: Biological Effects of heavy Ions From the Standpoint of Target Theory. *Adv. Space Res.*, vol. 6, no. 11, 1986, pp. 191–198.

Katz, R.; 1990a: "Cross Section," *Appl. Radiat. Isot.* 41:563–567.

Katz, R.; 1990b: "On the Normalized Yield (events/rad/dalton) of Biological Molecules Irradiated with Energic Heavy Ions," *Radiation Physics and Chemistry* in press.

Katz, R; Ackerson, R.; Homayoonfar, M.; and Sharma, S. C.; 1971: Inactivation of cells by heavy ion bombardment. *Radiat. Res.* 47:402–425.

Katz, R.; and Cucinotta, F. A.; 1991: "RBE vs. Dose for Low Doses of High LET Radiations," *Health Physics* 5:717–718.

Katz, R.; Dunn, D. E.; and Sinclair, G. L.; 1985: "Thindown in Radiobiology," *Radiat. Prot. Dosimetry* 13:281–284.

Katz, R.; and Huang, GuoRong; 1989: "Track Core Effects in Heavy Ion Radiolysis," *Physics and Chemistry* 33:345–349.

Katz, R.; Haung, G.; 1991: "Radiosensitivity Parameters for Cell Survival in Tradescantia and for Chromosome Aberrations in Chinese Hamster Cells," *Radiation Protection Dosimetry* 31:372–374.

Katz, R.; Loh, Kim Sum; Daling, Luo; and Huang, GuoRong; 1990: "An Analytic Representation of the Radial Distribution of Dose from Energetic Heavy Ions in Water, Si, LiF, NaI, and SiO_2" *Radiation Effects and Defects in Solids* 114:15–20.

Katz, R.; and Kobetich, E. J.; 1968: "Response of NaI(T1) to Energetic Heavy Ions," *Phys. Rev.* 170:397–400.

Katz, R.: and Kobetich, E. J.; 1969: "Particle Tracks in Emulsion," *Phys. Rev.* 186:344–351.

Katz, R.; Sharma, S. C.; and Homayoonfar, M.; 1972: "The Structure of Particle Tracks," in *Topics in Radiation Dosimetry*, ed. F. A. Attix, Academic Press, New York.

Katz, R.; Sinclair, G. L.; and Waligorski, M. P. R.; 1986: "The Fricke Dosimeter as a 1-hit Detector," *Nucl. Tracks Radiat. Meas.* 11:301–307.

Katz, R.; and Wesely, S.; 1991: "Cross Sections for Single and Double Strand Breaks in SV-40 Virus in E_0 Buffer after Heavy Ion Irradiation. Experiment and Theory," Radiation and Environmental Biophysics 30:81–85.

Katz, R.; and Varma, M. N.; 1990: "Radial Distribution of Dose," *to be published.*

Katz, R.; and Zachariah, R.; 1991: E. Coli B modeled as a 1-hit detector. In: *Radiation Research, a Twentieth Century Perspective*, J. D. Chapman, W. C. Dewey, and G. F. Whitmore, eds. Acadamic Press, San Diego, p. 117.

Kellerer, A. M.; and Brenot, J.; 1973: Nonparametric determination of modifying factors in radiation action. *Radiat. Res.* 56:28–29.

Silberberg, R.; Tsao, C. H.; and Shapiro, M. M.; 1976: *Semiempirical Cross Sections. In: Spallation Nuclear Reactions and Their Applications* (B. S. P. Shen and M. Merker, Eds.) pp. 49–81. Dardrecht Publishing Co., Boston, MA.

Skarsgard, L. D.; Kihlman, B. A.; Parker, L.; Pujara, C. M.; and Richardson, S.; 1967: Survival chromosome abnormalities, and recovery in heavy-ion and X-irradiated mammalian cells. *Radiat. Res. Suppl.* 7:208–221.

Todd, P.; 1967: Heavy ion inactivities of cultured human cells. *Radiat. Res. Suppl.* 7:196–207.

Townsend, Lawrence W.; Nealy, John E.; Wilson, John W.; and Simonsen, Lisa C.; 1990: *Estimates of Galactic Cosmic Ray Shielding Requirements During Solar Minimum.* NASA TM-4167.

Underbrink, A. G.; Huczkowski, J.; Woch, B; Gedlek, E.; Cebulska-Wasilweska, A.; Litwiniszyn, M.; and Kasper, E.; 1978: "The Relationship of Different Somatic Mutations Induced by Neutrons and X Rays to Loss of Reproductive Integrity in Tradescantia Stamen Hairs," Report 1039/B, 1978, Institute of Nuclear Physics, Krakow (D).

Waligorski, M. P. R.; Hamm, R. N.; and Katz, R.; 1986: "The Radial Distribution of Dose Around the Path of a Heavy Ion in Liquid Water," *Nucl. Tracks Radiat. Meas.* 11:309–319.

Waligorski, M. P. R.; Loh, K. S.; and Katz, R.; 1987: "Inactivation of Dry Enzymes and Viruses by Energetic Heavy Ions," *Radiat. Phys. Chem.* 30:201–208.

Waligorski, M. P. R.; Danialy, G.; Loh,K. S.; and Katz, R.; 1989: "Response of the alanine dosimeter to charge particle and neutron irradiations," *Applied Radiation and Isotopes* 40:923–933.

Waligorski, M. P. R.; Sinclair, G. L.; and Katz, R.; 1987: "Radiosensitivity Parameters for Neoplastic Transformation in C3H10T1/2 Cells," *Radiat. Res.* 111:424–437.

Wilson, J. W.; 1777a: "Analysis of the Theory of High Energy Transport." NASA TN D-8381.

Wilson, J. W.; 1977b: Depth dose relations for heavy ion beams. Va. J. of Sci., vol. 28:136–138.

Wilson, J. W.; 1983: "Heavy Ion Transport in the Straight Ahead Approximation." NASA TP-2178.

Wilson, J. W., et al. 1989: BRYNTRN: A Baryon Transport Computer Code. NASA TM-4037 (1988), also NASA TP-2887 (1989).

Wilson, J. W.; Lamkin, S. L.; 1976: Perturbation theory for charged-particle transport in one dimension. *Nucl. Sci. & Eng.* 57:292–299.

Wilson, J. W.; Townsend, L. W.; and Khan, F.; 1989: Evaluation of highly ionizing components in high-energy nucleon radiation fields. *Health Phys.* 57:717–724.

Wilson, J. W.; Townsend, L. W.; and Badavai, F. F.; 1984: Galactic HZE propagation through the earth's atmosphere. *Radiat. Res.* 109:173–183.

Yang, T. C.; Craise, L. M.; and Tobias, C. A.; 1985: "Neoplastic Cell Transformation by Heavy Charged Particles," *Radiat. Res. Suppl.* 104:S177–S187.

Ziegler, J. F.; 1980: *Stopping Cross-Sections for Energetic Ions in All Elements.* Pergamon Press, New York, 1980.

CONCEPTS OF MICRODOSIMETRY AND THEIR APPLICABILITY TO RADIATION PROTECTION PROBLEMS IN MANNED SPACE MISSIONS

Joachim Breckow

TÜV Rheinland (Technical Inspection Organization)
Dept. of Radiation Protection
P.O.Box 10 17 50
D-5000 Köln 91
Germany

ABSTRACT

A distinct and characteristic feature of ionizing radiation is its discontinuous interaction with matter. The energy imparted to the exposed material is due to discrete random events with stochastic fluctuating energy depositions. These fluctuations result in a highly non-uniform microdistribution of the transmitted energy and the subsequent radiation products. The current dosimetric quantities, such as the absorbed dose, D, or the Linear Energy Transfer, LET, are merely the statistical averages of the actual energy deposits and may deviate from these by orders of magnitude. The deviations are most substantial for small volumes, for small doses, or for densely ionizing radiations.

Biological effects of ionizing radiation depend on the energy deposited in volumes as small as individual cells or subcellular structures. Differences in the biological effectiveness of different types of ionizing radiation at equal values of absorbed dose can only be understood if the microstructure of the different spatial energy-deposition patterns are taken into account.

The characteristic feature of the ambient radiation field in space with its broad spectrum of different radiation quality makes it neccessary to take account of the stochastical nature of energy deposition. Microdosimetry provides the concepts that are required to quantify the microscopic distributions and the spatial correlation of energy deposits in small volumes and for radiations of different quality. Microdosimetric approaches have led to a deeper understanding of the underlying primary mechanisms of the interaction of ionizing radiation with biological structures.

INTRODUCTION

All ionizing radiations produce biological effects by the same basic mechanisms which result in excitations and ionizations in the irradiated tissue. Furthermore, the yields of the primary radiation products are essentially the same for the various types of radiations. Nevertheless, there are considerable differences in the biological effectiveness of ionizing radiations even at equal values of absorbed dose. It is generally agreed that these differences are related to the different spatial distributions of the primary radiation products within the irradited organism.

Biological Effects and Physics of Solar and Galactic Cosmic Radiation, Part A, Edited by C.E. Swenberg *et al.*, Plenum Press, New York, 1993

In radiation protection the characteristic biological effectiveness, the "quality", of a radiation type is taken into account in terms of the Linear Energy Transfer (LET) which charactarizes the mean density of energy loss of a charged particle along its track. However, the distribution of energy imparted along the particle track is by far not homogeneous and exhibits clusters and local concentrations of ionizations. This inhomogenity and the resulting random fluctuations of energy imparted are disregarded by the LET concept.

Furthermore, in order to assess quality factors for unknown radiation fields one needs to determine the LET distributions. For mixed and/or time-varying radiation fields measurements can hardly be performed. Calculations in terms of LET, on the other hand, tend to result in extreme simplifications that may lead to inconsistencies with radiobiological findings.

The complex radiation environmment in spaceflights is due to a variety of very different radiation sources, including galactic rays, radiation originating from the sun, charged particles trapped by the earth's magnetic field or high-energy charged particles such as protons, α-particles and heavy ions. The fluences and energy spectra of these radiations within a spacecraft depend on altitude, orbit inclination, solar activities, spacecraft orientation or shielding conditions. A quantification of any particular radiation condition has to be based on measurements which can be done most accurately by means of microdosimetric techniques.

The characteristic features of radiation in space and its biological consequences are of special interest in radiobiology and they are also relevant for radiation protection. Some of these aspects may lead to a better assessment of radiation action if they are considered in terms of microdosimetry.

BASIC CONCEPTS AND QUANTITIES

Ionizing radiation transfers energy to the exposed materials in discrete random events. The energy imparted, ε, to the matter in a volume is the difference between the energy of ionizing radiation entering the volume and that emerging from it. The specific energy, z, is the quotient of the energy imparted and the mass m of the volume (for a review see e.g.: ICRU, 1983, or Kellerer, 1985):

$$z = \varepsilon / m \tag{1}$$

Specific energy in a volume may be due to one or more energy deposition events. An event is the energy deposition by an ionizing particle and/or its secondaries, i.e., by energy deposits that are statistically dependent.

The specific energy is a random variable. Under a specified irradiation condition and for a given microdosimetric reference volume one can never predict the value of z. One can, however, state a probability distribution of the values of specific energy. Usually one considers spherical volume elements. The distribution of z depends on the size of the volume, on the type of the radiation and of the absorbed dose, D. The distribution is designated by $f(z;D)$ for multi-events and by $f_1(z)$ for single events. According to the usual definition of a probability density $f(z;D)\cdot dz$ is the probability that the specific energy is between z and z+dz. The specific energy has the same unit, Gy, as the absorbed dose which is merely the statistical mean of the stochastically deposited energy per unit mass, i.e., of the specific energy:

$$D = \bar{z} = \int z \cdot f(z; D)\, dz \tag{2}$$

The absorbed dose loses its meaning when applied to microscopic regions where the statistical fluctuations of imparted energy can be large.

The mean number of events at an absorbed dose, D, is called the mean event frequency, n. It is the ratio of the mean of the multi-event distribution, $\bar{z}$, and the mean of the single-event distributuion, $\bar{z}_1$:

$$n = \bar{z} / \bar{z}_1 \qquad \Rightarrow \qquad \bar{z} = D = n \cdot \bar{z}_1 \tag{3}$$

The lineal energy, y, is the quotient of the energy imparted, ε, to the volume of interest and the mean chord length, $\bar{l}$, in that volume:

$$y = \varepsilon / \bar{l} \tag{4}$$

The mean chord length $\bar{l}$ is the average length of the straight-line segments that result when the reference volume is randomly traversed by straight particle tracks from a uniform, isotropic field. For a convex volume the mean chord length is equal to 4 times its volume divided by its surface (for spherical volumes with diameter d: $\bar{l}=2/3 \cdot d$). The distributions of lineal energy, f(y), are analogeous to the distributions of specific energy, but they relate only to energy increments produced by single events.

The quantity lineal energy has been conceived as the microdosimetric analogue to the Linear Energy Transfer (LET) which is defined as L=dE/dx, where dE is the mean energy lost by a charged particle in electronic collisions along an element dx of its trajectory. Both quantities lineal energy and LET are conventionally expressed in the same unit, keV/μm. Measurements that aim at the determination of LET distributions determine, in actuality, the distribution of lineal energy. Due to energy-loss straggling, i.e., the clustered deposition of energy in the volume by δ-rays or groups of δ-rays, and other stochastic factors these distributions can differ substantially from the distributions of LET. For electrons, for example, the application of the LET concept never permits a direct prediction of energy deposition on a microscopic scale.

For heavy ions, on the other side, there are site sizes and particle energies for which the LET concept predicts adequately the energy deposition. But even in this case the LET is of limited value since it permits no statement on the energy distribution within the sites, although this distribution can differ substantially for particles that have the same LET but different velocities.

THE MEANING OF MICRODOSIMETRIC QUANTITIES

Microdosimetric quantities, such as specific energy, are required whenever one considers regions small enough that the relative fluctuations of energy deposition are substantial. Fig.1 gives two scatter diagrams that were obtained by the simulation of energy depositions in small volumes. The diagrams indicate the distribution of specific energies, z, that result at various levels of absorbed dose, D; they refer to Co-γ-radiation (left panel) and 14 MeV neutrons (right panel) for spherical microscopic regions of 8 μm diameter. One recognizes that there are very substantial deviations of the specific energy from its mean value, the absorbed dose D=$\bar{z}$ (diagonales of the diagram, dotted lines). At a given absorbed dose the specific energy may be zero or it may exceed the expectation value by orders of magnitude. With respect to biological targets, this means that the knowledge of absorbed dose may permit no statement on energy actually imparted to individual cells or subcellular structures. The stochastic nature of specific energy becomes more prominent with decreasing site size, with decreasing absorbed dose, and with increasing ionization density of the radiation.

In the right part of each panel of Fig.1, indicated as the "High Dose" region, nearly each imparted energy is due to multiple energy-deposition events and correlates to a dose-dependent distribution f(z;D); the mean event frequency, n, is much larger than 1. In the case of an irradiated ensemble of volumes this is equivalent to the statement that in almost all volumes out of the ensemble an energy deposition takes place; the fraction of volumes affected is about 100% (dashed lines).

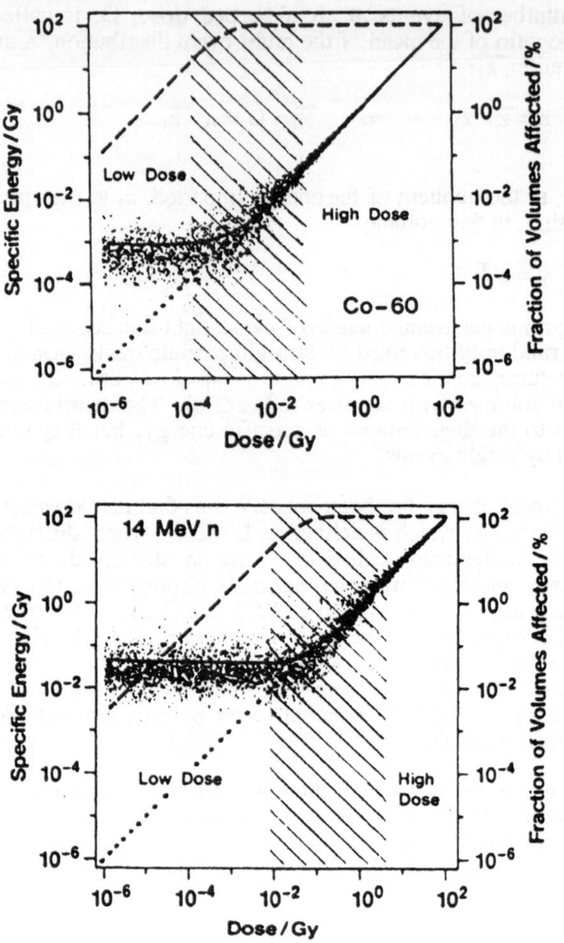

Figure 1. Scatter Diagrams of Simulated Distributions of Specific Energy at Various Levels of Absorbed Dose. The Simulations Refer to Co-γ-radiation (Upper Panel) and 14 MeV Neutron Radiation (Lower Panel) for Spherical Regions of 8 μm Diameter . The Fraction of Volumes Affected is Given by the Dashed Lines. The Mean of the Specific Energy is Represented by the Diagonales (Dotted Lines). The Mean of the Specific Energy due to All Non-zero Events is Given by the Solid Lines. (After Booz and Feinendegen, 1988)

In the left part of each panel, indicated as the "Low Dose" region, the energy imparted is either due to only single events or there are no energy depositions at all. The mean event frequency is much smaller than 1 and the fraction of volumes affected is lower than 100%. The energy imparted to an affected volume is exclusively determined by the single-event distribution $f_1(z)$ and it is independent on absorbed dose. The single-event distribution, in turn, is closely linked to the temporal and spatial correlation of energy depositions and determines the "quality" of a given radiation field. Thus, the induced hazard to an affected cell depends on the characteristic radiation quality, whereas the number of affected cells depends on dose. This fact is very fundamental for the understanding of the effects of low-dose radiation.

In most cases, one deals typically with low doses and low dose rates. Therefore, it is unnecessary to deal with the functions f(z;D), especially if densily ionizing radiations as HZE particles are considered. Instead it is sufficient to consider the distribution of increments z which are produced in single events of energy deposition. The functions $f_1(z)$ contain the essential information, while f(z;D) need hardly be invoked. It is unlikely that the nucleus of a cell is traversed - if at all - by more than one HZE particle.

Distributions of specific energy, as represented in Fig.1, can be useful tools in the biophysical analysis of cellular radiation effects. Specifically, they provide meaningful information on the event frequencies at the stated absorbed dose in specified regions, for example in the nucleus of a cell. They also indicate the possible energy concentrations that can occur in subcellular sites even at very small doses. In a further step microdosimetric considerations may be expanded to concepts that are linked more closely to the structure of charged particle tracks and to the non-homogeneous spatial distributions of energy which they produce in the cell.

CHARACTERISTICS OF TRACK STRUCTURE

Fig. 2 exemplifies the spatial patterns of energy deposition by a diagram that superimposes electron and proton tracks with a schematic representation of a chromatin fiber. The dots represent ionizations in the charged particle tracks.

Sparsely ionizing radiation, for instance fast electrons, produce ionizations with fairly wide spacing (Fig.2). However, they produce also a large number of δ-rays (energy clusters) that contribute to a considerable extent to the imparted energy.

Densely ionizing radiations, for example high-energy heavy ions (HZE particles), deliver energy mainly in the very close proximity of the particle track. The pattern of energy loss in matter consists of a central region of very high (and relatively homogeneous) ionization density and a much larger laterally extended region of lower ionization density with a considerable degree of stochastic fluctuation. Due to energetic δ-rays with energies up to several hundred keV a considerable amount of energy may be deposited in distances from the primary track that are as large as many hundred μm. In effect, one could say that each particle produces not only one track, but additionally many electron tracks with electron-track ends of high ionization density in far distances from the main track.

The resulting high degree of spatial inhomogenity of the energy deposition is illustrated in Fig.3 which shows radial dose distributions for 600-MeV/amu Fe-ions. The dotted curve and the dashed curve are from two track structure models based on the LET

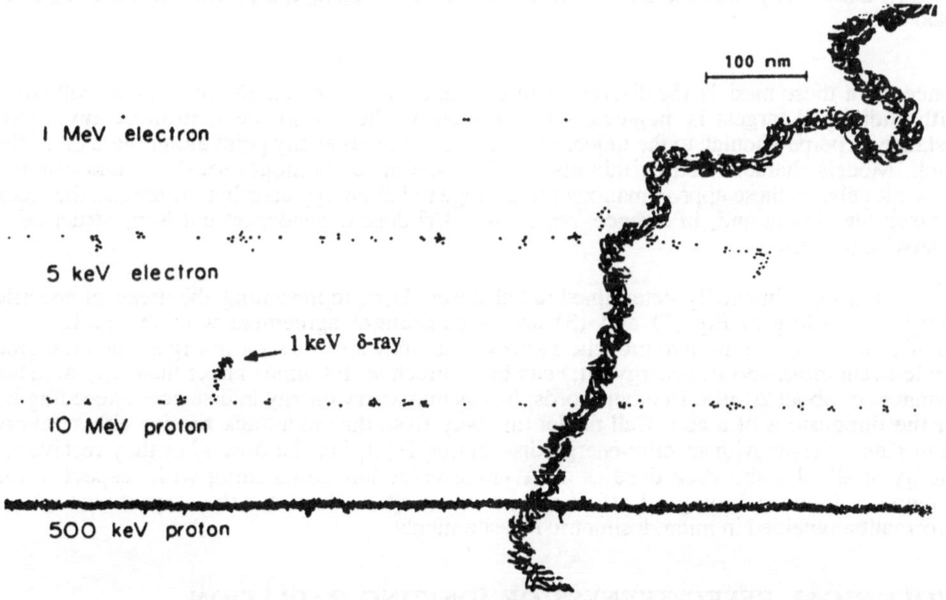

Figure 2. Schematic Representation of Electron and Proton Tracks of the Specified Energies Superimposed on a Schematic Chromatin Fiber that Illustrates Molecular Dimensions. (From Breckow and Kellerer, 1991)

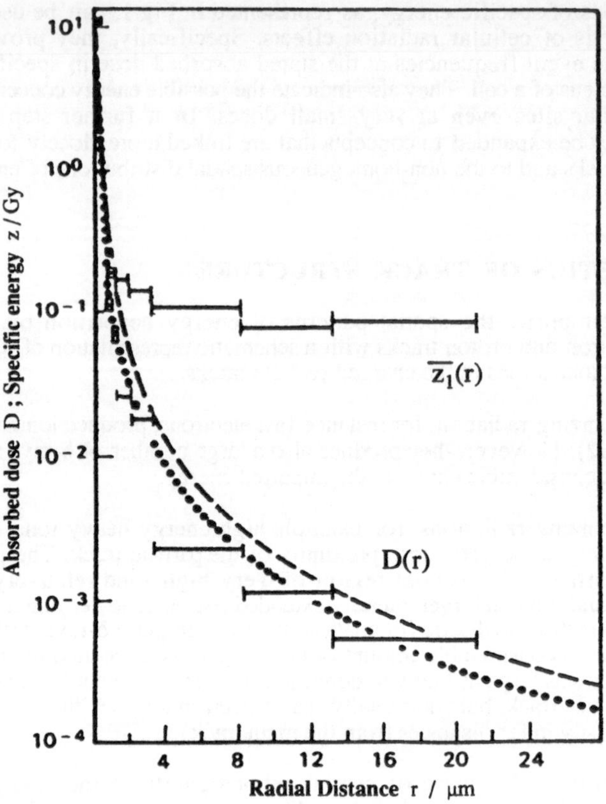

Figure 3. Radial Dose Distributions in a 600 MeV/amu Fe-ion Beam. Microdosimetric Measurements in a 1.3 μm-diameter Site are Compared with Dose Calculations Based on Two Homogeneous Track Structure Models (Dashed Curve: Chatterjee and Schäfer, 1976; Dotted Curve: Zhang et al., 1985). (After Metting et al., 1988)

concept. In these models the discrete nature of energy loss by an ion in random collisions with individual targets is neglected. Consequently, the energy deposition at any radial distance, r, perpendicular to the trajectory is also continuous at any point along the axis of the track. Models characterized by this discription are termed "homogeneous" or "amorphous" track models. In these approximations the average radial energy distribution around the track is taken into accout and, in essence, results in a $1/r^2$ dose-dependence, but δ-ray structure is otherwise neglected.

The experimentally determined radial doses, D(r), representing the mean of specific energy according to Eqs.(2) and (3) are in substantial agreement with the radial dose calculations based on the two models. However, as may also be seen in Fig.3, the measured single-event mean specific energy $\bar{z}_1(r)$ can be as much as 100 times larger than D(r) at radial distances of about 20 μm. In other words, δ-rays may carry energy in distances exceeding by far the dimensions of a cell. Cell nuclei far away from the main track receive either energy from single events with specific-energy distribution $f_1(z)$, i.e. the δ-rays, or they receive no energy at all; the absorbed dose as an average value has no meaning with respect to the potentially induced biological effect in the cells. This illustrates the importance of the information obtained in microdosimetric measurements.

BIOLOGICAL EFFECTIVENESS OF IONIZING RADIATION

Let e(z) be the "microscopic" yield of radiation products within a specified sensitive biological site. This yield determines the subsequent biological effect and is a function of the specific energy in the site. The "macroscopic" effect probability, E(D), is then the integral over the effect function, e(z), weighted by the distribution of z at a given absorbed dose, D (Kellerer and Rossi, 1972, 1978):

$$E(D) = \int e(z) \cdot f(z;D)\, dz \tag{5}$$

The effect function, e(z), reflects the basic molecular interactions of radiation products which lead to the relevant biological lesions, such as DNA alterations, mutations, etc. Radiation induced mutations, for example, are known to be to a considerable extent DNA deletions which result from two double-strand breaks (Ward, 1988). There are a variety of mechanisms where sublesions are formed by two energy transfers and where they interact to form a lesion. The sublesions could be molecular alterations in DNA, for example, single-strand breaks in close proximity; the "interaction" would then be misrepair due to the interference of excision repair of the two single-strand breaks. The sublesions could also be alterations on a more complex level, for example, pairs of double-strand breaks in chromosomal structures that may form chromosome aberrations by misrepair.

It was one of the fundamental ideas of the so-called "Dual Radiation Action" originally introduced by Kellerer and Rossi (1972) to take account of these pairwise interactions of radiation products forming biological lesions. If the yield of a particular cellular damage is due to such an "Interaction of pairs of energy transfers" it is a process of second order and, therefore, is proportional to the square of the specific energy within the sensitive site:[1]

$$e(z) = z^2 \tag{6}$$

Thus, the average yield, E(D), will be proportional to the mean square, $\overline{z^2}$, of the specific energy, i.e., to the second momentum of the distribution f(z;D):

$$E(D) = \int z^2 \cdot f(z;D)\, dz = \overline{z^2} \tag{7}$$

It was one of the central statements in the theorie of "Dual Radiation Action" that the mean square of a distribution f(z;D) can be expressed through a function of absorbed dose and the mean values of the corresponding single-event distribution $f_1(z)$:

$$\begin{aligned} E(D) = \overline{z^2} &= \overline{z_1^2} / \overline{z_1} \cdot D + D^2 \\ &= \overline{z}_D \cdot D + D^2 \end{aligned} \tag{8}$$

The ratio of the mean square, $\overline{z_1^2}$, and the mean, $\overline{z}_1$, of the single-event distribution is called the dose-mean specific energy, $\overline{z}_D$. It represents one of the fundamental microdosimetric quantities and gives, like the second momentum of any distribution, the variability of the quantity under consideration. In a single value, it incorporates the information about the temporal and spatial correlation in an energy-deposition process. Therefore, it is characteristic and specific for any given radiation. Since, as stated above, the energy-deposition pattern characterized by $\overline{z}_D$ determines the biological effectiveness, the dose-mean specific energy may serve as a measure for the quality of a given radiation.

$\overline{z}_D$ can either be measured or it can be calculated by means of Monte-Carlo simulations of charged particle tracks which have become a common method for the determination of [1]microdosimetric parameters (e.g.: Paretzke, 1983; Wilson et al., 1988). More subtle microdosimetric considerations lead to functions that describe the distribution of distances between the energy deposits. By these so-called proximity functions the spatial proximity of energy transfers in a particle track is represented. Once a single proximity function for a given radiation field has been determined, $\overline{z}_D$- or corresponding quantities - for biological sites of

[1]Other relationships for e(z) are conceivable, but may lead to inconsistent statements. For example, if e(z) is directly proportional to z the biological effect probability E(D) would be proportional to absorbed dose but independent from radiation quality; the RBE of all radiations would be equal to one.

any shape and any size can readily be computed (Kellerer and Chmelevsky, 1975; Kellerer, 1985; Breckow and Kellerer, 1991).

Eq.(8) is the original formula for the well-known linear-quadratic dose-effect relationship. For low doses and particularly for densely ionizing radiation it is sufficient to consider only the linear term:

$$E(D) \sim \overline{y}_D \cdot D \tag{9}$$

The effect probability, E(D), is here given in terms of lineal energy $\overline{y_D} = m/\overline{l} \cdot \overline{z_D}$ which is closely linked to the (weighted) mean of LET utilized in radiation protection.

MICRODOSIMETRY IN RADIATION PROTECTION

Many important application of microdosimetric techniques refer to the exploration of unknown or inadequately known radiation fields or to the monitoring for changes in radiation quality in time-varying radiation fields (Kellerer and Rossi, 1984; Lindborg et al., 1985; Breckow et al., 1988). Beyond its practical use, however, microdosimetry has by now found perhaps its most important application in radiation protection - somewhat against original expectations (Zaider and Brenner, 1985; Zaider and Rossi, 1989).

In radiation protection one discriminates between stochastic effects, such as hereditary damage and radiation cancerogenesis, and deterministic effects, such as skin damage or most prenatal malformations. It is an objective of radiation protection to avoid deterministic effects entirely. One assumes that this objective can be reached because they depend on damage to a multiplicity of cells in a tissue with a threshold, beyond the effect is totally absent.

Stochastic effects, on the other hand, can not be avoided entirely. It is assumed that they result from damage to individual cells and that they exhibit linear dose dependences at small doses. Therefore, the aim of radiation protection can merely be to minimize radiation exposures and to reduce thereby the risk of stochastic effects. This assumption of linear dependences for stochastic effects of low doses is crucial and specific to the risk considerations for ionizing radiations (see: ICRP, 1991).

The probability of radiation hazards depends, as stated above, on the density of energy imparted and its degree of clustering, i.e. on the radiation quality. To take account for the different effectiveness of different types of ionizing radiation the quantity dose equivalent, H, is utilized in radiation protection. It is equal to absorbed dose, D, times a qualitay factor, Q, which is defined in terms of LET, L:

$$H = Q(L) \cdot D \tag{10}$$

The value of L depends on the energy of the particles representing the radiation field. In the usual case of a distribution of energy in a mixed radiation field one deals with a distribution of L, f(L), in the exposed material. In order to characterize the quality of a radiation by a single quality factor one usually uses the weighted mean of the LET distrubtion:

$$\overline{Q} = 1/\overline{L} \cdot \int Q(L) \cdot L \cdot f(L)\, dL \tag{11}$$

Thus, in order to assess quality factors for unknown radiation fields one needs to determine the LET distributions. For mixed and/or time-varying radiation fields measurements can hardly, or only with restrictions, be performed. Calculations in terms of LET, on the other hand, tend to result in extreme simplifications, but nevertheless they are very widely utilized.

The LET concept can, at best, provide a crude characterization of the charged particle

tracks that occur in the exposed medium. Several features that are essential in describing charged particle tracks are disregarded in the LET concept. In particular, these are:
- The finite range of the particles and the change of LET along the track
- The lateral extension of the particle tracks due to the finite range of δ-rays
- The statistical fluctuation of energy loss along the particle track, termed the energy-loss straggling.

Therefore, it is not suprising that the present definition of the quality factor gives only a rough approximation of what a "real" radiation quality could be termed. In particular, this is true for the relative biological effectiveness of neutrons that is condiderably underestimated by the present definition of the quality factor. Lineal energy has several advantages over LET:
- It can be directly measured (in contrast to LET which usually is calculated)
- It is a quantity related to energy actually deposited in a site (LET relates to energy lost by the passing particle) and therefore would be expected to be better correlated with any biological effect
- It is a stochastic quantity and thus contains no inherent averaging.

On account of these reasons one could wish to define a quality factor in terms of the stochastic analogue of the LET, i.e. the lineal energy, y (Zaider and Brenner, 1985; ICRU, 1986; Kellerer and Hahn, 1988):

$$H = Q(y) \cdot D \qquad (12)$$

In a mixed radiation field a mean quality factor can be obtained analogous to the determination of the mean LET value according to Eq.(11):

$$\overline{Q} = 1/\overline{y} \cdot \int Q(y) \cdot y \cdot f(y)\, dy \qquad (13)$$

Since f(y) can be directly measured, $\overline{Q}$ is readily be obtained if one has an appropriate analytic function for Q(y). At this stage, Eqs.(12) and (13) are purely formal equivalences to Eqs.(10) and (11). Let, however, the quality factor Q(y) be simply proportional to the lineal energy y, i.e. Q(y)=y, it then follows:

$$\overline{Q} = 1/\overline{y} \cdot \int y^2 \cdot f(y)\, dy = \overline{y^2}/\overline{y} = \overline{y}_D$$
$$\Rightarrow \quad H = \overline{y}_D \cdot D \qquad (14)$$

From the comparison of Eq.(14) with Eq.(9) it is readily seen that a definition of dose equivalent in terms of lineal energy would lead to a closed and consistent figure of the risk for stochastic radiation effects and the underlying basic microscopic interactions. This figure includes in an analytical form considerations of stochastic effects that depend on dose, on the one hand, and on dose-independent radiation quality, on the other hand.

The linear relationship of Q(y) covers not the entire range of y. Models for Q(y) propose a linear increase up to about 100 keV/μm and decreasing values for higher y due to saturation effects (ICRU, 1986). This is, partly, also taken into account by the LET concept. However, if one thinks merely in terms of LET, it remains unclear why the quality factor according to its present definition, should be constant below 3.5 keV/μm before it increases with LET. In microdosimetric terms, however, proportionality is also valid in the low lineal-energy region. It is readily understood that for sparsely ionizing radiations a considerable contribution to the energy imparted is due to δ-rays with the consequence of much more dense energy depositions than in the "primary" particle track that alone is included by the LET. Therefore, the difference between y and LET increases with decreasing LET. There have been several proposals how to link, for practical purposes, the LET concept to microdosimetric quantities. An approximation that applies for the majority of practical applications is given by the relation (Kellerer and Hahn, 1988):

$$y = 9/8 \cdot L + \Delta \qquad (15)$$

The latter term represents the contribution due to δ-rays and is in the order of less than 1 keV/μm. For radiation protection purposes, one can use the equality Q(y)=Q(L) for y(L) according to Eq.(15). The numerical differences caused by the use of y spectra instead of LET spectra are of little concern in radiation protection applications.

For small values of L the δ-ray related additive term, Δ, dominates the contribution to y. For heavy ions Δ may considerably exceed 1 keV/μm (Metting et al., 1988). However, it is still small against L, so that for heavy ions a proportionality of y and L may be assumed.

In 1986 a liaison group of the International Commission on Radiological Protection (ICRP) and the International Commission on Radiation Units and Measurements (ICRU) published recommendations for conceptional changes in view of a possible revision of the quality factors for ionizing radiations (ICRU, 1986). The proposal for a revision addressed two separate issues. One issue concerned the numerical values of the quality factor, in particular for neutrons and heavy ions, the other concerned the formal definition of the quality factor in terms of lineal energy instead of LET.

Recently, the ICRP has published a fundamental report with the Commission's recommendations for changes and modifications of a variety of quantities and concepts (ICRP, 1991). The most relevant changes refer to revised assessments of stochastic radiation risks, to the conceptional framework of radiation protection and to the system of dose limits. Apart from the new risk estimates and the dose-limit recommendations the ICRP report introduces a new definition on the quality factor as a result of the lasting and sometimes controversal debates initiated by the report of the ICRP-ICRU joint task group.

In the new ICRP recommendations the definition of the quality factor is retained in terms of LET, whereas the analytic form of Q(L) has now changed. The following specified Q-L relationship is adopted (see Fig.4):

LET / keV·μm^{-1}	Q (L)
< 10	1
10-100	$0.32 \cdot L - 2.2$
>100	$300 / \sqrt{L}$

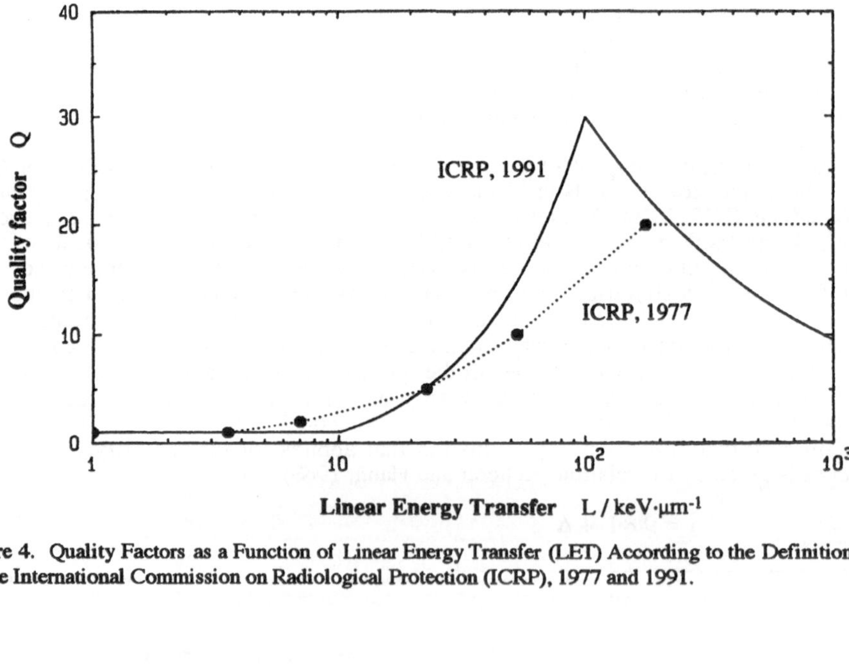

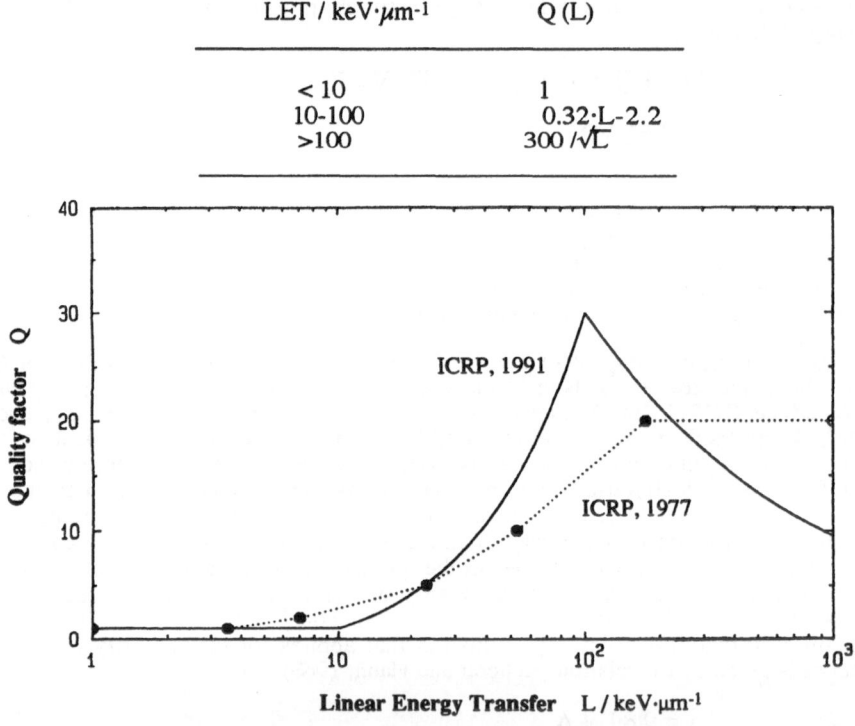

Figure 4. Quality Factors as a Function of Linear Energy Transfer (LET) According to the Definitions Given by the International Commission on Radiological Protection (ICRP), 1977 and 1991.

With a decreasing quality factor for LET>100 keV/μm the ICRP takes account for the reduced effectiveness of heavy ions. Thus, for LET values exceeding 225 keV/μm the quality factor now is smaller than according to the former definition where it was 20 for LET>175 keV/μm.

Parallel to the analytic Q-L relationship the ICRP selects radiation weighting factors, w_R, based on a review of the biological information and a variety of exposure conditions. This procedure leads to a new quantity in radiation protection called the equivalent dose, $H_{T,R}$.

Although in the new ICRP recommendations the quality factor has been given in terms of LET, the actual form of the Q(L) relationship is guided by the corresponding microdosimetric functions. It has than, for practical reasons, be transformed to a function of LET. However, it is emphasized by the ICRP that nevertheless both systems may be applied (ICRP, 1991, p. 81): "Since the mean lineal energy represents discrete energy deposition, it is in principle more meaningful than linear energy transfer (LET) as the physical quantity to be used in the specification of radiation quality. Although this characteristic of lineal energy is directly measurable, L has been used in most of the existing practical radiation protection calculations. Therefore, Q will be given here as a function of L although the Commission recognises that the use of lineal energy is also possible."

Measurements of microdosimetric quantities in spaceflights

During the past decades in the US and Soviet manned space missions a large number of dosimetric measurements have been made. The complexity of radiation in space and, in addition, the various shielding conditions in the spacecrafts raises a number of problems mainly with respect to the accurate determination of energy spectra and LET spectra. As stated above, the directly measurable quantity is the lineal energy, y, instead of LET. Microdosimetric measurements that aim at the determination of dose (specific energy) and ionization density (lineal energy) yield a direct information of what amount and what density of energy is actually deposited in a specified site of small volume.

Microdosimetric measurements have been made on some space missions (for a review see e.g.: Benton and Parnell, 1988). During the French-Soviet space mission "Aragatz" on the Soviet orbital station "Mir" in 1988/89, for example, a low-pressure tissue-equivalent proportional counter with simulated site size of about 3.5 μm has been utilized for such measurements. Specific energies as well as lineal energies in the range from some keV/μm up to greater than 1.000 keV/μm have been recorded. (Nguyen et al., 1990). From these values absorbed dose rates, $\dot{D}$, quality factors, Q, and dose equivalent rates, $\dot{H}$ have been calculated. For this purpose, lineal energies had to be transformed to LET in order to obtain quality factors according to the present definition (however, not yet on the basis of the very recent ICRP recommendations). Some of the results are shown in Fig.5. Quality factors due to LET greater than 3.5 keV/μm are considered separately and are indicated as QHL. The mean quality factor, Q, for the entire LET range is 1.8 and QHL is 7.4.

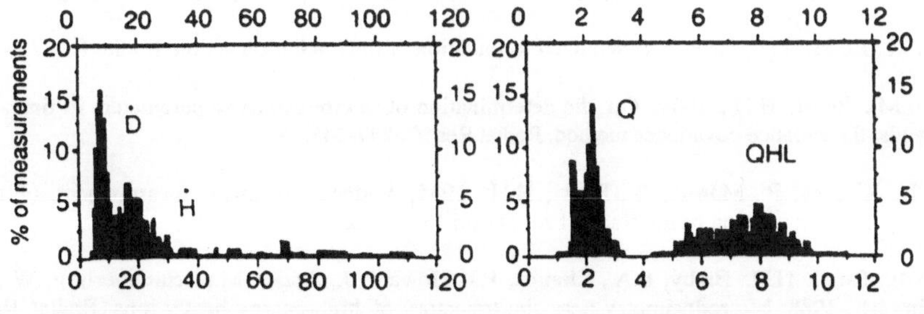

Figure 5. Distributions of Dose Rates and Quality Factors Measured on Board Soviet Orbital Station "Mir". $\dot{D}$: Absorbed Dose Rate (μGy/h); $\dot{H}$: Dose Equivalent Rate (μS_v/h); Q: Total Quality Factor; QHL: Quality Factor Due to High LET Radiation (>3.5 keV/μm). (After Nguyen et al., 1990)

The narrowness of the distributions of the absorbed dose rate and the quality factor compared with the broadness of the $\hat{H}$ and QHL distributions indicate that a considerable fraction of the absorbed dose is due to low LET radiations. Its percentage is almost 90%. On the other hand, due to the higher quality factor of high LET radiation it contributes more than half to the total dose equivalent. Thus, although radiation in a spacecraft is mainly due to low LET radiations the probability of a radiation induced biological hazard in man is assumed to be due in equal proportions by low and high LET radiations.

REFERENCES

Benton, E.V., Parnell, T.A., 1988, Space radiation dosimetry on US and Soviet manned missions. NATO ASI Series A: Life Sciences, 154: 729-794.

Booz, J., Feinendegen, L.E., 1988, A microdosimetric understanding of low-dose radiation effects. Int J Radiat Biol 53: 13-21.

Breckow, J., Kellerer, A.M., 1991, Concepts of microdosimetry and their applicability to DNA studies. In: Fielden, E.M., O'Neill, P. (eds.) The Early Effects of Radiation on DNA. NATO ASI Series H 54: 163-178.

Breckow, J., Wenning, A., Roos, H., Kellerer, A.M., 1988, The variance-covariance method: microdosimetry in time-varying low dose-rate radiation fields. Radiat Environ Biophys 27: 247-259.

Chatterjee, A., Schäfer, H.J., 1976, Microdosimetric structure of heavy ion tracks in tissue. Radiat Environ Biophys 13: 215-227.

ICRP, 1977, Recommendations of the International Commission on Radiological Protection. Report 26, Pergamon Press, Oxford.

ICRP, 1991, Recommendations of the International Commission on Radiological Protection. Report 60, Pergamon Press, Oxford.

ICRU, 1983, Microdosimetry. International Commission on Radiation Units and Measurements, Report 36, Bethesda, MD.

ICRU, 1986, The Quality Factor in Radiation Protection. International Commission on Radiation Units and Measurements, Report 40, Bethesda, MD.

Kellerer, A.M., 1985, Fundamentals of microdosimetry. In: Kase, K.R., et al. (eds) The Dosimetry of Ionizing Radiation. Academic Press, Orlando FL: 78-163.

Kellerer, A.M., Chmelevsky, D., 1975, Concepts of microdosimetry. III. Mean values of the microdosimetric distributions. Radiat Environ Biophys 12: 321-335.

Kellerer, A.M., Hahn, K., 1988, Considerations on a revision of the quality factor. Radiat Res 114: 480-488.

Kellerer, A.M, Rossi, H.H., 1972, The theory of dual radiation action. Curr Top Radiat Res Q 8: 85-158.

Kellerer, A.M., Rossi, H.H., 1978, A generalized formulation of dual radiation action. Radiat Res 75:471-488.

Kellerer, A.M., Rossi, H.H., 1984, On the determination of microdosimetric parameters in time-varying radiation fields: the variance-covariance method. Radiat Res 97: 237-245.

Lindborg, L., Kliauga, P., Marino, S., Rossi, H.H., 1985, Variance-covariance measurements of the dose mean lineal energy in a neutron beam. Radiat Prot Dosim 13: 347-351.

Metting, N.F., Rossi, H.H., Braby, L.A., Kliauga, P.J., Howard, J., Zaider, M., Schimmerling, W., Wong, M., Rapkin, M., 1988, Microdosimetry near the trajectory of high-energy heavy ions. Radiat Res 116: 183-195.

Nguyen, V.D., Bouisset, P., Akatov, Y.A., Petrov, V.M., Kozlova, S.B., Siegrist, M., Zwilling, J.F., 1990,

Measurements of quality factors and dose equivalents with CIRCE inside the Soviet space station MIR. Radiat Prot Dosim 31: 377-382.

Paretzke, H.G., 1983, Concepts of charged particle track structures. 8th Proc Symp Microdos, Jülich: 67-77.

Ward, J.F., 1988, DNA damage produced by ionizing radiation in mammalian cells: identities, mechanisms, and repairability. Prog_Nucleic Acid Res Mol Biol 35: 95-125.

Wilson, W.E., Metting, N.F., Paretzke, H.G., 1988, Microdosimetric aspets of 0.3- to 20-MeV proton tracks. Radiat Res 115: 389-402.

Zaider, M., Brenner, D.J., 1985, On the microdosimetric definition of quality factors. Radiat Res 103: 302-316.

Zaider, M., Rossi, H.H., 1989, Estimation of the quality factor on the basis of multi-event microdosimetric distributions. Health Physics 56/6: 885-892.

Zhang, C., Dunn, D.E., Katz, R., 1985, Radial distribution of dose and cross-sections for the inactivation of dry enzymes and viruses. Radiat Prot Dosim 13: 215-218.

THEORETICAL ANALYSIS OF HEAVY ION ACTION ON CELLS: MODEL-FREE APPROACHES, CONSEQUENCES FOR RADIATION PROTECTIONS

Jürgen Kiefer
Strahlenzentrum der Justus-Liebig-Universität
Giessen, Germany

ABSTRACT

The complexity of the space radiation field precludes a comprehensive coverage of possible biological effects taking into account all particle properties and biological endpoints. The estimation of radiation risks, however, demands the establishment of "risk coefficients" for all non-negligible contributions. In order to cope with this task theoretical analyses leading eventually to the discovery of general underlying principles may be helpful and presumably the only workable approach. The paper describes in a rather elementary manner the particular aspects how heavy particle interactions lead to biological effects at the cellular level with emphasis on inactivation and mutation induction. It is shown that the killing ability of single particles increase with charge and mass but never reaches unity. This means that there is always a chance for mutants or transformants to survive to give rise to late effects at the genetic and somatic level. The mutagenic - and presumably also the carcinogenic - potential also increases with charge and mass. A realistic hazard evaluation has to take into account not only the specific properties of a particular particle but also its range in the human body as well as secondary products formed on its way by degradation and nuclear reactions.

INTRODUCTION

The radiation field in space consists mainly of high energy particles and is quite different from the terrestrial background. In order to assess the possible hazard to astronauts during long-term missions into deeper space these special properties have to be taken into account. It will never be possible to establish a database for risk estimation as it already exists for radiation protection on earth for the space situation. One has, therefore, recur to a large extent to theoretical models in order to translate the existing knowledge which is essentially based on human data obtained from gamma-exposure to the application in space. A knowledge of the space field is, of course, indispensable for this exercise but by far not sufficient. The biological properties of the structured components are very different from those of electromagnetic radiations and considerable gaps still exist in our knowledge. Theoretical approaches, carefully checked by properly designed experiments, form an important - and in fact necessary - part to achieve the goal of a reliable risk estimation. It is an interesting sideline that similar problems exist in the field of microelectronics. The size of circuit elements approach that of cellular structures so that it may be hoped that theoretical methods originally developed in radiation biology might eventually find their way also into the domain of technical applications. This means that understanding of radiation action on cells does not only receive its justification from problems of manned space flight but may eventually turn out to be equally important for the control of unmanned satellites.

Biological Effects and Physics of Solar and Galactic Cosmic Radiation, Part A, Edited by C.E. Swenberg *et al.*, Plenum Press, New York, 1993

This paper will start with a few considerations on the interactions of heavy ions with matter demonstrating that "traditional" quantities like "dose" or "linear energy transfer" (LET) are only of limited value. The problem of "late damage" which - at the cellular level - is connected with mutations and transformations is then considered. Finally a few critical remarks about the current schemes of radiation protection guidelines and their applicability to the space situations will be addressed.

THE PHYSICAL CASE: SPECIAL PROPERTIES OF HEAVY IONS

Heavy charged particles deposit energy upon interaction with matter in a localized and structural manner there by creating regions of high ionization density. Classically this is quantitatively described by the concept of "Linear Energy Transfer" (LET) defined as the "energy locally imparted to the medium per unit travelling distance". Its relationship to "dose" D is given by

$$D = L \cdot \Phi / \rho \qquad (1)$$

where L stands for LET, Φ for ion fluence and ρ for the density of the medium.

The special mentioning of the word locally implies that the ion's path may not be viewed as a very narrow "pencil beam" but rather like a broad alley. This is due to the fact that by the initial interaction with the atomic shell electrons of sometimes considerable kinetic energy (usually termed "penumbra electrons") are liberated, These delocalized electrons are able to travel large distances. Another aspect which needs to be taken into account are the ranges of the ions in the human body.

For the sake of illustration three ions will be taken as representatives, namely He, C and Fe all of which are important components of galactic cosmic radiation (Simpson 1983). Figure 1 depicts the relationship between ion energy - expressed in MeV per atomic mass unit u - and linear energy transfer LET for water as absorbing medium. One sees that very high values can be reached but they are found with low energies and decline rapidly with ion speed. For risk assessment the relationship between LET and range which is given in Fig. 2 is presumably more meaningful. But even here it is clear that for ranges larger than a few centimetres the energy deposition is still very substantial. A more detailed consideration has the microscopical pattern of interaction take into account which is governed by the radial spread around the ion's trajectory. A measure for this is the "penumbra radius" indicating the largest distance electrons may travel perpendicularly to the ion path's central axis. There are a

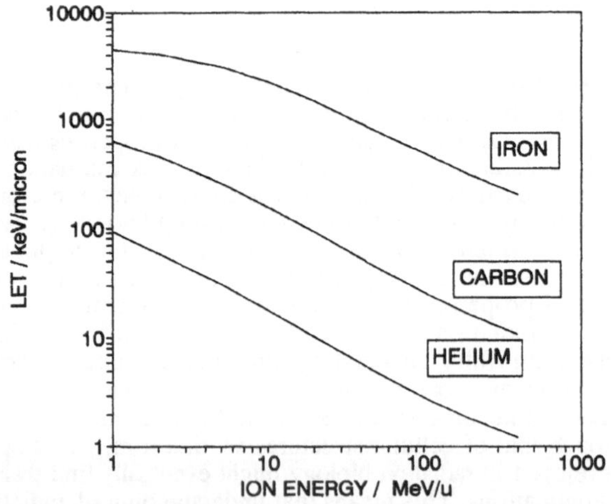

Figure 1. Linear Energy Transfer (LET) Dependence on Ion Energy for Helium, Carbon and Iron-ions. Calculated by Kiefer (1987).

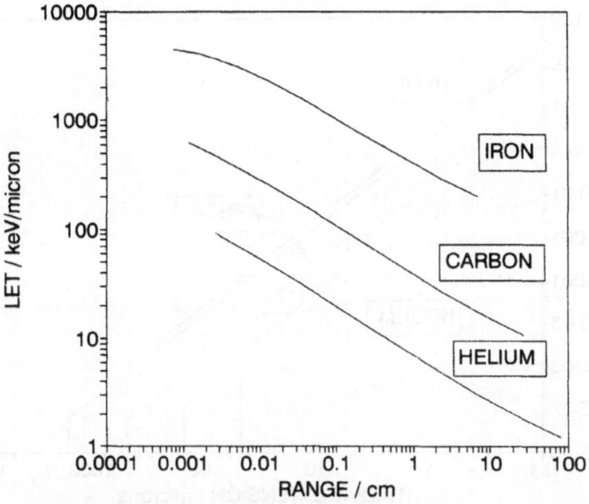

Figure 2. LET versus Ion Range. Calculated According to Kiefer (1987).

number of models in the literature for its calculation (see Kiefer and Straaten, 1986 and Kiefer and Kost 1988 for further discussion and references) which, however, agree in the assumption that r_p depends only the specific ion energy, i.e. ions of the same speed have identical penumbra extensions. (This is not taken to mean that the track structure is quantitatively the same, they differ in the number of electrons, see below). One current model (Kiefer and Straaten 1986) assumes

$$r_p = 0.0616\ E^{1.7} \quad (2)$$

(E denotes the ion specific energy in MeV/u, r_p in μm, see Figure 3), a relationship which seems to be compatible with experimental results for $E > \approx 1$ MeV/u.

The energy deposited within the penumbra cone depends on the effective ion charge Z^* according to the relation (Kiefer and Straaten 1986)

$$\rho_e = 1.25 Z^{*2} / (\ \rho\ \beta^2\ r^2\ ) \quad (3)$$

where ρ_e is the "energy density" within the track (i. e. the "dose" for infinitesimally small

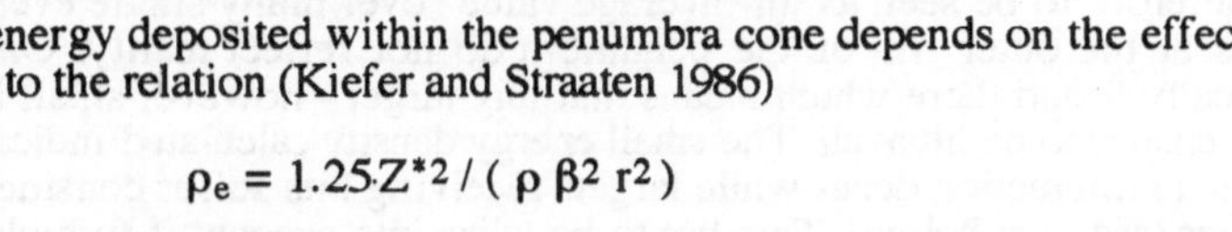

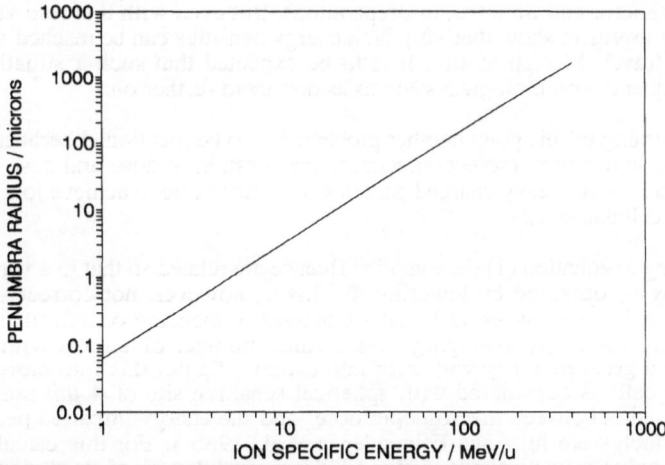

Figure 3. Penumbra Radii as a Function of Ion Energy. Calculated According to Kiefer and Straaten (1986).

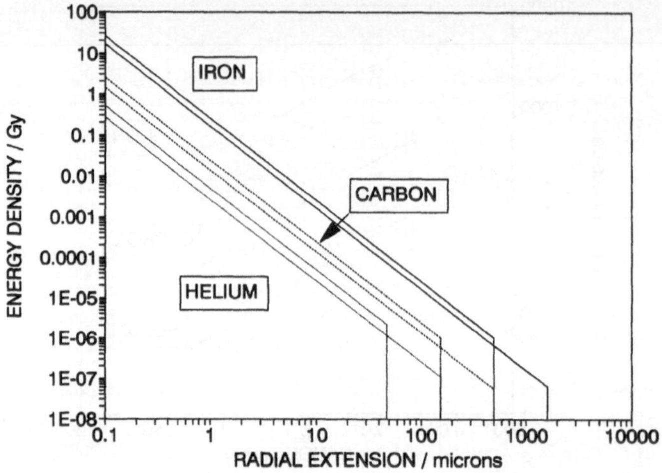

Figure 4. Track Structure of Some Ions: Helium 50 and 100 MeV/u, Carbon 100 and 200 MeV/u, Iron 200 and 400 MeV/u. Calculated According to Kiefer and Straaten (1986).

targets, measured in Gy), ρ the density of the absorbing medium, β the ion velocity relative to that of light in vacuo and r the radial distance from the track centre. Figure 4 depicts equation (3) for a few selected cases.

A few comments are necessary here to avoid oversimplifying conclusions: The model calculations are based on the so-called "amorphous track model" which assumes a continuous transfer of energy within the region of interest and neglects hence entirely the quantum nature of interaction processes. While such an approximation is already highly questionable for macroscopic situations it breaks down completely for the microscopic case.

ρ_e has therefore to be seen as an "average value" over many single events. Also the very low values at the outer rim of the penumbra do not reflect reality. Only very few electrons are usually found there which means that any target - however small it might be - has only a small chance to be hit at all. The small energy density calculated indicates only that at most positions no interaction occur while targets receiving hits suffer considerable higher energy depositions (see also below). This has to be taken into account if formula (3) is used to compute the distribution of energy depositions in targets of specified size. Such an improvement can in principle easily be achieved, work is in progress to develop a more realistic model (Kiefer and Straaten, in preparation). But even with this reservation in mind Figure 4 is quite useful to show that very high energy densities can be reached when charged heavy particles travel through matter. It is to be expected that such a situation may have profound consequences on biological systems as discussed further on.

Before turning to this point another problem has to be mentioned because it also plays an important role in radiation protection, namely the question of dose and dose rate. One can categorically state: With heavy charged particles it is impossible to achieve low doses or low dose rates at the cellular level.

According to equation (1) dose and ion fluence are related so that in a very formal way small doses may be obtained by lowering Φ This is, however, not correct since at small fluences only very few targets are really hit while most of them are completely spared so that small "doses" are found by averaging over a small number of targets with high energy depositions and a great majority with zero interactions. To put this into more quantitative terms a "model cell" is considered with spherical sensitive site of 4 μm radius. Figure 5 illustrates the relation between macroscopic"dose" and the energy imparted per mass unit in those entities which were hit (see Feinendegen et al. 1985). For this calculations which should be taken only for its illustrative value LET was used despite of its obvious limitations. If the energy imparted per mass unit per single traversal through the sensitive site is called d_1

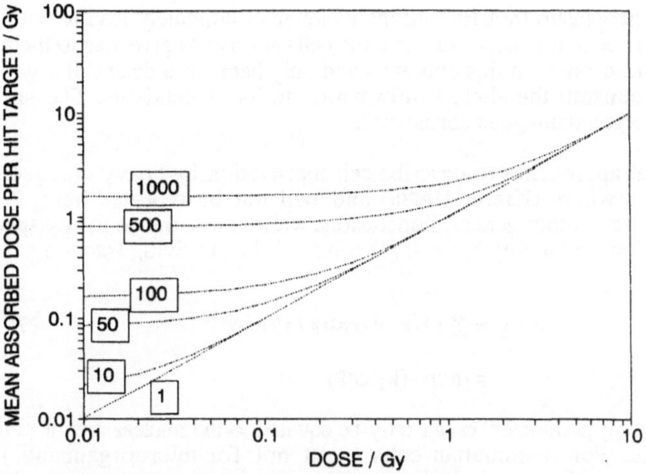

Figure 5. Relation between Macroscopic Dose and Mean Energy Deposition by Particles of Various LET (as Indicated) in a Hit Cells ("Event Size") with a Cross Section of 100 μm².

one has

$$d_1 = L / (\rho\sigma). \qquad (4)$$

The mean number of traversals n is obviously

$$n = D / d_1 = \sigma\Phi. \qquad (5)$$

Since the number of hits follows a Poisson-distribution the fraction of cells with at least one hit $p_{>1}$ is given by

$$p_{>1} = 1 - e^{-n} = 1 - \exp(D\rho\sigma/L). \qquad (6)$$

It is immediately clear from this that for any given target the lower limit of "dose" is given by d_1, quite irrespective how small the fluence might be. This statement is pertinent for the space situation where ion fluences are low, i. e. few cells will be hit but in each case where one occurs rather amount of energy will be transferred.

It should be pointed out at the end of this section that the approach described is in fact a very simple one and rather superficial. It neglects largely the variation of energy deposition at the molecular level which is very important for the understanding of the different effectiveness of radiation types. This the realm of "microdosimetry" and dealt with in greater depth else where in this volume (see the article by Breckow). Some model considerations have been given at the previous symposium (Kiefer 1988a) to which the reader is referred as well as to a collection of papers by different authors addressing especially the question of "Quantitative mathematical models in radiation biology" (Kiefer 1988b). The purpose of the present contribution is to demonstrate in simple terms problems and possible solutions encountered when one is dealing with the biological action of heavy charged particles. Reference should be made at this stage to some "classical papers" where these concepts were first developed and which are the basis for most of the thoughts described here (see Butts and Katz (1967); Kellerer and Rossi (1972, 1978)). Additional references can be found in the symposium proceedings quoted above (Kiefer 1988b).

THE BIOLOGICAL CASE: MUTATION INDUCTION BY HEAVY IONS

The cellular basis for late effects after low doses of ionizing radiations is the induction of mutations and neoplastic transformations which may ultimately lead to malformations and cancer. For this to occur it is necessary that the cells survive to give rise to the proliferation of the originally induced ones. In this context - and only here - is a dead cell a good cell because it is not able to transmit the altered information to its descendants. The situation is quite different if acute organ damage is considered.

Theoretical approaches to describe cell inactivation by heavy charged particles have been reviewed elsewhere (Kiefer 1988a) and will not be repeated here. It is, however, possible to draw a few rather general conclusions without recurrence to any specific model. If k_1 is the inactivation probability by a single traversal the surviving fraction y may be written as

$$y = \Sigma\, (1-k_1)^i\, (\sigma\Phi)^i / i!\; e^{-\sigma\Phi} \qquad (7)$$

$$= \exp -(k_1\, \sigma\Phi)$$

where i is the running parameter. σ can only be equated to the nuclear area if penumbra effects can be neglected. For mammalian cells (but not for microorganisms) this can be approximately assumed for not too heavy particles. The inactivation cross section σ_i is then

$$\sigma_i = \kappa\, \sigma\Phi \qquad (8)$$

This is equivalent to saying that σ_i equals the cross section of the sensitive site if the killing probability per particle traversal is unity - an obvious conclusion! If σ is known k_1 may be calculated from survival curves. Some data obtained for V79 Chinese hamster cells ($\sigma = 110$ μm^2 , Geard 1980) are displayed in Figure 6. One sees that the inactivation probability increases with LET but it does not reach unity even for very heavy ions.

Mutation induction can be treated in an analogous way. Let M be the number of surviving mutants (commonly called the "mutation frequency") and m_1 the probability to induce a mutation by a single traversal. One has then

$$M = \frac{\Sigma\; i\, m_1\, (1-k_1)^i (\sigma\Phi)^i / i!\; e^{-\sigma\Phi}}{\Sigma\, (1-k_1)^i\, (\sigma\Phi)^i / i!\; e^{-\sigma\Phi}} \qquad (9)$$

$$= m_1\, (1-k_1)\sigma\Phi \qquad (10)$$

This is again an obvious conclusion: the mutation frequency vanishes if the cell is killed by a single traversal. The "mutation induction cross section" σ_m is given by

$$\sigma_m = m_1\, (1-\kappa_1)\, \sigma \qquad (11)$$

If one is interested in fundamental processes equation (10) may be used to estimate the mutagenic potential of different particles. To illustrate this m_1 is plotted versus LET in Figure 7. One notes that the effectivity increases with LET and levels off at very high values. This quite different from the behaviour of σ_m which decreases with LET. The reason is, of course, immediately clear: this is due to the increased killing efficiency so that the probability for surviving mutants becomes progressively smaller. For radiation protection purposes it is, however, significant to state that it is different from zero - even with uranium ions.

CONSEQUENCES FOR RADIATION PROTECTION

It has been shown above that the mutagenic potential of HZE-particles is very high indeed, and the same holds presumably true for their carcinogenic power. At the same time the killing efficiency is smaller than unity which means that there is always a finite probability for a mutant or transformed cell to survive and thus give rise to tumours. This constitutes a severe risk for astronauts and has to be taken into account when formulating radiation guidelines. The present regulation based on dose, equivalent dose and radiation factors has only limited applicability. The particle fluences involved are quite small and calculating

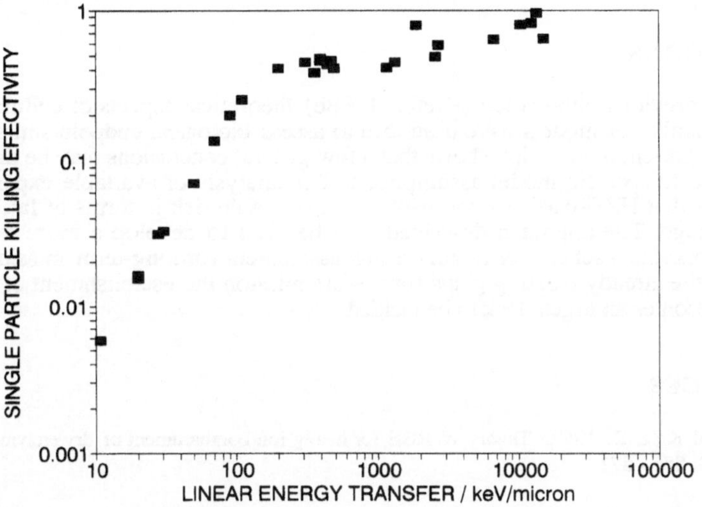

Figure 6. Inactivation Probability for single Traversals in V79 Chinese Hamster Cells. Based on Data from Thacker et al. (1979) and Kranert et al. (1990).

macroscopic parameters might lead to grossly misleading risk estimates. This has been realized before (Curtis, this issue) and prompted the introduction of "fluence-related risk coefficients". This approach is definitely better than the conventional one but it necessiates inclusion of the fact that the spectrum of particles impinging on the body surface suffers considerable changes and degradations. Theoretically this can be taken into account by calculating all possible organ doses as postulated in the protection recommendations. For practical purposes this is usually only performed for a number of "key organs", e.g. the blood forming system, so that those which lie between the surface and the bone marrow are neglected. It is, however, quite clear that no part of the body is spared from the potential carcinogenic radiation risk. This means that the fluence-related hazard estimate has to be based not only on the impinging particle itself but also on its penetration power and possibly secondary reaction products.

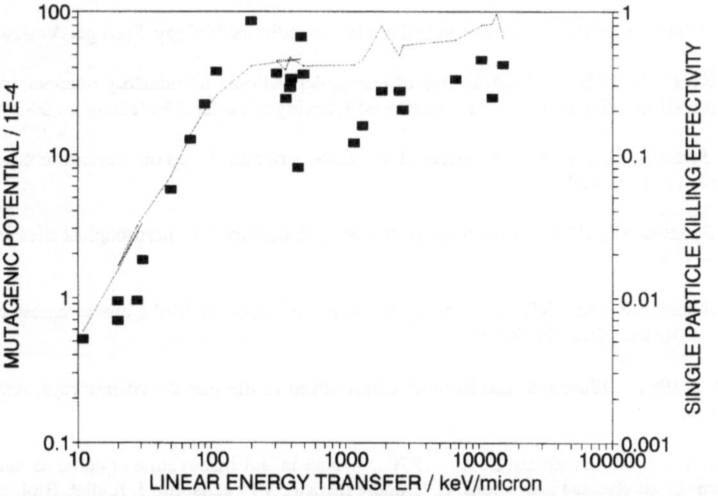

Figure 7. Mutagenic Potential m_1 of Charged Particles versus LET for V79 Cells (Squares, Left Ordinate). The Dotted Line is Taken from the Data in Fig. 6 for Comparison (Right Ordinate). Results Based on Thacker et al. (1979) and Kranert et al. (1990) Data.

CONCLUSIONS

In a previous publication (Kiefer, 1988b) theoretical aspects of cellular heavy ion action of a number of models were described to assess biological endpoints in a quantitative manner. The present manuscript shows that a few general conclusions may be drawn without any reference to specific model assumptions. The analysis of available experimental data demonstrates that HZE-particles constitute a considerable risk in terms of late genetic and somatic damage. The approach described may be used to develop a more realistic - and hopefully workable - scheme of radiation risk assessment for long-term missions into deep space. With the already existing plans for a Mars mission the establishment of a permanent lunar base becomes an urgent task to be tackled.

REFERENCES

Butts, J. J. and Katz, R. 1967, Theory of RBE for heavy ion bombardment of dry enzymes and viruses. Radiat. Res. 30: 855 - 871.

Feinendegen, L. E. 1985, Microdosimetric approach to the analysis of cell responses at low dose and low dose rate. Radiat. Prot. Dosimetry 13: 299-306.

Geard, C. R., 1980, Initial changes in cell cycle progression of chinese hamster V-79 cells induced by high-LET charged particles. Radiat Res , 83.

Kellerer, A. M. and Rossi, H. H., 1972, The theory of dual radiation action. Curr. Top. Radiat. Res. Quarterly 8: 85-158.

Kellerer, A. M. and Rossi, H. H., 1978, A generalized formulation of dual radiation action. Radiat. Res. 75: 471-488.

Kiefer, J. , 1987, Semi-empirical calculations for stopping powers and effective charge of heavy ions in water. GSI Scientific Report 1987, p.217.

Kiefer, J., 1988a, Heavy ion effects on cells: an approach to theoretical understanding.in "Terrestrial Space Radiation and Its Biological Effects" (P. D. McCormack, C. E. Swenberg, H. Bücker, eds.), Plenum Press, p. 117-127.

Kiefer, J. (ed.), 1988b, Quantitative mathematical models in radiation biology, Springer-Verlag, Heidelberg.

Kiefer, J. and Kost, M., 1988, Finestructures of energy deposition - introductory remarks, in: Quantitative mathematical models in radiation biology (J. Kiefer, ed.), Springer-Verlag, Heidelberg, p. 29-39.

Kiefer, J. and Straaten, H. , 1986, A model of ion track structure based on classical collision dynamics. Phys. Med. Biol. 11: 1201-1209.

Kiefer, J. and Straaten, H., 1987, Calculations of energy deposition in microscopical sites by heavy ions. GSI-Report 87-11, A7.

Kranert, T., Schneider, E. and Kiefer, J., 1990, Mutation induction in V79 Chinese hamster cells by very heavy ions. Int. J. Radiat, Biol. 58: 975-987.

Simpson, J. A. , 1983, Elemental and isotopic composition of the galactic cosmic rays. Ann. Rev. Nucl. Part. Sci. 33: 323.

Thacker, J., Stretch, A. and Stephens, M.A. , 1979, Mutation and inactivation of cultured mammalian cells exposed to beams of accelerated heavy ions. II. Chinese hamster V79 cells. Int. J. Radiat. Biol. 36: 137-148.

MATHEMATICAL MODELS OF LESION INDUCTION AND REPAIR IN IRRADIATED CELLS

S. Kozubek+* and G. Horneck*

+Institut of Biophysics, 61265 Brno 12, CSFR
*DLR Institut für Aerospace, 5000 Köln 90, Germany

The interaction of ionizing radiation with biological objects gives rise to two opposite processes. The first one comprises the destructive action of radiation on biologically important molecules, with the formation of radicals, their interaction and the fixation of biologically significant damage. The second one concerns the elimination of these injuries by biological processes accompanied by fixation of final molecular changes and their macroscopic manifestation. The kinetics of these two processes are quite different. Physical and chemical reactions are terminated within the first milliseconds, whereas early steps of the cellular repair run at least in the range of seconds. DNA strand breaks will be considered as biologically significant lesions that can be defined and detected.

The initial "black box" of the biological effects of ionizing radiation can be split into two boxes (box 1 and box 2), where the output from the first box is also the input into the second one. Box 1 includes the most of physical and chemical processes; box 2 includes mainly biological processes and will be called further biological box. The cellular damage fixed before the enzymatic repair starts will be called "biologically significant lesions" (b.s.l.). Such separation seems to be a logical step to a better understanding of biological radiation effects.

Both boxes can be investigated independently if we know what types of cellular lesions are important for what biological effect (if we could measure and induce them). If we could measure them, box 1 could be investigated. If we could produce them independently, box 2 could be investigated. The main problem is the identification of b.s.l. Three main methods are used to reach this aim: a) direct measurements of molecular changes in irradiated cells; b) predictions from energy deposition in target molecules by means of computation; c) predictions through the biological box by modelling.

Whereas the first two methods provide a menu of lesions that might be potentially significant for the fate of an irradiated cell, the third method is a way to predict b.s.l. through a biological box. The biological box joints b.s.l. with the investigated endpoint by enzymatic repair processes. Recent information on these processes is reflected in repair models. In such models usually some initial lesions are postulated and linked by theory to some endpoint macroscopically measured in the experiment. Some properties of initial lesions can be then established, which should facilitate the search for b.s.l. The problem remains whether the model is correct. The correctness of the model cannot be proved; it can, however, be substantially verified. The possibilities of model construction and verification will be discussed.

Let us consider the biological box. The input of this box are b.s.l. of different types. Their production rates depend on the conditions of irradiation. For example, in anoxic conditions in the presence of protectors or sensitizers the induction rates can be different. Also the ratios among different lesion types can vary depending on the conditions during irradiation. Therefore, we can modify the input of the biological box by modifying the conditions during irradiation. On the other hand, the output can be determined if a set of different endpoints are investigated. The combination of the modification of initial lesions (IL) with determination of final lesions (FL) give us substantial information about the structure of the biological box.

To reduce the number of unknown parameters in mathematical models, the separation of the above mentioned two main boxes can be done. The following basic axioms can be accepted: 1) Ionizing radiation produces different classes of cellular damage (IL), the ratios of which depend on the conditions during radiation. 2) The production rates of different types of IL are the same for all endpoints. 3) The repair processes do not depend on the conditions during irradiation.

The separation in the two boxes and consequently also the axioms can be controlled experimentally. To fulfil the second axiom it is necessary in the experiment to use only isogenic strains or strains with very similar genetic organisation. The third axiom can be fulfilled if the treatment before and during irradiation does not influence cellular repair processes. This is the case in most experiments and it can be tested in a parallel experiment where the conditions are changed just before irradiation.

Onsetting repair leads to the formation of FL. Frequently assumed random repair can be described by some fixation probability for each IL and each FL. Each type of IL can contribute to each type of FL. The production rate of FL is then given by the equation:

$$\beta_{j,c} = \sum_{(i)} F_{i,j} \times N_{i,c} \qquad (1)$$

where $\beta_{j,c}$ is the production rate of the j-th type of FL in c-th type of radiation conditions, $N_{i,c}$ is the production rate of i-th type of IL in c-th type of radiation conditions, and $F_{i,j}$ is the corresponding fixation probability (the probability that i-th type of IL will be fixed to j-th type of FL). The Equation 1 can be written in the matrix form: $\boldsymbol{\beta} = \mathbf{F.N}$. The properties of the IL (their induction rates in different conditions) are given by the matrix **N**, the structure of the biological box is reflected in the matrix **F**.

An important question is the relation between FL and observed endpoint. Different endpoints can be investigated: survival, mutation induction, etc. Random repair should lead to random distribution of FL in a cell population and therefore to exponential survival curves and linear mutation induction dose dependence:

$$S = \exp(-\beta_l x D) \qquad (2)$$

$$M = \beta_m x D \qquad (3)$$

where β_l or β_m are the production rates of the lethal or mutagenic FL, respectively. In fact, some kind of interaction between FL can lead to the shoulder in the cell survival curve, to quadratic or linear-quadratic dose dependences of cell killing or mutation induction. Generally for some endpoint (G) we have some dose response function:

$$G_{j,c} = f(-\beta_{j,c} x D) \qquad (4a)$$

or

$$\mathbf{G} = \mathbf{f}(\boldsymbol{\beta} \mathbf{x} \mathbf{D}) = \mathbf{f}(\mathbf{F} \mathrm{x} \mathbf{N} \mathrm{x} \mathbf{D}) \qquad (4b)$$

Equations 4 provides the possibility to link measurable quantities (S and M) to unknown parameters **N** and **F**.

The number of free parameters can now be calculated. For various conditions during irradiation different production rates are obtained. For p different conditions and m different types of IL we have **p x m** fixation probabilities. For different endpoints we have different fixation probabilities; for q endpoints we have **q x m** parameters. Together we have **(q + p) x m** parameters. It can be easily shown that there are only **(q + p - 1) x m** independent parameters.

The maximum number of data points (when all endpoints are measured for all conditions) is **q x p**. The problem can be mathematically solved if we have more experimental points than free parameters:

$$(q + p - 1) \times m < q \times p. \qquad (5)$$

According to this condition the following table has been calculated.

Table 1. The number of types of IL that can be determined for a given number of radiation conditions and a given number of endpoints investigated

NUMBER of END-POINTS (q)	**NUMBER OF RADIATION CONDITIONS (p)**							
	2	**3**	**4**	**5**	**6**	**7**	**8**	**9**
2	-	-	-	-	-	-	-	-
3	-	-	2	2	2	2	2	2
4	-	2	2	2	2	2	2	3
5	-	2	2	2	3	3	3	3
6	-	2	2	3	3	3	3	3
7	-	2	2	3	3	3	4	4
8	-	2	2	3	3	3	4	4
9	-	2	3	3	3	4	4	4

If a certain number of types of IL should be considered, we need data for some minimal number of radiation conditions and endpoints. For example, to determine parameters for 2 types of IL we need at least 4 different conditions and 3 endpoints or 3 conditions and 4 endpoints. To determine 3 types of IL we need at least 5 conditions and 6 endpoints or 4 conditions and 9 endpoints.

The numerical analysis is not simple although it can be performed by means of minimizing procedure such as using a least square fit criteria. The sum of squares per degree of freedom needs to be calculated. The number of experimental conditions and endpoints examined should be substantially greater than the minimum number given in Table 1. The assumptions inherent in the model can be checked and different hypotheses can also be tested. The number of types of cellular damage can be determined. The properties of b.s.l. are predicted by **N** and the structure of the biological box is reflected in the matrix **F**.

CELL KINETICS AND TRACK STRUCTURE

John W. Wilson, F. A. Cucinotta, and J. L. Shinn

NASA Langley Research Center

Hampton, VA, U.S.A.

ABSTRACT

A major uncertainty in shield requirements for deep-space missions is establishing biological risk for high charge and energy (HZE) exposure. Estimates of biological risk in space requires an understanding of the relationship of ground-based biological experiments with intense particle beams to the low exposure rates in the space environment. We have examined the relation of a (relatively) general cell kinetic model to the track structure theory of Katz and determined repair coefficients from the experiments of Yang et al. as a means of predicting biological response to low dose-rate exposure in the deep-space environment. The model provides repair dependent relative biological effectiveness (RBE's) which agree well with values found in ion exposure experiments and makes predictions which could be tested in future laboratory studies. The model seems to provide the necessary requirement of relating laboratory response data to space exposure conditions with the exception of the gravity environment effects.

INTRODUCTION

Before sending men into deep space for more than several months, there are many issues concerning radiation safety of the high energy and charge ions (HZE) of the galactic background radiation (Grahn, 1973) which must be resolved. The preponderance of human data available on radiation exposure is from (mainly) γ-ray exposure data obtained from the nuclear weapons of World War II (BEIR V). These are augmented by biological experiments for various radiation types in search of a means of extrapolating the human risk data for γ-rays to any arbitrary radiation field (Sinclair, 1985). This is the origin of the quality factor (Q) based on experimentally observed relative biological effectiveness (RBE) for various radiation exposure types. Central to the development of this method of estimating radiation protection requirements is the assumption that the experimental RBE reaches a maximum value as the delivered dose approaches zero independent of the dose rate at which the experiment was performed (ICRU, 1986). Such an assumption depends on the biological repair rates and repair efficiencies and there is evidence that the maximum RBE may not be practically achievable in laboratory experiments for some biological systems (Wilson and Cucinotta, 1991). Furthermore, very large RBE values have already been observed for some biological systems (Merriam et al., 1984; Thomson, Williamson and Grahn, 1989) that if implemented into this conservative protection methodology based on Q then the implementation of deep space travel would be difficult to achieve. Clearly, the resolution of these issues is of the utmost importance to the future of NASA's manned space program.

Biological Effects and Physics of Solar and Galactic Cosmic Radiation, Part A, Edited by C.E. Swenberg *et al.*, Plenum Press, New York, 1993

The present work is an outgrowth from a request by the Life Sciences Branch to develop methods of extrapolation of laboratory experiments to long duration space exposure. Most of that work was accomplished with disregard to the role of biological repair mechanisms (Wilson et al., 1990; Cucinotta et al., 1991) which would have a limited role in most laboratory experiments but is of critical importance in the low level space environment (17 μGy/hr). To address this issue we implemented a cell kinetic model based on a simple analogy with chemical kinetics (Frost and Pearson, 1962) motivated by the belief that the ultimate cell repair mechanisms are chemical in nature (Wilson and Cucinotta, 1991). At the same time, our awareness of the complexity of rather simple chemical systems (Wilson et al., 1984) especially those excited by ionizing radiation (Wilson, DeYoung, and Harris, 1979) gives us pause in approaching such a complicated chemical system as a living cell where thousands of chemical species are inhomogenously mixed even prior to radiation exposure. Clearly, there must be a great simplification in approaching this problem if any analytic expression is to be derived concerning radiation response.

There are essentially three types of kinetic models which have been used in past studies (Tobias, 1985; Goodhead, 1985; Scott and Ainsworth, 1980). The first two models evaluate the average number of lesions per cell and assume nonlinear kinetic terms as the source of sigmoid behavior in the survival curve. The repair-misrepair (RMR) model (Tobias, 1985) assumes a linear repair kinetic term to remove lesions (sublesions) from the cell, while a binary misrepair kinetic term produces lethal lesions appearing as lack of biological survival. A later version of the model assumes that sublesions can be "fixed," presumably as the cell progresses (Curtis, 1986). The binary misrepair term has been shown to be relatable to the dual action response model (Curtis, 1986) and the multitarget theory of Katz et al. (1981). A second nonlinear kinetic model assumes that repair enzymes are depleted in the repair process so that at large exposure levels the repair rates decrease to zero (saturated repair) as the repair enzyme pool is depleted (Goodhead, 1985). The unrepaired damage is assumed to be "fixed" at cell cycle progression when inactivation is expressed. Although the mechanisms are substantially different, both models describe well the sigmoid behavior and dose-rate-dependent response of mammalian cell cultures. Linear kinetic models (called state vector models) have found application in the literature (Scott and Ainsworth, 1980) and are adaptable to inclusion of cell environmental effects (Crawford-Brown and Hoffmann, 1990). The state vector models are closely related to the multihit model (Casarett, 1968). The main success of the two nonlinear models and the state vector model is for x-ray exposures. There is no clear development of these models to include track structure effects in heavy ion exposures.

It is well known that nonlinear processes dominate at high power densities in chemical processing (for example, Wilson, 1980; Wilson and Lee, 1980). Two and three body recombination processes are well known examples of nonlinear processes (Wilson, DeYoung and Harris, 1979). Chemical combinations of reactive species are present even at the lowest power levels where they tend to dominate and often follow linear kinetic equations (Wilson and Lee, 1980). High density power levels are locally present with the passage of high LET particles even at low exposure rates so that chemical products at high power with low LET radiation are similar to those produced by even low exposure levels of high LET radiation (Charlesby, 1967). Such facts have long been known to radiation chemist for high LET neutron environments at nuclear reactors. Similar nonlinear processes are related to the columnar recombination in ion chamber and scintillator detectors resulting in reduced detector efficiency. Any viable radiation model must account for the high power densities within particle tracks but the nonlinear time scale within the track chemistry is (very) short compared to the time scale of subsequent biological repair mechanisms giving hope that the repair kinetics may yet be describable by a linear kinetic model (Ngo et al., 1990) in which nonlinear processes are ascribed to mainly track structure effects. The veracity of such an approach would lie with the observed biological response under varied exposure conditions and the role of modeling would be to help define critical experiments. It is our hope that the low dose and low dose rate inherent in most space exposure can be adequately described by linear repair kinetics with nonlinear behavior confined to track structure effects as motivated by the above considerations.

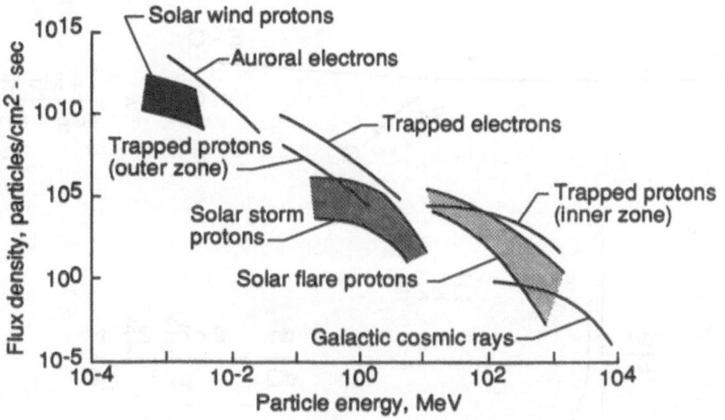

Figure 1. Space-radiation environment.

There is ample evidence that radiation injury to cell membranes and cytoplasmic material is only effective at doses in excess of tens of Gy (Casarett, 1968). Such high exposures are associated with early lethality and of little relevance to normal space exposure. The normal doses in space are much less than 10 Gy and relevant biological effects are expected to occur many years after exposure. There will be no directly observable changes in tissue systems during or immediately after the flight. Rather subtle changes in cell chemistry have occurred which will not be fully expressed until much later in the life of the cell line or the individual. It is believed that such changes are related to changes in the DNA structure. This will be the assumption of the present work.

A multitarget model with track structure derived by Katz et al. (1971) has been quite successful in describing track-structure-dependent phenomena in biological cell systems (Waligorski et al., 1987). The model considers the simple physical arrangements of sensitive sites required for some observable to be manifest and a physical model for the energy deposited around the path of a moving charged particle. The effects of charged-particle irradiation are correlated with that of gamma-ray irradiation by assuming that the response in sensitive sites near the particle's path is part of a larger system irradiated with gamma rays at the same dose (Katz et al., 1971). In the Katz model, mechanistic assumptions are avoided. The parameterization of the response to gamma rays provides for calibration of a biological systems response, as well as a transfer function for describing heavy ion effects. The main criticism of the model is its inability to predict repair-dependent phenomena (Curtis, 1986; Lett et al., 1989) and failure to achieve a maximum RBE at low exposure.

In a previous report (Wilson and Cucinotta, 1991), we presented a simple phase dependent repair model in which track structure effects were added through the use of the Katz formalism. Repair coefficients were estimated from the experiments of Yang et al. (1989) on stationary G_1 mouse cells in which varying amounts of repair in G_1-phase was allowed before cell cycling. Highly efficient repair was demonstrated for G_1-phase for light ions while high energy ^{56}Fe exposures showed little repair in good agreement with the kinetic model.

INTERACTIONS AND KINETIC PROCESSES

The energetic particles in space consists of mainly atomic constituents covering a very broad energy spectrum and flux values as shown in figure 1 (Wilson, 1978). The particles themselves are small ($\approx 10^{-13}$ cm) but are electrically charged resulting in a long-range force component. A casual look at condensed matter reveals mostly the structure of the electron clouds which contain only 0.05 percent of the mass but occupy virtually all the space within the material. Embedded within these electron clouds are the atomic nuclei whose dimensions

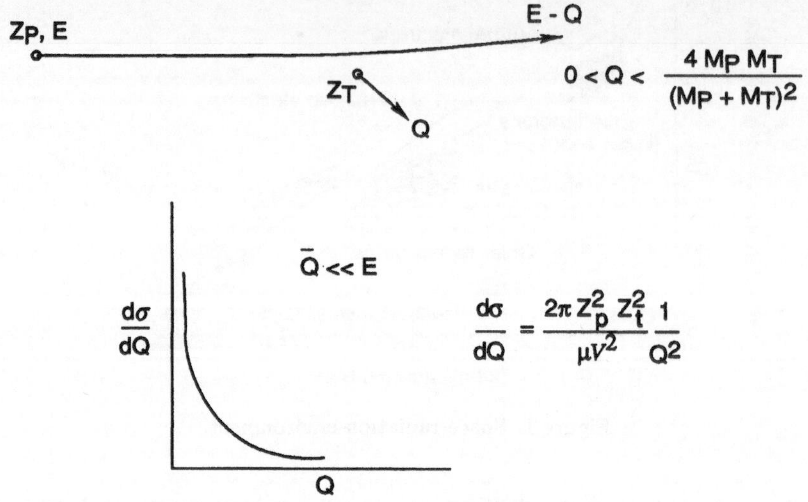

Figure 2. Coulomb scattering.

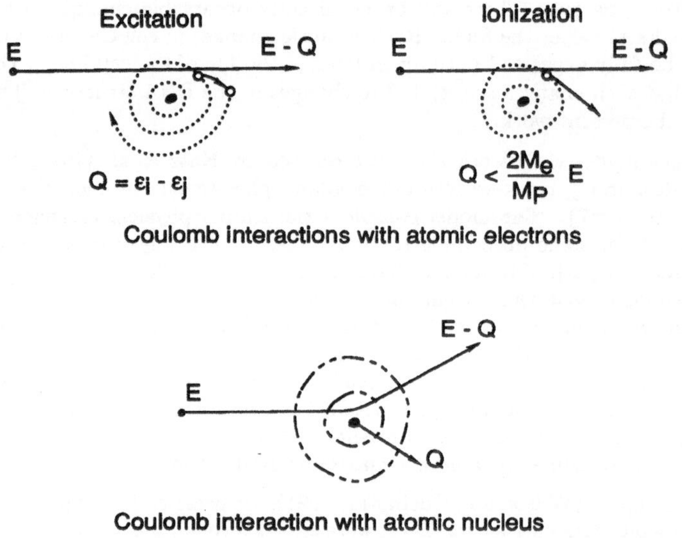

Figure 3. Schematic of Coulomb interactions with atomic electrons and atomic nucleus.

are 10^{-5} times smaller than the complete atom but contain 99.95 percent of the mass of the atom. Clearly an energetic particle passing through such a material will mainly interact with the electrons in the cloud and seldom strike a nucleus.

The dominate energy transfer process is energy loss through ionization, that is, a collision between the incoming charged particle (whether it is a proton, electron, or heavy ion) and the orbital electrons of the shielding material (fig. 2). They interact through Coulomb scattering, and the energy transferred from an ion of energy E and charge Z_p to a target particle of charge Z_t is labeled Q. The cross section σ has an inverse Q^2 dependence, and therefore the energy transfer is usually quite small. In the figure, μ is reduced mass for the projectile target system of masses M_P and M_T.

When the target is an electron bound in an atomic orbital, there are two options of either producing excitation when specific energy transfers ($\epsilon_i - \epsilon_j$, where ϵ_i and ϵ_j denote atomic energy levels) are made or ionization where the energy transferred must be greater than the ionization potential (fig. 3). The cross section is related to this energy transfer and goes like the inverse of Q^2. Another important process, especially for incident electrons, is Coulomb interaction with the atomic nucleus which results in multiple scattering effects. These multiple scattering effects are important for electron diffusion within the media.

The cross sections for secondary electrons produced from impacts of ions with atoms as described in figures 2 and 3 are shown in figure 4. This figure shows curve fits to the experimental data for 1 and 5 MeV proton impact (Manson et al., 1975), and the inverse Q^2 dependence above about 20 eV for the secondary electron energy is clearly evident. The corrections below 20 eV are due to binding effects which can only be treated quantum mechanically. The electron is actually bound in an atom, and these binding effects become important when the energy transfer is on the order of the binding energy. This type of data is important in giving the lateral spread of the energy from the track as the particle passes through a material.

There are added degrees of freedom when atoms are bound into molecular systems. Shown in figure 5 is a collection of data for N_2 molecules, which we chose as a typical molecule mainly because we could find the most data for it. This molecule has been under extensive investigation because of its importance to high power lasers and atmospheric phenomena. Vibrational excitation is important for electron energies below about 10 eV. Once the electronic excitation or ionization threshold is exceeded, the cross sections become heavily dominated by those two processes alone. In about one half the cases, ionization results in dissociation; and according to the data we have been able to collect, most molecules undergoing electronic excitation result in dissociation. There are, however, considerable differences in the dissociation cross section for these two processes as shown in figure 5. These differences are probably due to the small number of molecular states observed in the experiments. The molecular excitation cross section will probably change as further experiments are performed and the total electronic excitation cross section will probably show the same energy dependence as the ionization cross section at high energy. The data are taken from Schulz (1976), Cartwright et al. (1977), Köllman (1975), and Wight, Van der Wiel, and Brion (1976). The problem of molecular binding effects is difficult to treat using quantum theory but local plasma models have shown some success in treating both the molecular binding problem (Wilson and Kamaratos, 1981; Kamaratos, 1982; Xu, Khandelwal, and Wilson, 1984a and 1984b; Xu et al., 1984) as well as condensed phase effects (Wilson et al., 1984; Xu, Khandelwal, and Wilson, 1985).

Although most collisions in the material are with orbital electrons, the rare nuclear collisions are of importance because of the large energy transferred in the collision and the generation of new energetic particles. This process of transferring kinetic energy into new secondary radiations occurs through several different processes, such as direct knockout of nuclear constituents, resonant excitation followed by particle emission, pair production, and possible coherent effects within the nucleus. Through these processes, a single particle incident on the material may attenuate through energy transfer to electrons of the media or generate a multitude of secondaries causing an increase in exposure (transition effect). Which process dominates depends on energy, particle type, and material composition. This development of cascading particles is depicted in figure 6 as a relative comparison between high-energy proton and α-particle cascades in the Earth's atmosphere. Note the similarities displayed in figure 6 for individual reaction events and the nuclear-star events as seen in nuclear emulsion.

Neutrons and γ-rays are produced in local shield material and by local manmade sources such as nuclear power reactors. The neutrons interact through nuclear reactions similar to energetic protons whereby secondary charged particles are produced. The γ-rays interact through three main processes. The photoelectric cross section above ionization threshold, Compton scattering above tens of thousand electron volts and pair production above 1 MeV. The neutron induced nuclear stars are highly ionizing local events. The γ-ray produced

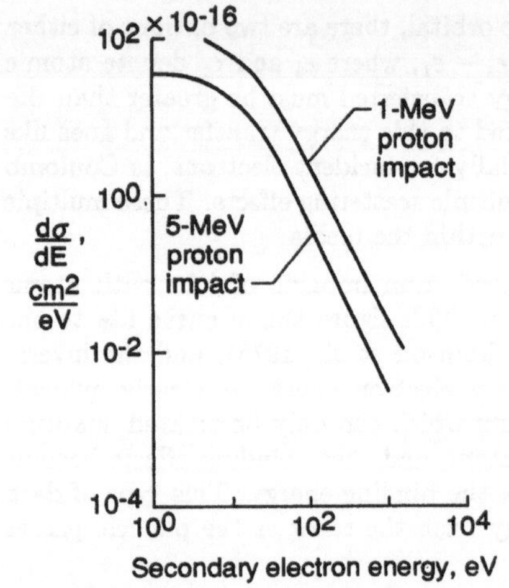

Figure 4. Secondary electron production spectra from proton impact with helium.

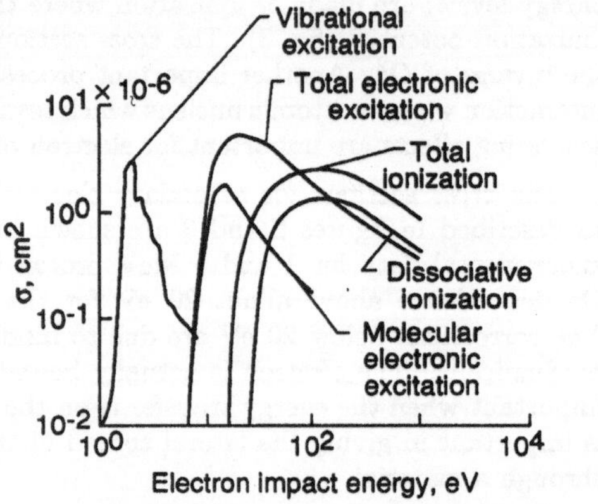

Figure 5. Electron impact cross section with N_2 molecules.

secondary electrons are broadly dispersed throughout the media giving a rather uniform distribution of ionization and excitation events.

The initiating events occur on the time scale of the passing ionizing particle ($\sim 10^{-14}$ sec). Even a free thermal target particle would drift less than 10^{-5} μm (0.1Å) in this time period. Clearly, diffusion and chemical reaction are precluded in this time period. The initiating events produce ions and free radicals within the media distributed in space according to the nature of the particle initiating the event. Whether the event is initiated by a neutral particle or passing ion, the secondary electrons ultimately dominate in producing the nascent chemical products. The source distribution (in space and energy) of the electrons is intimately connected to the initiating event and largely determines the initial distribution of ions and radicals (Rustgi et al., 1988). To a first approximation the distribution of ions and radicals is related to the average energy deposit first studied by Schaefer (1952) and extensively investigated by Katz and coworkers (for example, Katz et al., 1971).

The organic molecules of a living cell are suspended in water so that the radiolysis of water is one key to understanding mechanism of radiation injury. Without details we simply note that a principle product is the dissociation of H_2O into hydrogen and hydroxyl radicals (Casarett, 1968).

$$H_2O \rightarrow H + OH$$

Subsequent events depend on the density of such radicals and the other molecules present. At high power densities, peroxide and hydrogen formation

$$OH + OH \rightarrow H_2O_2$$

$$H + H \rightarrow H_2$$

are in competition with recombination

$$H + OH \rightarrow H_2O$$

In the presence of dissolved oxygen the peroxyl radical is formed as

$$H + O_2 \rightarrow HO_2$$

which cycles to form hydrogen peroxide

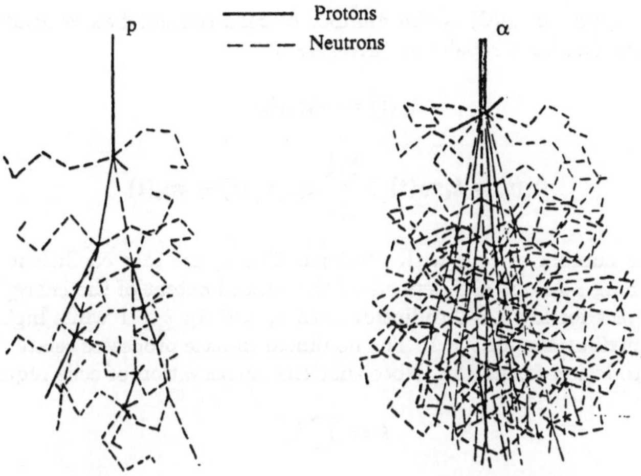

Figure 6. Cascade development in matter.

$$HO_2 + HO_2 \rightarrow O_2 + H_2O_2$$

The peroxides are highly active chemical (oxydizing) agents which are biologically damaging. The density of hydroxyl radicals and peroxides are undoubtedly related to damage to nearby organic molecules. Within the energy deposit are also organic molecules which may likewise be disassociated and form peroxyl radicals which may further form chemical reactions. Although we may not know specifically how a given structure within the DNA is altered it is easy to imagine that the degree of alteration of the DNA by either direct interaction or a chemical product is related to the local energy density associated with the initial event.

We will not attempt to describe the processes by which the DNA is damaged by the radicals and chemical agents described above but simply note that the direct reaction of these species with the organic molecules is a linear kinetic process. Furthermore, the enzyme repair of such damage is also expected to follow a linear kinetic description (Ngo et al., 1990). Even in spite of these facts, we would try linear kinetic modeling on the general principle of "keeping it simple" unless required to do otherwise according to experimental evidence. Nonlinear kinetics are of course important if sufficient energy is given to the cell in a sufficiently short time but such exposure levels are not assumed important to space radiation protection.

A SIMPLE LINEAR KINETIC SURVIVAL MODEL

Whether the DNA is altered by the direct interaction of the passing particle or chemical agents produced in the surrounding medium we may think in terms of chemical change at a specific site along the DNA chain which we term a lesion. Due to the complexity of the DNA molecule it is clear that many different types of lesions may occur and may exist along the DNA strand simultaneously. Such lesions are those we have termed the nascent cellular lesions (Wilson and Cucinotta, 1991). These nascent lesions are formed soon after the passage of the ionizing particle and are now dealt with by the slower cell repair processes (Ngo et al., 1990). Radiation injury to a DNA site is related to the probability of an initiating ionization/excitation event at or near the site and the probability that a chemical lesion results. The probability of the initiation is related to the absorption cross section and the probability of forming a lesion is related to chemical kinetics within the energy deposit and rates of diffusion. The lesion formation rates are proportional to the ionizing particle flux and depend on particle type. If $n_0(0)$ is the number of cells present before irradiation then the cell kinetic equations (without repair) are given as

$$\dot{n}_0(t) = -kn_0(t) \tag{1}$$

$$\dot{n}_i(t) = k_i n_0(t) + \sum_{j=1}^{i-1} k_{i-j} n_j(t) - kn_i(t) \tag{2}$$

where $n_i(t)$ are the number of cells with i lesions. The k_i are rate coefficients (cross section times flux) for forming i lesions and depend on the spatial extent of the energy deposit (track structure). We expect for low LET radiation that $k_i \approx 0$ for $i > 1$ while high LET particles have significant contributions for $i > 1$. The nonlinear kinetic processes relate the distribution of energy deposit to lesion formation. Note that the conservation of cells requires

$$k = \sum_i k_i \tag{3}$$

The distribution of nascent lesions in the cell population after a time t is

$$n_0(t) = n_0 e^{-kt} \tag{4}$$

$$n_1(t) = n_0 k_1 t \, e^{-kt} \tag{5}$$

$$n_2(t) = n_0 \left[k_2 t + \frac{1}{2}(k_1 t)^2 \right] e^{-kt} \tag{6}$$

$$n_3(t) = n_0 \left[k_3 t + \frac{2}{2!} k_1 t k_2 t + \frac{1}{3!}(k_1 t)^3 \right] e^{-kt} \tag{7}$$

and similarly for higher order terms where $n_0 \equiv n_0(0)$. It is likely that these nascent lesions are chemical alterations yet to be stabilized by the cell repair processes.

If in subsequent reactions the DNA is restored to its original state then the cell has been repaired. Some altered status after repair is called misrepair and the cell undergoes clonogenic death (Casarett, 1968). We assume a simple form for the linear repair kinetics as

$$\dot{n}_0(t) = \sum_{i=1}^{\infty} \alpha_{ri} n_i(t) \tag{8}$$

$$\dot{n}_i(t) = -\alpha_i n_i(t) \tag{9}$$

where α_{ri} and α_i are repair rate coefficients. The balance of misrepaired cells are

$$\dot{n}_d(t) = \sum_{i=1}^{\infty} \alpha_{mi} n_i(t) \tag{10}$$

where α_{mi} is the misrepair rate coefficient. Conservation of number of cells requires $\alpha_i = \alpha_{ri} + \alpha_{mi}$. The subsequent repaired, unrepaired, and misrepaired states are given by

$$n_0(t) = n_0(0) + \sum_{i=1}^{\infty} n_i(0) \frac{\alpha_{ri}}{\alpha_i} \left(1 - e^{-\alpha_i t}\right) \tag{11}$$

$$n_i(t) = n_i(0) e^{-\alpha_i t} \tag{12}$$

$$n_d(t) = \sum_{i=1}^{\infty} n_i(0) \frac{\alpha_{mi}}{\alpha_i} \left(1 - e^{-\alpha_i t}\right) \tag{13}$$

where $n_0(0)$ and $n_i(0)$ represents in equations (11) through (13) and in subsequent equations the lesion distribution at the start of the repair period. After the repairs are complete ($\alpha_i t \gg 1$) the final populations are given as

$$n_0(\infty) = n_0(0) + \sum_{i=1}^{\infty} \frac{\alpha_{ri}}{\alpha_i} n_i(0) \tag{14}$$

$$n_d(\infty) = \sum_{i=1}^{\infty} \frac{\alpha_{mi}}{\alpha_i} n_i(0) \tag{15}$$

which are dependent on the initial distribution of lesions and the repair efficiencies. At 100 percent repair efficiency

$$n_0(\infty) = n_0(0) + \sum_{i=1}^{\infty} n_i(0) \equiv n_0 \tag{16}$$

so that the population is fully restored to its initial number n_0 prior to exposure. In practice, relatively efficient repair is found for $i < m_d$ with nearly complete misrepair for higher numbers of lesions ($i \geq m_d$). The final population is then

$$n_0(\infty) = n_0(0) + \sum_{i=1}^{m_d - 1} n_i(0) \tag{17}$$

Examination of $n_0(\infty)$ for a short γ-ray exposure period t with the aid of equations (4) to (7) yields

$$n_0(\infty) \approx n_0 \left[1 - \frac{1}{m_d!} (k_1 t)^{m_d} \right] \tag{18}$$

with $k_i = 0$ for $i > 1$. This solution shows typical sigmoid response at low dose. If however $k_{m_d} \neq 0$ as expected for HZE exposure then

$$n_0(\infty) \simeq n_0(1 - k_{m_d} t) \tag{19}$$

exhibiting a linear slope response with no shoulder. Clearly a broad range of solution is available as a function of radiation type (k_i) and the repair efficiencies.

The cellular repair mechanisms depend on the overall status of the cell chemistry. The chemical processes in the cell vary greatly throughout the cell cycle. For example, the cytoplasm viscosity is normally quite high and suddenly drops to near zero just before mitosis. Such shifts in physical properties are affected through the cell chemistry and reflects tremendous variability. More subtle are the changes prior to synthesis as well as throughout the synthesis phase. Such variations are clearly seen in different experimental protocols with the same cell line and radiation source. These considerations will be key to determination of the coefficients in the above kinetic equations.

Sigmoid Survival Curves

If any particle is capable of forming single lesions it is the photon. We further assume the photon to form only single lesions. The lesion distribution for a pulse of photons of duration t_r is then

$$n_0(t_r) = n_0 e^{-k t_r} \tag{20}$$

$$n_i(t_r) = n_0 \frac{1}{i!} (k_1 t_r)^i \, e^{-k t_r} \tag{21}$$

Allowing for complete repair subsequent to exposure yields

$$\frac{n_0(\infty)}{n_0} = 1 - \sum_{i=1}^{\infty} \frac{\alpha_{mi}}{\alpha_i} \frac{1}{i!}(k_1 t_r)^i \, e^{-kt_r} \tag{22}$$

The response is proportional to the exposure to the m_d power where m_d is the lowest i for which $\alpha_{mi} \neq 0$. Generally, there could be a small linear term present if $0 < \alpha_{m1}/\alpha_1 \ll 1$. Such a response would appear linear-quadratic at low exposure. Note that a linear term could also arise from $k_{m_d} > 0$ for γ-rays as well.

Exponential Survival Curves

The exposure with high energy iron ions is near exponential and would be accommodated if the $k_i \approx 0$ for all $i < m_d$. The lesion distribution would then be

$$n_0(t_r) = n_0 e^{-kt_r} \tag{23}$$

$$n_i(t_r) = 0 \tag{24}$$

$$n_{m_d}(t_r) = k_{m_d} t_r e^{-kt_r} \tag{25}$$

The survival is given as

$$\frac{n_0(\infty)}{n_0} = 1 - \frac{\alpha_{m1m_d}}{\alpha_{1m_d}} k_{m_d} t_r e^{-kt_r} - \frac{\alpha_{m2m_d}}{\alpha_{2m_d}} \frac{1}{2!}(k_{m_d} t_r)^2 e^{-kt_r} - \ldots \tag{26}$$

showing linear-quadratic dependence at low exposure. If the repair efficiency is zero for $i \geq m_d$ then the response is given as

$$\frac{n_0(\infty)}{n_0} = e^{-k_{m_d} t_r} \tag{27}$$

as is usually observed for iron beam exposures.

Discussion

In summarizing our results to this point, we have suggested that track structure effects and the associated fast nonlinear kinetic processes contribute to the source terms for nascent lesions within the cell DNA. That the subsequent repair kinetics are much slower and are represented by linear repair processes (Ngo et al., 1990). The sigmoid behavior in survival curves is related to repair efficiency and the exponential survival curves of HZE particles result from track structure effects for which multiple lesions are formed in a single ion passage. We now consider means of evaluating the kinetic parameters through relation to a track structure model and repair dependent experiments.

TRACK STRUCTURE MODEL

The cellular track model of Katz et al. (1971) attributes biological damage from energetic ions to the secondary electrons (delta rays) produced along the ion's path. The effects caused by energetic ions are correlated with those of gamma rays by assuming the response in sensitive sites near the ion's path is part of a larger system irradiated with gamma rays at the same dose. The response due to ion effects is then approximately related to the gamma-ray response and the delta-ray dose surrounding the ion's path. For a multitarget response with target number m, the inactivation of cells by gamma rays is assumed to follow a multitarget distribution reflecting the random accumulation of sublethal damage, with a radiosensitivity parameter D_o.

For the inactivation of cells by ions, two modes are identified: "ion-kill" which corresponds to intratrack effects and "gamma-kill" which corresponds to intertrack effects. Here, the ion-kill mode is unique to ions corresponding to single particle inactivation of cells described by the cross section σ. The inactivation cross section for a sensitive site whose response to radiation is ahistoric is determined as

$$\sigma = \int_0^\infty 2\pi r dr (1 - e^{-\overline{D}/D_0})^m \tag{28}$$

where $\overline{D}$ is the average dose at the sensitive site from the ion's delta rays. The evaluation of the cross section is separated by Katz et al. (1971) into a so-called grain-count regime, where inactivation occurs randomly along the path of the particle, and into the so-called track-width regime, where many inactivations occur and are said to be distributed like a "hairy-rope." In the grain-count regime, σ may be parameterized as

$$\sigma = \sigma_0 (1 - e^{-Z^{*2}/\kappa\beta^2})^m \tag{29}$$

where σ_0 is the plateau value of the cross section, the effective charge number is given by

$$Z^* = Z(1 - e^{-125\beta/Z^{2/3}}) \tag{30}$$

and κ is a parameter related to the radius of the sensitive site, a_0, by

$$D_0 a_0^2/\kappa \cong 2 \times 10^{-7} \text{ erg/cm} \tag{31}$$

The transition from the grain-count regime to the track-width regime is observed to take place at a value of $Z^{*2}/\kappa\beta^2$ of about 4 at lower values we are in the grain-count regime and at higher values the track-width regime.

The fraction of the cells damaged in the ion-kill mode is $P = \sigma/\sigma_0$ and note that in the track-width regime $\sigma > \sigma_0$ and it is assumed that $P = 1$. The track model assumes that a fraction of the ion's dose, $(1 - P)$, acts cumulatively with that for other particles to inactivate cells in the gamma-kill mode. The surviving fraction of a cellular population $n_0(\infty)$, whose response parameters are m, D_0, and κ or a_0 after irradiation by a fluence of particles F, is then written

$$\frac{n_0(\infty)}{n_0} = \pi_i \times \pi_\gamma \tag{32}$$

where

$$\pi_i = e^{-\sigma F} \tag{33}$$

is the ion-kill survival probability and

$$\pi_\gamma = 1 - \left(1 - e^{-D_\gamma/D_0}\right)^m \tag{34}$$

is the gamma-kill survival probability. The gamma-kill dose fraction is

$$D_\gamma = (1 - P)D \tag{35}$$

where D is the absorbed dose. Note this division into ion-kill and gamma-kill also divides our track into regions where the fast nonlinear kinetics are expected to dominate (ion-kill) and a region where the fast linear kinetics are expected to be more important (gamma-kill).

The RBE at a specific survival level is given by

$$\text{RBE} = D_x/D \tag{36}$$

where

$$D_x = -D_0 \ln\left\{1 - \left[1 - n_0(\infty)/n_0\right]^{1/m}\right\} \tag{37}$$

is the x-ray dose at which this level is obtained. Equations (28) through (37) represent the cellular track model for monoenergetic particles. We must now consider the relationship of the kinetic model to the Katz model.

Physics and Kinetics of Cell Injury

The Katz model is formulated on the basis of physical arguments about track structure, geometric arrangement of sensitive (chemical bond) sites, the size of the cell nucleus, and energy thresholds for changes in the cell molecules. In practice, the Katz parameters $(m, D_0, \sigma_0, \kappa)$ are determined from biological experiments for a given cell system and experimental protocol. The degree to which cell repair is reflected in the final parameters is uncertain, but the effects of differing experimental protocols on the Katz parameters are well known and in someway reflect repair mechanisms. We will attempt to better define the relationship of repair to the Katz model parameters within the context of the present repair kinetic model.

In the Katz model, it is assumed that electromagnetic radiations form single lesions with an efficiency related to D_0 and generally more than one lesion ($m_d \geq 2$) is required to express the biological effect (cell death in the present case). We assume that stationary G_1 phase cells show near complete repair of lesion multiples less than m_d. If the cells are irradiated with γ-rays in stationary G_1 and are held in this phase until repair is complete, then the surviving population is found to be

$$n_0(\infty) \approx n_0(t_r) + \sum_{i=1}^{m_d-1} n_i(t_r) \tag{38}$$

assuming maximum repair in stationary G_1 (i.e., $\frac{\alpha_{r_i}}{\alpha_i} \approx 1$ for $i < m_d$). Equation (38) allows us to relate the k_i coefficients to the corresponding Katz parameters of equations (28) to (35) as applied to the appropriate experimental protocol (namely, G_1 exposure followed by complete G_1 repair). In the kinetic model, n_0 is the initial number of G_1 cells, and equation (38) is rewritten as

$$\begin{aligned}\frac{n_0(\infty)}{n_0} &\approx e^{-kt_r} + \sum_{i=1}^{m_d-1} k_i t_r e^{-kt_r} + \frac{1}{2!}k_1 t_r \sum_{i=1}^{m_d-1} k_i t_r e^{-kt_r} \\ &+ \frac{1}{3!}(k_1 t_r)^2 \sum_{i=1}^{m_d-2} k_i t_r e^{-kt_r} + \ldots\end{aligned} \tag{39}$$

According to the Katz model, a $m_d = 3$ system has a γ-ray response given by

$$\frac{n_0(\infty)}{n_0} \approx 1 - \left(1 - e^{-D_\gamma/D_0}\right)^3 \approx 1 - \left(\frac{D_\gamma}{D_0}\right)^3 \tag{40}$$

which is matched to equation (39) if

$$k_1 t_r \cong 6^{\frac{1}{3}} D_\gamma / D_0 \tag{41}$$

$$k_m t_r \approx 0 \quad (m > 1) \tag{42}$$

as is appropriate for γ-rays. Similarly, the remaining terms in equation (39) can be determined from the remaining Katz terms by noting for strictly ion kill kinetics

$$k_3 t_r \cong \sigma F \tag{43}$$

$$k_2 t_r \cong 0 \tag{44}$$

Requiring k_2 to be zero results from our matching of equation (39) at low dose. There may be nonzero values of k_2 but they cannot be strictly determined in the present form of Katz's theory. Although the k_i's may reflect both physical and chemical processes because of their empirical nature, we assume here that they are most clearly identified with the physical processes discussed by Katz. We now examine means by which repair rates can be estimated at least for some experimental cell systems.

Three-target Repair/Misrepair Systems

The above can be applied to an approximate three-target system as

$$\frac{n_0(\infty)}{n_0} \approx \left[1 + \frac{\alpha_{r_1}}{\alpha_1} 6^{\frac{1}{3}} \frac{D_\gamma}{D_0} + \frac{\alpha_{r_2}}{\alpha_2} \frac{6^{\frac{2}{3}}}{2} \frac{D_\gamma^2}{D_0^2}\right] e^{-\sigma F - 6^{\frac{1}{3}} D_\gamma / D_0} \tag{45}$$

where D_γ, D_0, and σF are related to the usual Katz model for $m_d = 3$ and $\frac{\alpha_{r_1}}{\alpha_1}, \frac{\alpha_{r_2}}{\alpha_2}$ are the repair ratios for the once hit and twice-hit cells. Presumably, $\frac{\alpha_{r_1}}{\alpha_1} > \frac{\alpha_{r_2}}{\alpha_2}$. We take

$$\frac{\alpha_{r_2}}{\alpha_2} = \left(\frac{\alpha_{r_1}}{\alpha_1}\right)^p \tag{46}$$

in the present analysis and expect p to be 2 or greater. In the limit of vanishing dose where RBE is presumably maximum

$$\mathrm{RBE}_m \approx 1 - \frac{\sigma}{\sigma_0} + 6^{-\frac{1}{3}} D_0 \frac{\alpha_1 \sigma}{\alpha_{m_1} L} \tag{47}$$

which is unbound for small α_{m1}. The RBE in the Katz model is found to increase with ion dose as $D^{-1+1/m}$ (Cucinotta et al. 1991, Katz and Cucinotta 1991) so that no maximum is achieved. A similar dependence on dose is found here at higher exposure levels than assumed in equation (47), however, misrepair prevents a one-to-one correspondence especially at low dose where a maximum RBE is achieved in the kinetics model for $\alpha_{m_1} > 0$.

APPLICATION TO CELL SURVIVAL

The experiments of Yang et al. (1989) have utilized contact stabilized mouse cells C3H10T1/2 in the stationary G_1 phase. In one set of experiments, the cells were held in the G_1 phase for 24 hours before separation and introduction into a nutrient medium to stimulate growth (delayed plating). A second series of cells was immediately plated and thus greatly altered the cell kinetics by progression towards the synthesis cycle (S phase) soon after exposure. It is well known (Sinclair, 1968) that the early G_1 phase is efficient in cell repair while the early S phase is mistake prone (Radman et al., 1981). We assume the stationary G_1 phase repair ratio $\frac{\alpha_{r_1}}{\alpha_1}$ is near maximum, while the accident-prone early S phase has a significant rate of misrepair. Furthermore, cell survival of the mouse cell is shown by Katz to be a three-target system, and even higher rates of misrepair are expected from the doubly injured cell ($p \gg 2$), especially later in the cell cycle.

The Katz parameters (see table 1) for the delayed experiments (Waligorski et al., 1987) are used directly to estimate $\sigma F, D_0, D_\gamma$ with assumed $\frac{\alpha_r}{\alpha} = 1$ and provides a good fit, as expected, to Yang et al.'s delayed plating data. Good agreement is found for the immediately plated cells by taking $p = 6$ and $\frac{\alpha_{r_1}}{\alpha_1} = 0.7$ (for the exponential population). The results are shown in figure 7. The figure is arranged in the order of increasing LET, and the sigmoid behavior associated with multitarget phenomena is apparent for the lighter ions. The sigmoid behavior disappears at higher LET as the repair processes become less effective and the ion-kill mechanism of Katz dominates.

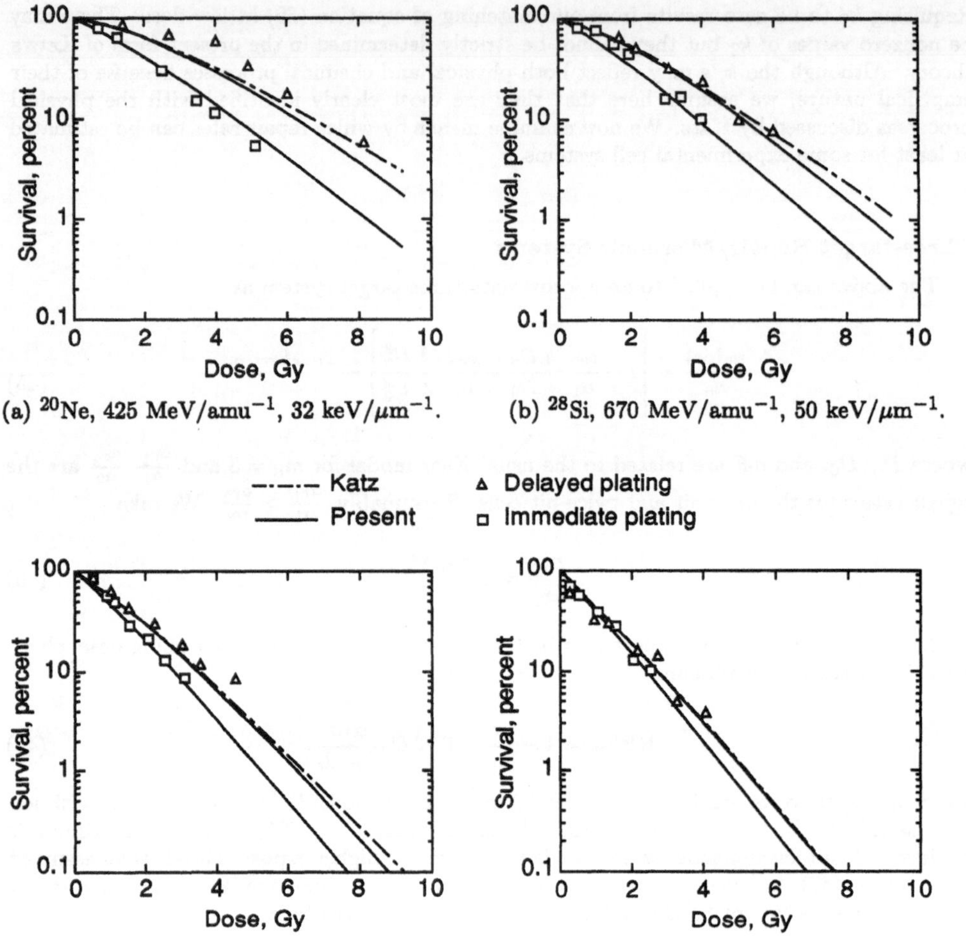

(a) ^{20}Ne, 425 MeV/amu^{-1}, 32 keV/μm^{-1}.

(b) ^{28}Si, 670 MeV/amu^{-1}, 50 keV/μm^{-1}.

(c) ^{28}Si, 320 MeV/amu^{-1}, 82 keV/μm^{-1}.

(d) ^{56}Fe, 600 MeV/amu^{-1}, 190 keV/μm^{-1}.

Figure 7. Cell survival of C3H10T 1/2 for delayed plating (Δ) and immediate plating ($\square$). The dash-dot curves are Katz model values while the present model values are the full curves.

Comparisons of calculated and measured RBE values for several ions are shown in table 2 at survival levels of 10 percent and 50 percent for immediate plating conditions. The agreement with experiment is very good except for the U ion. Here, we have not taken track-width effects in the thin down region into account. The maximum RBE value given by equation (47) with $\alpha_{ri}/\alpha_i = 0.7$ is also shown in table 2. We note that for the delayed plating experiments, no maximum RBE is predicted in the kinetic model (assumed $\alpha_{m1} = 0$), as well as in the Katz model (Katz and Cucinotta, 1991).

TABLE 1. KATZ PARAMETERS FOR CELL SURVIVAL USED IN THE PRESENT TRACK STRUCTURE REPAIR/MISREPAIR MODEL

	σ_0, cm^2	κ	m	D_0, Gy
C3H10T 1/2	5×10^{-7}	750	3	2.8

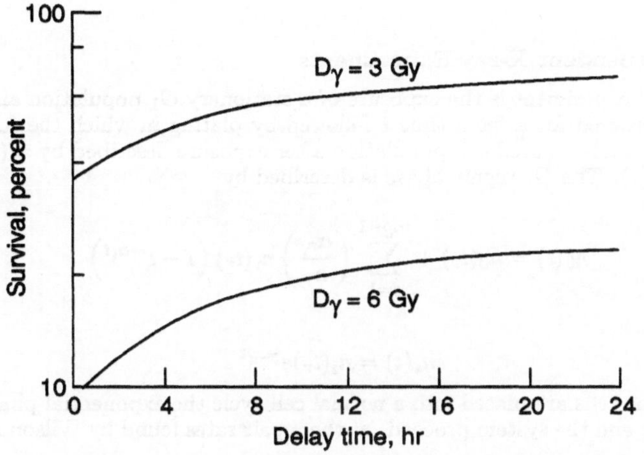

Figure 8. Percent survival at two exposure levels as a function of G_1 delay time before plating.

TABLE 2. RBE FOR SURVIVAL OF C3H10T 1/2 CELLS (IMMEDIATE PLATING)

		10%	50%	Maximum
Radiation	LET*	Experiment (Theory)	Experiment (Theory)	Theory
x-rays		1.00 (1.00)	1.00 (1.00)	1.00
C-12	10	1.00 (1.03)	1.00 (1.02)	1.10
Ne-20	32	1.50 (1.29)	1.56 (1.38)	1.71
Si-28	50	1.50 (1.52)	1.67 (1.67)	2.28
Si-28	82	2.23 (2.10)	3.00 (2.51)	3.73
Ar-40	140	2.30 (2.60)	3.00 (3.15)	5.00
Fe-56	192	2.20 (2.50)	3.10 (3.08)	4.87
Fe-56	286	2.00 (2.35)	3.00 (2.99)	4.68
Fe-56	475	1.62 (1.65)	2.72 (2.11)	3.31
U-238	1860	0.88 (0.43)	1.20 (0.55)	0.86

*LET in units of keV/μm

Although the present results are encouraging, there is a fuller range of protraction experiments to which the model is to be compared. Furthermore, other biological endpoints must yet be added and further tested against experimental observation.

Repair Rate Dependent X-ray Experiments

Another useful experiment is the exposure of a stationary G_1 population and to allow G_1 phase repair to proceed for a fixed time t followed by plating in which the full cell cycle is promoted. The initially injured cell population after exposure described by $n_i(t_r)$ is given by equations (4) to (7). The G_1 repair phase is described by

$$n_0(t) = n_0(t_r) + \sum_{i=1}^{m_d-1} \left(\frac{\alpha_{r_i}}{\alpha_i}\right) n_i(t_r) \left(1 - e^{-\alpha_i t}\right) \tag{48}$$

and

$$n_i(t) = n_i(t_r) e^{-\alpha_i t} \tag{49}$$

If after a time t the cells are placed into a normal cell cycle the exponential phase repair rates are quite different and the system proceeds at the repair rates found by Wilson and Cucinotta (1991) as

$$n_0(\infty) = n_0(t_r) + \sum_{i=1}^{m_d-1} \left(\frac{\alpha_{r_i}}{\alpha_i}\right) n_i(t_r) \left(1 - e^{-\alpha_i t}\right) + \sum_{i=1}^{m_d-1} \left(\frac{\alpha'_{r_i}}{\alpha'_i}\right) n_i(t_r)\, e^{-\alpha_i t} \tag{50}$$

where t remains as the G_1 repair period and α'_{r_i} and α'_i are the repair rate coefficients for an exponential population. Results are shown in figure 8 as a function of G_1 delay for two x-ray exposure levels of 3 and 6 Gy.

Another approach to study G_1 repair rates is to use fractionated exposures of a G_1 population. The initial exposure followed by a G_1 repair period of length t results in a cell population after repair of

$$n_0(t) = n_0(t_r) + \sum_{i=1}^{m_d-1} \left(\frac{\alpha_{r_i}}{\alpha_i}\right) n_i(t_r) \left(1 - e^{-\alpha_i t}\right) \tag{51}$$

and

$$n_i(t) = n_i(t_r) e^{-\alpha_i t} \tag{52}$$

A subsequent second exposure of equal duration t_r results in a new population

$$n'_0(t_r) = n_0(t)\ e^{-k t_r} \tag{53}$$

$$n'_1(t_r) = n_1(t)\ e^{-k t_r} + k_1\ t_r\ n_0(t)\ e^{-k t_r} \tag{54}$$

$$n'_2(t_r) = n_2(t)\ e^{-k t_r} + n_1(t)\ k_1\ t_r\ e^{-k t_r} + \frac{1}{2}\ n_0(t)\ k_1^2\ t_r^2\ e^{-k t_r} \tag{55}$$

which if plated immediately after exposure yields

$$n'_0(\infty) = n'_0(t_r) + \sum_{i=1}^{m_d-1} \left(\frac{\alpha'_{r_i}}{\alpha_i}\right) n'_i(t_r) \tag{56}$$

These results are compared to the variable repair and fractionated exposure experiments of Yang et al. (1989) in figure 9. The agreement is excellent.

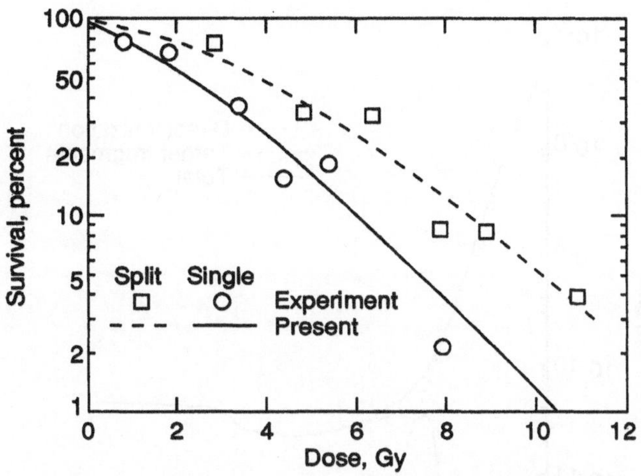

Figure 9. Comparison of present theory with Yang's experiments for single and fractionated exposure.

Target Fragments in Proton Induced Kinetics

The target fragmentation fields are found in closed form in terms of the collision density (Wilson, 1977), because the fragment ions are of relatively low energy. Away from any interfaces, the target fields are in local equilibrium and may be written as

$$\phi_\alpha(x, E_\alpha; E_j) = \frac{1}{S_\alpha(E_\alpha)} \int_{E_\alpha}^{\infty} \frac{d\Sigma_{\alpha j}(E', E_j)}{dE'} \phi_j(x, E_j) dE' \tag{57}$$

where the subscript α labels the target fragment type, $S_\alpha(E)$ the stopping power, and E_α and E_j are in units of MeV.

The particle fields of the projectiles and target fragments determine the level and type of radiation damage for the endpoint of interest. The relationship between the fields and the cellular response is now considered within the Katz cellular track model.

The ion-kill term now contains a projectile term (Cucinotta et al., 1991b) as well as a target fragment term as

$$(\sigma F) = \sigma_j(E_j)\phi_j(x, E_j) + \sum_\alpha \int_0^\infty dE_\alpha \phi_\alpha(x, E_\alpha; E_j)\sigma_\alpha(E_\alpha) \tag{58}$$

while the corresponding gamma-kill dose becomes

$$\begin{aligned} D_\gamma = {} & [1 - P_j(E_j)]S_j(E_j)\phi_j(x, E_j) \\ & + \sum_\alpha \int_0^\infty dE_\alpha [1 - P_\alpha(E_\alpha)]S_\alpha(E_\alpha)\phi_\alpha(x, E_\alpha; E_j) \end{aligned} \tag{59}$$

Use of equation (57) allows one to define an effective cross section as

$$\sigma_j^*(E_j) = \sigma_j(E_j) + \sum_\alpha \int_0^\infty dE_\alpha \frac{\sigma_\alpha(E_\alpha)}{S_\alpha(E_\alpha)} \int_{E_\alpha}^\infty dE' \frac{d\Sigma_{\alpha j}(E', E_j)}{dE'} \tag{60}$$

The first term of equation (60) is caused by the direct ionization of the media by the passing ion of type j. The second term results from target fragments produced in the media.

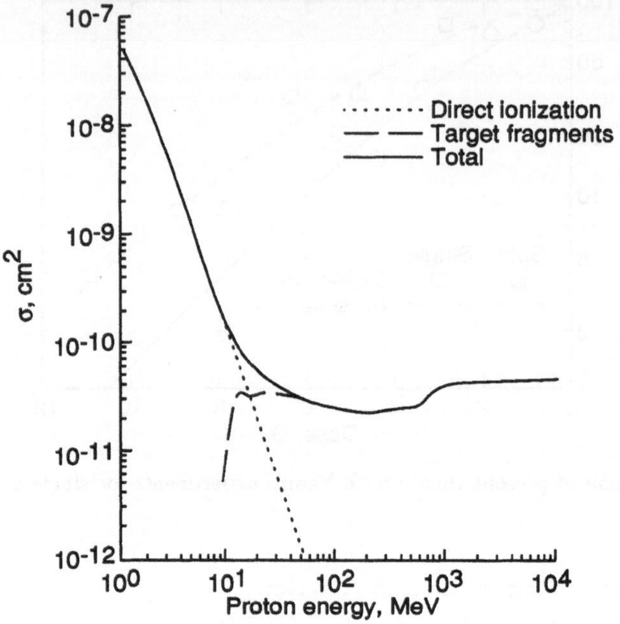

Figure 10. Katz cross section for protons in C3H10T 1/2 cells.

The Katz (Waligorski et al., 1987) cellular parameters for survival of C3H10T1/2 fit to the experiments of Yang et al. (1985) as given in table 1 are used to evaluate target fragment contributions according to equations (59) and (60). General agreement with the measured RBE values (Waligorski et al., 1987) was found using these parameter sets. The single-particle inactivation cross section neglecting target fragmentation of equation (60) is shown in figure 10 for cell death as a function of the energy, MeV, of the passing proton. The target fragmentation contribution [second term of equation (60)] has been evaluated and also shown in figure 10. For protons, the effect of the target fragments (dashed line, the second term of equation (60)) dominates over the proton direct ionization (dotted line) at high energy. For high-LET particles (low energy), the direct ionization dominates and target fragmentation effects become negligible. The effects of target fragments on the gamma-kill dose equation (59) are small (Cucinotta et al., 1991b) and are neglected here. The effective cross section is now used to study the repair capability of the cell for target fragment induced lesions. We have calculated the immediate and delayed plating response including target fragment contributions and compare these with results in which target fragments are neglected (see figure 11).

Results for 10 MeV proton exposures are shown in figure 11a. The response curves are characteristic of x-ray exposures and target fragments play a small role at this energy. Exposures at 50 MeV and 100 MeV clearly display target fragment effects (figs. 11b and 11c) but are beyond our ability to measure in biological experiments. Target fragment effects are quite large at 1,000 MeV and show in figure 11d clearly a reduced capability of the cell to repair fragment induced lesions. When target fragments are neglected the response curves are nearly those expected for x-ray exposures.

Remarks

The multilesion track structure model described herein agrees well with the available experimental data for C3H10T 1/2 cells. The lack of repair capability of the cell for target fragment induced damage by high energy protons is predicted by the model. We must await further experiments to confirm these predictions.

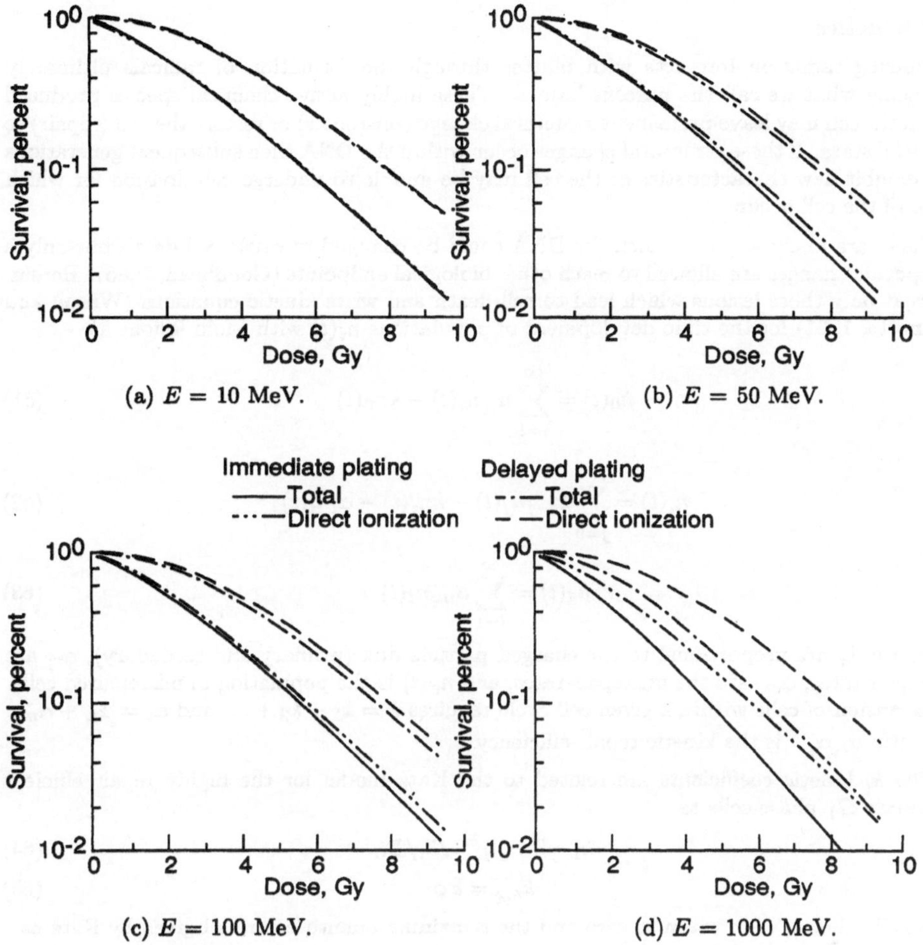

Figure 11. Target fragmentation effects on repair processes in proton exposures.

CELL KINETIC EFFECTS ON RBE

Radiation exposure limits and the associated biological risks are mainly estimated from human exposures at the two nuclear weapon sites of WWII and were predominately due to γ-rays (BEIRV, 1990) although the neutron component is still uncertain. The quality factor Q is defined as an extrapolation factor for risk estimation for radiations of different quality and is a judgment from estimates of RBE_m as the low dose limit for selected biological endpoints (ICRU, 1986; Sinclair, 1985). It is assumed that RBE_m is dose rate independent and is achieved in controlled laboratory experiments usually with single exposure of the biological specimens. Cell cultures are playing an increasing role in determining RBE_m since the large populations required for low dose response is more easily achieved. An added advantage of cell culture studies is that the role of the effects of radiation quality may be better understood (Katz and Cucinotta, 1991) and the role of cell kinetics can be studied more directly in these simpler biological systems (Yang, et al., 1989). The limitation of the cell radiation studies is that the relationship of *in vitro* response to *in vivo* response is not fully understood. In the present section we use a simple kinetic model to shed some light on the use of cell culture derived RBE's and their possible implication for tissue systems within an organism.

Cell Kinetics

Ionizing radiation interacts with matter through the formation of radicals ultimately producing what we call the nascent lesions. These highly active chemical species produced within the cell may leave permanent structural change (misrepair) or restore the cell (repair) to its initial state. If these structural changes occur within the DNA then subsequent generations may exhibit new characteristics or the cell may be unable to undergo cell division for which death of the cell occurs.

There are many ways in which the DNA could be changed to cause cell death but only a few specific changes are allowed to reach other biological endpoints (Goodhead, 1985). Herein, we treat only those lesions which lead to cell death and write kinetic equations (Wilson and Cucinotta, 1991) for the time development of populations $n_i(t)$ with i-fold lesions as

$$\dot{n}_0(t) = \sum_{i=1}^{\infty} \alpha_{r_i} n_i(t) - k n_0(t) \tag{61}$$

$$\dot{n}_i(t) = \sum_{j=0}^{i-1} k_{i-j} n_j(t) - k n_i(t) - \alpha_i n_i(t) \tag{62}$$

$$\dot{n}_d(t) = \sum_{i=1}^{\infty} \alpha_{m_i} n_i(t) \tag{63}$$

where the k_i are proportional to the charged particle flux (primary and secondary), α_{r_i} are the repair rates, α_{m_i} are the misrepair rates, and $n_d(t)$ is the population of misrepaired cells. Conservation of cells within a given cell cycle requires $k = k_1 + k_2 + \ldots$ and $\alpha_i = \alpha_{r_i} + \alpha_{m_i}$. The ratio $\alpha_{r_i} \alpha_i^{-1}$ is the kinetic repair efficiency.

The k_i kinetic coefficients are related to the Katz model for the highly repair efficient stationary G_1 phase cells as

$$k_1 = (m_d!)^{\frac{1}{m_d}} \dot{D}_\gamma / D_0 \tag{64}$$

$$k_{m_d} = \sigma\phi \tag{65}$$

where all other k_i's are taken as zero and the remaining quantities are all given by Katz as

$$\dot{D}_\gamma = \left(1 - \frac{\sigma}{\sigma_0}\right) L\phi \tag{66}$$

where ϕ is the local charged particle flux (primary and secondary), L is their corresponding LET, σ is approximated using the Katz model, equations (29) to (31).

The repair coefficients are found to be cell phase dependent and the stationary G_1-phase repair efficiencies are near maximum for $i < m_d$ and near zero otherwise. The exponential population showed relatively high single lesion repair efficiency and much lower multiple lesion repair efficiencies (see table 3) in analyzing the repair dependent experiments of Yang, et al. (1989). As examples, the G_1 repair enhanced exposures (delayed plating) and exponential

TABLE 3. SURVIVAL REPAIR RATES (h^{-1}) AND REPAIR EFFICIENCIES

	G_1 Phase			Exponential Phase		
i	1	2	$> m$	1	2	$> m$
α_i	.25	.125	$< .08$	.25	.125	$< .08$
$\alpha_{r_i} \alpha_i^{-1}$	$> .97$	$> .84$	~ 0	.7	.118	~ 0

phase repair exposures are compared to the present results in figure 7 for various ions (Wilson and Cucinotta, 1991) and with x-ray fractionated exposures (Wilson, Cucinotta, and Shinn, 1991) in figure 8. We will use this model to study the functional dependence of RBE at low total dose for G_1 phase and exponential phase repair processes.

Low Dose Rate Exposures

We consider now a special solution of equations (61) to (63) for an exposure field with a low constant dose rate ($\alpha_i \gg k_j$ for all i, j). At low dose rates the populations of cells with lesions can be approximated as

$$n_1(t) \simeq k_1 n_0(t)/\alpha_1 \tag{67}$$

$$n_2(t) \simeq k_1^2 n_0(t)/\alpha_1\alpha_2 \tag{68}$$

$$n_3(t) \simeq (k_1^3/\alpha_1\alpha_2\alpha_3 + k_3/\alpha_3) n_0(t) \tag{69}$$

In the case of low total exposure $n_0(t)$ may be taken as constant and the accumulation of misrepaired cells is written as

$$\frac{n_m(t)}{n_0} \simeq \frac{\alpha_{m_1}}{\alpha_1} 6^{\frac{1}{3}} \frac{(1-P)D}{D_0} + \frac{\alpha_{m_2}}{\alpha_2} 6^{\frac{2}{3}} \frac{(1-P)^2 \dot{D} D}{D_0^2 \alpha_1} + \frac{\alpha_{m_3}}{\alpha_3} 6 \frac{(1-P)^3 \dot{D}^2 D}{D_0^3 \alpha_1 \alpha_2} + \frac{\alpha_{m_3}}{\alpha_3} \frac{\sigma}{L} D \tag{70}$$

where $\dot{D}$ is the dose rate and $P = \sigma/\sigma_0$. In the case of an exponential population $\frac{\alpha_{m_1}}{\alpha_1} \simeq 0.3$ so that the first term is always dominant over the second and third term for very low dose rate exposures ($\dot{D}\alpha_i^{-1} \ll D_0$). The RBE is found to be

$$\mathrm{RBE}_m = 1 - P + 6^{-\frac{1}{3}} \frac{\alpha_{m_3}}{\alpha_3} \frac{\alpha_1}{\alpha_{m_1}} \frac{\sigma}{L} D_0 \tag{71}$$

as was found for our earlier result (Wilson and Cucinotta, 1991). If the repair efficiency of G_1 phase is highly efficient $\left(\frac{\alpha_{m_1}}{\alpha_1} \ll \frac{\dot{D}}{\alpha_i D_0}\right)$ then the higher order terms of equation (70) cannot be ignored in determining RBE for which there are important dose rate dependent factors whenever $\dot{D} \gg \alpha_i D_0 \approx 0.01$ Gy min^{-1}. At much lower dose rates ($\dot{D} \ll 0.01 \frac{\alpha_{m_1}}{\alpha_1}$ Gy min^{-1}.) then the RBE$_m$ given by equation (71) is obtained. A parameter study using the data in figure 7 shows $\frac{\alpha_{m_1}}{\alpha_1} < 0.03$ corresponding to 97 percent repair efficiency as noted in table 3. Taking this as the lower limit on G_1 repair efficiency the dose rate dependent low-dose limit of RBE is shown in figure 12. Although the exponential population RBE$_m$ is easily achieved for all the ions shown in figure 12, the G_1 population RBE shows a strong dose rate dependence with the RBE$_m$ reached at effectively zero dose rate. Clearly RBE$_m$ will be difficult to measure experimentally.

Results and Discussion

Values of RBE$_m$ are shown for the exponential population in figure 13 and table 2 according to the parameters in tables 1 and 3. Also shown in table 2 are RBE values measured by Yang, et al. at two exposure levels (corresponding to 10 and 50 percent survival levels). It might be surmised that a slight increase in repair efficiency for the exponential population may be appropriate according to table 2. The RBE for ^{4}He ions (Bettega, et al., 1990) shown as the single datum in figure 13 was measured at 0.01 Gy exposure. A 10 percent increase in $\alpha_{m_1}\alpha_1^{-1}$ would bring the theory and the ^{4}He datum into agreement. There are no low dose and low

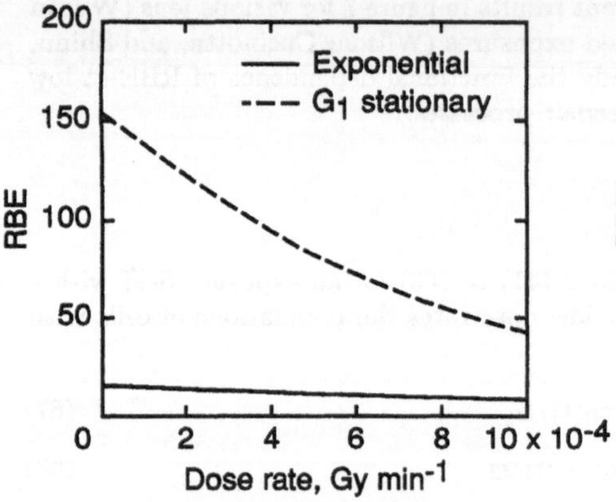

(a) ^{1}H, 95 keV μm^{-1}.

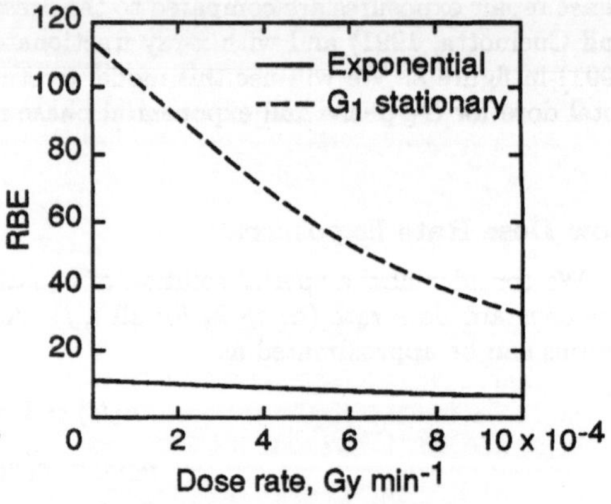

(b) ^{4}He, 101 keV μm^{-1}.

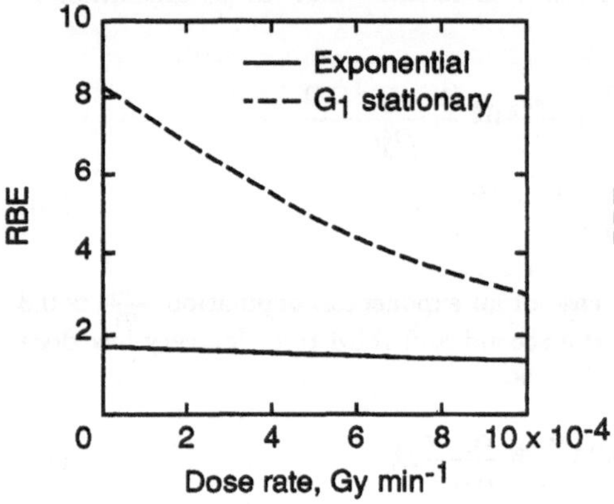

(c) ^{20}Ne, 32 keV μm^{-1}.

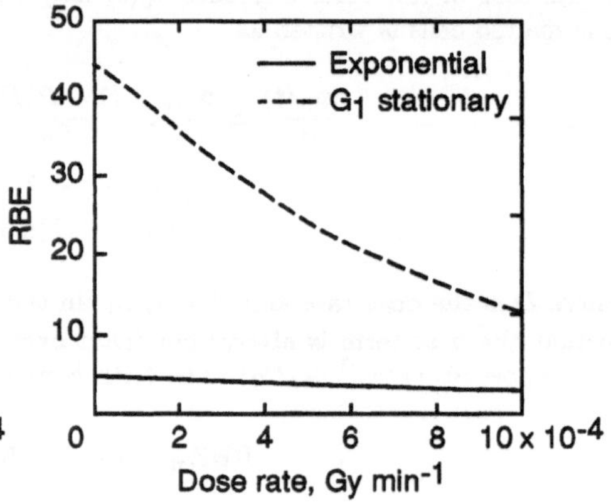

(d) ^{56}Fe, 192 keV μm^{-1}.

Figure 12. Predicted RBE for G_1 stationary and exponential population as a function of dose rate for various ion types.

dose rate measurements with which to make comparison for the G_1 population. Values greatly in excess of those given for the exponential population are expected. The low dose rate RBE measurements would be helpful in establishing the G_1 phase repair efficiencies.

Considering that highly differentiated tissues consist mainly of G_1 phase cells at any instant of time one might argue that the relevant RBE's for the mouse would be the stationary G_1 phase values which have not yet been measured and are difficult to estimate from current G_1 phase studies. The exponential phase RBE_m then appears as a lower limit on the relevant RBE values. Clearly the relevance of RBE_m of the exponential population to the mouse cannot be adequately resolved until the relationship of cell culture experiments to tissue response is better understood.

GENERALIZED LINEAR KINETIC MODEL

We assume at low total dose (<10 Gy) that cell injury is through damage on specific loci along the DNA strands and such loci are related to some characteristic of the cell. The nascent lesions are assumed to be chemically active species which directly involve the locus of interest. We label the loci as l and the number of lesions within the loci as i. Each locus will

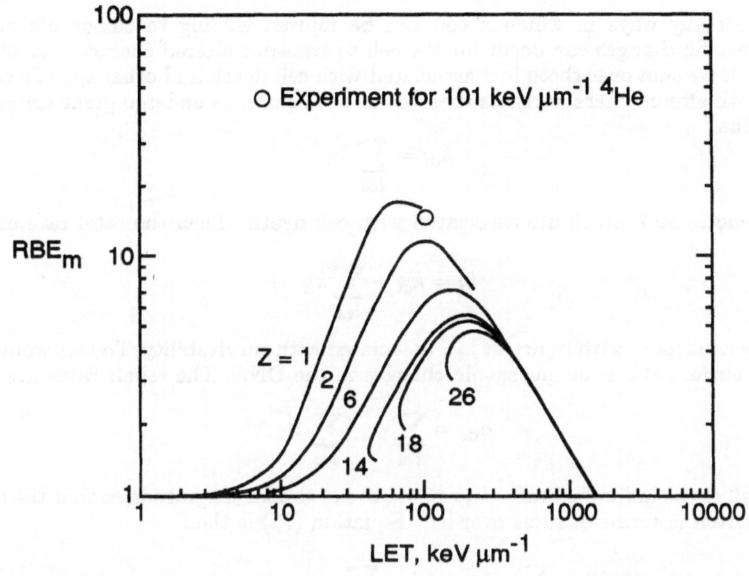

Figure 13. RBE$_m$ for cell survival of a C3H10T 1/2 exponential population. The (○) is the value measured by Bettega, et al. (1990) for low energy ^{4}He ions at 0.01 Gy.

have lesion generation rates and repair rates similar to our multilesion model discussed in the previous sections. The initial uninjured population $n_0(0)$ then develops in time as

$$\dot{n}_0(t) = \sum_{li} \alpha_{rli}\, n_{li}(t) - kn_0(t) \tag{72}$$

where $n_{li}(t)$ is the number of cells with one loci l damaged with i lesions and α_{rli} is the corresponding repair rate coefficient and k is the rate at which damage is induced in the population. The α_{rli} represent the rates associated with a complex series of events resulting in the final repair of the locus. We assume a corresponding misrepair rate coefficient α_{mli} for which the highly chemically active species (radicals) have been stabilized but the corresponding locus is left in a misrepaired state

$$\dot{n}_l(t) = \sum_i \alpha_{mli} n_{li}(t) + \sum \alpha_{rl'i'} n_{ll'i'}(t) - kn_l(t) \tag{73}$$

and results in a permanent change in the state of the cell. The number of cells with l, i satisfy similar equations

$$\dot{n}_{li}(t) = \sum_{l'i'} \alpha_{rl'i'} n_{lil'i'}(t) + k_{li} n_0(t) + \sum_{j=1}^{i-1} k_{li-j} n_{lj}(t) - kn_{li}(t) - \alpha_{li} n_{li}(t) \tag{74}$$

where the collision induced lesion rate coefficients satisfy

$$k = \sum_{li} k_{li} \tag{75}$$

with

$$\alpha_{li} = \alpha_{rli} + \alpha_{mli} \tag{76}$$

and $n_{lil'i'}(t)$ are the cells with lesions at two loci and similar rate equations.

There are many ways in which a cell can be injured leading to clonogenic death and only a few specific changes can occur for the cell to transmit altered character to subsequent generations. If we sum over those loci associated with cell death and other specific target loci associated with altered characteristics then the above equations undergo great simplification. First we define

$$k_{di} = \sum_{l\epsilon d} k_{li} \tag{77}$$

where $l\epsilon d$ denotes all l which are associated with cell death. Then the total rate of induced injury is

$$k_i = k_{di} + \sum_{l\epsilon s} k_{li} \tag{78}$$

where $l\epsilon s$ are associated with injury at loci associated with survivability. The $l\epsilon s$ would include neoplastic transformations or mutagenic changes in the DNA. The repair rates are similarly defined as

$$\alpha_{di} = \sum_{l\epsilon d} \alpha_{li} / \sum_{l\epsilon d} 1^l \tag{79}$$

We assume all of the individual rates k_{li} and α_{li} are of similar magnitude so that the equations may be rewritten in terms of sums over $l\epsilon d$. Equation (72) is then

$$\dot{n}_0(t) = \sum_i \alpha_{rdi} n_{di}(t) + \sum_{l,i} \alpha_{rli} n_{li}(t) - k n_0(t) \tag{80}$$

where the second sum is over all l which do not contribute to d. Similarly equation (73) becomes

$$\dot{n}_d(t) = \sum_i \alpha_{mdi} n_{di}(t) + \sum_i \alpha_{li} n_{lid}(t) + \sum_{i'} \alpha_{mdi'} n_{ldi'}(t) - k_s n_d(t) \tag{81}$$

and

$$\dot{n}_l(t) = \sum_i \alpha_{mli} n_{li}(t) + \sum_{i'} \alpha_{rdi'} n_{ldi'}(t) - k_d n_l(t) \tag{82}$$

with equation (74) given by

$$\dot{n}_{di}(t) = \sum_{l'i'} \alpha_{rl'i'} n_{l'i'di}(t) + k_{di} n_0(t) + \sum_{j=1}^{i-1} k_{di-j} n_{dj}(t) - k n_{di}(t) - \alpha_{di} n_{di}(t) \tag{83}$$

and

$$\dot{n}_{li}(t) = \sum_{i'} \alpha_{rdi'} n_{lidi'}(t) + k_{li} n_0(t) + \sum_{j=1}^{i-1} k_{li-j} n_{lj}(t) - k n_{li}(t) - \alpha_{li} n_{li}(t) \tag{84}$$

We further give the time development for cells with more than one loci injured as

$$\begin{aligned} \dot{n}_{lidi'}(t) = k_{li} n_{di'}(t) + k_{di'} n_{li}(t) + \sum_{j=1}^{i-1} k_{li-j} n_{ljdi'}(t) + \sum_{j'=1}^{i-1} k_{di'-j'} n_{lidj'}(t) \\ - k n_{lidi'}(t) - \alpha_{li} n_{lidi'}(t) - \alpha_{di'} n_{lidi'}(t) \end{aligned} \tag{85}$$

where the cells $n_{lil'i'}(t)$ are small in number compared to $n_{lidi'}(t)$ and are neglected. The corresponding misrepair equations are

$$\dot{n}_{lid}(t) = \sum_{i'} \alpha_{mdi'} n_{lidi'}(t) + k_{li} n_d(t) + \sum_{j=1}^{i-1} k_{li-j} n_{ljd}(t) - k_s n_{lid}(t) - \alpha_{li} n_{lid}(t) \tag{86}$$

$$\dot{n}_{ldi'}(t) = \sum_i \alpha_{mli} n_{lidi'}(t) + k_{di'} n_l(t) + \sum_{j'=1}^{i'-1} k_{di'-j'} n_{ldj'}(t) - k_d n_{ldi'}(t) - \alpha_{di'} n_{ldi'}(t) \quad (87)$$

Note that

$$k_s = \sum_{\substack{l \epsilon s \\ i}} k_{li} \quad (88)$$

$$k_d = \sum_{\substack{l \epsilon d \\ i}} k_{li} \quad (89)$$

are the total injury rate coefficients for altered cells and cell death respectively. In the following we will treat only neoplastic transformation and clonogenic death. Other biological endpoints are easily treated in the present formalism. We will now use the repair kinetic studies of Yang et al. and the Katz formalism to evaluate the kinetic coefficients for the C3H10T1/2 system.

Katz Model for Transformations

As before in the Katz model, there are four cellular parameters to describe the response of the cells for a given biological endpoint, two of which (m, the number of targets per cell, and D_0, the characteristic x-ray dose) are extracted from the response of the cellular system to x and γ irradiation. The other two (σ_0 interpreted by Katz as the cross-sectional area of the cell nucleus within which the sensitive sites are located and κ, a measurement of the size of the sensitive site) are found principally from cell assays after track-segment irradiations with energetic, charge particles.

As discussed before the capacity of cells to accumulate sublethal damage, two modes of injury are identified: "ion-kill" (corresponding to intratrack effects) and "gamma-kill" (corresponding to intertrack effects). When the passage of a single ion damages cells, the ion-kill mode occurs. In the grain-count regime, the fraction of cells damage in the ion-kill mode is taken as $P = \sigma/\sigma_0$, where σ is the single-particle injury cross section and P is the probability of damage in the ion-kill mode. The track model assumes that a fraction of the ion dose, $(1-P)$, acts cumulatively with that from other particles to injure cells in the gamma-kill mode. The untransformed fraction of a cellular population $n_0(\infty)$, whose response parameters are m, D_0, σ_0, and κ after irradiation by a fluence of particles F, is written

$$\frac{n_0(\infty)}{n_0} = \pi_i \times \pi_\gamma \quad (90)$$

where

$$\pi_i = e^{-\sigma F} \quad (91)$$

is the ion-kill injury probability and

$$\pi_i = 1 - \left(1 - e^{-D_\gamma/D_0}\right)^m \quad (92)$$

is the gamma-kill injury probability. The gamma-kill dose fraction is

$$D_\gamma = (1-P)D \quad (93)$$

where D is the absorbed dose as in the case of cell inactivation. The single-particle injury cross section σ is given by equation (29) as in cell inactivation. The transformation yield, Y_t is written as

$$Y_t = 1 - \pi_i \times \pi_\gamma \quad (94)$$

It is the way in which transformation experiments are performed that one observes not the fraction of uninjured cells or even the transformation yield (the number of transformations per cell at risk). Rather survival experiments run concurrently and among the cell colonies (survivors) that are observed, a fraction are transformed colonies. The fraction of surviving colonies which are transformed is called the transformation frequency, F_t. In our computational model, we find directly the transformation yield which is relatable to transformation frequency and survival fraction S as

$$Y_t = SF_t = (1 - \pi_i \times \pi_\gamma) \tag{95}$$

At low dose levels the survival fraction approaches unity so that

$$Y_t \approx F_t \tag{96}$$

To compensate, we allow for changes in the four Katz parameters as fit to frequency data. We now reconsider the relationship of the Katz model to the kinetics model.

Physics and Kinetics of Cell Injury

We again define the relationship of repair to the Katz model parameters within the context of the present repair kinetic model. We use the same approach as used in the cell survival model. We first consider the impulsive exposure of a cell population for a short time period t_r after which the cell population is given by

$$n_0(t_r) = n_0(0)e^{-kt_r} \tag{97}$$

$$n_{d1}(t_r) = k_{d1}t_r n_0(0)e^{-kt_r} \tag{98}$$

$$n_{d2}(t_r) = \left(k_{d2}t_r + \frac{1}{2!}k_{d1}^2 t_r^2\right) n_0(0)e^{-kt_r} \tag{99}$$

$$n_{d3}(t_r) = \left(k_{d3}t_r + \frac{2}{2!}k_{d2}k_{d1}t_r^2 + \frac{1}{3!}k_{d1}^3 t_r^3\right) n_0(0)e^{-kt_r} \tag{100}$$

$$n_{l1}(t_r) = k_{l1}t_r n_0(0)e^{-kt_r} \tag{101}$$

$$n_{l2}(t_r) = \left(k_{l2}t_r + \frac{1}{2!}k_{l1}^2 t_r^2\right) n_0(0)e^{-kt_r} \tag{102}$$

$$n_{l1d1}(t_r) = \frac{2}{2!}k_{l1}k_{d1}t_r^2 n_0(0)e^{-kt_r} \tag{103}$$

$$n_{l1d2}(t_r) = \left(\frac{2}{2!}k_{l1}k_{d2}t_r^2 + \frac{3}{3!}k_{d1}^2 k_{l1}t_r^3\right) n_0(0)e^{-kt_r} \tag{104}$$

$$n_{l2d1}(t_r) = \left(\frac{2}{2!}k_{d1}k_{l2}t_r^2 + \frac{3}{3!}k_{l1}^2 k_{d1}t_r^3\right) n_0(0)e^{-kt_r} \tag{105}$$

and similarly for higher order terms. After the exposure period the system is allowed to repair in the G_1 stationary phase where repair efficiencies are near maximum. The repair kinetics subsequent to exposure are described by

$$n_{lidi'}(t') = e^{-(\alpha_{li}+\alpha_{di'})t}\, n_{lidi'}(t_r) \tag{106}$$

$$n_{di}(t) = e^{-\alpha_{di}t}n_{di}(t_r) + \sum_{l'i'} \frac{\alpha_{rl'i'}}{\alpha_{l'i'}}\left[e^{-\alpha_{di}t} - e^{-(\alpha_{l'i'}+\alpha_{di})t}\right] n_{l'i'di}(t_r) \tag{107}$$

$$n_{li}(t) = e^{-\alpha_{li}t}n_{li}(t_r) + \sum_{i'} \frac{\alpha_{rdi'}}{\alpha_{di'}}\left[e^{-\alpha_{li}t} - e^{-(\alpha_{li}+\alpha_{di'})t}\right] n_{lidi'}(t_r) \tag{108}$$

$$n_{lid}(t) = \sum_{i'} \frac{\alpha_{mdi'}}{\alpha_{di'}} \left[e^{-\alpha_{li}t} - e^{-(\alpha_{li}+\alpha_{di'})t} \right] n_{lidi'}(t_r) \tag{109}$$

$$n_{ldi}(t) = \sum_{i'} \frac{\alpha_{mli'}}{\alpha_{li'}} \left[e^{-\alpha_{di}t} - e^{-(\alpha_{li'}+\alpha_{di})t} \right] n_{li'di}(t_r) \tag{110}$$

$$\begin{aligned} n_d(t) = {} & \sum_i \frac{\alpha_{mdi}}{\alpha_{di}} \left(1 - e^{-\alpha_{di}t}\right) n_{di}(t_r) \\ & + 2 \sum_{l'i'} \frac{\alpha_{mdi'}}{\alpha_{di'}} \frac{\alpha_{mli}}{\alpha_{li}} \left[\left(1 - e^{-\alpha_{di'}t}\right) - \left(\frac{\alpha_{di'}}{\alpha_{li} + \alpha_{di'}}\right) \left(1 - e^{-(\alpha_{li}+\alpha_{di'})t}\right) \right] n_{l'i'di}(t_r) \\ & + \sum \frac{\alpha_{mdi'}}{\alpha_{di'}} \left[\left(1 - e^{-\alpha_{li}t}\right) - \left(\frac{\alpha_{li}}{\alpha_{li} + \alpha_{di'}}\right) \left(1 - e^{-(\alpha_{li}+\alpha_{di'})t}\right) \right] n_{lidi'}(t_r) \end{aligned} \tag{111}$$

$$\begin{aligned} n_l(t) = {} & \sum_{i'} \frac{\alpha_{mli}}{\alpha_{li}} \left(1 - e^{-\alpha_{li}t}\right) n_{li}(t_r) \\ & + \sum_i \alpha_{mli} \sum_{i'} \frac{\alpha_{rdi'}}{\alpha_{di}} \left[\frac{1 - e^{-\alpha_{li}t}}{\alpha_{li}} - \frac{1 - e^{-(\alpha_{li}+\alpha_{di'})t}}{\alpha_{li} + \alpha_{di'}} \right] n_{lidi'}(t_r) \\ & + \sum \frac{\alpha_{mli}}{\alpha_{li}} \frac{\alpha_{rdi'}}{\alpha_{di'}} \left[1 - e^{-\alpha_{di'}t} - \left(\frac{\alpha_{di'}}{\alpha_{li} + \alpha_{di'}}\right) \left(1 - e^{-(\alpha_{li}+\alpha_{di'})t}\right) \right] n_{lidi'}(t_r) \end{aligned} \tag{112}$$

The number of uninjured cells is

$$\begin{aligned} n_0(t) = {} & n_0(t_r) + \sum \frac{\alpha_{rdi'}}{\alpha_{di'}} \left(1 - e^{-\alpha_{di'}t}\right) n_{di}(t_r) + \sum \frac{\alpha_{rli}}{\alpha_{li}} \left(1 - e^{-\alpha_{li}t}\right) n_{li}(t_r) \\ & + \sum \frac{\alpha_{rdi'}}{\alpha_{di'}} \frac{\alpha_{rli}}{\alpha_{li}} \left[1 - e^{-\alpha_{di'}t} - \left(\frac{\alpha_{di'}}{\alpha_{di'} + \alpha_{li}}\right) \left(1 - e^{-(\alpha_{di'}+\alpha_{li})t}\right) \right] n_{lidi'}(t_r) \\ & + \sum \frac{\alpha_{rli}}{\alpha_{li}} \frac{\alpha_{rdi'}}{\alpha_{di'}} \left[1 - e^{-\alpha_{li}t} - \left(\frac{\alpha_{li}}{\alpha_{li} + \alpha_{di'}}\right) \left(1 - e^{-(\alpha_{di'}+\alpha_{li})t}\right) \right] n_{lidi'}(t_r) \end{aligned} \tag{113}$$

and the number of survivors is

$$\begin{aligned} n_s(t) = {} & n_0(t_r) + \sum \left(1 - e^{-\alpha_{li}t}\right) n_{li}(t_r) + \sum \frac{\alpha_{rdi'}}{\alpha_{di'}} \left(1 - e^{-\alpha_{di'}t}\right) n_{di'}(t_r) \\ & + \sum \frac{\alpha_{rdi'}\alpha_{rli}}{\alpha_{di'}\alpha_{li}} \left[1 - e^{-\alpha_{di}t} - \left(\frac{\alpha_{di'}}{\alpha_{di'} + \alpha_{li}}\right) 1 - e^{-(\alpha_{li}+\alpha_{di'})t} \right] n_{lidi}(t_r) \\ & + \sum \frac{\alpha_{rdi'}}{\alpha_{di'}} \left[1 - e^{-\alpha_{li}t} - \left(\frac{\alpha_{li}}{\alpha_{li} + \alpha_{di'}}\right) \left(1 - e^{-(\alpha_{li}+\alpha_{di'})t}\right) \right] n_{lidi'}(t_r) \end{aligned} \tag{114}$$

If the cells are allowed to complete their repair then the number of survivors is

$$n_s(\infty) = n_0(t_r) + \sum n_{li}(t_r) + \sum \frac{\alpha_{rdi'}}{\alpha_{di'}} n_{di'}(t_r) + \sum \frac{\alpha_{rdi'}}{\alpha_{di'}} \left(\frac{\alpha_{rli} + \alpha_{di'}}{\alpha_{li} + \alpha_{di'}}\right) n_{lidi'}(t_r) \tag{115}$$

and the number of transformed cells is

$$n_l(\infty) = \sum \frac{\alpha_{mli}}{\alpha_{li}} n_{li}(t_r) + \sum \frac{\alpha_{mli}}{\alpha_{li}} \frac{\alpha_{rdi'}}{\alpha_{di'}} n_{lidi'}(t_r) \tag{116}$$

We now look to the Katz model for guidance in estimating k's. We again use the G_1 stationary phase parameters for which m_l and m_d are determined by Katz and we assume

$$\alpha_{mdi} \approx 0 \qquad i \leq m_d \tag{117}$$

$$\alpha_{mli} \approx 0 \qquad i \leq m_l \tag{118}$$

Expanding the Katz model in the low dose limit and equating to the lowest surviving terms in equations (115) and (116) results in

$$k_{dm_d} t_r = \sigma_d F \tag{119}$$

$$k_{d1} t_r = (m_d!)^{\frac{1}{m_d}} \frac{D_{\gamma d}}{D_{0d}} \tag{120}$$

$$k_{lm_l} t_r = \sigma_l F \tag{121}$$

$$k_{l1} t_r = (m_l!)^{\frac{1}{m_l}} \frac{D_{\gamma l}}{D_{0l}} \tag{122}$$

where $\sigma_d, m_d, D_{\gamma d}, D_{0d}$ are the usual Katz functions and parameters for cell death and $\sigma_l, m_l, D_{\gamma l}, D_{0l}$ correspond to the values for cell transformation. The α_{di} and α_{rdi} will be taken from our earlier study of cell survival and the α_{li} and α_{rli} will be found by analysis of the data of Yang et al.

The series solutions given by equations (115) and (116) converge slowly at high dose and it is convenient to perform some of the summations in closed form. For example, equation (115) may be written as

$$n_s(\infty) = n_0 e^{-k_d t_r} + \sum_{i'=1}^{m_d-1} \frac{\alpha_{rdi'}}{\alpha_{di'}} \left[n_{di'}(t_r) + \sum_{i=1}^{\infty} \left(\frac{\alpha_{rli} + \alpha_{di'}}{\alpha_{li} + \alpha_{di'}} \right) n_{lidi'}(t_r) \right] \tag{123}$$

with corresponding values for equation (116) as

$$\begin{aligned} n_l(\infty) = \left(e^{k_s t_r} - 1\right) & \left[e^{-kt_r} n_0 + \sum_{i'=1}^{m_d-1} \frac{\alpha_{rdi'}}{\alpha_{di'}} n_{di'}(t_r) \right] \\ & - \sum_{i=1}^{m_l-1} \frac{\alpha_{rli}}{\alpha li} \left[n_{li}(t_r) + \sum_{i'=1}^{m_d-1} \frac{\alpha_{rdi'}}{\alpha_{di'}} n_{lidi'}(t_r) \right] \end{aligned} \tag{124}$$

These expressions apply to both the delayed plating experiments ($\alpha_r \approx \alpha$) as well as the immediate plating experiments of Yang et al. provided the appropriate α's are used.

Delayed Plating Studies

Yang and coworkers have utilized the following protocol, G_1 stationary cells are given a high dose rate exposure. G_1 stationary repair kinetics are utilized by retaining the population in G_1 phase for 24 hours before plating and scoring the cell modifications. Exponential phase repair kinetics are envoked by plating the cells immediately following exposure. The scoring of these protocols relate to the repair efficiency of an exponential population. Experiments with variable delay times allows evaluation of the repair rates. The delayed plating and immediate plating experiments are described by equations (115) and (116) with repair coefficients for G_1 stationary and exponential populations respectively. Description of the variable delay experiment require further development.

The cell populations subsequent to exposure are described by equations (106) to (112) where t is the time period. If at time t (delay time) the cells are plated then the final populations are given as

$$n_s(\infty) = n_0(t) + \sum n_{li}(t) + \sum \frac{\alpha'_{rd'}}{\alpha'_{di'}} n_{di'}(t) + \sum \frac{\alpha'_{rdi'}}{\alpha'_{di'}} \left(\frac{\alpha'_{rli} + \alpha'_{di}}{\alpha'_{li} + \alpha'_{di}} \right) n_{lidi}(t) \tag{125}$$

$$n_l(\infty) = n_l(t) + \sum_i \frac{\alpha'_{mli}}{\alpha'_{li}} n_{li}(t) + \sum_{ii'} \frac{\alpha'_{mli}}{\alpha'_{li}} \frac{\alpha'_{rdi'}}{\alpha'_{di'}} n_{lidi'}(t) \tag{126}$$

where the α coefficients are the G_1 stationary values and α' coefficients are the exponential phase values. The coefficients for cell death from our earlier studies are given in table 3. We have used the transformation studies of Yang et al. to determine the transformation repair coefficients using the Katz parameters in table 4a. The results for x-ray exposures are shown in figure 14. The original Katz parameters were fit to transformation frequency whereas the coefficients in our model relate to transformation yield. As a result we made some adjustments in the Katz parameters in arriving at our results in figure 14. It is difficult to ascribe the quality of fit to Yang's data due to scatter in the experiments. What is certain is that $m_l = 3$ is the only value consistent with the immediate plating data. The corresponding results are shown in figures 15 and 16 for various ions. The scattering in the transformation data limits our ability to evaluate the model.

CELL CYCLE RADIATION SURVIVAL MODEL

Living cells are found to proceed through a series of events leading to cell division referred to as the cell cycle. There are two significant events denoted by S-phase (synthesis of DNA material) and M-phase (cell division). These phases are separated by two gaps called G_1 (following mitosis) and G_2 (following S and preceding M). The cell cycle may be limited by the physical/chemical environment, interaction with adjacent cells, or available nutrients. Indeed, the growth of specialized tissues in complex organisms is controlled by cell contact interaction and exchange of growth controlling chemical compounds (Allen 1962).

The role of repair in radiobiological response was elegantly presented by Fritz-Niggli (1988). The (early) S- and M-phases appear accident prone for which G_1 and G_2 are instrumental in making repairs. Evidence of these facts lie in the following observations. First, the errors of the S- and M-phases are normally repaired, otherwise life would not exist (Fritz-Niggli,

TABLE 4a. KATZ C3H10T1/2 CELL PARAMETERS

	σ_0, cm^2	k	m	D_0, GY
Survival	5×10^{-7}	750	3	2.8
Transformation	7×10^{-11}	475	3	117

TABLE 4b. TRANSFORMATION REPAIR RATES (h^{-1}) AND REPAIR EFFICIENCY

	G_1 Phase			Exponential		
i	1	2	≥3	1	2	≥3
α_i	.25	.125	≤.08	.25	.125	≤.08
$\alpha_{r_i}\alpha_i^{-1}$	1.0	1.0	0.0	.99	.70	0.0

1988). Second, radiation injury sustained in (early) S- and M-phases is more likely to end the cell line than injury received in the (early) G_1- and G_2-phase (especially early G_1 and late G_2, Sinclair, 1968). The cell cycle progression can be blocked (delayed) in G_1 or G_2 by injury sustained in that phase until the injury is repaired (Mitchison, 1971)). These simple facts alone provide insight as to the biological response of more complex organisms.

A tissue from a complex organism exhibits a distribution of cells over various phases. The highly differentiated tissues are predominantly stationary G_1 and are well known to be radiation resistant. That (stationary) G_1 repair systems are highly efficient would seem necessary to preserve complex organisms. Stem cell tissues have significant populations of M- and especially S-phase cells and are in part responsible for acute radiation syndrome in higher animals. Immature individuals are more sensitive than adults and the embryo is most sensitive of all. Clearly, a viable model of radiation response must account for the varying repair kinetics for the differing cell phases and the distribution of tissue cells within the cell cycle.

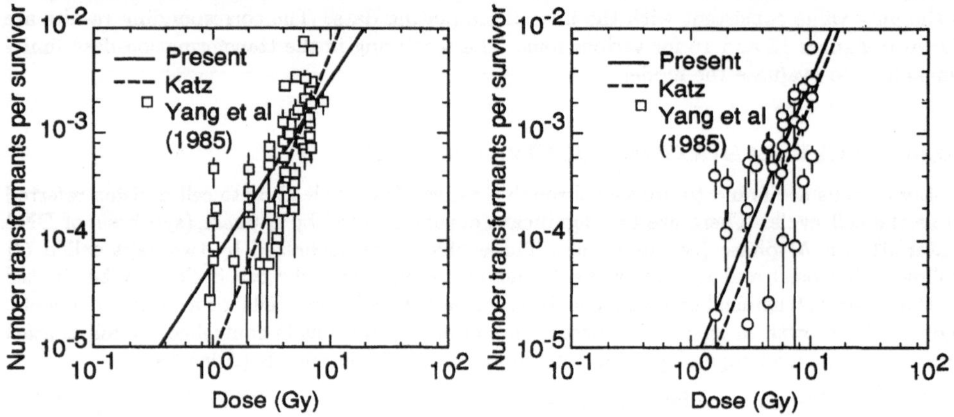

(a) Immediate plating x-ray. (b) Delayed plating x-ray.

Figure 14. Transformation frequencies for x-ray exposures.

Ultimately, the radiation response of an organism or tissue is determined by the biological processes of individual cells. It is the specific molecular structures and their physical interaction with passing ionizing radiation which initiates the response mechanisms, but the cell's ability to repair such physical insult is a primary determinant of cell sensitivity to ionizing radiation. In that the progression of cell chemistry is controlled by cellular environmental factors, individual cell response is governed in part by external factors for which there is some experimental control. Indeed, the environmental factors existing within tissue systems are ultimately related to carcinogenic response. From this viewpoint, cell repair kinetics and the relation to cell environment is of fundamental importance in understanding radiation effects in biological systems.

In the present section, we show how to develop a more comprehensive model of the cell kinetics. The present model lacks age dependence within the cycle. We still rely on the Katz model for a description of the physics of the track structure. The cell kinetics are represented by an unbounded set of coupled linear differential equations describing multiples of lesions within the cell. The kinetic coefficients in the model are to be determined from repair dependent cell response data.

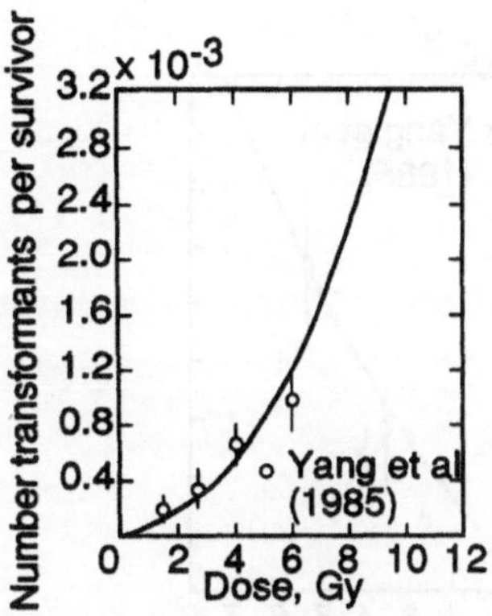

(a) Delayed plating Ne^{20} (425 MeV/n).

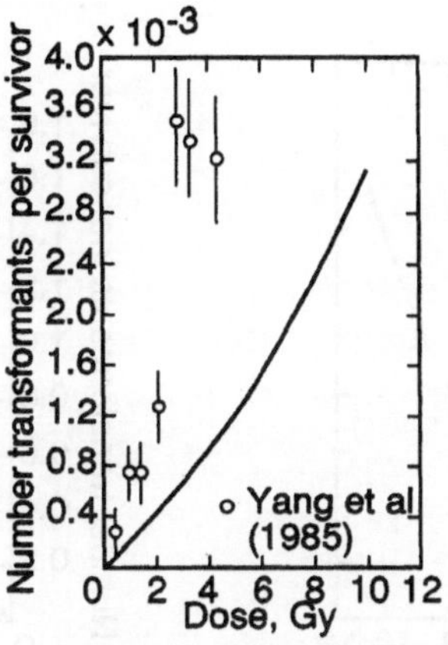

(b) Delayed plating Si^{28} (320 MeV/n).

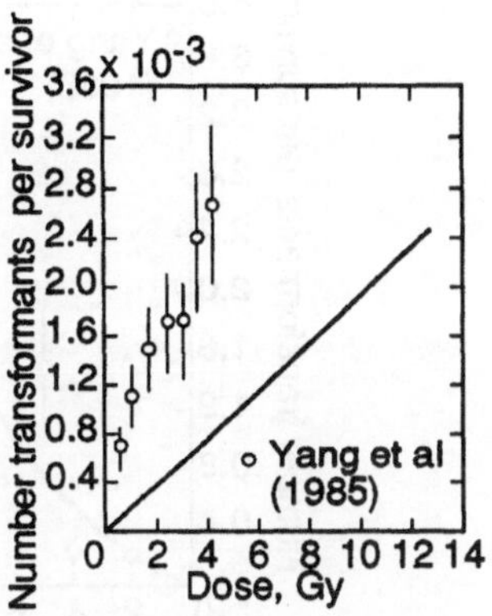

(c) Delayed plating Si^{28} (670 MeV/n).

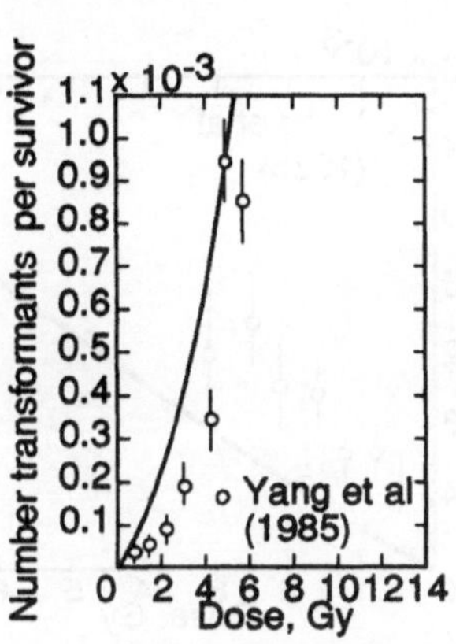

(d) Delayed plating Ar^{40} (330 MeV/n).

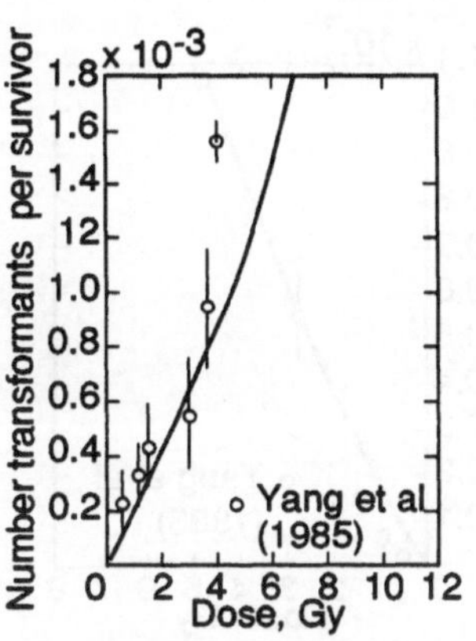

(e) Delayed plating Fe^{56} (600 MeV/n).

Figure 15. Transformation frequencies for various ions. Delayed plating.

Cell Cycle Kinetics

Cell cultures of higher animal cells cycle in approximately 24 hours. The cell populations n_1, n_2, n_3, n_4 represent the phases G_1, S, G_2, M and satisfy the approximate equations

$$\dot{n}_1 = 2n_4/\tau_4 - n_1/\tau_1 \tag{127}$$

$$\dot{n}_2 = n_1/\tau_1 - n_2/\tau_2 \tag{128}$$

$$\dot{n}_3 = n_2/\tau_2 - n_3/\tau_3 \tag{129}$$

$$\dot{n}_4 = n_3/\tau_3 - n_4/\tau_4 \tag{130}$$

The time spent in mitosis, τ_4, is quite small compared to the other phases so that n_4 is in near equilibrium with n_3 so that the approximate equations can be used

$$\dot{n}_1 = 2n_3/\tau_3 - n_1/\tau_1 \tag{131}$$

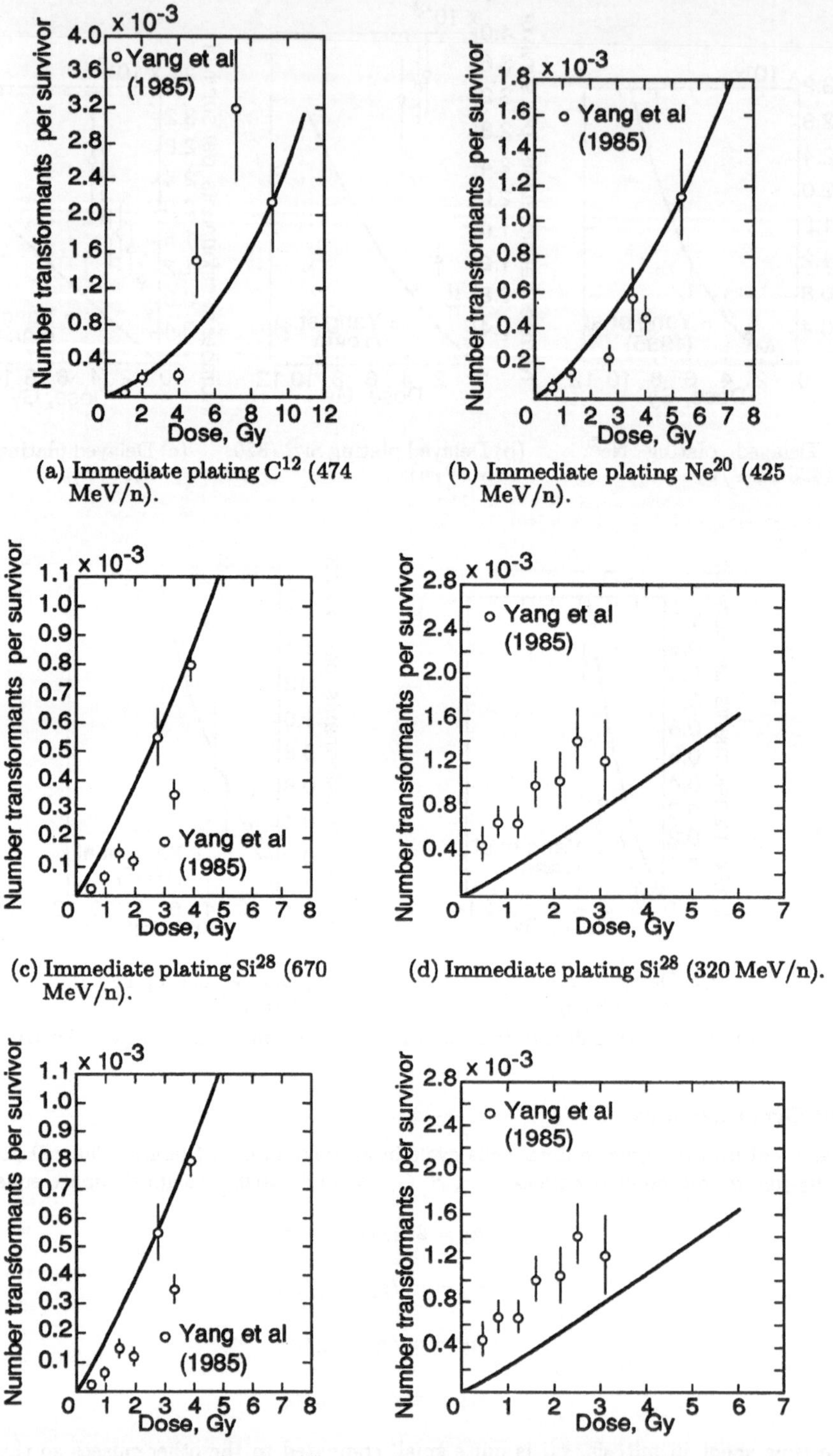

(a) Immediate plating C^{12} (474 MeV/n).

(b) Immediate plating Ne^{20} (425 MeV/n).

(c) Immediate plating Si^{28} (670 MeV/n).

(d) Immediate plating Si^{28} (320 MeV/n).

(e) Immediate plating Ar^{40} (330 MeV/n).

(f) Immediate plating Fe^{56} (300 MeV/n).

Figure 16. Transformation frequencies for various ions. Immediate plating.

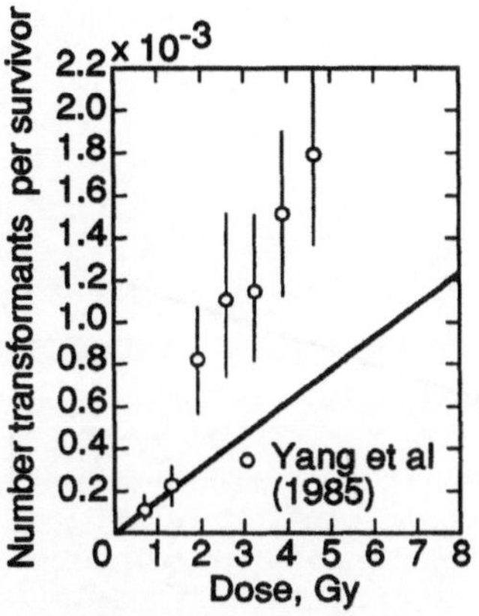

(g) Immediate plating Fe^{56} (400 MeV/n).

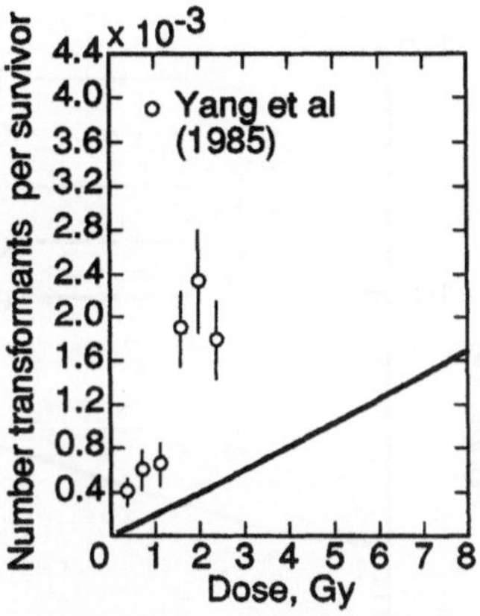

(h) Immediate plating Fe^{56} (600 MeV/n).

Figure 16. Concluded.

$$\dot{n}_2 = n_1/\tau_1 - n_2/\tau_2 \tag{132}$$

$$\dot{n}_3 = n_2/\tau_2 - n_3/\tau_3 \tag{133}$$

Furthermore, these three phases are sometimes of near equal duration so that $\tau = \tau_1 \approx \tau_2 \approx \tau_3$. One may then show that for an initial G_1 population that

$$n_1 = n_0\, e^{-t/\tau} + n_0\, \frac{1}{3}\left(\frac{t}{\tau}\right)^3 e^{-t/\tau} + n_0\, \frac{2}{3\cdot 4\cdot 5\cdot 6}\left(\frac{t}{\tau}\right)^6 e^{-t/\tau} + \ldots \tag{134}$$

$$n_2 = n_0\, \frac{t}{\tau}\, e^{-t/\tau} + n_0\, \frac{1}{3\cdot 4}\left(\frac{t}{\tau}\right)^4 e^{-t/\tau} + \ldots \tag{135}$$

$$n_3 = n_0\, \frac{1}{2}\left(\frac{t}{\tau}\right)^2 e^{-t/\tau} + n_0\, \frac{1}{3\cdot 4\cdot 5}\left(\frac{t}{\tau}\right)^5 e^{-t/\tau} + \ldots \tag{136}$$

The total cell population is given by

$$\begin{aligned} n_{\text{tot}} = n_0 + n_0 \sum_j \frac{(2^j - 1)}{(3j)!}\left(\frac{t}{\tau}\right)^{3j} e^{-t/\tau} + n_0 \sum_j \frac{(2^j - 1)}{(3j+1)!}\left(\frac{t}{\tau}\right)^{3j+1} e^{-t/\tau} \\ + n_0 \sum_j \frac{(2^j - 1)}{(3j+2)!}\left(\frac{t}{\tau}\right)^{3j+2} e^{-t/\tau} \end{aligned} \tag{137}$$

We note that the present cell phase model is never fully synchronized resulting from the lack of an age variable within each phase. This is not a serious limitation since real cells loose their synchronization after about two cycles. More importantly, cell repair efficiency varies within a given phase which must be treated in a comprehensive model. Equations (130) to (136) will be used herein to describe the cell phase kinetics for exponential populations. The time development of an exponential population is shown in figure 17. The first generation phase populations are shown separately as is the second generation G_1 population. Growth limited populations require $\tau_1 \gg \tau_2 + \tau_3$ and a corresponding generalization of equation (137).

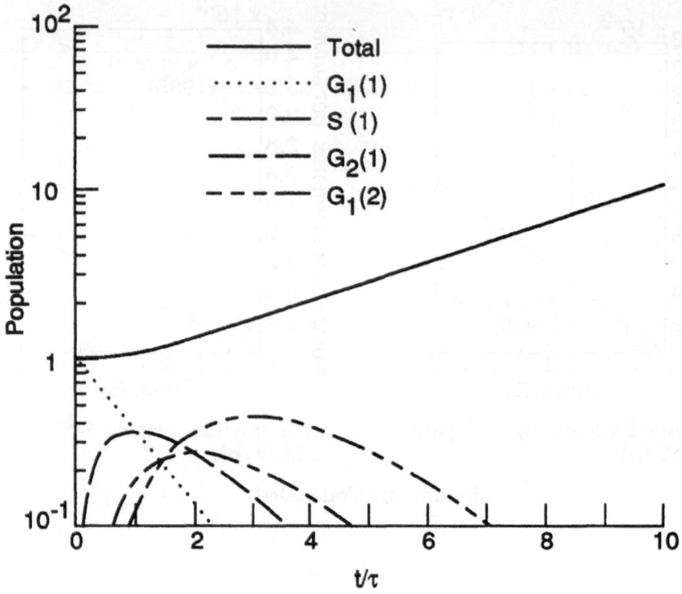

Figure 17. Population distribution for cell cycle model.

Tissue systems are noted for their low degree of mitotic activity obtained in the above model for $\tau_1 \gg \tau_2 \simeq \tau_3 = \tau$. This leaves the n_2 and n_3 populations in near equilibrium with the n_1 population as

$$n_2 \simeq \tau_2 n_1/\tau_1 \tag{138}$$

$$n_3 \simeq \tau_3 n_1/\tau_1 \tag{139}$$

and the solution to equation (131) is

$$n_1(t) \simeq n_1(0)\, e^{t/\tau_1} \tag{140}$$

At any given time S-, G_2- and M-phase cells will be very few as a result of $\tau/\tau_1 \ll 1$. If τ_2, τ_3 are the same as for the exponential population above the doubling time for n_1 given by equation (140) is on the order of one year. We now consider the coupling of the cell phase kinetics with the cell repair kinetics as we have used before.

RADIATION INJURY AND REPAIR

We define $n_{si}(t)$ as the number of cells in phase labeled by s which have received a number i of radiation-induced lesions at any time, t, in the original population of cells, n_{so} $(t = 0)$. Lacking a mechanistic foundation, we will consider an average lesion and neglect differences in the types that will occur. More important will be an accounting of the number of lesions that do occur. We also assume herein that cell death is independent of where the lesions occur; the treatment of other biological endpoints will require the differentiation among specific loci within the DNA and was dealt with in the previous sections.

We assume that these lesions are subsequently chemically stabilized and may either restore the cell to its initial structure or leave permanent changes which alter the cell function. The kinetics of cellular repair will, to a large degree, determine the time development of the initial population of radiation-produced lesions. Survival curves are based on measurements long after the induction of lesions by radiation and are functions of both the track structure of the radiations and the repair kinetics. The progression in time of the cellular populations

$n_{si}(t)$ is assumed to be governed by a coupled set of linear first-order differential equations reflecting the losses and gains by radiation injury and repair. The rate at which radiation-induced lesions are produced is denoted k_i and is assumed to be cell phase independent. The k_i scale linearly with the flux of ionizing radiations and will depend on particle type, i.e., track structure. The mean lesion repair rate (averaged over loci with i-fold lesions) is denoted α_{si} and the number of i-fold lesions repaired per unit time is then $\alpha_{si}\, n_{si}$. The equations within a given cell phase which determine the change in the lesion populations with time are given by

$$\dot{n}_{so} = \sum_{i=1}^{\infty} \alpha_{sr_i}\, n_{si} - k\, n_{so} \tag{141}$$

$$\dot{n}_{si} = \sum_{j=0}^{i-1} k_{i-j}\, n_{sj} - k\, n_{si} - \alpha_{si}\, n_{si} \tag{142}$$

where the subscript i of n_{si} denotes the multiplicity of (chemical) lesions. We allow for lesion misrepair in the model by the time evolution equation for misrepaired cell m as

$$\dot{m}_s = \sum_{i=1}^{\infty} \alpha_{smi}\, n_{si} \tag{143}$$

Conservation of cell number dictates $\alpha_{si} = \alpha_{sri} + \alpha_{smi}$ and $k = k_1 + k_2 + \ldots$ within a given cell phase. In equations (140) and (141), we have assumed the rates k_i are independent of possible previous lesions; however, this restriction could be lifted if necessary. The lesions are chemically active species neutralized by enzyme activity and radical recombination at rate constants α_{si}. The α_{sri} and α_{smi} are repair and misrepair rates for the lesion to be restored (α_{sr}) or permanently injured (α_{sm}). This is not the most general model but reasonably represents the essential kinetics for cell survival.

The effects of radiation injury on the cell cycle kinetics has been reviewed by Casarett (1968) and the implications will now be modeled. Cell populations with no lesions (including those which are repaired) undergo the usual cell cycle. We assume injured cells progress to G_2 where repairs are completed (G_2 block) before mitosis. The appropriate equations are given as

$$\dot{n}_{10} = 2\, n_{30}/\tau_3 - n_{10}/\tau_1 - k\, n_{10} + \sum_{i=1} \alpha_{1ri}\, n_{1i} \tag{144}$$

$$\dot{n}_{1i} = -\beta_i\, n_{1i}/\tau_1 + \sum_{j=0}^{i-1} k_{i-j}\, n_{1j} - k n_{1i} - \alpha_{1i}\, n_{1i} \tag{145}$$

$$\dot{n}_{20} = n_{10}/\tau_1 - n_{20}/\tau_2 - k\, n_{20} + \sum_{i=1} \alpha_{2ri}\, n_{2i} \tag{146}$$

$$\dot{n}_{2i} = \beta_i\, n_{1i}/\tau_1 - \gamma_i n_{2i}/\tau_2 + \sum_{j=0}^{i-1} k_{i-j}\, n_{2j} - k\, n_{2i} - \alpha_{2i}\, n_{2i} \tag{147}$$

$$\dot{n}_{30} = n_{20}/\tau_2 - n_{30}/\tau_3 - k\, n_{30} + \sum_{i=1} \alpha_{3ri}\, n_{3i} \tag{148}$$

$$\dot{n}_{3i} = \gamma_i n_{2i}/\tau_2 + \sum_{j=0}^{i-1} k_{i-j}\, n_{3j} - k\, n_{3i} - \alpha_{3i}\, n_{3i} \tag{149}$$

$$m_s = \sum_{i=1} \alpha_{smi}\, n_{si} \tag{150}$$

where m_s are the number of misrepairs in a given cell phase S and are assumed to terminate the cell. The total number of cell deaths is $m = m_1 + m_2 + m_3$. In practice there is a G_1 block with a repair efficiency different from the G_1 stationary phase. This must be dealt with using a G_1 age dependent repair rate beyond the scope of the present work.

In the above equations are many parameters which must be determined. The τ_s parameters related to cell cycle times are characteristic of the cells and their environment. We will assume τ_2 and τ_3 are fixed by cell properties but τ_1 is easily affected by environmental factors. The k_i relate to the formation rates of nascent lesions and are linear in particle flux. The k_i are functions of track structure. The α's are functions of the cell repair chemistry and determine how the nascent lesions are ultimately resolved by the cell. In previous work we have determined the k coefficients using the Katz parameters and assumptions concerning the G_1 repair efficiencies. We will show how these methods apply to the present formalism.

Physics and Kinetics of Cell Injury

In the Katz model, it is assumed that electromagnetic radiations form single lesions with an efficiency related to D_o and generally more than one lesion ($m \geq 2$) is required to express the biological effect (cell death in the present study). It will be useful to first consider the case of irradiation in stationary G_1-phase where misrepair effects are small. If the cells are irradiated in G_1 and are held ($\tau_1 \to \infty$, τ_2, $\tau_3 \ll \tau_1$) in this phase until repair is complete, then the surviving population is from equations (143) and (144) as

$$n_{10}(\infty) = n_{10}(t_r) + \sum_{i=1}^{m-1} \left(\frac{\alpha_{1ri}}{\alpha_{1i}}\right) n_{1i}(t_r) \tag{151}$$

where assuming maximum repair in G_1 implies $\frac{\alpha_{ri}}{\alpha_i} \approx 1$ for $i < m$. Equation (151) allows us to relate the k_i coefficients to the corresponding Katz parameters as applied to the appropriate experimental protocol (namely, G_1 exposure followed by complete G_1 repair). In the kinetic model, $n_{10}(0)$ is the initial number of G_1 cells, and equation (151) is rewritten assuming maximum repair efficiency as

$$\frac{n_{10}(\infty)}{n_{10}(0)} = e^{-kt_r} + \sum_{i=1}^{m-1} k_i\, t_r\, e^{-kt_r} + \frac{1}{2!} k_1\, t_r \sum_{i=1}^{m-1} k_i\, t_r\, e^{-kt_r}$$

$$+ \frac{1}{3!}(k_1\, t_r)^2 \sum_{i=1}^{m-2} k_1\, t_r\, e^{-kt_r} + \ldots \tag{152}$$

We consider first the example of an $m = 3$ system. According to the Katz model, an $m = 3$ system has a γ-ray response given by

$$\frac{n_{10}(\infty)}{n_{10}(0)} \cong 1 - \left(1 - e^{-D_\gamma/D_o}\right)^3 \cong 1 - \left(\frac{D_\gamma}{D_o}\right)^3 \tag{153}$$

For γ-rays, only single-step transitions are allowed. This assumption leads to the identification applied to equation (152) if

$$k_1\, t_r \cong 6^{\frac{1}{3}} D_\gamma / D_o \tag{154}$$

$$k_m\, t_r \approx 0 \quad (m > 1) \tag{155}$$

as is appropriate for γ-rays as was assumed for inactivation in equations (41) and (42). For ions, m-step transitions may occur through multiple δ-ray production leading to a non-zero k_m. The correspondence between the k_m and the cross section σ is then determined from the remaining Katz terms by noting for strictly ion kill kinetics

$$k_3\ t_r \cong \sigma F \tag{156}$$

$$k_2\ t_r \cong 0 \tag{157}$$

Requiring k_2 to be zero results from our matching of equation (151) and equation (34) at low dose. There may be non-zero values of k_2, but they cannot be strictly determined from the present form of Katz's theory. Radiation rates k_n for $i \gg m$ are also not considered in the Katz model. Although the k_i's may reflect both physical and chemical processes because of their empirical nature, we assume here that they are most clearly identified with the physical processes discussed by Katz. We now examine means by which repair rates can be estimated at least for some experimental cell systems.

Impulsive Exposures

The typical exposure protocol uses an intense particle beam in which exposure times are short compared to the cell cycle period and the cell repair times. The distribution of nascent lesions are then given for each phase of the cell cycle as

$$n_{s0}(t_r) = n_{s0}(0)\ e^{-kt_r} \tag{158}$$

$$n_{s1}(t_r) = n_{s0}(0)\ k_1\ t_r\ e^{-kt_r} \tag{159}$$

$$n_{s2}(t_r) = n_{s0}(0)\left[\frac{1}{2!}\ k_1^2\ t_r^2\ e^{-kt_r} + k_2\ t_r\ e^{-kt_r}\right] \tag{160}$$

$$n_{s3}(t_r) = n_{s0}(0)\left[\frac{1}{3!}\ k_1^3\ t_r^3\ e^{-kt_r} + k_1\ k_2\ t_r^2\ e^{-kt_r} + k_3\ t_r\ e^{-kt_r}\right] \tag{161}$$

and similarly for higher order terms. Equations (158) to (161) may be taken as the initial populations for the cell system described by equations (141) to (150) with the corresponding k_i coefficients taken as zero. A general solution is found as follows.

We assume a solution of the form ($i > 0$)

$$n_{si}(t) = \sum_{j=0}\left[a_{sij}\ t^j\ e^{-\nu_{1i}t} + b_{sij}\ t^j\ e^{-\nu_{2j}t} + c_{sij}\ t^j\ e^{-\nu_{3j}t}\right] \tag{162}$$

The initial conditions are met as

$$a_{1i0} = n_{1i}(0) \tag{163}$$

$$b_{2i0} = n_{2i}(0) \tag{164}$$

$$c_{3i0} = n_{3i}(0) \tag{165}$$

with all other a_{sio}, b_{sio}, c_{sio} being zero and

$$\nu_{1i} = \beta_i\ \tau_1^{-1} + \alpha_{1i} \tag{166}$$

$$\nu_{2i} = \gamma_i\ \tau_2^{-1} + \alpha_{2i} \tag{167}$$

$$\nu_{3i} = \alpha_{3i} \tag{168}$$

The remaining coefficients of equation (162) are found to satisfy recurrence relations which allows evaluation of the sums term-by-term. The recurrence relations for each cell phase are distinct with the following valid for G_1-phase ($s = 1$)

$$(j+1)a_{1ij+1} = 0 \tag{169}$$

$$(j+1)b_{1ij+1} = (\nu_{2i} - \nu_{1i})\, b_{1ij} + \gamma_i\, b_{2ij}/\tau_2 \tag{170}$$

$$(j+1)c_{1ij+1} = (\nu_{3i} - \nu_{1i})\, c_{1ij} + \gamma_i\, c_{2ij}/\tau_2 \tag{171}$$

The corresponding S-phase relations ($s = 2$) are

$$(j+1)a_{2ij+1} = (\nu_{1i} - \nu_{2i})\, a_{2ij} + \beta_i\, a_{1ij}/\tau_1 \tag{172}$$

$$(j+1)b_{2ij+1} = \beta_i\, b_{1ij}/\tau_1 \tag{173}$$

$$(j+1)c_{2ij+1} = (\nu_{3i} - \nu_{2i})\, c_{2ij} + \beta_i\, c_{1ij}/\tau_1 \tag{174}$$

with similar G_2-phase relations ($s = 3$)

$$(j+1)a_{3ij+1} = (\nu_{1i} - \nu_{3i})\, a_{3ij} + \gamma_i\, a_{2ij}/\tau_2 \tag{175}$$

$$(j+1)b_{3ij+1} = (\nu_{2i} - \nu_{3i})\, b_{3ij} + \gamma_i\, b_{2ij}/\tau_2 \tag{176}$$

$$(j+1)c_{3ij+1} = \gamma_i\, c_{2ij}/\tau_2 \tag{177}$$

In writing the solution in the form of equation (162) we have found the repair solutions to run in three separate cycles according to the initial population distributions prior to exposure. Clearly a G_1 stationary population cycles through the 'a' coefficients only with the b's and c's all identically zero. Similarly, a synchronized S-phase exposed population cycles according to the 'b' coefficient recurrence relations while the 'c' coefficients are exclusively related to exposure in G_2-phase. Clearly an exponential population excites all three recurrence relations according to the equilibrium distribution of the cell population.

First Generation Kinetics

According to our general kinetic equations (141) to (150), injured cells are blocked in G_2 until repairs are complete. The repair kinetics in the first generation of injured cells determine the fate of the cell. This is not in complete accordance with experimental observation in which one or two cell divisions are relatively common before the cell line is terminated. The recurrence relations among the a, b, and c coefficients may be solved in closed form to generate the cycle dependent repair processes.

a. G_1 repair cycle. We first consider the repair cycle of G_1-phase injured cells using recurrence relations given by equations (169), (172) and (175). The G_1 repair cycle is found to be given by

$$n_{1i}(t) = n_{1i}(0)\; e^{-\nu_{1i}t} \tag{178}$$

$$n_{2i}(t) = n_{1i}(0)\; \frac{\beta_i}{\tau_1} \left[\frac{e^{-\nu_{2i}t} - e^{-\nu_{1i}t}}{\nu_{1i} - \nu_{2i}} \right] \tag{179}$$

$$n_{3i}(t) = n_{1i}(0)\; \frac{\beta_i}{\tau_1}\, \frac{\gamma_i}{\tau_2}\; e^{-\nu_{1i}t} \sum_{j=2}^{\infty} \frac{t^j}{j!} \sum_{\ell=0}^{j} (\nu_{1i} - \nu_{2i})^{\ell}\, (\nu_{1i} - \nu_{3i})^{j-\ell} \tag{180}$$

$$m_1(t) = \sum_{i>0} n_{1i}(0)\; \frac{\alpha_{m1i}}{\nu_{1i}} \left(1 - e^{-\nu_{1i}t} \right) \tag{181}$$

$$m_2(t) = \sum_{i>0} n_{1i}(0)\ \frac{\beta_i}{\tau_1}\frac{\alpha_{m2i}}{(\nu_{1i}-\nu_{2i})}\left[\frac{1}{\nu_{2i}}\left(1-e^{-\nu_{2i}t}\right) - \frac{1}{\nu_{1i}}\left(1-e^{-\nu_{1i}t}\right)\right] \tag{182}$$

$$m_3(t) = \sum_{i>0} n_{1i}(0)\ \frac{\beta_i\ \gamma_i\ \alpha_{m3i}}{\tau_1\ \tau_2\ \nu_{1i}}\sum_{j=2}^{\infty}\left[1-\frac{\Gamma(j+1,\nu_{1i}t)}{j!}\right]\sum_{\ell=0}^{j}\frac{(\nu_{1i}-\nu_{2i})^{\ell}\ (\nu_{1i}-\nu_{2i})^{j-\ell}}{\nu_{1i}^{j}} \tag{183}$$

The above are solutions for cycling cell systems and approach the stationary G_1 repair kinetics for $\tau_1 \to \infty$ with τ_2 and τ_3 finite. Such a limit corresponds to the delayed plating experiments of Yang et al. (1989). The stationary phase solutions are given as

$$n_{1i}(t) = n_{1i}(0)\ e^{-\alpha_{1i}t} \tag{184}$$

$$m_1(t) = \sum_{i>0} n_{1i}(0)\ \frac{\alpha_{m1i}}{\alpha_{1i}}\left(1-e^{-\alpha_{1i}t}\right) \tag{185}$$

with $n_{2i}(t)$, $n_{3i}(t)$, $m_2(t)$, and $m_3(t)$ all zero. If complete repair is allowed in G_1-phase as in Yang et al. (1989) delayed plating experiments then the misrepaired population is

$$m(\infty) = \sum_{i>0}\ n_{1i}(0)\ \frac{\alpha_{m1i}}{\alpha_{1i}} \tag{186}$$

as we found earlier (Wilson and Cucinotta, 1991). The misrepair ratios were found near zero for $m < 3$ as required for the sigmoid shape in the x-ray survival data and recovery of the Katz model at low dose.

Herein we assume G_1 and G_2 misrepair is near zero so that the cycling G_1 exposures have total misrepair given by

$$m(\infty) \simeq \sum_{i>1}\ n_{1i}(0)\ \frac{\beta_i}{\tau_1}\frac{\alpha_{m2i}}{\nu_{1i}\nu_{2i}} \tag{187}$$

which should match our earlier results for the immediately plated exposures as

$$\frac{\beta_1}{\tau_1}\frac{\alpha_{m21}}{\nu_{11}\nu_{21}} = 0.3 \tag{188}$$

$$\frac{\beta_2}{\tau_1}\frac{\alpha_{m22}}{\nu_{12}\nu_{22}} = 0.88 \tag{189}$$

We have no means for evaluating the β_i and γ_i at this time and so we assume the cell cycle times of injured cells are the same as the unexposed population (i.e., $\beta_i = \gamma_i = 1$). The corresponding misrepair coefficients are then found to be

$$\alpha_{m21} = 0.3\tau_2^{-1}\,(1+\alpha_{11}\ \tau_1)\,(1+\alpha_{21}\ \tau_2) \tag{190}$$

$$\alpha_{m22} = 0.88\tau_2^{-1}\,(1+\alpha_{12}\ \tau_1)\,(1+\alpha_{22}\ \tau_2) \tag{191}$$

where the τ_1 and τ_2 are those associated with the Yang et al. experiments ($\tau_1 \simeq 4$ hr and $\tau_2 \simeq 10$ hr resulting in $\alpha_{m21}/\alpha_{21} \simeq\ 0.84$ and $\alpha_{m22}/\alpha_{22} \simeq 1$).

b. S-repair cycle. Cells injured in the S-phase cycle are described through the b coefficient recurrence relations as given by equations (170), (173), and (176). The solutions for the S-repair cycle is given as

$$n_{1i}(t) = 0 \tag{192}$$

$$n_{2i}(t) = n_{2i}(0)\ e^{-\nu_{2i}t} \tag{193}$$

$$n_{3i}(t) = n_{2i}(0)\ \frac{\gamma_i}{\tau_2}\left[\frac{e^{-\nu_{3i}t} - e^{-\nu_{2i}t}}{\nu_{2i}-\nu_{3i}}\right] \tag{194}$$

$$m_1(t) = 0 \tag{195}$$

$$m_2(t) = \sum_{i>0} \frac{\alpha_{m2i}}{\nu_{2i}} n_{2i}(0) \left(1 - e^{-\nu_{2i}t}\right) \tag{196}$$

$$m_3(t) = \sum_{i>0} \frac{\alpha_{m3i}}{(\nu_{2i} - \nu_{3i})} \frac{\gamma_i}{\tau_2} n_{2i}(0) \left[\frac{1}{\nu_{3i}}\left(1 - e^{-\nu_{3i}t}\right) - \frac{1}{\nu_{2i}}\left(1 - e^{-\nu_{2i}t}\right)\right] \tag{197}$$

The total number of misrepaired cells following a single impulsive exposure is

$$m(\infty) = \sum_{i>0} \left[\frac{\alpha_{m2i}}{\nu_{2i}} + \frac{\alpha_{m3i}}{\nu_{2i}\nu_{3i}} \frac{\gamma_i}{\tau_2}\right] n_{2i}(0) \tag{198}$$

Since in our current understanding $\tau_1\nu_{1i}\beta_i^{-1} > 1$ we see that the misrepair rate of the S-phase injury is higher than the G_1-phase injury cycle. This is not surprising since repair during the G_1-phase is very efficient.

c. G_2-repair cycle. The G_2-repair cycle is simplified by the complete G_2 block assumed in the present model. The c coefficient recurrence relations (171), (174), and (177) give solutions

$$n_{1i}(t) = n_{2i}(t) = 0 \tag{199}$$

$$n_{3i}(t) = n_{3i}(0)\, e^{-\nu_{3i}t} \tag{200}$$

$$m_1(t) = m_2(t) = 0 \tag{201}$$

$$m_3(t) = \sum_{i>0} \frac{\alpha_{m3i}}{\nu_{3i}} n_{3i}(0) \left(1 - e^{-\nu_{3i}t}\right) \tag{202}$$

The total misrepair fraction is near zero since the G_2-phase repair efficiency is assumed near unity and shows similar survival characteristics as the stationary G_1-phase repair kinetics.

d. Exponential population repair kinetics. Exposure of an exponential population is described as a superposition of the previous solutions. The initial irradiated cell population is the equilibrium distribution in which

$$n_s(0) = \frac{\tau_s}{\sum_{s'} \tau_{s'}} n_0(0) \tag{203}$$

where $n_0(0)$ is the total cell population. The initial distribution of injured cells are given by equations (158) to (161). Subsequent to exposure, the misrepair is described by equations (178) to (202) and the total misrepair population is given as

$$\begin{aligned} m(\infty) &\approx \sum_{i>0} \frac{1}{\tau_1} \frac{\alpha_{m2i}}{\nu_{1i}\nu_{2i}} n_{1i}(0) + \sum_{i>0} \frac{\alpha_{m2i}}{\nu_{2i}} n_{2i}(0) \\ &= \sum_{i>0} \left(\frac{1}{\tau_c\nu_{1i}} + \frac{\tau_2}{\tau_c}\right) \frac{\alpha_{2mi}}{\nu_{2i}} n_i(0) \end{aligned} \tag{204}$$

where

$$\tau_c = \sum_s \tau_s \tag{205}$$

$$n_i(0) = \sum_s n_{si}(0) \tag{206}$$

e. Tissue systems repair kinetics. The main characteristic of tissue systems represented in the present model is the excessive time spent in the G_1-phase. The cell distribution is then approximately

$$n_1 = \frac{\tau_1}{\tau_c} n_0 \simeq n_0 \tag{207}$$

$$n_2 = \frac{\tau_2}{\tau_c} n_0 \simeq \frac{\tau_2}{\tau_1} n_0 \tag{208}$$

$$n_3 = \frac{\tau_3}{\tau_c} n_0 \simeq \frac{\tau_3}{\tau_1} n_0 \tag{209}$$

Repair within the G_1 population after exposure is given by equations (184) and (185) so that

$$m_1(\infty) = \sum_{i>0} \frac{\alpha_{m1i}}{\alpha_{1i}} n_{1i}(0) \tag{210}$$

The repair for S-phase exposures are

$$m_2(\infty) = \sum_{i>0} \frac{\alpha_{m2i}}{\nu_{2i}} n_{2i}(0) \tag{211}$$

$$m_3(\infty) = \sum_{i>0} \frac{1}{\tau_2} \frac{\alpha_{m3i}}{\nu_{2i}\nu_{1i}} n_{2i}(0) \tag{212}$$

while G_2 injury results in

$$m_3(\infty) = \sum_{i>0} \frac{\alpha_{m3i}}{\nu_{3i}} n_{3i}(0) \tag{213}$$

Due to the assumed efficient repair in G_2 we see that $m_3(\infty)$ given by equations (212) and (213) are inferior to $m_2(\infty)$ of equation (211). The total misrepair is then

$$m(\infty) = \sum_{i>0} \left[\frac{\alpha_{m1i}}{\alpha_{1i}} + \frac{\tau_2}{\tau_1} \frac{\alpha_{m2i}}{\nu_{2i}} \right] n_{2i}(0) \tag{214}$$

In the event that $\frac{\tau_2}{\tau_1} \ll \frac{\alpha_{m1i}}{\alpha_{1i}}$ then the population response is similar to G_1-phase stationary populations as is usually the case.

Discussion

With the formalism of cell injury and repair within the cell cycle presented in this last section there is some hope of deriving a more complete description of the cell kinetic problem. However, a more complete model must first treat the age dependence within the phase. For example, the G_1 phase must be treated as an early G_1 and late G_1 as a minimum requirement. This we know since the G_1 block repair efficiency is similar to the S-phase repair efficiency and not the G_1 stationary repair efficiency. With such a complete model one may turn to the issue of cell progression within a tissue system. Such issues are critical for understanding biological response.

REFERENCES

Allen, J. M.; 1962: *The Molecular Control of Cellular Activity.* McGraw-Hill, Inc. New York.

Bettega, D.; Calzolari, P.; Ottolenghi A.; and Tallone Lombardi, L.; 1990: Oncogenic Transformation Induced by High and Low LET Radiations. *Radiat. Prot. Dosim.* vol. 31, pp. 279–283.

BEIR V; 1990: *Health Effects of Exposure to Low Levels of Ionizing Radiation*, Washington, D.C.; National Academy Press.

Casarett, A. P.; 1968: *Radiation Biology*, Prentice-Hall, Englewood Cliffs, NJ.

Cartwright, D. C.; Trajmar, S.; Chutjian, A.; and Williams, W.; 1977: Electron Impact Excitation of the Electronic States of N_2. II. Integral Cross Sections at Incident Energies From 10 to 50 eV. *Phys. Review A*, vol. 16, no. 3, pp. 1041–1051.

Charlesby, A.; 1967: Radiation Mechanisms in Polymers, In *Irradiation of Polymers*, R. F. Gould, ed., American Chemical Society, Washington, D.C., pp. 1–21.

Crawford-Brown, D. J.; and Hoffmann, W.; 1990: A Generalized State-vector Model for Radiation-induced Cellular Transformation. *Int. J. Radiat. Biol.*, vol. 57, pp. 407–423.

Cucinotta, F. A.; Katz, R.; Wilson, J. W.; Townsend, L. W.; Nealy J. E.; and Shinn, J. L.; 1991a: *Cellular Track Model of Biological Damage to Mammalian Cell Cultures From Galactic Cosmic Rays.* NASA TP-3055.

Cucinotta, F. A.; Katz, R.; Wilson, J. W.; Townsend, L. W.; Shinn, J. L.; and Hajnal, F.; 1991b: Biological Effectiveness of High Energy Protons: Target Fragmentation. *Radiat. Res.*, vol. 127, pp. 130–137.

Curtis, S. B.; 1986: Lethal and Potentially Lethal Lesions Induced by Radiation—A Unified Repair Model. *Radiat. Res.*, vol. 106, pp. 252–270.

Fritz-Niggli, H.; 1988: The Role of Repair Processes in Cellular and Genetic Response to Radiation. In: *Terrestrial Space Radiation and It's Biological Effects*, Edited by P. D. McCormack, C. E. Swenberg and H. Bücker, Plenum Press, New York.

Frost, A. A.; and Pearson, R. G.; 1962: *Kinetics and Mechanisms.* John Wiley and Sons, NY.

Goodhead, D. T.; 1985: Saturable Repair Models of Radiation Action in Mammalian Cells. *Radiat. Res.*, vol. 104, pp. S58–S67.

Grahn, D.; 1973: HZE Effects in Manned Spaceflight. Space Science Board, National Academy of Science, Washington, DC.

International Commission on Radiological Units and Measurements; 1986: *The Quality Factor in Radiation Protection.* Bethesda, MD: ICRU; ICRU Report 40.

Kamaratos, E.; Chang, C. K.; Wilson, J. W.; and Xu, Y. J.; 1982: Valence Bond Effects on Mean Excitation Energies for Stopping Power in Metals. *Phys. Lett.*, vol. 92A, no. 7, pp. 363–365.

Katz, R.; Ackerson, B.; Homayoonfar, M.; and Sharma, S. C.; 1971: Inactivation of Cells by Heavy Ion Bombardment. *Radiat. Res.*, vol. 47, pp. 402–425.

Katz, R.; and Cucinotta, F. A.; 1991: RBE vs. Dose for Low Doses of High-LET Radiations. *Health Physics*, vol. 60, pp. 717–718.

Köllman, K.; 1975: Dissociative Ionization of H_2, N_2, and CO by Electron Impact—Measurements of Kinetic Energy, Angular Distributions, and Appearance Potentials. *Int. J. Mass Spectrom. & Ion Phys.*, vol. 17, pp. 261–285.

Lett, J. T.; Cox, A. B.; and Story, M. D.; 1989: The Repair in the Survival of Mammalian Cells From Heavy Ion Irradiation: Approximation to the ideal case of target theory. *Adv. Space Res.* vol. 9, pp. 99–104.

Manson, S. T.; Toburen, L. H.; Madison, D. H.; and Stolterfoht, N.; 1975: Energy and Angular Distribution of Electrons Ejected From Helium by Fast Protons and Electrons: Theory and Experiment. *Phys. Review A*, vol. 12, third ser., no. 1, pp. 60–79.

Merriam, G. R.; Worgul, B. V.; Medvedovsky, C.; Zaider, M.; and Rossi, H.; 1984: Accelerated Heavy Particles on the Lens. *Radiat. Res.*, vol. 98, pp. 129–140.

Mitchison, J. M.; 1971: *The Biology of the Cell Cycle.* Cambridge University Press, Great Britian.

Ngo, F. Q. H.; Xian-Li, J.; Kalvakolanu, I.; Roberts, W.; Blue, J.; and Higgins, J.; 1990: Basic Biological Investigations of Fast Neutrons. *International Colloquium on Neutron Radiation Biology*, Rockville, MD, Nov. 5–7, 1990.

Radman, M.; Dohert, C.; Courguignon, M. F.; Doubleday, O. P.; and Letcomte, P.; 1981: High Fidelity Devices in the Reproduction of DNA in *Chromosome Damage and Repair*, E. Seeberg and K. Kleppe, eds., Plenum Press, New York.

Rustgi, M. L.; Pandy, L. N.; Wilson, J. W.; Long, S. A. T.; and Zhu, G.; 1988: Distribution of Energy in Polymers Due to Incident Electrons and Protons. *Radiat. Effects*, vol. 105, pp. 303–311.

Schaefer, Hermann J.; 1952: Exposure Hazards From Cosmic Radiation Beyond the Stratosphere and in Free Space. *J. Aviation Med.*, vol. 23, no. 4, pp. 334–344.

Schulz, George J.; 1976: A Review of Vibrational Excitation of Molecules by Electron Impact at Low Energies. *Principles of Laser Plasmas*, George Bekefi, ed., John Wiley & Sons, Inc., pp. 33–88.

Scott, B. R.; and Ainsworth, E. J.; 1980: State Vector Model for Life Shortening in Mice After Brief Exposures to Low Doses of Ionizing Radiation. *Math. Biosci.*, vol. 49, pp. 185–205.

Sinclair, W. K.; 1968: Cyclic X-ray Responses in Mammalian Cells in Vitro. *Radiat. Res.*, vol. 33, pp. 620–643.

Sinclair, W. K.; 1985: Experimental RBE Values of High LET Radiations at Low Doses and the Implications for Quality Factor Assignments. *Rad. Prot. Dos.*, vol. 13, pp. 319–326.

Tobias, C. A.; 1985: The Repair-Misrepair Model in Radiobiology: Comparison to Other Models. *Radiat. Res.*, vol. 104, pp. S77–S95.

Waligorski, M. P. R.; Sinclair, G. L.; and Katz, R.; 1987: Radiosensitivity Parameters for Neoplastic Transformations in C3H10T 1/2 CELLS. *Radiat. Res.*, vol. 111, pp. 424–437.

Wight, G. R., Van der Wiel, M. J.; and Brion, C. E.; 1976: Dipole Excitation, Ionization and Fragmentation of N_2 and CO in the 10–60 eV Region. *J. Phys. B: At. Mol. Phys.*, vol. 9, no. 4, pp. 675–689.

Wilson, J. W.; 1977a: *Analysis of the Theory of High-Energy Ion Transport.* NASA TN D-8381.

Wilson, J. W.; 1980: Nuclear-Induced Xe-Br* Photolytic Laser Model. *Appl. Phys. Lett.*, vol. 37, no. 8, pp. 695–697;.

Wilson, J. W.; and Cucinotta, F. A.; 1991: *Cellular repair/misrepair track model.* Washington, DC; NASA, Report TP-3124.

Wilson, J. W.; Cucinotta, F. A.; and Shinn, J. L.; 1991: *Multiple Lesion Track Structure Model.* Washington, DC.; NASA, Report TP- .

Wilson, J. W.; DeYoung, R. J.; and Harries, W. L.; 1979: Nuclear-Pumped ^{3}He-Ar Laser Modeling. *J. Appl. Phys.*, vol. 50, no. 3, pt. I, pp. 1226–1235.

Wilson J. W.; and Lee, J. H.; 1980: Modeling of a Solar-Pumped Iodine Laser. *Virginia J. Sci.*, vol. 31, pp. 34–38.

Wilson, J. W.; Lee, Y.; Weaver, W. R.; Humes, D. H.; and Lee, Ja H.; 1984: *Threshold Kinetics of a Solar-Simulator-Pumped Iodine Laser.* NASA TP-2241.

Wilson, J. W.; and Kamaratos, E.; 1981: Mean Excitation Energy for Molecules of Hydrogen and Carbon. *Phys. Lett.*, vol. 85A, no. 1, pp. 27–29.

Xu, Y. J.; Khandelwal, G. S.; and Wilson, J. W.; 1984b: Low-Energy Proton Stopping Power of N_2, O_2, and Water Vapor, and Deviations From Bragg's Rule. *Phys. Review A*, vol. 29, third ser., no. 6, pp. 3419–3422.

Xu, Y. J.; Khandelwal, G. S.; and Wilson, J. W.; 1985: Proton Stopping Cross Sections of Liquid Water Vapor. *Phys. Review A*, vol. 32, third ser., no. 1, pp. 629–636.

Yang, T. C.; Craise, L. M.; Mei, M.; and Tobias, C. A.; 1989: Neoplastic Cell Transformation by High LET Radiation: Molecular Mechanisms. *Adv. Space Res.*, vol. 9, pp. 131–140.

INDEX